harr

Springer Series in
SOLID-STATE SCIENCES 135

Springer
Berlin
Heidelberg
New York
Hong Kong
London
Milan
Paris
Tokyo

Physics and Astronomy

ONLINE LIBRARY

springeronline.com

Springer Series in
SOLID-STATE SCIENCES

Series Editors:
M. Cardona   P. Fulde   K. von Klitzing   R. Merlin   H.-J. Queisser   H. Störmer

The Springer Series in Solid-State Sciences consists of fundamental scientific books prepared by leading researchers in the field. They strive to communicate, in a systematic and comprehensive way, the basic principles as well as new developments in theoretical and experimental solid-state physics.

126 **Physical Properties of Quasicrystals**
Editor: Z.M. Stadnik

127 **Positron Annihilation in Semiconductors**
Defect Studies
By R. Krause-Rehberg and H.S. Leipner

128 **Magneto-Optics**
Editors: S. Sugano and N. Kojima

129 **Computational Materials Science**
From Ab Initio to Monte Carlo Methods
By K. Ohno, K. Esfarjani, and Y. Kawazoe

130 **Contact, Adhesion and Rupture of Elastic Solids**
By D. Maugis

131 **Field Theories for Low-Dimensional Condensed Matter Systems**
Spin Systems and Strongly Correlated Electrons
By G. Morandi, P. Sodano, A. Tagliacozzo, and V. Tognetti

132 **Vortices in Unconventional Superconductors and Superfluids**
Editors: R.P. Huebener, N. Schopohl, and G.E. Volovik

133 **The Quantum Hall Effect**
By D. Yoshioka

134 **Magnetism in the Solid State**
By P. Mohn

135 **Electrodynamics of Magnetoactive Media**
By I. Vagner, B.I. Lembrikov, and P. Wyder

136 **Nanoscale Phase Separation and Colossal Magnetoresistance**
The Physics of Manganites and Related Compounds
By E. Dagotto

137 **Quantum Transport in Submicron Devices**
A Theoretical Introduction
By W. Magnus and W. Schoenmaker

138 **Phase Separation in Soft Matter Physics**
Micellar Solutions, Microemulsions, Critical Phenomena
By P.K. Khabibullaev and A.A. Saidov

139 **Optical Response of Nanostructures**
Microscopic Nonlocal Theory
By K. Cho

140 **Fractal Concepts in Condensed Matter Physics**
By T. Nakayama and K. Yakubo

141 **Excitons in Low-Dimensional Semiconductors**
Theory, Numerical Methods, Applications
By S. Glutsch

Series homepage – springer.de

Volumes 1–125 are listed at the end of the book.

I.D. Vagner   B.I. Lembrikov   P. Wyder

Electrodynamics of Magnetoactive Media

With 97 Figures

Springer

Professor I.D. Vagner
MPI für Festkörperforschung
Hochfeld-Magnetlabor CNRS
25, Avenue des Martyrs
38042 Grenoble CX 09, France
E-mail: vagner@labs.polycnrs-gre.fr

Holon Academic Institute of Technology
Golomb Street 52
58102 Holon, Israel
E-mail: vagner_i@hait.ac.il

Professor P. Wyder
MPI für Festkörperforschung
Hochfeld-Magnetlabor CNRS
25, Avenue des Martyrs
38042 Grenoble CX 09, France
E-mail: wyder@labs.polycnrs-gre.fr

Dr. B.I. Lembrikov
Holon Academic Institute of Technology
Golomb Street 52
58102 Holon, Israel
E-mail: borisle@hait.ac.il

Series Editors:

Professor Dr., Dres. h. c. Manuel Cardona
Professor Dr., Dres. h. c. Peter Fulde*
Professor Dr., Dres. h. c. Klaus von Klitzing
Professor Dr., Dres. h. c. Hans-Joachim Queisser

Max-Planck-Institut für Festkörperforschung, Heisenbergstrasse 1, D-70569 Stuttgart, Germany
* Max-Planck-Institut für Physik komplexer Systeme, Nöthnitzer Strasse 38
  D-01187 Dresden, Germany

Professor Dr. Roberto Merlin
Department of Physics, 5000 East University, University of Michigan
Ann Arbor, MI 48109-1120, USA

Professor Dr. Horst Störmer
Dept. Phys. and Dept. Appl. Physics, Columbia University, New York, NY 10027 and
Bell Labs., Lucent Technologies, Murray Hill, NJ 07974, USA

ISSN 0171-1873

ISBN 3-540-43694-4 Springer-Verlag Berlin Heidelberg New York

Library of Congress Cataloging-in-Publication Data

Vagner, Israel D.
Electrodynamics of magnetoactive media / I. Vagner, B.I. Lembrikov, P.R. Wyder. p. cm. – (Springer series in solid-state sciences, ISSN 0171-1873 ; 135) Includes bibliographical references and index. ISBN 3-540-43694-4 (acid-free paper) 1. Electrodynamics. 2. Magnetic fields. I. Lembrikov, B. I. (Boris I.), 1950- II. Wyder, P. (Peter), 1936- III. Title. IV. Series.
QC631.V325 2004
537.6–dc22   2003061743

This work is subject to copyright. All rights are reserved, whether the whole or part of the material is concerned, specifically the rights of translation, reprinting, reuse of illustrations, recitation, broadcasting, reproduction on microfilm or in any other way, and storage in data banks. Duplication of this publication or parts thereof is permitted only under the provisions of the German Copyright Law of September 9, 1965, in its current version, and permission for use must always be obtained from Springer-Verlag. Violations are liable for prosecution under the German Copyright Law.
Springer-Verlag is a part of Springer Science+Business Media
springeronline.com
© Springer-Verlag Berlin Heidelberg 2004
Printed in Germany
The use of general descriptive names, registered names, trademarks, etc. in this publication does not imply, even in the absence of a specific statement, that such names are exempt from the relevant protective laws and regulations and therefore free for general use.

Typesetting and production: PTP-Berlin Protago-TeX-Production GmbH, Berlin
Cover concept: eStudio Calamar Steinen
Cover production: *design & production* GmbH, Heidelberg

Printed on acid-free paper   SPIN: 10876330   57/3141/YU - 5 4 3 2 1 0

Preface

Our objective was primarily to consider in a separate treatise from the general point of view a theory of as many electrodynamic phenomena in a magnetic field as possible. The choice of material was determined by both the absence of such a book and the scientific interests of the authors. From the very beginning, however, we felt it necessary to include the fundamentals of electrodynamics that are required for the thorough analysis of particular processes. We believe that it is convenient for a reader to find in the same book a consistent review of some special fields in physics and a complete set of theoretical instruments that are necessary for the clear understanding of more advanced parts of the book.

There exists a number of excellent textbooks and monographs describing the problems of classical electrodynamics in general and its applications to continuous media. We have to acknowledge, for example, the following fundamental books: *Electrodynamics* by A. Sommerfeld [1], *The Classical Theory of Fields* by L.D. Landau and E.M. Lifshitz [2], *Electromagnetic Theory* by J.A. Stratton [3], and *Electrodynamics of Continuous Media* by L.D. Landau and E.M. Lifshitz [4]. This list is certainly not exhaustive. However, to our knowledge, a book specifically covering the theory of electrodynamic phenomena in a magnetic field has not yet been written.

This book is suitable for third year undergraduate students, graduate students, experimentalists specialized in the magnetic field activities and engineers occupied in high-tech industries who wish to update their knowledge in physics. The first group of possible readers can limit themselves to the chapters containing the fundamentals of classical electrodynamics and the sample problems, which illustrate the applications of the formalism developed in these chapters. Graduate students and engineers will be interested in the special part of the book where the electrodynamics of complex media in magnetic fields is exposed. The aim of this part of the book is twofold. First, the brief and compact review of electrodynamic phenomena in plasma, liquid crystals, superconductors subject to a magnetic field provided with the detailed derivations and pictures delivers the necessary scientific information about the phenomena and the main quantitative results. Secondly, one can follow the general procedures of electromagnetic field calculation in such a way that these methods can be applied to the solution of practically important problems.

The book is constructed as follows. In the first chapter we introduce the Maxwell equations in the general form as a basis for all further analysis. It should be noted that elsewhere in the book the MKS system of units is used. According to the most common point of view, we write the Maxwell equations a priori as the set of postulates adequately describing the well-established empirical laws. The Maxwell equations are presented also in a relativistic 4-dimensional form, which appeared to be necessary in the special part of the book. Then, in each section of the book we try to solve the Maxwell equations for any particular case, sometimes along with the equations of motion of a particle, or a continuous medium as a whole. In many cases these equations of motion can be obtained by virtue of the Lagrangian formalism. For this reason, we construct the total Lagrangian containing the terms describing an electromagnetic field, charged particles and their interaction.

The second chapter concerns magnetostatics. In the context of the book, it plays a key role, in particular, since its results are often used in the subsequent chapters. Besides the basic equations relating the magnetic field characteristics such as a vector potential, magnetic induction and current density, we provide the detailed derivation of a magnetic field of some typical current configurations.

In the third chapter the quasi-static electromagnetic field is considered. Here we concentrate on the phenomena in a slowly varying magnetic field such as eddy currents, skin effect and electromagnetic induction. We also consider a circuit with a variable current and introduce an impedance as its characteristic.

The fourth chapter describes the electromagnetic wave propagation in a vacuum, anisotropic media, dispersive media and guiding systems. Here we introduce the dielectric permittivity tensor and show that it is basic in determination of the propagation regime, type of waves, etc. We consider the essential topics such as plane waves, light reflection at a boundary between two media, types of waves and their propagation directions in anisotropic media, magnetic-optical effects, conditions of light absorption, the Kramers and Kronig's relations between the real and imaginary parts of the dielectric permittivity, running wave solutions for hollow and dielectric wave-guides and standing waves in cavity resonators.

The first four chapters compose the general part of the book. A reader with a limited background knowledge in this field can use it as a manual in classical electrodynamics.

Chapters 5–9 concern the electrodynamics of complicated systems where the interaction with a medium is taken into account. The comprehensive study of each particular problem can be found in separate excellent monographs. We, however, tried our best to include the most significant and, at the same time, methodically instructive examples of the classical electrodynamics application.

We start with charged particle behavior in all kinds of external field combinations: in static electric and magnetic fields separately, in combined electric and magnetic fields, both uniform and spatially inhomogeneous, and in time dependent fields (Chap. 5) using here the knowledge obtained in the three first chapters. For example, we derive the equations of motion of the charged particle by using the Lagrangian formalism. The appropriate form of the Maxwell equations is used for the field description. The results presented in Chap. 5 are the basis for the study of electrodynamic effects in plasma. Generally, plasma represents an ensemble of charged particles of different types. The motion of charged particles in external electric and magnetic fields is usually accompanied by so-called instabilities, i.e., an excitation of collective oscillations and waves. In Chap. 6 the basics of the theory of instabilities are considered. Then the quasi-static instabilities in solid state plasma subject to external electric and magnetic fields are analyzed by virtue of this formalism.

Electromagnetic waves in plasma are analyzed in Chap. 7. We do not pretend to give an exhaustive analysis of phenomena in plasma, since this is an enormous field of physics and explored in a great number of specialized books and papers. We present only the typical cases illustrating the propagation of electromagnetic waves in strongly anisotropic dispersive medium in the presence of a magnetic field. The chapter contains a description of the general approach to the plasma investigation and the results for several types of waves in one- and two-component magneto-active plasma. In accordance with the purpose of this book, we emphasize the analysis of helicons as predominantly magnetic waves. The study of helicons also permits the demonstration of non-locality effects, or the so-called spatial dispersion. In this chapter we mainly use the formalism developed in Chap. 4 for electromagnetic wave propagation in anisotropic dispersive media.

In Chaps. 8 and 9 we consider the peculiarities of electrodynamics of the media with the specific types of long range ordering: liquid crystals and superconductors that possess the characteristic dynamic variable – a so-called order parameter. In both cases the adequate description of the electrodynamic phenomena requires the simultaneous solution of the Maxwell equations and the equations of motion for the order parameter. The latter are usually obtained by virtue of the Lagrangian type formalism. The difference between the two cases is that the liquid crystal ordering is a purely classical phenomenon that can be explained in the framework of the classical thermodynamics, while the existence of a super-conducting state is based on a quantum phenomenon of electron pairing.

In Chap. 8 we gradually analyze the magnetostatics, electrodynamic instabilities and some specific optical effects in liquid crystals. Once more we should underline that the ideas described in Chaps. 2–5 are applied. For example, liquid crystals can serve as an instructive model for the analysis of light wave propagation in a strongly anisotropic, spatially inhomogeneous medium.

In Chap. 9 the electrodynamics of superconductors is considered. We discuss the London theory, the Ginzburg–Landau theory and the Josephson effect with some applications. The study follows the same scheme: magnetostatics, the effects in variable fields and wave processes. Superconductivity is a striking example of a macroscopic manifestation of the purely quantum effect. It is shown that using a macroscopic order parameter that results from the microscopic theory as a starting point, the consistent phenomenological electrodynamics of superconductors can be constructed. We underline that a magnetic field plays a crucial role in superconductor behavior.

We are grateful to W. Joss, T. Maniv, the late A.S. Rozhavsky, Yu. Pershin and P.C.E. Stamp for the fruitful discussions and constructive help at different stages of the writing. We especially appreciate the contribution of Ju H. Kim and G. Kventsel who actively participated in the planning of the book and proposed some topics for Chap. 5, *Charged Particle in Electromagnetic Field*, at the first stage of the work. We are grateful to A. Kaplunovsky for the preparation of the electronic version of the manuscript.

Grenoble, Holon
July 2003

Israel D. Vagner
Boris I. Lembrikov
Peter Wyder

Contents

1 Fundamentals of Electrodynamics 1
   1.1 Charges and Fields 1
   1.2 Maxwell Equations 4
        1.2.1 Maxwell Equations in the Differential Form 4
        1.2.2 Maxwell Equations in the Integral Form 7
        1.2.3 Field Potentials 9
   1.3 Relativistic Properties of the Maxwell Equations 13
        1.3.1 Lorentz Transformation of Coordinates 13
        1.3.2 Four-Dimensional Form of the Maxwell Equations 15
        1.3.3 Lorentz Transformation
              of Electromagnetic Field 4-Tensor 19
   1.4 Lagrangian Formalism 22
        1.4.1 Action Functional 22
        1.4.2 Field Equations 27

2 Magnetostatics 29
   2.1 Fundamentals of Magnetostatics 29
        2.1.1 Constant Magnetic Field 29
        2.1.2 Magnetic Moment of a System of Currents 36
        2.1.3 Energetic Characteristics of a Static Magnetic Field .. 37
        2.1.4 Forces in a Magnetic Field 40
   2.2 Magnetic Field of a Given Current Distribution 45
        2.2.1 Magnetic Field of Linear Currents 45
        2.2.2 Magnetic Field of a Contour with a Current 49
        2.2.3 A Cylinder in an External Magnetic Field 52
        2.2.4 Magnetic Field of a Solenoid 58
        2.2.5 Magnetic Field of a Torus 63
        2.2.6 Magnetic Field of a Rotating Charged Sphere 64

3 Quasi-static Electromagnetic Field 68
   3.1 Conductors in a Variable Magnetic Field 68
        3.1.1 Basic Equations 68
        3.1.2 Eddy Currents 71
        3.1.3 Skin Effect 73

3.2　Quasi-steady Current Circuit 75
　　　3.3　Electromagnetic Induction Law 78

4　Electromagnetic Waves 82
　　　4.1　Electromagnetic Waves in a Dielectric..................... 82
　　　4.2　Plane Waves 85
　　　4.3　Monochromatic Plane Waves.............................. 87
　　　4.4　Polarization of a Monochromatic Plane Wave 89
　　　4.5　Reflection and Refraction of Electromagnetic Waves 92
　　　4.6　Electromagnetic Waves in Anisotropic Media................ 98
　　　　　4.6.1　Plane Wave in an Anisotropic Medium 98
　　　　　4.6.2　Optical Properties of Uniaxial Crystals.............. 105
　　　　　4.6.3　Birefringence in an Electric Field 108
　　　　　4.6.4　Magnetic-Optical Effects 109
　　　4.7　Electromagnetic Waves in Dispersive Media................ 114
　　　　　4.7.1　Dispersion of the Dielectric Permittivity............. 114
　　　　　4.7.2　Field Energy in Dispersive Media.................... 118
　　　　　4.7.3　The Kramers and Kronig's Relations................. 121
　　　　　4.7.4　Dispersion Relation
　　　　　　　　for a Monochromatic Plane Wave................... 128
　　　4.8　Electromagnetic Wave Propagation in Guiding Systems 130
　　　　　4.8.1　Hollow Wave-Guides 130
　　　　　4.8.2　Rectangular Waveguides 134
　　　　　4.8.3　Energy Absorption in Waveguides 138
　　　　　4.8.4　Optical Dielectric Waveguides...................... 140
　　　　　4.8.5　Cavity Resonators 145

5　Charged Particle in Electromagnetic Field 152
　　　5.1　Equations of Motion 152
　　　5.2　Energy of Particles and Fields............................ 153
　　　5.3　A Charged Particle Motion in Static Uniform Fields 156
　　　　　5.3.1　Static Uniform Magnetic Field 156
　　　　　5.3.2　Adiabatic Invariance 159
　　　　　5.3.3　Static Uniform Electric Field 161
　　　　　5.3.4　Combined Static Uniform Electric
　　　　　　　　and Magnetic Fields 162
　　　5.4　Hall Effect in Semiconductors 164
　　　5.5　A Charged Particle Motion in Static Non-uniform
　　　　　Magnetic Fields 169
　　　　　5.5.1　Gradient Drift 170
　　　　　5.5.2　Centrifugal Drift 171
　　　　　5.5.3　Magnetic Mirrors................................... 174
　　　5.6　A Charged Particle Motion
　　　　　in Static Non-uniform Electric Field 180
　　　5.7　A Charged Particle Motion in Time Dependent Fields....... 182

		5.7.1	Time-Dependent Uniform Magnetic Field............ 182

        5.7.1 Time-Dependent Uniform Magnetic Field............ 182
        5.7.2 Static Magnetic
               and Uniform Time-Dependent Electric Field 186

6 Current Instabilities 193
6.1 General Approach 193
6.2 Criteria of Instability Classification 195
        6.2.1 Criteria of Absolute and Convective Instability 195
        6.2.2 Criterion of a Spatial Amplification 201
6.3 Analysis of Dispersion Equation for Two Coupled Waves 204
6.4 Transformation of Instabilities 207
6.5 Instabilities in Solid State Plasma 210
        6.5.1 Parameters of Solid State Plasma 210
        6.5.2 Quasineutral Oscillations in Semiconductors 213
        6.5.3 Gradient Instability with Two Types of Carriers...... 217
        6.5.4 Instability in a Semiconductor
               with Three Types of Carriers 224
        6.5.5 Parametric Excitation of the Instability 226
        6.5.6 Helicoidal Instability in Semiconductors 229

7 Waves in Plasma 236
7.1 Hydrodynamic Model 236
7.2 Waves in One Component Electron Plasma 239
        7.2.1 General Dispersion Relations 239
        7.2.2 Conductivity Tensor in a Local Approximation 243
        7.2.3 Plasmons.. 245
        7.2.4 Ordinary and Extraordinary Electromagnetic Waves .. 250
        7.2.5 Helicons... 253
7.3 Waves in Two Component Plasma 258
        7.3.1 Ion Acoustic Waves................................ 258
        7.3.2 Electrostatic Ion Waves 261
        7.3.3 Alfven Waves 264
        7.3.4 Magnetosonic Waves 266
7.4 Excitation of Transverse Waves
        in Indium Antimonide 268
7.5 Kinetic Theory 273
        7.5.1 The Boltzmann–Vlasov Equation 273
        7.5.2 Landau Damping................................... 277
        7.5.3 Permittivity Tensor in a Non-local Case 282
        7.5.4 Helicon Dispersion in a Non-local Case 288

8 Electrodynamics of Liquid Crystals 293
8.1 Classification and Fundamental Properties
        of Liquid Crystals 293

8.2 Liquid Crystals in a Static Magnetic Field 297
    8.2.1 Continuum Theory of Liquid Crystals 297
    8.2.2 Nematic Liquid Crystal in a Static Magnetic Field.... 300
    8.2.3 Cholesteric Liquid Crystal
         in a Static Magnetic Field 309
    8.2.4 Helfrich–Hurault Effect
         in Smectic A Liquid Crystals 313
8.3 Electrohydrodynamic Instabilities in Nematic
    Liquid Crystals 315
    8.3.1 Domain Formation in an External Electric Field...... 315
    8.3.2 General System of Equations 317
    8.3.3 Excitations with a DC Electric Field................ 321
    8.3.4 Excitation with AC Electric Field 323
8.4 Excitation of Second Sound in a Smectic A Liquid Crystal ... 326
8.5 Electromagnetic Wave Propagation in a Cholesteric
    Liquid Crystal 331

9 Electrodynamics of Superconductors...................... 337
9.1 Superconducting Current 337
9.2 The Meissner Effect and Penetration Depth................ 341
9.3 Magnetic Flux in a Superconducting Ring 348
9.4 Examples of a DC Regime in a Superconductor 351
    9.4.1 Junction between a Normal Conductor
         and a Superconductor............................. 351
    9.4.2 Electrostatic Analogy 353
    9.4.3 Inductance of a Superconducting Thin Strip 355
9.5 Phenomena in a Superconductor under AC Regime 357
    9.5.1 The Phenomenological Two-Fluid Model 357
    9.5.2 Surface Impedance 360
    9.5.3 Superconducting Transmission Lines
         and Microwave Cavities 361
9.6 The Ginzburg–Landau Theory 366
    9.6.1 The Ginzburg–Landau Equations................... 366
    9.6.2 Examples of Ginzburg–Landau Theory Application ... 369
    9.6.3 Surface Energy at the Boundary
         between Normal Conductor and Superconductor 373
9.7 Type II Superconductors 377
    9.7.1 Mixed State 377
    9.7.2 London Model of the Mixed State 379
    9.7.3 Vortex Energy 381
    9.7.4 Vortex Lattice 382
    9.7.5 Upper Critical Field 383
    9.7.6 Vortex Motion 385

	9.8	The Josephson Effect 388
		9.8.1 The Josephson Relations 388
		9.8.2 Spatial Variation of the Phase Difference in a Magnetic Field 391
		9.8.3 Current Dependence on a Magnetic Field 394
		9.8.4 Wave Equation for a Josephson Junction 395

A The List of Notations 401

B Formulae of Vector Analysis 404
   B.1 Vector Operations in the Cartesian Coordinates 404
   B.2 Vector Operations in Cylindrical Coordinates 405
   B.3 Vector Operations in Spherical Coordinates 405
   B.4 Vector Analysis Identities................................. 406

C The Physical Constants 408

D The MKS System of Units 409

References 417

Index 421

1 Fundamentals of Electrodynamics

1.1 Charges and Fields

Electrodynamics as a scientific discipline is based primarily on the concept of electric charges as a specific physical quantities and interaction between them. The existence of electric charges is an experimentally established fact. They can be produced, for example, by rubbing a sample of amber. The observed phenomena of attraction, repulsion, heat generation are explained as a result of charges which have been produced [1]. An electric charge can be either positive, or negative. It can be measured and has its own dimension. In the MKS system of units which is used in this book the unit of an electric charge is *coulomb*. The charge of an electron is, as it is well known, negative, and its absolute value e is 1.60×10^{-19} coulomb.

Interaction between electric charges can be described by virtue of the concept of a field of forces. One can say that any electric charge creates around itself a field, and other electric charges placed in such a field are affected by some force [2]. In classical physics the introduction of a field is a method of description of the charged particles interaction as a specific physical phenomenon. From the point of view of the theory of relativity, the forces acting on a charged particle at a given moment of time are determined by the configuration of all other charged particles in previous moments of time due to the finite velocity of light [2]. As a result, a change of a position of any electric charge influences other electric charges only after some finite time interval. The field exists as itself, and we must consider an interaction between electric charges as a two stage process. First an electric charge interacts with a field, and then the field interacts with another electric charge [2]. Such a field is called electromagnetic one. Electromagnetic field consists of an electric field and a magnetic field. Below it will be shown that in the relativistic case they cannot exist separately. However, in the non-relativistic limit, i.e. when a velocity of the reference frame is small in comparison with the vacuum light velocity c, the electric and magnetic fields can be considered separately. It is convenient to define the characteristic quantities of the both fields for this case. Then these definitions will be generalized for the relativistic case. We start with the electric field intensity $\boldsymbol{E}$.

Definition 1. *The electric field intensity is the mechanical force exerted in an electric field on an infinitesimally small test body, divided by the charge of the test body* [1].

$$\mathbf{E} = \frac{\text{FORCE}}{\text{CHARGE}}. \tag{1.1}$$

The dimension of the electric field intensity in the MKS system of units is newton/coulomb.

Consider now the line integral between two points A and B:

$$\int_A^B E_l \, dl = \int_A^B \mathbf{E} \cdot d\mathbf{l}, \tag{1.2}$$

where E_l is the projection of the vector $\mathbf{E}$ on the direction of the line element vector $d\mathbf{l}$; the integrand in the right-hand side of (1.2) is the scalar product of the both vectors.

Definition 2. *The integral (1.2) is called the voltage V* [1]:

$$V = \int_A^B \mathbf{E} \cdot d\mathbf{l}. \tag{1.3}$$

The dimension of the voltage is according to (1.3)

$$(\text{newton} \cdot \text{m}) / \text{coulomb} = \frac{\text{joule}}{\text{coulomb}}. \tag{1.4}$$

The unit of voltage (1.4) is called volt.

In potential fields the voltage (1.3) does not depend on an integration path, being determined only by the integration limiting points A and B. In such a case voltage is simply equal to the difference of potential between these two points: $V = V_{AB}$.

Definition 3. *The electric displacement or electric induction vector $\mathbf{D}$ can be introduced as follows. Imagine that the charge creates the field by a kind of "excitation" of the surrounding medium. The electric induction vector $\mathbf{D}$ can be used as a measure of such an "excitation". For a single point charge e the flux of the vector $\mathbf{D}$, i.e. the density of "$\mathbf{D}$ lines" leaving e uniformly in all directions is given by* [1]:

$$\oint_S D_n \, ds = e, \tag{1.5}$$

where ds is an element of an arbitrary surface S surrounding the charge e. For example, for a spherical surface of radius R we get from (1.5):

$$4\pi R^2 D = e. \tag{1.6}$$

For arbitrary charge distribution, including a continuous one, (1.5) becomes [1]:

$$\oint_S D_n ds = \sum_i e_i, \tag{1.7}$$

where the right-hand side is the algebraic sum of positive and negative charges. The dimension of D is

$$\frac{\text{CHARGE}}{\text{AREA}} = \frac{\text{coulomb}}{\text{m}^2}, \tag{1.8}$$

which is essentially different from the electric field intensity dimension (1.1).

This definition of D, however, is not unique. It must be completed with the relation between the vectors D and E which will be obtained below both in general form and for each particular case. In the simplest case of an isotropic medium and linear relation between D and E it turned out that the "lines" of D are identical with the lines of force defined by the vector E [1]. The dimensionality analysis shows that the time derivative $\partial D/\partial t$ dimension coincides with the dimension of the electric current density j:

$$\frac{\text{CHARGE}}{\text{AREA} \cdot \text{TIME}} = \frac{\text{coulomb}}{\text{m}^2 \cdot \text{s}} = \frac{\text{ampere}}{\text{m}^2}, \tag{1.9}$$

which permits the generalization of the concept of electric current. The quantity of the total current j^{tot} introduced by Maxwell consists of the conduction current (1.9) and the so-called displacement current $\partial D/\partial t$ [1,4]:

$$j^{\text{tot}} = j + \frac{\partial D}{\partial t}. \tag{1.10}$$

Consider now the quantities characterizing the magnetic field. The magnetic field exerts a mechanical force on a magnetic pole which is at the beginning assumed to be isolated. Quantitatively the magnetic pole can be evaluated by virtue of the Ampère's relation between current and magnetism. For example, the magnetic field of a plane circulating current I about the area S is, at a great distance from I, equal to the field of a bar magnet placed normal to S at I, with the moment [1]

$$\mu_m = IS. \tag{1.11}$$

On the other hand, the moment (1.11) of the magnet can be expressed in terms of the pole strength P and pole separation l

$$\mu_m = Pl. \tag{1.12}$$

Comparison of (1.11) and (1.12) permits the definition of the pole strength and its dimension [1]:

$$P = \frac{IS}{l}, \tag{1.13}$$

with the dimension coulomb · m/s. Once the pole strength is defined we can define the magnetic field induction $\boldsymbol{B}$.

Definition 4. *The magnetic induction $\boldsymbol{B}$ is the mechanical force exerted by the magnetic field on the magnetic pole divided by the pole strength [1]:*

$$\boldsymbol{B} \equiv \frac{\text{FORCE}}{\text{POLE STRENGTH}}. \tag{1.14}$$

The dimension of the magnetic induction $\boldsymbol{B}$ is given by

$$\frac{\text{newton} \cdot \text{s}}{\text{coulomb} \cdot \text{m}}.$$

The direction of the vector $\boldsymbol{B}$, varying from point to point, is represented by the form of the magnetic lines of force.

Definition 5. *The magnetic field intensity $\boldsymbol{H}$ can be identified as the magnetic pole strength per unit area of the surface enclosing the magnetic moment [1]:*

$$\boldsymbol{H} = \frac{\text{POLE STRENGTH}}{\text{AREA}}. \tag{1.15}$$

The dimension of the magnetic field intensity $\boldsymbol{H}$ is given by

$$\frac{\text{coulomb}}{\text{m} \cdot \text{s}} = \frac{\text{ampere}}{\text{m}}.$$

In isotropic media the directions of $\boldsymbol{H}$ and $\boldsymbol{B}$ coincide, and the lines of $\boldsymbol{H}$ are identical with the lines of force of the vector $\boldsymbol{B}$.

1.2 Maxwell Equations

1.2.1 Maxwell Equations in the Differential Form

The Maxwell equations in the rationalized MKS system of units read [3]

$$\text{div}\,\boldsymbol{B} = 0 \tag{1.16}$$

$$\operatorname{curl} \boldsymbol{E} = -\frac{\partial \boldsymbol{B}}{\partial t} \tag{1.17}$$

$$\operatorname{div} \boldsymbol{D} = \rho \tag{1.18}$$

$$\operatorname{curl} \boldsymbol{H} = \boldsymbol{j} + \frac{\partial \boldsymbol{D}}{\partial t}. \tag{1.19}$$

They are the generalization of the set of empirical laws established in the 18-th and 19-th centuries, and governing the properties of electric $\boldsymbol{E}(\boldsymbol{r},t)$ and magnetic $\boldsymbol{H}(\boldsymbol{r},t)$ field intensities, electric displacement (electric induction) $\boldsymbol{D}(\boldsymbol{r},t)$ and magnetic induction $\boldsymbol{B}(\boldsymbol{r},t)$ in the medium with the charge density of $\rho(\boldsymbol{r},t)$ and the current density of $\boldsymbol{j}(\boldsymbol{r},t)$. The physical meaning of these equations may be explained as follows:

1) Equation (1.18) is actually a sophisticated form of Coulomb law, stating that the polarization produced at the point $\boldsymbol{r}$ by the point charge q placed at the point $\boldsymbol{r}$: $\rho(\boldsymbol{r},t) = q\delta(\boldsymbol{r}-\boldsymbol{r}')$ is directed along the vector $\boldsymbol{r}$.

2) Equation (1.16) when compared to (1.18) tells us that there are no magnetic charges (it is impossible to created a positive pole of a magnet without creating a negative one).

3) Equation (1.17) demonstrates Faraday's law, according to which a changing magnetic induction produces an electric field.

4) Equation (1.19) is Ampere's law determining the magnetic field created by electric current $\operatorname{curl} \boldsymbol{H} = \boldsymbol{j}$, combined with Maxwell's additional term

$$\frac{\partial \boldsymbol{D}}{\partial t}$$

(displacement current) stating (in analogy with (1.17)) that changing electric fields produce magnetic fields. Equations (1.16)–(1.19) are macroscopic, and the quantities appearing there are assumed to be the result of averaging over physically infinitesimal elements of volume thus ignoring the variations on a microscopic level, [4, 5]. The system of (1.16)–(1.19) should be completed with the so-called material equations which establish the relation between the fields $\boldsymbol{E}$, $\boldsymbol{H}$ and induction vectors $\boldsymbol{D}$, $\boldsymbol{B}$, respectively which have been defined independently in the previous section. The material equations are especially important in the MKS form of the Maxwell equations since in this case the dielectric constant ε and permeability μ possess their characteristic dimensionality, and even in vacuum they differ from unity. In general case both quantities are second rank tensors and may, in principle, depend on coordinates, time and in non-linear approximation on fields. However, we will limit our study with linear homogeneous media. In the majority of problems to be considered a medium is also assumed to be isotropic, and we begin with the simplest case where the dielectric constant and permeability are scalars independent on time and spatial coordinates. The material equations have the form:

$$\boldsymbol{D} = \varepsilon \boldsymbol{E}; \quad \boldsymbol{B} = \mu \boldsymbol{H}, \tag{1.20}$$

where

$$\varepsilon = \varepsilon_0 \varepsilon^r; \quad \mu = \mu_0 \mu^r. \tag{1.21}$$

The dielectric constant of vacuum $\varepsilon_0 = 10^{-9}/(36\pi) \approx 8.854 \times 10^{-12}$ farad/m, the permeability of vacuum $\mu_0 = 4\pi \times 10^{-7} \approx 1.257 \times 10^{-6}$ henry/m. The factors ε^r, μ^r are the dimensionless relative permittivity (dielectric constant) and relative permeability, respectively. It should be noted that the velocity of light in vacuum c is expressed in terms of ε_0, μ_0 as follows

$$c = \frac{1}{\sqrt{\varepsilon_0 \mu_0}} = 2.997 \times 10^8 \text{ m/s} \approx 3 \times 10^8 \text{ m/s}.$$

In vacuum, where both the relative permittivity ε^r and permeability μ^r are unity the field inductions and intensities differ only by the constant numerical factors ε_0, μ_0 determined by the choice of units. The situation in a medium is much more complicated. In order to clarify the material equations we must introduce the additional vectors $\boldsymbol{P}$ and $\boldsymbol{M}$ of the electric and magnetic polarization respectively which describe the medium response to the external field. They can be defined as follows [3]:

$$\boldsymbol{P} = \boldsymbol{D} - \varepsilon_0 \boldsymbol{E}, \quad \boldsymbol{M} = \frac{1}{\mu_0} \boldsymbol{B} - \boldsymbol{H}. \tag{1.22}$$

Obviously, both polarizations vanish in vacuum. Substitution of (1.22) into (1.18) and (1.19) makes it possible to exclude $\boldsymbol{D}$ and $\boldsymbol{H}$ from the Maxwell equations. Equations (1.16) and (1.17) contain only $\boldsymbol{E}$ and $\boldsymbol{B}$ while (1.18) and (1.19) take the form [3]:

$$\text{curl}\boldsymbol{B} - \varepsilon_0 \mu_0 \frac{\partial \boldsymbol{E}}{\partial t} = \mu_0 \left(\boldsymbol{j} + \frac{\partial \boldsymbol{P}}{\partial t} + \text{curl}\boldsymbol{M} \right) \tag{1.23}$$

and

$$\text{div}\boldsymbol{E} = \frac{1}{\varepsilon_0} (\rho - \text{div}\boldsymbol{P}). \tag{1.24}$$

Equations (1.23) and (1.24) show that the medium subject to electromagnetic field can be replaced by the equivalent distributions of intrinsic charges ρ_{int} and currents $\boldsymbol{j}_{\text{int}}$ which have the form

$$\rho_{\text{int}} = -\text{div}\boldsymbol{P}, \quad \boldsymbol{j}_{\text{int}} = \frac{\partial \boldsymbol{P}}{\partial t} + \text{curl}\boldsymbol{M}. \tag{1.25}$$

In isotropic media the polarization vectors are parallel to the corresponding field intensity vectors. It is known from the well established experimental results that in moderate fields where the linear approximation is valid and in the absence of ferromagnetism and ferroelectricity the polarization vectors are proportional to the field intensities [3]:

$$\boldsymbol{P} = \chi_e \varepsilon_0 \boldsymbol{E}, \ \boldsymbol{M} = \chi_m \boldsymbol{H}. \tag{1.26}$$

The electric susceptibility χ_e and the magnetic susceptibility χ_m of a medium are dimensionless factors which are related to the relative permittivity and permeability as follows:

$$\chi_e = \varepsilon^r - 1, \ \chi_m = \mu^r - 1. \tag{1.27}$$

In an anisotropic medium both susceptibilities represent the second rank tensors. It can be easily shown that the electric polarization $\boldsymbol{P}$ is the electric dipole moment per unit volume of a medium, and the magnetic polarization, or magnetization $\boldsymbol{M}$ is the magnetic dipole moment per unit volume of a medium.

To complete the description of the overall structure of Maxwell's equation two important points should be added. Firstly, while the first pair of these equations ((1.16), (1.17)) is homogeneous and in this sense self-sufficient, the second pair ((1.18), (1.19)) is inhomogeneous, thus requiring a knowledge of charge $\rho(\boldsymbol{r},t)$ and current $\boldsymbol{j}(\boldsymbol{r},t)$ densities. The latter quantities are determined by equations of motion (in general quantum, but for a large number of problems classical to a good approximation) of charges in the presence of electromagnetic field $\boldsymbol{E}, \boldsymbol{H}$. Thus Maxwell's equations become coupled with the equations of motion and the whole set of equations should be solved simultaneously in a self-consistent manner. Usually at the intermediate stage of calculation the dependencies of ρ and $\boldsymbol{j}$ on $\boldsymbol{D}$ and $\boldsymbol{H}$ are evaluated, determining the material equations[1] (1.20). Moreover, since ρ and $\boldsymbol{j}$ are averaged quantities, a statistical average should be incorporated; this condition gives rise to the temperature dependence in material equations.

Secondly, it may easily be shown that the second pair of Maxwell's equations satisfies the continuity equation (which is a differential form of the law of charge conservation)

$$\frac{\partial \rho}{\partial t} + \mathrm{div}\boldsymbol{j} = 0 \tag{1.28}$$

irrespective of the concrete form of material equations or, in other words, in any medium.

1.2.2 Maxwell Equations in the Integral Form

The Maxwell equations can be represented in the integral form which is important in applications. In order to obtain these relations we use two well

---
[1] Sometimes to avoid the solution of equations of motion, the material equations are simply postulated, and are then called phenomenological equations.

known theorems of the vector analysis: the Stokes theorem and the Gauss theorem. The Stokes theorem reads that a vector function $\boldsymbol{f}(x,y,z)$ which is continuous with its first derivatives everywhere on the surface S and on the closed contour C limiting this surface obeys the following relationship:

$$\int_C \boldsymbol{f}(x,y,z) \cdot d\boldsymbol{l} = \int_S \operatorname{curl}\boldsymbol{f}(x,y,z) \cdot d\boldsymbol{s}, \tag{1.29}$$

where $d\boldsymbol{l}$ is the element of the contour C length tangential to the contour at each point, and $d\boldsymbol{s}$ is the area element of the surface S representing the positive normal with respect to the surface S. The integral in the left-hand side of (1.29) is called a circulation of the vector function $\boldsymbol{f}(x,y,z)$.

The Gauss theorem permits the transformation of an integral over a closed volume V into an integral over the surface S limiting this volume. If a vector function $\boldsymbol{f}(x,y,z)$ meets the conditions mentioned above everywhere in the volume V and on its surface S then the following expression takes place:

$$\int_S \boldsymbol{f}(x,y,z) \cdot d\boldsymbol{s} = \int_V \operatorname{div}\boldsymbol{f}(x,y,z) \, dV. \tag{1.30}$$

The integral in the left-hand side of (1.30) is called the flux of the vector function $\boldsymbol{f}(x,y,z)$ through the surface S.

The application of the Gauss theorem to the Maxwell equation (1.16) yields:

$$\int_V \operatorname{div}\boldsymbol{B}\,dV = \int_S \boldsymbol{B} \cdot d\boldsymbol{s} = \Phi = 0, \tag{1.31}$$

where Φ is the flux of the magnetic induction vector $\boldsymbol{B}$, or magnetic flux through the closed surface S. This Maxwell equation in the integral form claims that the total magnetic flux through the closed surface is zero.

Being applied to the Maxwell equation (1.18) the Gauss theorem shows that the total flux of the electric induction vector $\boldsymbol{D}$ through the closed surface S is equal to the total electric charge Q contained in the volume V limited by S. Indeed,

$$\int_V \operatorname{div}\boldsymbol{D}\,dV = \int_S \boldsymbol{D} \cdot d\boldsymbol{s} = \int_V \rho\,dV = Q. \tag{1.32}$$

Integrate both sides of the Maxwell equation (1.17) over the closed surface S traversing the magnetic field $\boldsymbol{B}$. We get:

$$\int_S \operatorname{curl}\boldsymbol{E} \cdot d\boldsymbol{s} = -\frac{\partial}{\partial t}\int_S \boldsymbol{B} \cdot d\boldsymbol{s}. \tag{1.33}$$

1.2 Maxwell Equations

Application of the Stokes theorem to the left-hand side of (1.33) yields the circulation of the electric field intensity $\boldsymbol{E}$ along the closed contour C limiting the surface S.

$$\int_S \mathrm{curl}\boldsymbol{E} \cdot d\boldsymbol{s} = \int_C \boldsymbol{E} \cdot d\boldsymbol{l}. \tag{1.34}$$

The circulation of the electric field along the closed contour is called the electromotive force (e.m.f.) in this contour [2]. The integral in the right-hand side of (1.33) is the magnetic flux Φ through the surface S as it is seen from the comparison of (1.31) and (1.33). So we obtain:

$$\int_C \boldsymbol{E} \cdot d\boldsymbol{l} = -\frac{\partial \Phi}{\partial t}. \tag{1.35}$$

The similar procedure applied to the Maxwell equation (1.19) gives for the circulation of the magnetic field intensity $\boldsymbol{H}$

$$\int_S \mathrm{curl}\boldsymbol{H} \cdot d\boldsymbol{s} = \int_C \boldsymbol{H} \cdot d\boldsymbol{l} = \int_S \boldsymbol{j} \cdot d\boldsymbol{s} + \frac{\partial}{\partial t} \int_S \boldsymbol{D} \cdot d\boldsymbol{s}. \tag{1.36}$$

The first integral in the right-hand side of (1.36) is the total conduction current I flowing in the contour C. The second term in the

$$\int_S \boldsymbol{j} \cdot d\boldsymbol{s} = I. \tag{1.37}$$

The second term in the right-hand side of (1.36) is the so-called displacement current which exists only in the time dependent field. Finally we have:

$$\int_C \boldsymbol{H} \cdot d\boldsymbol{l} = I + \frac{\partial}{\partial t} \int_S \boldsymbol{D} \cdot d\boldsymbol{s}. \tag{1.38}$$

Equations (1.31), (1.32), (1.35) and (1.38) are the Maxwell equations in the integral form. They can be especially useful in the problems with highly symmetrical bodies where the surface and contour integrals can be easily calculated in a closed form. Some problems of such a type will be analyzed below.

1.2.3 Field Potentials

The complete system of the Maxwell equations considered above along with the appropriate boundary conditions allows for a complete determination of both electromagnetic fields and current densities, without introducing any additional concepts and objects. However, it is certainly permissible to play correct mathematical games with Maxwell's equations, though their fruitfulness may not be immediately apparent. In this sense, the first pair of Maxwell equations would seem particularly attractive to anyone who has enjoyed his or her undergraduate course in vector calculus. Indeed, when compared with identity div curl$q$ = 0, where q is an arbitrary vector, (1.16) at once suggests that the magnetic field vector $B(r, t)$ may be written as a curl of another vector, $A(r, t)$, called the vector potential:

$$B = \operatorname{curl} A. \tag{1.39}$$

Then (1.17) may be rewritten as

$$\operatorname{curl} E + \frac{\partial}{\partial t} \operatorname{curl} A \equiv \operatorname{curl} \left[E + \frac{\partial A}{\partial t} \right] = 0. \tag{1.40}$$

Now, comparing (1.40) with another identity

$$\operatorname{curl} \operatorname{grad} q = 0$$

we see that the expression in square brackets in this equation may be written as a grad of some scalar function, say $(-\varphi(r, t))$

$$E + \frac{\partial A}{\partial t} = -\nabla \varphi(r, t),$$

or

$$E = -\frac{\partial A}{\partial t} - \nabla \varphi(r, t). \tag{1.41}$$

Thus, instead of the two vector fields E and B (six independent functions of r, t) we have invented two potentials, vector and scalar (four independent functions), which, if known, determine the six components of electromagnetic field via (1.39) and (1.41). Actually, our gain is even more significant: in terms of these potentials the first two Maxwell equations are satisfied automatically, and we are left with the problem of solving the second pair (four equations), to determine four potential functions.

Before we proceed this way, let us make the following important observation: relationships (1.39), (1.41) allow for a unique determination of the fields through the potentials, but the opposite is not true – different potentials may result in the same fields! Indeed, (1.39) indicates that the two vector potentials, A and A', give rise to the same magnetic field B if

$$A' = A + \nabla \chi, \tag{1.42}$$

where $\chi(r, t)$ is an arbitrary scalar function.

Turning now to (1.41) we observe that if the transformation $A \to A'$ (1.42) is accompanied by the appropriate transformation

$$\varphi' = \varphi - \frac{\partial \chi}{\partial t} \tag{1.43}$$

then the electric field $\boldsymbol{E}$ remains unchanged as well.

Transformations (1.42), (1.43) are called gauge transformations and $\boldsymbol{A}$ and φ are sometimes (in quantum field theory always) called gauge potentials. Using this terminology, we can say that (1.39), (1.41) and (1.42), (1.43) indicate that electromagnetic field $\boldsymbol{E}, \boldsymbol{B}$ is invariant with respect to the gauge transformation.

Two conclusions – one of great fundamental importance, the other of a corresponding technical force – may be drawn from our consideration of the gauge properties of Maxwell equations. Both conclusions are based on the fact that the electromagnetic potentials $\boldsymbol{A}$ and φ have been invented through (1.39), (1.41) for the sake of convenience, as a pure mathematical device; only fields $\boldsymbol{E}$ and $\boldsymbol{B}$ have physical significance, being experimentally measurable quantities and affecting the motion of charged particles (e.g. in classical mechanics via the Lorentz force). Hence, first, any equation in physics (classical and quantum field equations and equations of motion, material equations etc.) containing electromagnetic potentials has to be gauge invariant – e.g. has to remain unchanged (or lead to the same physical results) under gauge transformations (1.42), (1.43). The requirement of gauge invariance of any physical theory is at least as crucial as the requirement of its Lorentz-invariance. Secondly, we are allowed to make use of the freedom to change the electromagnetic potentials in the framework of gauge transformations and choose a proper gauge (a particular $\chi(\boldsymbol{r},t)$) to simplify either certain equations or calculations of certain physical properties.

Now we have all the required tools to proceed in rewriting Maxwell's equations in terms of electromagnetic potentials. Substituting (1.39), (1.41) into (1.18), (1.19), respectively, and taking into account the material (1.20) and the definitions (1.21) we obtain for a homogeneous, isotropic and linear medium

$$\Delta\varphi - \mu\varepsilon\frac{\partial^2 \varphi}{\partial t^2} = -\frac{\rho}{\varepsilon} - \frac{\partial}{\partial t}\left(\operatorname{div}\boldsymbol{A} + \mu\varepsilon\frac{\partial \varphi}{\partial t}\right) \tag{1.44}$$

$$\Delta\boldsymbol{A} - \mu\varepsilon\frac{\partial^2 \boldsymbol{A}}{\partial t^2} = -\mu\boldsymbol{j} + \operatorname{grad}\left(\operatorname{div}\boldsymbol{A} + \mu\varepsilon\frac{\partial \varphi}{\partial t}\right), \tag{1.45}$$

where Δ is the Laplace operator (Laplacian):

$$\Delta = \frac{\partial^2}{\partial x^2} + \frac{\partial^2}{\partial y^2} + \frac{\partial^2}{\partial z^2}.$$

Let us use the freedom of performing the gauge transformation and choose such an $\chi(\boldsymbol{r},t)$ that the transformed potentials satisfy the so-called Lorentz condition (Lorentz gauge):

$$\operatorname{div} \boldsymbol{A} + \mu\varepsilon \frac{\partial \varphi}{\partial t} = 0. \tag{1.46}$$

With such a gauge, (1.44), (1.45) take a considerably simplified decoupled form:

$$\Box \varphi \equiv \left[\Delta - \mu\varepsilon \frac{\partial^2}{\partial t^2}\right] \varphi = -\frac{\rho}{\varepsilon} \tag{1.47}$$

$$\Box \boldsymbol{A} \equiv \left[\Delta - \mu\varepsilon \frac{\partial^2}{\partial t^2}\right] \boldsymbol{A} = -\mu \boldsymbol{j}. \tag{1.48}$$

Equations (1.47), (1.48) are the standard inhomogeneous wave equations, intensively studied in mathematical physics; the differential operator

$$\Box = \Delta - \mu\varepsilon \frac{\partial^2}{\partial t^2} \equiv \frac{\partial^2}{\partial x^2} + \frac{\partial^2}{\partial y^2} + \frac{\partial^2}{\partial z^2} - \mu\varepsilon \frac{\partial^2}{\partial t^2} \tag{1.49}$$

is known as D'Alambert's operator. In the case of wave propagation in vacuum using the relation between the light velocity c and the quantities ε_0, μ_0 mentioned above we can rewrite the D'Alambert's operator (1.49) in the more traditional form

$$\Box = \frac{\partial^2}{\partial x^2} + \frac{\partial^2}{\partial y^2} + \frac{\partial^2}{\partial z^2} - \frac{1}{c^2} \frac{\partial^2}{\partial t^2}. \tag{1.50}$$

For homogeneous isotropic media where the relative dielectric constant ε_r and permeability μ_r differ from unity the vacuum light velocity c is changed to the quantity

$$c_m = \frac{c}{\sqrt{\varepsilon_r \mu_r}} < c,$$

which has a meaning of a light velocity in a medium.

Equations (1.47), (1.48) and the Lorentz condition (1.46) still do not fix the unique form of electromagnetic potentials $\boldsymbol{A}$ and φ. It is easy to check that if the given $\boldsymbol{A}(\boldsymbol{r}, t)$ and $\varphi(\boldsymbol{r}, t)$ are the solutions of (1.46) and (1.47), (1.48) then $\boldsymbol{A}$ and φ, related to $\boldsymbol{A}$ and φ by gauge transformations (1.42), (1.43) are their solutions too, provided that $\chi(\boldsymbol{r}, t)$ satisfies the equation

$$\Box \chi(\boldsymbol{r}, t) = 0 \tag{1.51}$$

(which means that χ may be an arbitrary function of the argument $(\boldsymbol{r} - ct)$).

Another widely used gauge is the so-called Coulomb gauge, defined by the condition

$$\operatorname{div} \boldsymbol{A} = 0. \tag{1.52}$$

If this condition is imposed, (1.44), (1.45) take the form

$$\Delta \varphi = -\frac{\rho}{\varepsilon} \tag{1.53}$$

$$\Box \boldsymbol{A} = -\mu \boldsymbol{j} + \mu \varepsilon \frac{\partial}{\partial t} (\mathrm{grad} \varphi). \tag{1.54}$$

These equations are effectively decoupled too, since (1.53) can be solved independently; after substituting its solution in (1.54), the latter takes the standard form of an inhomogeneous wave equation. As in the case of the Lorentz gauge, (1.52) and (1.53), (1.54) do not determine the potentials uniquely; gauge transformation (1.42), (1.43), with $\chi(\boldsymbol{r}, t)$ satisfying the Laplace equation

$$\Delta \chi(\boldsymbol{r}, t) = 0 \tag{1.55}$$

leaves (1.52), (1.53), (1.54) unchanged. Solutions of (1.55) are well known – they are spherical harmonics.

In the quantum mechanical description of the electron motion in constant magnetic field $\boldsymbol{B}$ two special cases of the Coulomb gauge are widely used: (a) the Landau gauge

$$\boldsymbol{A}_L = \{-By, 0, 0\} \tag{1.56}$$

(b) the symmetric gauge

$$\boldsymbol{A}_{sym} = \frac{1}{2} \{\boldsymbol{B} \times \boldsymbol{r}\}. \tag{1.57}$$

Both gauges are related by gauge transformation (1.42) with a function

$$\chi = \frac{1}{2} B x y \tag{1.58}$$

being proportional to the second rank spherical harmonic d_{xy}.

1.3 Relativistic Properties of the Maxwell Equations

1.3.1 Lorentz Transformation of Coordinates

Experiment shows that the two fundamental principles are valid [2, 3].

1. The first principle is the so-called relativity principle. According to this principle, all the laws of nature are identical in all inertial systems of reference, i.e. in such systems where a moving body which is not acted upon by external forces proceeds with the constant velocity. The equations expressing the laws of nature are invariant with respect to transformations

of coordinates and time from one inertial system to another. This means that the equation describing any law of nature written in terms of coordinates and time in different inertial reference systems has one and the same form.
2. The second principle reads that the velocity of light c in a free space is the universal constant which is independent on a reference system.

We try to find the formula of transformation from an inertial reference frame K to a system K′ moving relative to K with velocity v along the z-axis. The so-called Lorentz transformation establishes the relativistic transformation as a consequence of the requirement that it leave the interval between events

$$dS^2 = c^2 dt^2 - dx^2 - dy^2 - dz^2 \tag{1.59}$$

invariant. If we use the quantity $ict = \tau$ the interval between events can be considered as the distance between the corresponding pair of world points (x_1, y_1, z_1, t_1) and (x_2, y_2, z_2, t_2) in a four-dimensional system of coordinates. Here $i = \sqrt{-1}$ is the imaginary unity. Consequently, one may say that the required transformation must leave unchanged all distances in the four-dimensional (x, y, z, τ) space [2]. Such transformations consist only of parallel displacements and rotations of the coordinate system. The displacement of the coordinate system parallel to itself can be ignored, since it leads only to a shift in the origin of the space coordinates and a change in the time reference point. Thus the required transformation must be expressible mathematically as a rotation of the four-dimensional (x, y, z, τ) coordinate system. Every rotation in the four-dimensional space can be resolved into six rotations, in the planes $xy, zy, xz, \tau x, \tau y, \tau z$. For the sake of generality we introduce the four-vector denotion

$$x = x_1, \ y = x_2, \ z = x_3, \ ict = \tau = x_4. \tag{1.60}$$

Then the rotation transformation can be written as follows [3]:

$$x'_j = \sum_{k=1}^{4} a_{jk} x_k; \quad (j = 1, 2, 3, 4), \tag{1.61}$$

where the determinant of transformation $\det \|a_{jk}\|$ is equal to unity, the coordinates with a prime belong to the moving reference system, and

$$\sum_{j=1}^{4} a_{ji} a_{jk} = \delta_{ik}; \quad (j = 1, 2, 3, 4). \tag{1.62}$$

The first three of these rotations transform only the space coordinates; they correspond to the usual space rotations [2]. For the sake of definiteness consider a rotation in the $x_3 x_4$ plane; under this, the y- and x-coordinates do not change. Then we have [3]

1.3 Relativistic Properties of the Maxwell Equations

$$x_1 = x_1', \quad x_2 = x_2', \quad x_3' = a_{33}x_3 + a_{34}x_4, \quad x_4' = a_{43}x_3 + a_{44}x_4. \tag{1.63}$$

Our purpose is to calculate the finite components of the transformation tensor $a_{11}, a_{22}, a_{33}, a_{34}, a_{43}, a_{44}$. The conditions (1.62) give the following relations between the coefficients a_{jk}

$$a_{34}^2 + a_{44}^2 = 1, \quad a_{33}^2 + a_{43}^2 = 1, \quad a_{33}a_{34} + a_{43}a_{44} = 0. \tag{1.64}$$

The results (1.64) and the unitary character of the transformation yield [3]:

$$a_{11} = 1, \quad a_{22} = 1, \quad a_{33} = a_{44}, \quad a_{34} = -a_{43}. \tag{1.65}$$

Consider the motion of the reference frame K′origin $x_3' = 0$ [2]. This implies the condition

$$\frac{x_3}{x_4} = \frac{z}{ict} = -i\frac{v}{c} = -\frac{a_{34}}{a_{33}}. \tag{1.66}$$

Combining (1.64), (1.65) and (1.66) we finally obtain [3]:

$$a_{33} = a_{44} = \frac{1}{\sqrt{1 - v^2/c^2}}; \quad a_{34} = -a_{43} = i\frac{v}{c}\frac{1}{\sqrt{1 - v^2/c^2}}. \tag{1.67}$$

Returning to the spatial coordinates and time we write using the results obtained

$$x' = x, \quad y' = y, \quad z' = \frac{z - vt}{\sqrt{1 - v^2/c^2}}, \quad t' = \frac{1}{\sqrt{1 - v^2/c^2}}\left(t - \frac{v}{c^2}z\right) \tag{1.68}$$

This result is known to be the *Lorentz transformation* [2,3]. It is seen from (1.68) that the inverse transformation to the reference system K can be achieved immediately by changing in these equations v to $-v$. It is also easy to see from (1.68) that on making the transition to the limit $c \to \infty$ and classical mechanics, the Lorentz transformation goes over into the Galileo transformation [2,3].

$$x' = x, \quad y' = y, \quad z' = z - vt, \quad t' = t.$$

1.3.2 Four-Dimensional Form of the Maxwell Equations

Let us now examine the transformation properties of fields and currents under the Lorentz transformation, as an essential step in the construction of Lagrangian formalism. For this purpose we must represent them in a 4-dimensional form. Starting with (1.47), (1.48) we first point out that the D'Alambert operator (1.49) is Lorentz invariant (that is, it behaves like a scalar). The quantities $ic\rho$ and $\boldsymbol{j}$ form a 4-vector $\boldsymbol{J}^{(4)}$. Indeed, multiplying every component dx^i of the 4-vector $\{d\boldsymbol{r}, icdt\}$ by the total charge dq contained in the volume element dV ($dq = \rho dV$), one obtains

1 Fundamentals of Electrodynamics

$$dq dx^i = \rho dV dx^i = dV dt \frac{dx^i}{dt} \rho \ldots \ldots (i = 1, 2, 3, 4). \tag{1.69}$$

Since dq (the charge) and $dV dt$ (of the volume element in 4-space) are scalars and dx^i is a 4-vector, then $\rho dx^i/dt$ is a 4-vector as well. This vector (whose components are $\{\boldsymbol{j}, i\rho c\}$) is called the 4-current. Thus, the right-hand sides of (1.47), (1.48) form a 4-vector $\boldsymbol{J}^{(4)}$ and (1.47), (1.48) are Lorentz-invariant when and only when φ and $\boldsymbol{A}$ form a 4-vector $\boldsymbol{A}^{(4)} = \{\boldsymbol{A}, i\varphi/c\}^2$; this vector is called a 4-potential. We proceed with the construction of the Lorentz-invariant relationships between 4-vectors of the potential and current. We start with the Maxwell equations (1.16) and (1.19) choosing the vectors of electric induction $\boldsymbol{D}$ and magnetic field intensity $\boldsymbol{H}$ as the variables. The explicit form of these equations in the Cartesian coordinates yields:

$$\sum_{k=1}^{4} \frac{\partial G_{jk}}{\partial x_k} = J_j; \quad (j = 1, 2, 3, 4), \tag{1.70}$$

where G_{jk} are 6 independent components of the antisymmetric second rank tensor which has the form

$$G_{jk} = \begin{vmatrix} 0 & H_3 & -H_2 & -icD_1 \\ -H_3 & 0 & H_1 & -icD_2 \\ H_2 & -H_1 & 0 & -icD_3 \\ icD_1 & icD_2 & icD_3 & 0 \end{vmatrix}. \tag{1.71}$$

The pair of homogeneous Maxwell equations (1.17) and (1.18) yields the second pair of variables $\boldsymbol{E}$ and $\boldsymbol{B}$ that compose the other antisymmetric second rank tensor F_{jk}

$$F_{jk} = \begin{vmatrix} 0 & B_3 & -B_2 & -\frac{i}{c}E_1 \\ -B_3 & 0 & B_1 & -\frac{i}{c}E_2 \\ B_2 & -B_1 & 0 & -\frac{i}{c}E_3 \\ \frac{i}{c}E_1 & \frac{i}{c}E_2 & \frac{i}{c}E_3 & 0 \end{vmatrix}. \tag{1.72}$$

Equations (1.17) and (1.18) now can be rewritten as follows

$$\frac{\partial F_{ij}}{\partial x_k} + \frac{\partial F_{ki}}{\partial x_j} + \frac{\partial F_{jk}}{\partial x_i} = 0, \quad (i, j, k = 1, 2, 3, 4). \tag{1.73}$$

It is seen from expressions (1.71), (1.72) that the real components of the tensors correspond to the magnetic field while the imaginary components are associated with the electric field. This feature can be expressed in the form of the axial vectors equivalent to the antisymmetric tensors

---

[2] An additional justification of φ being the zero (time) component of 4-potential will be presented when the Langranian formalism is considered.

$$^2\boldsymbol{F} = \left(\boldsymbol{B}, -\frac{i}{c}\boldsymbol{E}\right), \quad ^2\boldsymbol{G} = (\boldsymbol{H}, -ic\boldsymbol{D}). \tag{1.74}$$

Such a definition permits the alternative formulation of the field equations which is based on the dual system of variables. Namely, we can write

$$^2\boldsymbol{F}^* = \left(-\frac{i}{c}\boldsymbol{E}, \boldsymbol{B}\right), \quad ^2\boldsymbol{G}^* = (-ic\boldsymbol{D}, \boldsymbol{H}), \tag{1.75}$$

or in the explicit form

$$^2F^*_{jk} = \begin{vmatrix} 0 & -\frac{i}{c}E_3 & \frac{i}{c}E_2 & B_1 \\ \frac{i}{c}E_3 & 0 & -\frac{i}{c}E_1 & B_2 \\ -\frac{i}{c}E_2 & \frac{i}{c}E_1 & 0 & B_3 \\ -B_1 & -B_2 & -B_3 & 0 \end{vmatrix} \tag{1.76}$$

and

$$^2G^*_{jk} = \begin{vmatrix} 0 & -icD_3 & icD_2 & H_1 \\ icD_3 & 0 & -icD_1 & H_2 \\ -icD_2 & icD_1 & 0 & H_3 \\ -H_1 & -H_2 & -H_3 & 0 \end{vmatrix} \tag{1.77}$$

Then, the field equations take the form

$$\sum_{k=1}^{4} \frac{\partial F^*_{jk}}{\partial x_k} = 0, \quad (j = 1, 2, 3, 4) \tag{1.78}$$

$$\frac{\partial G^*_{ij}}{\partial x_k} + \frac{\partial G^*_{ki}}{\partial x_j} + \frac{\partial G^*_{jk}}{\partial x_i} = J_l, \quad (i, j, k, l = 1, 2, 3, 4). \tag{1.79}$$

Expressions (1.75)–(1.79) have an essential disadvantage in comparison with expressions (1.70)–(1.74) since the components F^*_{jk} and G^*_{jk} being reduced to the three-dimensional vectors yield the axial vector $\boldsymbol{E}$ and the polar vector $\boldsymbol{B}$ while the opposite case is true. Consequently, the 4-tensors F_{jk} and G_{jk} should be considered as the basic ones for the natural form of the Maxwell equations in the 4-vector notation (1.70) and (1.73). We should add to this system the continuity (1.28) which in the 4-vector notation takes the form

$$\sum_{k=1}^{4} \frac{\partial J_k}{\partial x_k} = 0. \tag{1.80}$$

Taking into account (1.39), (1.41) we can express the basic 4-tensor components F_{jk} in terms of the 4-potential $\boldsymbol{A}^{(4)}$ in the following way

$$F_{jk} = \frac{\partial A^{(4)}_k}{\partial x_j} - \frac{\partial A^{(4)}_j}{\partial x_k}; \quad (j, k = 1, 2, 3, 4). \tag{1.81}$$

18   1 Fundamentals of Electrodynamics

Unlike the CGS system of units where the Maxwell equations contain only two field variables, E and H, the Maxwell equations in the MKS system of units contain four field vectors E, D, H, B. In order to express all these quantities in terms of the 4-potential $A^{(4)}$ keeping the symmetry of denotions we need to introduce some kind of a symmetric relation between F_{jk} and G_{jk}. For this purpose we introduce the symmetric tensor γ_{jk} such that

$$\gamma_{jk} = \frac{1}{\mu} \text{ if } j,k = 1,2,3, \quad \gamma_{jk} = \varepsilon c^2 \text{ if } j \text{ or } k = 4, \text{ and } \gamma_{44} = \mu \varepsilon^2 c^4 \tag{1.82}$$

Then we have

$$G_{jk} = \gamma_{jk} F_{jk} \tag{1.83}$$

$$\sum_{k=1}^{4} \gamma_{jk} \frac{\partial F_{jk}}{\partial x_k} = J_j; \quad (j = 1,2,3,4) \tag{1.84}$$

and taking into account (1.81) we finally obtain

$$\sum_{k=1}^{4} \frac{\partial^2}{\partial x_k^2} \left(\gamma_{jk} A_k^{(4)} \right) = -J_j; \quad (j = 1,2,3,4) \tag{1.85}$$

The Lorentz condition (Lorentz gauge) in the 4-vector notation takes the form

$$\sum_{k=1}^{4} \frac{\partial}{\partial x_k} \left(\gamma_{jk} A_k^{(4)} \right) = 0, \quad (j = 1,2,3,4) \tag{1.86}$$

In vacuum the tensor γ_{jk} reduces to a scalar μ_0^{-1}. Then the field equations are simplified and coincide with the symmetric CGS form except for the factor μ_0:

$$\sum_{k=1}^{4} \frac{\partial^2 A_j^{(4)}}{\partial x_k^2} = -\mu_0 J_j, \quad \sum_{k=1}^{4} \frac{\partial A_k^{(4)}}{\partial x_k} = 0, \quad (j = 1,2,3,4). \tag{1.87}$$

Some formal difference between the 4-vector denotion of the Maxwell equations in CGS and MKS systems of units does not seem occasional or artificial. Actually, in the case of CGS system of units it is assumed implicitly from the very beginning that the electromagnetic field described by the vectors E and H exists in a homogeneous isotropic space with dielectric constant and magnetic permeability equal to unity, i.e. in vacuum. As a result, the analysis of the electromagnetic field in a continuous medium with specific polarization and magnetic characteristics requires inevitably the introduction of the vectors $\boldsymbol{D}$ and $\boldsymbol{B}$ by means of material equations and the generalization of the original system of equations. The MKS version of the Maxwell equations appears to be more general and consistent since it contains all necessary variables $\boldsymbol{E}, \boldsymbol{H}, \boldsymbol{D}, \boldsymbol{B}$ which makes it possible to consider the electromagnetic field in vacuum as a particular case where $\mu_r = \varepsilon_r = 1$. The definition of the vacuum light velocity c by means of the material constants of vacuum ε_0 and μ_0 results in the complete coincidence of both forms of notation in the situation of a free space without currents and charges.

1.3.3 Lorentz Transformation of Electromagnetic Field 4-Tensor

Now we proceed with the Lorentz transformation of the 4-vectors of current $\boldsymbol{J}^{(4)}$ and $\boldsymbol{A}^{(4)}$ and the 4-tensors of the electromagnetic field F_{ij} and G_{ij}. Using the relationships (1.67) and the explicit forms of the 4-vectors $\boldsymbol{J}^{(4)}$ and $\boldsymbol{A}^{(4)}$ we write [3]

$$J'_x = J_x, \ J'_y = J_y, \ J'_z = \frac{(J_z - v\rho)}{\sqrt{1 - v^2/c^2}} \quad (1.88)$$

$$\rho' = \frac{(\rho - J_z v/c^2)}{\sqrt{1 - v^2/c^2}}$$

and

$$A'_x = A_x, \ A'_y = A_y, \ A'_z = \frac{(A_z - \varphi v/c^2)}{\sqrt{1 - v^2/c^2}} \quad (1.89)$$

$$\varphi' = \frac{(\varphi - v A_z)}{\sqrt{1 - v^2/c^2}}.$$

The inverse transformation for the current has the form

$$J_z = \frac{(J'_z + v\rho')}{\sqrt{1 - v^2/c^2}}; \ \rho = \frac{(\rho' + J'_z v/c^2)}{\sqrt{1 - v^2/c^2}} \quad (1.90)$$

and a similar expression takes place for the 4-vector of the field potential $\boldsymbol{A}^{(4)}$. In the classical limit $c \to \infty$ equations (1.90) reduce to the obvious conditions

$$J_z = (J'_z + v\rho'), \ \rho = \rho' \quad (1.91)$$

For example, it is instructive to check directly the invariance of the current $\boldsymbol{J}^{(4)}$ with respect to the Lorentz transformation. Indeed, we have

$$\left(\boldsymbol{J}^{(4)'}\right)^2 = J'^2_x + J'^2_y + J'^2_z + (ic\rho')^2 = \quad (1.92)$$

$$= J^2_x + J^2_y + \frac{1}{1 - v^2/c^2} \left[(J_z - v\rho)^2 + (ic)^2 \left(\rho - \frac{v}{c^2} J_z\right)^2\right] \equiv \left(\boldsymbol{J}^{(4)}\right)^2$$

The next step is to define the Lorentz transformation relationships for the 4-tensors of electromagnetic field F_{ik} and G_{ik}. It is known that the tensor components transform as follows [2,3]:

$$F'_{il} = \sum_{j=1}^{4} \sum_{k=1}^{4} a_{ij} a_{lk} F_{jk}, \quad (i, l = 1, 2, 3, 4), \quad (1.93)$$

where a_{lk} are the components (1.65). The inverse transformation has the form

$$F_{jk} = \sum_{i=1}^{4} \sum_{l=1}^{4} a_{ij} a_{lk} F'_{il}, \quad (j, k = 1, 2, 3, 4). \quad (1.94)$$

Substituting into (1.93) a_{lk} (1.65), (1.67) and F_{jk} in the explicit form we obtain for the components of $^2\boldsymbol{F'}$ [3]:

$$F'_{12} = a_{11} a_{22} F_{12} = F_{12} \quad (1.95)$$

$$F'_{13} = a_{11} a_{33} F_{13} + a_{11} a_{34} F_{14} = \frac{1}{\sqrt{1 - v^2/c^2}} \left(F_{13} + i\frac{v}{c} F_{14} \right) \quad (1.96)$$

$$F'_{14} = a_{11} a_{43} F_{13} + a_{11} a_{44} F_{14} = \frac{1}{\sqrt{1 - v^2/c^2}} \left(F_{14} - i\frac{v}{c} F_{13} \right) \quad (1.97)$$

$$F'_{23} = a_{22} a_{33} F_{23} + a_{22} a_{34} F_{24} = \frac{1}{\sqrt{1 - v^2/c^2}} \left(F_{23} + i\frac{v}{c} F_{24} \right) \quad (1.98)$$

$$F'_{24} = a_{22} a_{43} F_{23} + a_{22} a_{44} F_{24} = \frac{1}{\sqrt{1 - v^2/c^2}} \left(F_{24} - i\frac{v}{c} F_{23} \right) \quad (1.99)$$

$$F'_{34} = (a_{33} a_{44} - a_{34} a_{43}) F_{34} = F_{34}. \quad (1.100)$$

The components of the 4-tensor G_{ik} can be presented analogously. Returning to the three-dimensional field components $\boldsymbol{E}$ and $\boldsymbol{B}$ we find

$$B'_x = \frac{1}{\sqrt{1 - v^2/c^2}} \left(B_x + \frac{v}{c^2} E_y \right), \quad E'_x = \frac{1}{\sqrt{1 - v^2/c^2}} (E_x - vB_y) \quad (1.101)$$

$$B'_y = \frac{1}{\sqrt{1 - v^2/c^2}} \left(B_y - \frac{v}{c^2} E_x \right), \quad E'_y = \frac{1}{\sqrt{1 - v^2/c^2}} (E_y + vB_x) \quad (1.102)$$

1.3 Relativistic Properties of the Maxwell Equations

$$B'_z = B_z, \ E'_z = E_z. \tag{1.103}$$

We can easily generalize the relationships (1.101)–(1.103) decomposing the electric field and the magnetic induction vectors into the components parallel and perpendicular to direction of the reference frame motion. Then we find that the parallel, or longitudinal components $E_\parallel$ and $B_\parallel$ are conserved and do not depend on the reference frame while the perpendicular, or transverse components $E_\perp$ and $B_\perp$ undergo the Lorentz transformation

$$E_\parallel = E'_\parallel, \ B_\parallel = B'_\parallel, \ E'_\perp = \frac{(\boldsymbol{E} + [\boldsymbol{v} \times \boldsymbol{B}])_\perp}{\sqrt{1 - v^2/c^2}}, \ B'_\perp = \frac{(\boldsymbol{B} - \frac{1}{c^2}[\boldsymbol{v} \times \boldsymbol{E}])_\perp}{\sqrt{1 - v^2/c^2}}. \tag{1.104}$$

In the classical limit $v \ll c$ equations (1.104) take the form

$$E_\parallel = E'_\parallel, \ B_\parallel = B'_\parallel, \ B'_\perp = B_\perp, \ E'_\perp = E_\perp + [\boldsymbol{v} \times \boldsymbol{B}]_\perp. \tag{1.105}$$

The second pair of the fundamental variables $\boldsymbol{H}$ and $\boldsymbol{D}$ transform in a similar way. Namely, it can be shown that

$$H'_\parallel = H_\parallel, \ D'_\parallel = D_\parallel, \ H'_\perp = \frac{(\boldsymbol{H} - [\boldsymbol{v} \times \boldsymbol{D}])_\perp}{\sqrt{1 - v^2/c^2}}, \ D'_\perp = \frac{(\boldsymbol{D} + \frac{1}{c^2}[\boldsymbol{v} \times \boldsymbol{H}])}{\sqrt{1 - v^2/c^2}}. \tag{1.106}$$

The important conclusion of the analysis carried out is that neither an electric field $\boldsymbol{E}$, nor a magnetic one $\boldsymbol{B}$ do not exist separately. The structure of the fundamental 4-tensors $^2\boldsymbol{F} = (\boldsymbol{B}, -i\boldsymbol{E}/c)$ and $^2\boldsymbol{G} = (\boldsymbol{H}, -ic\boldsymbol{D})$ components depends entirely on the reference frame velocity $\boldsymbol{v}$. For example, in the fixed coordinate system it is possible to measure the purely magnetic field. But in the moving coordinate system the electrostatic component $\boldsymbol{E} = [\boldsymbol{v} \times \boldsymbol{B}]$ inevitably appears. In general, the Lorentz-invariance of the Maxwell equations means that if the vectors $\boldsymbol{E}, \boldsymbol{B}, \boldsymbol{H}$ and $\boldsymbol{D}$ determine the electromagnetic field in the coordinate system X, then the Maxwell equations in the coordinate system X' moving with the velocity $\boldsymbol{v}$

$$\text{curl}' \boldsymbol{E}' + \frac{\partial \boldsymbol{B}'}{\partial t'} = 0, \ \text{div}' \boldsymbol{B}' = 0 \tag{1.107}$$

$$\text{curl}' \boldsymbol{H}' - \frac{\partial \boldsymbol{D}'}{\partial t'} = \boldsymbol{J}', \ \text{div}' \boldsymbol{D}' = \rho'$$

are also satisfied. Here the primed operators refer to the spatial coordinates of the system X', and the vectors $\boldsymbol{E}', \boldsymbol{B}', \boldsymbol{H}'$ and $\boldsymbol{D}'$ are determined by (1.101)–(1.104), (1.106).

The material equations defining the connection between the vectors $\boldsymbol{D}$ and $\boldsymbol{E}$, $\boldsymbol{B}$ and $\boldsymbol{H}$, respectively, do not conserve passing from one reference frame to another. The variation of macroscopic parameters ε, μ, σ can be attributed to the change of a moving medium structure.

The material equations defining the connection between the vectors $\boldsymbol{D}$ and $\boldsymbol{E}$, $\boldsymbol{B}$ and $\boldsymbol{H}$, respectively, do not conserve passing from one reference frame to another. The variation of macroscopic parameters ε, μ, σ can be attributed to the change of a moving medium structure.

Finally, we need to determine the possible field invariants with respect to the Lorentz transformation besides the scalars $(\boldsymbol{J}^{(4)'})^2$ and $(\boldsymbol{A}^{(4)})^2$. It is known that the scalar products of two second rank tensors S_{jk} and T_{jk} are invariant with respect to rotations [3]:

$$\sum_{j=1}^{4}\sum_{k=1}^{4} S_{jk}T_{jk} = \text{invariant}, \quad \sum_{j=1}^{4}\sum_{k=1}^{4} S_{jk}T_{kj} = \text{invariant}. \quad (1.108)$$

In vacuum, taking the fundamental 4-tensors ${}^2\boldsymbol{F} = (\boldsymbol{B}, -i\boldsymbol{E}/c)$ and ${}^2\boldsymbol{G} = (\boldsymbol{H}, -ic\boldsymbol{D})$ and relations (1.20), (1.21) $\boldsymbol{B} = \mu_0 \boldsymbol{H}$, $\boldsymbol{D} = \varepsilon_0 \boldsymbol{E}$ we compose according to expression (1.108) invariants which reduce to the invariant $\boldsymbol{E}\cdot\boldsymbol{H}$, and to the quantity $(\mu_0 H^2 - \varepsilon_0 E^2)$ which differs from the second invariant $(H^2 - E^2)$ in the CGS system of units due to the presence of the vacuum permitivity ε_0 and the vacuum permeability μ_0.

1.4 Lagrangian Formalism

1.4.1 Action Functional

The relativistic equation of motion of a free particle can be obtained from the principle of the least action which states that for each mechanical system there exists a certain integral S, called the action, possessing a minimum value for the actual motion, so that its variation δS is zero [2]. The action integral S_p for a free material particle, i.e. a particle not under the influence of any external force, must not depend on the choice of reference system, that is must be invariant under Lorentz transformation. Consequently, it must depend on a scalar. It is clear that the integrand must be a differential of the first order. The only scalar of this kind that can be constructed for a free particle is the product of the interval dS and some constant α_p characterizing the particle. For a free particle the action must have the form

$$S_p = -\alpha_p \int_a^b dS, \quad (1.109)$$

where the integration is carried out along the so-called world line of the particle between the two particular events of the arrival of the particle at the initial position and at the final position at definite times t_1 and t_2. It can be shown that the constant α_p must be positive [2]: $\alpha_p > 0$. The action integral can be represented as an integral with respect to time taking into account that

$$dS = \sqrt{c^2 dt^2 - dx^2 - dy^2 - dz^2} = cdt\sqrt{1 - \frac{v^2}{c^2}}, \quad (1.110)$$

where v is the particle velocity:

$$v^2 = \frac{dx^2 + dy^2 + dz^2}{dt^2}. \quad (1.111)$$

Then we get

$$S_p \equiv \int_{t_1}^{t_2} \mathcal{L}_p dt, \quad (1.112)$$

where the so-called Lagrangian function of the mechanical system $\mathcal{L}_p$ has the form [2]

$$\mathcal{L}_p = -\alpha_p c \sqrt{1 - \frac{v^2}{c^2}}. \quad (1.113)$$

We find the constant α_p using the transition to the classical limit $c \to \infty$ where $\mathcal{L}_p$ must go over into the classical expression

$$\mathcal{L}_p = \frac{mv^2}{2}. \quad (1.114)$$

Expanding $\mathcal{L}_p$ in powers of a small parameter v/c, omitting the constant part $\alpha_p c$, and neglecting terms of the higher order in v/c we obtain

$$\mathcal{L}_p = \frac{mv^2}{2} \approx \frac{\alpha_p v^2}{2c}, \quad \alpha_p = mc \quad (1.115)$$

and

$$\mathcal{L}_p = -mc^2 \sqrt{1 - \frac{v^2}{c^2}}, \quad S_p = -mc \int_a^b dS. \quad (1.116)$$

If we have a system consisting of a number of charged particles (for instance, electrons) interacting with an electromagnetic field, the total action will contain 3 terms

$$S = S_p + S_{\text{int}} + S_f, \quad (1.117)$$

where the first S_p stands for free particles

$$S_p = -\sum_i m_i c \int dS_i,$$

the second term S_{int} for the interaction of particles with a field, and the last term S_f for the action of the free field. The form of the last two terms can be guessed with high confidence, almost determined, through the use of very general considerations. Starting with the interaction term, we require that it fulfills the following conditions:

1. being a Lorentz-invariant scalar.
2. being linear in fields, so as to guarantee that the variational equations of motion will contain the force which is linear in fields (like the Coulomb or the Lorentz force).
3. being gauge invariant at least in the sense that the gauge transformation will not affect variation results (the equations of motion and the field equations).

The simplest expression which meets all these requirements is

$$S_{\text{int}} = \alpha \sum_i \int A_i dx_i. \tag{1.118}$$

Equation (1.118) clearly fulfills the first two conditions; it fulfills the third one as well, since gauge transformation (1.42) results in adding to $\boldsymbol{A}$ a four-gradient term, which, when substituted in (1.118), integrates out, yields a constant and does not affect the variation. The constant α should be determined by comparing the resulting equations of motions with the experimentally established expression for Lorentz force; it appears to be

$$\alpha = e$$

in the MKS units and will be introduced into (1.118) from now on. Thus, finally,

$$S_{\text{int}} = e \sum_i \int A_i dx_i. \tag{1.119}$$

It is convenient at this point to postpone the construction of the free-field action and derive the explicit expressions for the fundamental functions, determining the motion of a charged particle in an electromagnetic field.

Equations (1.118) and (1.119) imply that if the electromagnetic field is treated as an external field, e.g. determined only by external sources, then the action of a single electron, whose motion is affected by the field but does not affect the field itself is given by

$$S = \int_a^b \left(-mcdS + e \sum_i A_i dx_i \right). \tag{1.120}$$

Substituting $A_i = \{\boldsymbol{A}, i\varphi/c\}$, $dx^i = \{d\boldsymbol{r}, icdt\}$ and using (1.110) we present (1.120) in the form

1.4 Lagrangian Formalism

$$S = \int_{t_1}^{t_2} \left(-mc^2 \sqrt{1 - \frac{v^2}{c^2}} + e\mathbf{A}\mathbf{v} - e\varphi \right) dt. \quad (1.121)$$

Thus, the Lagrangian of an electron is given by

$$\mathcal{L} = -mc^2 \sqrt{1 - \frac{v^2}{c^2}} + e\mathbf{A}\mathbf{v} - e\varphi. \quad (1.122)$$

The canonical momentum which in the context of particle motion in an electromagnetic field is usually called generalized momentum is equal to

$$\mathbf{P}_g = \frac{\partial \mathcal{L}}{\partial \mathbf{v}} = \frac{m\mathbf{v}}{\sqrt{1 - v^2/c^2}} + e\mathbf{A} \equiv \mathbf{p} + e\mathbf{A}, \quad (1.123)$$

where

$$\mathbf{p} = \frac{m\mathbf{v}}{\sqrt{1 - v^2/c^2}} \quad (1.124)$$

is the ordinary momentum in the absence of the field. Once the Lagrangian is known, we are able to construct the Hamiltonian $\mathcal{H}$ of the particle, via the relation

$$\mathcal{H} = \mathbf{v}\frac{\partial \mathcal{L}}{\partial \mathbf{v}} - \mathcal{L}.$$

Using (1.123) one obtains

$$\mathcal{H} = \frac{mc^2}{\sqrt{1 - v^2/c^2}} + e\varphi. \quad (1.125)$$

The Hamiltonian should be expressed in terms of generalized coordinates and momenta. Combining (1.123) and (1.125) we obtain

$$\left(\frac{\mathcal{H} - e\varphi}{c} \right)^2 = m^2 c^2 + (\mathbf{P}_g - e\mathbf{A})^2,$$

or

$$\mathcal{H} = \sqrt{m^2 c^4 + c^2 (\mathbf{P}_g - e\mathbf{A})^2} + e\varphi. \quad (1.126)$$

In the non-relativistic limit $v^2/c^2 \ll 1$ the results obtained reduce to

$$\mathcal{L} \simeq \frac{mv^2}{2} + e\mathbf{A}\mathbf{v} - e\varphi \quad (1.127)$$

$$\mathbf{p} \simeq m\mathbf{v} = \mathbf{P}_g - e\mathbf{A} \quad (1.128)$$

$$\mathcal{H} = \frac{1}{2m}(\mathbf{P}_g - e\mathbf{A})^2 + e\varphi. \quad (1.129)$$

In (1.127) and (1.129) the unimportant rest energy is omitted.

26   1 Fundamentals of Electrodynamics

Turning now to the action S_f of the free field, we require S_f to be:

1. a Lorentz invariant scalar.
2. bilinear in fields (or field potentials), to provide linear field equations as a result of variation.
3. gauge invariant.

We have invented two Lorentz-covariant objects describing the electromagnetic field – the 4-vector $\mathbf{A}$ and the 4-tensor F_{jk}. The only bilinear scalar which can be constructed from vector potential $\sum (A_i)^2$ is essentially not gauge-invariant and is thus unsuitable for our purposes. Since it has been shown that the electromagnetic field tensor F_{jk} is gauge-invariant, and since the only bilinear in field potentials scalar that can be constructed from F_{jk} is $\sum_{jk} (F_{jk})^2$, the free-field action S_f should have the form

$$S_f = \sum_{jk} \gamma \int (F_{jk})^2 \, dt dV, \quad dV = dxdydz. \tag{1.130}$$

The constant γ should be determined by equating the field equations derived from the variation of $S_f + S_{\text{int}}$ with respect to field potentials to Maxwell equations in the chosen system of units. Substituting the values of F_{jk} (1.72) into (1.130) and passing to the 4-dimensional denotion we obtain:

$$S_f = -\frac{i}{c}\gamma \int 2\left(B^2 - \frac{1}{c^2}E^2\right) d\Omega, \tag{1.131}$$

where the 4-dimensional volume element

$$d\Omega = icdxdydzdt = dxdydzd\tau.$$

The analysis shows that γ should be negative: $\gamma = -|\gamma| < 0$. Otherwise, in the case of rapid oscillations the term with E^2, i.e. with $(\partial A/\partial t)^2$ appearing in the integrand with a positive sign could make S_f a negative quantity with an arbitrarily large value. Choosing the MKS system of units we have

$$\gamma = -\frac{c^2 \varepsilon_0}{4}$$

and

$$S_f = \frac{c^2 \varepsilon_0}{2} \iint \left(\frac{1}{c^2}E^2 - B^2\right) dt dV. \tag{1.132}$$

Equation (1.132) implies that the Lagrangian $\mathcal{L}_f$ of the free electromagnetic field is given by

$$\mathcal{L}_f = \frac{c^2 \varepsilon_0}{2} \int \left(\frac{1}{c^2}E^2 - B^2\right) dV = \frac{1}{2}\int \left(\varepsilon_0 E^2 - \mu_0 H^2\right) dV. \tag{1.133}$$

Finally, we have in the 4-dimensional form

$$S = -\sum_i m_i c \int dS_i + e \sum_i \int A_i dx_i + \frac{i}{c}\frac{c^2\varepsilon_0}{4}\sum_{jk}\int (F_{jk})^2 d\Omega. \quad (1.134)$$

1.4.2 Field Equations

Now we derive the equations of the field using the expression (1.134) for action. Actually, we know the final results in advance — these should be Maxwell equations for the field; thus, the aim of our derivation is to make sure that our expressions for S_{int} and S_f are correct. We begin the derivation of the electromagnetic field equations by presenting the interaction term S_{int} in action (1.119) in somewhat different form. First, assuming that the charged particles are distributed in space with a charge density ρ, we replace the summation over the particles in (1.119) by integration over the space:

$$S_{\text{int}} = e\sum_i A_i dx_i = \int \rho dV \sum_i A_i dx_i.$$

Then, writing

$$dx_i = \frac{dx_i}{dt} dt$$

and recalling that

$$J_i = \rho \frac{dx_i}{dt}$$

is a vector of 4-current, we bring the interaction term to the form

$$S_{\text{int}} = -\frac{i}{c}\int \sum_j J_j A_j d\Omega; \quad d\Omega = icdtdV \quad (1.135)$$

and the action of the electromagnetic field interacting with the charged particles is obtained by the adding of S_{int} to the action of a free field (1.130), (1.131):

$$S = -\frac{i}{c}\sum_j \int J_j A_j d\Omega + \frac{i}{c}\frac{c^2\varepsilon_0}{4}\sum_{jk}\int (F_{jk})^2 d\Omega. \quad (1.136)$$

The field equations are derived by means of the principle of the least action $\delta S = 0$ that is $\delta S/\delta \boldsymbol{A} = 0$, where $\delta/\delta \boldsymbol{A}$ stands for a functional derivative. Taking into account that

$$\sum_{jk}\delta(F_{jk})^2 = 2\sum_{jk}F_{jk}\delta F_{jk},$$

we obtain

$$\delta S = -\frac{i}{c} \int \left[\sum_j J_j \delta A_j - \frac{c^2 \varepsilon_0}{2} \sum_{jk} F_{jk} \delta F_{jk} \right] d\Omega = 0,$$

or, since according to (1.81)

$$F_{jk} = \frac{\partial A_k}{\partial x_j} - \frac{\partial A_j}{\partial x_k}$$

we have

$$\delta S = -\frac{i}{c} \int \left[\sum_j J_j \delta A_j - \frac{c^2 \varepsilon_0}{2} \sum_{jk} \left(F_{jk} \frac{\partial}{\partial x_j} \delta A_k - F_{jk} \frac{\partial}{\partial x_k} \delta A_j \right) \right] d\Omega = 0.$$

Interchanging indices i and k in the second term, and using the relation $F_{jk} = -F_{kj}$, we arrive at

$$\delta S = -\frac{i}{c} \int \left[\sum_j J_j \delta A_j + c^2 \varepsilon_0 \sum_{jk} F_{jk} \frac{\partial}{\partial x_j} \delta A_j \right] d\Omega = 0.$$

Integrating the second term by parts and omitting the 'free' term since the field vanishes at infinity, we finally get

$$\delta S = -\frac{i}{c} \sum_j \int \left(J_j - c^2 \varepsilon_0 \sum_{k=1}^{4} \frac{\partial F_{jk}}{\partial x_k} \right) \delta A_j d\Omega = 0.$$

For the integral to vanish for arbitrary δA_j, the integrand should be equal to zero:

$$c^2 \varepsilon_0 \sum_{k=1}^{4} \frac{\partial F_{jk}}{\partial x_k} = J_j. \tag{1.137}$$

The relation (1.137) is the tensorial 4-dimensional form of the second pair of Maxwell's equations (1.18) and (1.19); we recall that the first pair of equations (1.16) and (1.17) follows simply from the definition of F_{jk}. Substituting into (1.137) the relationship (1.81) and taking into account that $c^2 \varepsilon_0 = \mu_0^{-1}$ one can see that the result (1.137) immediately coincides with the corresponding equations of the system (1.87). The direct substitution of components F_{jk} (1.72) into (1.137) yields the three-dimensional equations (1.18) and (1.19).

2 Magnetostatics

2.1 Fundamentals of Magnetostatics

2.1.1 Constant Magnetic Field

We start with the Maxwell equations for the case of the constant magnetic field. They relate the vectors of magnetic induction $\boldsymbol{B}$, magnetic field intensity $\boldsymbol{H}$ and an electric current density $\boldsymbol{j}$.

$$\mathrm{curl}\boldsymbol{H} = \boldsymbol{j} \tag{2.1}$$

$$\mathrm{div}\boldsymbol{B} = 0 \tag{2.2}$$

$$\boldsymbol{B} = \mu_0 \mu^r \boldsymbol{H} = \mu_0 \left(\boldsymbol{M} + \boldsymbol{H}\right), \tag{2.3}$$

where the vector $\boldsymbol{M}$ is a magnetization of a medium. It has the form

$$\boldsymbol{M} = \chi \boldsymbol{H}, \; \chi = \mu^r - 1. \tag{2.4}$$

The magnetic permeability μ^r may be greater or less than unity. The magnetic susceptibility χ may correspondingly be either positive or negative. In general case of an anisotropic medium both a magnetic permeability μ^r_{ik} and a magnetic susceptibility χ_{ik} are tensor quantities. The magnetic susceptibility of the great majority of media is very small because the magnetization of a non-ferromagnetic body is a relativistic effect, of an order of magnitude v^2/c^2, where v is the velocity of the electrons in the atoms.

The equation defining the magnetic field induction $\boldsymbol{B}$ by virtue of the vector potential $\boldsymbol{A}$ remains the same:

$$\mathrm{curl}\boldsymbol{A} = \boldsymbol{B} \tag{2.5}$$

which immediately yields

$$\mathrm{curl}\,\mathrm{curl}\boldsymbol{A} = \mu_0 \mu^r \boldsymbol{j}. \tag{2.6}$$

The type of the gauge transformation can be chosen for each specific case. Taking, for example, $\mathrm{div}\boldsymbol{A} = 0$, we get

$$\nabla^2 \boldsymbol{A} = -\mu_0 \mu^r \boldsymbol{j}. \tag{2.7}$$

Equations (2.1)–(2.5) do not contain explicitly the electric field induction $\boldsymbol{D}$ and intensity $\boldsymbol{E}$. In the static case these vectors are calculated separately using the electric charge distribution. The current density $\boldsymbol{j}$, in turn, is determined by the electric field intensity $\boldsymbol{E}$:

$$\boldsymbol{j} = \sigma \boldsymbol{E}. \tag{2.8}$$

However, in this chapter we limit the study with the constant magnetic field, or magnetostatics, assuming that the electric current density $\boldsymbol{j}$ is known. One can, therefore, conclude that the problems of magnetostatics can be solved without being referred to an electric field. For the further analysis it is useful to write (2.1) and (2.2) in the integral form:

$$\oint_L \boldsymbol{H} dl = I \tag{2.9}$$

and

$$\oint_S \boldsymbol{B} d\boldsymbol{s} = 0, \tag{2.10}$$

where I is the total electric current flowing through the cross-section of a conductor, L is the contour enclosing the linear current I, and S is the surface limiting the volume with the magnetic field. Here we used the well known theorems of the vector analysis. Equation (2.9) is also known as the Ampère's law.

Equations (2.1)–(2.10) describing the magnetic field should be completed with the appropriate boundary conditions. Let the surface S separates two media with the relative magnetic permeabilities μ_1^r and μ_2^r. Consider first the behavior of the magnetic field intensity tangential component $\boldsymbol{H}_t$ at the boundary between the two media. The geometry of the problem is shown in Fig. 2.1.

We integrate both sides of the Maxwell equation (2.1) on the surface S_1 limited by the rectangular contour L shown in Fig. 2.1 with two parallel sides of the length Δl_t in each medium and two sides of the length Δl_n traversing the boundary. The integral in the left-hand side of (2.1) can be transformed into the circulation of the vector $\boldsymbol{H}$ along the chosen contour L, as was mentioned above. Then we have

$$\oint_L \boldsymbol{H} dl = \int_{S_1} \boldsymbol{j} \cdot d\boldsymbol{s}. \tag{2.11}$$

In the limit $\Delta l_n \to 0$ the integral in the left-hand side of (2.11) over the sides of the contour L traversing the boundary vanishes. The right-hand side of

(2.11) also vanishes since the current density j is finite and the surface tends to zero when $\Delta l_n \to 0$. The remaining terms yield

$$(\boldsymbol{H} \cdot \boldsymbol{\tau}_1 + \boldsymbol{H} \cdot \boldsymbol{\tau}_2) = 0, \tag{2.12}$$

where $\boldsymbol{\tau}_{1,2}$ are the unit vectors directed along the lower and upper sides of the rectangular contour L, respectively. Clearly

$$-\boldsymbol{\tau}_1 = \boldsymbol{\tau}_2 = \boldsymbol{\tau}$$

and we finally obtain the condition of the continuity of the tangential component of the magnetic field intensity at the boundary between two media:

$$\boldsymbol{H}_{t1} = \boldsymbol{H}_{t2}. \tag{2.13}$$

In the particular case when the current density $j \to \infty$ the surface current density $\varkappa = j \Delta l_n$ can be introduced that remains constant as $\Delta l_n \to 0$. Then the condition (2.13) should be modified as follows:

$$\boldsymbol{H}_{t1} - \boldsymbol{H}_{t2} = \varkappa. \tag{2.14}$$

Derive now the boundary condition for the magnetic induction $\boldsymbol{B}$. We choose a thin transition layer where the permeabilities are varying rapidly although remaining to be continuous. We assume that the field vectors and their first derivatives are continuous functions of the coordinates and, in general case, the time in the transition layer as well as in the both media. Inside the layer we single out a small cylinder with the base surface Δs and the height Δl equal to the thickness of the transition layer. The axis of the cylinder $\boldsymbol{n}$ is perpendicular to the surface S as it is shown in Fig. 2.2.

We apply the Maxwell equation in the integral form (2.10) to the field in this cylinder in the limit of the infinitesimal height $\Delta l \to 0$. In such a case the induction $\boldsymbol{B}$ on the surface of the cylinder is assumed to be constant.

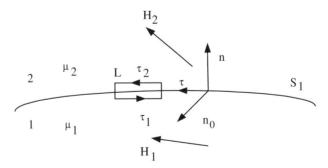

Fig. 2.1. The boundary conditions for the tangential component of the magnetic field intensity $\boldsymbol{H}$

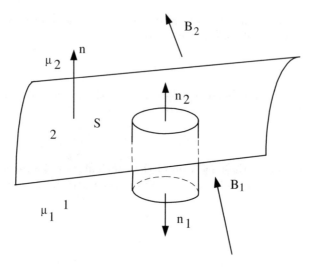

Fig. 2.2. The boundary conditions for the normal component of the magnetic induction $\boldsymbol{B}$

The contribution of the integral on the lateral surface vanishes due to the condition $\Delta l \to 0$, and (2.10) yields

$$(\boldsymbol{B} \cdot \boldsymbol{n}_1 + \boldsymbol{B} \cdot \boldsymbol{n}_2)\,\Delta s = 0 \tag{2.15}$$

where $\boldsymbol{n}_{1,2}$ are the normals to the cylinder base surfaces in each medium, respectively. Obviously,

$$-\boldsymbol{n}_1 = \boldsymbol{n}_2 = \boldsymbol{n}, \tag{2.16}$$

where $\boldsymbol{n}$ is the normal to the boundary surface S chosen in the positive direction. Combining (2.15) and (2.16) we obtain the condition of the continuity of the induction normal component B_n at the boundary between two media:

$$B_{1n} = B_{2n}. \tag{2.17}$$

Suppose that the current density components j_i are smooth functions of coordinates, i.e., they possess the first derivatives. Then applying the curl operator to both sides of (2.1) we obtain assuming the medium to be homogeneous, isotropic and non-magnetic:

$$\nabla^2 \boldsymbol{H} = -\mathrm{curl}\boldsymbol{j}, \tag{2.18}$$

where we have taken into account (2.2). Equation (2.18) is the vector analog of the Poisson equation, and it has the known general solution:

$$\boldsymbol{H}(\boldsymbol{r}) = \frac{1}{4\pi} \int_V \frac{\mathrm{curl}'\boldsymbol{j}(\boldsymbol{r}')}{|\boldsymbol{r} - \boldsymbol{r}'|}\,dV', \tag{2.19}$$

where r and r' are the radii-vectors of the observation point and the element of the current distribution, respectively. The prime means that the differentiation and integration are applied to the coordinates of the current distribution. The integrand in (2.19) can be simplified by using the theorems of the vector analysis. It can be shown directly that

$$\operatorname{curl}(\psi \boldsymbol{F}) = \psi \operatorname{curl} \boldsymbol{F} + [\operatorname{grad} \psi \times \boldsymbol{F}], \qquad (2.20)$$

where ψ is a scalar function. Substituting (2.20) into (2.19) with

$$\psi = \frac{1}{|\boldsymbol{r}-\boldsymbol{r'}|}, \quad \boldsymbol{F} = \operatorname{curl}' \boldsymbol{j}(\boldsymbol{r'})$$

we get

$$\boldsymbol{H}(\boldsymbol{r}) = \frac{1}{4\pi} \int_V \operatorname{curl}' \frac{\boldsymbol{j}(\boldsymbol{r'})}{|\boldsymbol{r}-\boldsymbol{r'}|} dV' - \frac{1}{4\pi} \int_V \left[\left(\operatorname{grad}' \frac{1}{|\boldsymbol{r}-\boldsymbol{r'}|}\right) \times \boldsymbol{j}(\boldsymbol{r'})\right] dV'. \qquad (2.21)$$

The first integral can be reduced to the one over the surface S:

$$\int_V \operatorname{curl}' \frac{\boldsymbol{j}(\boldsymbol{r'})}{|\boldsymbol{r}-\boldsymbol{r'}|} dV' = -\oint_S \frac{[\boldsymbol{j}(\boldsymbol{r'}) \times d\boldsymbol{s'}]}{|\boldsymbol{r}-\boldsymbol{r'}|}.$$

The surface integral vanishes since the surface S can be chosen in such a way that all currents are kept inside it. The integrand in the second integral can be transformed as follows:

$$\operatorname{grad}' \frac{1}{|\boldsymbol{r}-\boldsymbol{r'}|} = \frac{\boldsymbol{r}-\boldsymbol{r'}}{|\boldsymbol{r}-\boldsymbol{r'}|^3}.$$

The insertion of the latter result into (2.21) gives

$$\boldsymbol{H}(\boldsymbol{r}) = \frac{1}{4\pi} \int_V \frac{[\boldsymbol{j}(\boldsymbol{r'}) \times \boldsymbol{r_0}]}{|\boldsymbol{r}-\boldsymbol{r'}|^2} dV', \qquad (2.22)$$

where the unit vector $\boldsymbol{r_0}$ in the direction connecting the observation point $\boldsymbol{r}$ and the current distribution point $\boldsymbol{r'}$ has the form

$$\boldsymbol{r_0} = \frac{\boldsymbol{r}-\boldsymbol{r'}}{|\boldsymbol{r}-\boldsymbol{r'}|}. \qquad (2.23)$$

Relationship (2.23) possesses an important advantage being applicable for the piecewise continuous distribution of the current density $\boldsymbol{j}(\boldsymbol{r'})$ that occurs in many practically important cases. For instance, the linear current density is usually assumed to be constant inside the wire and zero outside of it.

Consider the particular case of the magnetic field caused by a so-called linear current, i.e. the current which flows in a conductor with a cross-section dimension sufficiently small as compared to the distance $|r - r'|$ to an observation point r. In such a case the quantity $|r - r'|$ is assumed to be constant when the integration is carried out over the conductor cross-section, $j = \text{const}$, and the total current I can be written as

$$I = jS, \qquad (2.24)$$

where S is the cross-section area. Then, taking into account that

$$j dV' = jS dl' = I dl',$$

where dl' is directed along the conductor, we obtain

$$H(r) = \frac{I}{4\pi} \oint \frac{[dl' \times r_0]}{|r - r'|^2} = \frac{I}{4\pi} \oint \frac{[dl' \times (r - r')]}{|r - r'|^3}. \qquad (2.25)$$

Relationship (2.25) is known as the Biot–Savart law.

The magnetic field of the given current distribution can be also expressed in terms of the vector potential A. Indeed, solving directly (2.7) for the unlimited space and the spatially confined current distribution we have

$$A(r) = \frac{\mu_0 \mu^r}{4\pi} \int_V \frac{j(r')}{|r - r'|} dV'. \qquad (2.26)$$

The validity of the solution (2.26) can be proved by the direct substitution into (2.7). The field induction is calculated by virtue of (2.5) using the result (2.26). In the case of the linear current (2.26) takes the form

$$A(r) = \frac{\mu_0 \mu^r I}{4\pi} \oint \frac{dl'}{|r - r'|}. \qquad (2.27)$$

Let us introduce a magnetic flux Φ which is the important characteristic of the magnetic field. It is defined as the flux of the magnetic induction vector B through the surface S:

$$\Phi = \int_S B ds. \qquad (2.28)$$

Obviously, the magnetic flux is a scalar quantity. Integrating both sides of (2.5) over the surface S and using the Stokes theorem we obtain that the magnetic flux through this surface is equal to the circulation of the vector potential:

$$\int_S B ds = \int_S \text{curl} A ds = \oint_L A dl = \Phi, \qquad (2.29)$$

where L is the contour enclosing the surface S.

Consider now the situation where currents are absent in some isolated volume V enclosed by the surface S. Then, in this region (2.1) and (2.9) become homogeneous. Any vector with zero curl can be presented as a gradient of a scalar function. Consequently, we can introduce a scalar potential $\mathcal{U}$ of the magnetic field in the region without currents:

$$\boldsymbol{H} = -\mathrm{grad}\,\mathcal{U}. \tag{2.30}$$

Substituting (2.30) into (2.2) we obtain that in a homogeneous medium with $\mu^r = \mathrm{const}$ the magnetostatic potential $\mathcal{U}$ obeys the Laplace equation similarly to the scalar potential of an electric field in a volume without electric charges:

$$\nabla^2 \mathcal{U} = 0. \tag{2.31}$$

The vector Poisson equation (2.18) in our case also reduces to the Laplace one.

$$\nabla^2 \boldsymbol{H} = 0. \tag{2.32}$$

In ferromagnetic media with a spontaneous magnetization $\boldsymbol{M}_{\mathrm{sp}}$ independent of an external magnetic field equation (2.3) takes the form:

$$\boldsymbol{B} = \mu_0 \mu^r \left(\boldsymbol{H} + \boldsymbol{M} \right)_{\mathrm{sp}}.$$

In such a case (2.2) becomes

$$\mathrm{div}\,\boldsymbol{B} = \frac{1}{\mu_0 \mu^r} \mathrm{div}\,\boldsymbol{M}_{\mathrm{sp}}$$

and the magnetostatic potential $\mathcal{U}$ satisfies the Poisson equation with some fictitious "charge" in the right-hand side:

$$\nabla^2 \mathcal{U} = \frac{1}{\mu_0 \mu^r} \mathrm{div}\,\boldsymbol{M}_{\mathrm{sp}} \tag{2.33}$$

and

$$\mathcal{U}(\boldsymbol{r}) = -\frac{1}{4\pi \mu_0 \mu^r} \int_V \frac{\mathrm{div}'\,\boldsymbol{M}_{\mathrm{sp}}(\boldsymbol{r}')}{|\boldsymbol{r} - \boldsymbol{r}'|} dV'. \tag{2.34}$$

In general, the magnetostatic problems can be successfully solved by virtue of the Laplace or Poisson equation for any closed volume that does not contain any current but may at the same time contain a constant magnet. However, it should be emphasized that the introduction of the magnetostatic potential is a rather formal method of calculation which is valid only under the strictly defined conditions, since the existence of magnetic charges ("monopoles") has yet to be proved.

2.1.2 Magnetic Moment of a System of Currents

Consider the vector potential (2.26) caused by the current distribution $\boldsymbol{j}(\boldsymbol{r}')$ concentrated in a fixed closed volume V which is situated far from the observation point P with the coordinates (x, y, z). We choose the origin inside V. The distance $R = |\boldsymbol{r} - \boldsymbol{r}'|$ between the current element $\boldsymbol{j}(\boldsymbol{r}')$ and the observation point P has the form:

$$R = \sqrt{(x-x')^2 + (y-y')^2 + (z-z')^2} \gg r', \qquad (2.35)$$

where the distance r' between the current element $\boldsymbol{j}(\boldsymbol{r}')$ and the origin is

$$r' = \sqrt{x'^2 + y'^2 + z'^2}.$$

The quantity $1/R$ can be expanded in powers of coordinates (x', y', z') in the vicinity of the origin as follows:

$$\frac{1}{R} = \frac{1}{r} - \sum_{i=1}^{3} x'_i \frac{\partial}{\partial x_i}\left(\frac{1}{R}\right)\Big|_{x'_i=0} + \frac{1}{2}\sum_{i=1}^{3}\sum_{k=1}^{3} x'_i x'_k \frac{\partial^2}{\partial x_i \partial x_k}\left(\frac{1}{R}\right)\Big|_{x'_i=0} - \cdots, \qquad (2.36)$$

where i, k stand for x, y, z, and $r = \sqrt{x^2 + y^2 + z^2}$. Substituting the expansion (2.36) into (2.26) we obtain:

$$\boldsymbol{A}(\boldsymbol{r}) = \frac{\mu_0 \mu^r}{4\pi}\frac{1}{r}\int_V \boldsymbol{j}(\boldsymbol{r}')\,dV' - \frac{\mu_0 \mu^r}{4\pi}\int_V \left(\boldsymbol{r}' \cdot \mathrm{grad}\left(\frac{1}{R}\right)\right)\boldsymbol{j}(\boldsymbol{r}')\,dV' +$$

$$+ \frac{\mu_0 \mu^r}{4\pi}\int_V \left\{\boldsymbol{r}' \cdot \mathrm{grad}\left(\boldsymbol{r}' \cdot \mathrm{grad}\left(\frac{1}{R}\right)\right)\right\}\boldsymbol{j}(\boldsymbol{r}')\,dV' - \cdots. \qquad (2.37)$$

Note that the grad operator is applied only to the coordinates (x, y, z). Consider the first integral in the series (2.37). The arbitrary current contribution can be divided into the closed tubes with infinitesimal cross-section da, the length element in the current direction $d\boldsymbol{r}'$ and current $I = jda$. Then, for each current tube the corresponding integral along the closed contour C of the tube would vanish:

$$\frac{\mu_0 \mu^r}{4\pi}\frac{1}{r}\int_V \boldsymbol{j}(\boldsymbol{r}')\,dV' = \frac{\mu_0 \mu^r}{4\pi}\frac{I}{r}\oint_C d\boldsymbol{r}' = 0.$$

The term of interest is the second one. The integrand can be transformed according to the well known formula of the vector algebra

$$[\boldsymbol{a} \times [\boldsymbol{b} \times \boldsymbol{c}]] = \boldsymbol{b}(\boldsymbol{a} \cdot \boldsymbol{c}) - \boldsymbol{c}(\boldsymbol{a} \cdot \boldsymbol{b}),$$

which yields

$$\left(\mathbf{r}' \cdot \mathrm{grad}\left(\frac{1}{R}\right)\right) d\mathbf{r}' = \frac{1}{2}\left[[\mathbf{r}' \times d\mathbf{r}'] \times \mathrm{grad}\left(\frac{1}{R}\right)\right] +$$

$$+ \frac{1}{2} d\left\{\left(\mathbf{r}' \cdot \mathrm{grad}\left(\frac{1}{R}\right)\right) \mathbf{r}'\right\}. \tag{2.38}$$

The second term in the right-hand side of (2.38) is a complete differential, and the integration of it over the closed contour gives zero. Then we have neglecting the higher order terms in (2.37):

$$\mathbf{A}(\mathbf{r}) = -\frac{\mu_0 \mu^r}{4\pi} I \int \mathbf{n} da \times \mathrm{grad}\left(\frac{1}{R}\right), \tag{2.39}$$

where we have taken into account that

$$\frac{1}{2}[\mathbf{r}' \times d\mathbf{r}'] = \mathbf{n} da,$$

and $\mathbf{n}$ is the unit vector perpendicular to the surface da. The quantity

$$\boldsymbol{\mu}_m = I \int \mathbf{n} da = \frac{I}{2} \oint_C [\mathbf{r}' \times d\mathbf{r}'] = \frac{1}{2} \int_V [\mathbf{r}' \times \mathbf{j}] dV' \tag{2.40}$$

is called the dipole magnetic moment of the system. The magnetization $\mathbf{M}$ of the system is determined as the magnetic moment per unit volume:

$$\mathbf{M} = \frac{d\boldsymbol{\mu}_m}{dV} = \frac{1}{2}[\mathbf{r}' \times \mathbf{j}]. \tag{2.41}$$

Equation (2.41) shows that the magnetization exists only when the current density possesses the component perpendicular to the radius vector $\mathbf{r}'$ which corresponds to the electric charge rotation around the origin. Let the electric charge has the density ρ and the linear velocity $\mathbf{v}$. Then the current density can be written as

$$\mathbf{j} = \rho \mathbf{v}$$

and the magnetization $\mathbf{M}$ takes the form

$$\mathbf{M} = \frac{1}{2}[\mathbf{r}' \times \rho \mathbf{v}]. \tag{2.42}$$

2.1.3 Energetic Characteristics of a Static Magnetic Field

The magnetic field energy density $\mathcal{W}^M$ and the energy W^M of the static magnetic field in the closed volume V have the form, respectively:

$$\mathcal{W}^M = \frac{1}{2}\mathbf{HB}, \quad W^M = \frac{1}{2}\int_V \mathcal{W}^M dV = \frac{1}{2}\int_V \mathbf{HB}\, dV. \qquad (2.43)$$

Substituting (2.5) into (2.43) we find:

$$W^M = \frac{1}{2}\int_V \mathbf{H}\,\mathrm{curl}\mathbf{A}\, dV. \qquad (2.44)$$

Applying to (2.44) the vector analysis identity

$$\mathbf{H}\,\mathrm{curl}\mathbf{A} = \mathbf{A}\,\mathrm{curl}\mathbf{H} + \mathrm{div}\,[\mathbf{A}\times\mathbf{H}]$$

and the Gauss theorem we obtain the sum of two integrals:

$$W^M = \frac{1}{2}\int_V \mathbf{j}\mathbf{A}\, dV + \frac{1}{2}\oint_S [\mathbf{A}\times\mathbf{H}]\, dS. \qquad (2.45)$$

The second integral in the right-hand side of (2.45) is the surface one, and it vanishes, since the current density and the vector potential decrease at large distances as $1/r^3$ and $1/r^2$, respectively, while the surface increases as r^2. Then (2.45) reduces to the simple expression

$$W^M = \frac{1}{2}\int_V \mathbf{j}\mathbf{A}\, dV, \qquad (2.46)$$

where the integration is carried out over the volume containing the currents only. The expression (2.46) shows that the magnetic energy is zero when the space is free of electric currents. In order to emphasize this fact we express the vector potential in terms of a current density according to (2.26). Insertion of this result into (2.46) yields:

$$W^M = \frac{\mu_0 \mu^r}{8\pi}\int_V\int_V \frac{\mathbf{j}(\mathbf{r})\,\mathbf{j}(\mathbf{r}')}{|\mathbf{r}-\mathbf{r}'|}\, dV'dV. \qquad (2.47)$$

Expression (2.47) can be rewritten as follows:

$$W^M = \frac{1}{2}\mathcal{L}I^2, \qquad (2.48)$$

where the coefficient $\mathcal{L}$ is called inductance. It has the form:

$$\mathcal{L} = \frac{\mu_0 \mu^r}{4\pi I^2}\int_V\int_V \frac{\mathbf{j}(\mathbf{r})\,\mathbf{j}(\mathbf{r}')}{|\mathbf{r}-\mathbf{r}'|}\, dV'dV. \qquad (2.49)$$

Obviously, the inductance does not depend on the total current in the system.

Consider now the system of volumes V_i with currents I_i. The total vector potential of this system is the sum of the vector potentials caused by each current:

$$\boldsymbol{A}^{\text{tot}} = \sum_{i=1}^{N} \boldsymbol{A}_i. \tag{2.50}$$

Combining (2.46) and (2.50) we obtain for the energy W^{tot} of the system of currents:

$$W^{\text{tot}} = \frac{1}{2} \sum_{k=1}^{N} \int_{V_k} \boldsymbol{j}\boldsymbol{A} dV_k = \frac{1}{2} \sum_{k=1}^{N} \sum_{i=1}^{N} \int_{V_k} \boldsymbol{j}\boldsymbol{A}_i dV_k. \tag{2.51}$$

Once more using (2.26) for each vector potential $\boldsymbol{A}_i$ we arrive at the following relationship:

$$W^{\text{tot}} = \frac{\mu_0 \mu^r}{8\pi} \sum_{k=1}^{N} \sum_{i=1}^{N} \int_{V_k} \int_{V_i} \frac{\boldsymbol{j}(\boldsymbol{r}_i)\,\boldsymbol{j}(\boldsymbol{r}_k)}{|\boldsymbol{r}_i - \boldsymbol{r}_k|} dV_k dV_i. \tag{2.52}$$

The terms in the sum (2.52) can be divided in two groups: the diagonal contributions W_{ii} and the non-diagonal ones with $i \neq k$. We write these terms in the form similar to (2.48):

$$W_{ii} = \frac{\mathcal{L}_i I_i^2}{2},\ W_{ik}_{i \neq k} = \frac{\mathcal{M}_{ik} I_i I_k}{2}, \tag{2.53}$$

where the so-called self-inductance $\mathcal{L}_i$ of the conductor is determined by expression (2.49) and the mutual inductance $\mathcal{M}_{ik}$ of conductors is:

$$\mathcal{M}_{ik} = \frac{\mu_0 \mu^r}{4\pi I_i I_k} \int_{V_k} \int_{V_i} \frac{\boldsymbol{j}(\boldsymbol{r}_i)\,\boldsymbol{j}(\boldsymbol{r}_k)}{|\boldsymbol{r}_i - \boldsymbol{r}_k|} dV_k dV_i. \tag{2.54}$$

We conclude that the energy of the system of currents can be expressed in terms of self-inductances and mutual inductances of conductors in a following way:

$$W^{\text{tot}} = \sum_{i=1}^{N} \frac{\mathcal{L}_i I_i^2}{2} + \sum_{k=1}^{N} \sum_{i=1, i \neq k}^{N} \frac{\mathcal{M}_{ik} I_i I_k}{2}. \tag{2.55}$$

The first sum in expression (2.55) represents the self-energy of the field, and the second one includes the mutual energy. Clearly, the magnetic field energy W^{tot} should be positive definite which results in positive value of self-inductance $\mathcal{L}_i > 0$ and a condition that $\mathcal{L}_i \mathcal{L}_k > \mathcal{M}_{ik}^2$. The calculation of the mutual induction essentially simplifies in the case of linear currents. Then (2.54) reduces to

$$\mathcal{M}_{ik} = \frac{\mu_0 \mu^r}{4\pi} \oint_{L_i} \oint_{L_k} \frac{dl_i dl_k}{|\boldsymbol{r}_i - \boldsymbol{r}_k|}. \tag{2.56}$$

The calculation of a self-inductance is a more difficult problem, since the linear conductor approximation results in the logarithmic divergence of the integral (2.49) due to the contribution from small values $|\boldsymbol{r} - \boldsymbol{r}'|$. In general, the finite thickness of a wire should be taken into account, and the appropriate approximations should be formulated for each particular case. Some practically important cases will be analyzed below.

Consider another form of the total energy of a system of linear currents. For this purpose we write equation (2.46) for the system of linear currents. It takes the form

$$W^M = \frac{1}{2} \sum_{i=1}^{N} I_i \oint_L \boldsymbol{A} dl_i. \tag{2.57}$$

The integrals in the right-hand side of (2.57) can be transformed into the surface integrals

$$\oint_{L_i} \boldsymbol{A} dl_i = \int_{S_i} \mathrm{curl}\boldsymbol{A} d\boldsymbol{s}_i = \int_{S_i} \boldsymbol{B} d\boldsymbol{s}_i, \tag{2.58}$$

where S_i is the surface enclosed by the contour L_i. The last integral in (2.58) is, by definition (2.28) the magnetic flux Φ_i through the circuit of the i-th current. Substituting (2.58) into (2.57) we finally obtain the connection between the magnetic energy and the magnetic flux.

$$W^M = \frac{1}{2} \sum_{i=1}^{N} I_i \Phi_i. \tag{2.59}$$

The comparison of (2.59) and (2.55) shows that the magnetic flux can be expressed in terms of the mutual inductance as follows

$$\Phi_i = \sum_{k=1}^{N} \Phi_{ik}, \quad \Phi_{ik} = \mathcal{M}_{ik} I_k, \quad \Phi_{ii} = \mathfrak{L}_i I_i, \tag{2.60}$$

where Φ_{ik} is the magnetic flux through the i-th contour created by the k-th current, Φ_{ii} is the magnetic flux through the i-th contour due to the current in this contour I_i itself.

2.1.4 Forces in a Magnetic Field

It is known from mechanics that the force $\boldsymbol{f} dV$ acting on the matter in a volume dV can be expressed in terms of the mechanical stress tensor σ_{ik}^M. The force density f_i has the form:

2.1 Fundamentals of Magnetostatics

$$f_i = \frac{\partial \sigma_{ik}^M}{\partial x_k}. \tag{2.61}$$

It can be shown that the stress tensor σ_{ik}^M in a fluid medium with a magnetic induction (2.3) can be written as follows:

$$\sigma_{ik}^M = -P_0(\rho, T)\delta_{ik} - \mu_0 \frac{H^2}{2}\left[\mu^r - \rho\left(\frac{\partial \mu^r}{\partial \rho}\right)_T\right]\delta_{ik} + \mu_0\mu^r H_i H_k, \tag{2.62}$$

where P_0, ρ and T are a pressure, a mass density and temperature, respectively. Substituting (2.62) into (2.61) we obtain the general expression for the force density acting on the matter in a magnetic field. It has the form:

$$\boldsymbol{f} = -\mathrm{grad}P_0 + \frac{1}{2}\mu_0 \mathrm{grad}\left[H^2\rho\left(\frac{\partial \mu^r}{\partial \rho}\right)_T\right] - \mu_0\frac{H^2}{2}\mathrm{grad}\mu^r - \frac{\mu_0\mu^r}{2}\mathrm{grad}H^2$$

$$+\mu_0\mu^r(\boldsymbol{H}\cdot\mathrm{grad})\boldsymbol{H}. \tag{2.63}$$

Here we take into account that

$$\mathrm{div}\boldsymbol{B} = \mathrm{div}\mu_0\mu^r\boldsymbol{H} = 0.$$

Using a well known formula of vector analysis and (2.1) we transform the last term in (2.63):

$$(\boldsymbol{H}\cdot\mathrm{grad})\boldsymbol{H} = \frac{1}{2}\mathrm{grad}H^2 - [\boldsymbol{H}\times\mathrm{curl}\boldsymbol{H}] = \frac{1}{2}\mathrm{grad}H^2 + \mu_0\mu^r[\boldsymbol{j}\times\boldsymbol{H}].$$

Then the force (2.63) takes the form:

$$\boldsymbol{f} = -\mathrm{grad}P_0 + \frac{1}{2}\mu_0\mathrm{grad}\left[H^2\rho\left(\frac{\partial \mu^r}{\partial \rho}\right)_T\right] - \mu_0\frac{H^2}{2}\mathrm{grad}\mu^r + \mu_0\mu^r[\boldsymbol{j}\times\boldsymbol{H}]. \tag{2.64}$$

The term $-\mathrm{grad}P_0$ is of no interest, and therefore it can be omitted. Usually, μ^r is very close to unity, and the last term in (2.64) gives the main contribution to the force in the presence of a conduction current. We finally obtain neglecting all small terms:

$$\boldsymbol{f} = \mu_0\mu^r[\boldsymbol{j}\times\boldsymbol{H}] = [\boldsymbol{j}\times\boldsymbol{B}]. \tag{2.65}$$

The total force $\boldsymbol{F}$ exerted by a magnetic field on conductor carrying a current is calculated by integration of (2.65) over the volume containing a system of currents:

$$\boldsymbol{F} = \int_V [\boldsymbol{j}\times\boldsymbol{B}]\,dV. \tag{2.66}$$

If the current density in (2.66) corresponds to the steady-state current I_1 in a wire, consideration similar to one that resulted in (2.25) allows for (2.66) to be written in the form

$$F = I_1 \int [d\boldsymbol{l}_1 \times \boldsymbol{B}]. \tag{2.67}$$

The magnetic field itself is produced by another current I_2 flowing through another wire with the length element $d\boldsymbol{l}_2$ and if both wires form loops, then, according to (2.25) the total force experienced by the first current is

$$\boldsymbol{F}_{12} = \frac{\mu_0 \mu^r I_1 I_2}{4\pi} \oint \oint \frac{[d\boldsymbol{l}_1 \times [d\boldsymbol{l}_2 \times \boldsymbol{R}_{12}]]}{R_{12}^3}. \tag{2.68}$$

Transforming the double vector product in (2.68) according to

$$[\boldsymbol{a} \times [\boldsymbol{b} \times \boldsymbol{c}]] = \boldsymbol{b}\,(\boldsymbol{a}\cdot\boldsymbol{c}) - \boldsymbol{c}\,(\boldsymbol{a}\cdot\boldsymbol{b})$$

we obtain for the integrand in (2.68)

$$\frac{[d\boldsymbol{l}_1 \times [d\boldsymbol{l}_2 \times \boldsymbol{R}_{12}]]}{R_{12}^3} = d\boldsymbol{l}_2 \frac{(d\boldsymbol{l}_1 \cdot \boldsymbol{R}_{12})}{R_{12}^3} - \frac{\boldsymbol{R}_{12}\,(d\boldsymbol{l}_1 \cdot d\boldsymbol{l}_2)}{R_{12}^3}.$$

The first term in the right-hand side of this expression involves a perfect differential in the integral over $d\boldsymbol{l}_1$ and gives no contribution to the integral providing that the path of integration is closed or extends to infinity. Hence, the expression for a force between two current loops takes a symmetric form

$$\boldsymbol{F}_{12} = -\frac{\mu_0 \mu^r I_1 I_2}{4\pi} \oint \oint \frac{\boldsymbol{R}_{12}\,(d\boldsymbol{l}_1 \cdot d\boldsymbol{l}_2)}{R_{12}^3}. \tag{2.69}$$

The force (2.67) can be also expressed in terms of the magnetic moment $\boldsymbol{\mu}_m$ (2.40). For this purpose we should change the integration along the contour L_1 with the integration over the surface S_1 replacing the length element $d\boldsymbol{l}_1$ of the linear current by the operator $[d\boldsymbol{s}_1 \times \text{grad}]$ containing the surface element $d\boldsymbol{s}_1$:

$$\oint_{L_1} [d\boldsymbol{l}_1 \times \boldsymbol{B}] = \int_{S_1} [[d\boldsymbol{s}_1 \times \text{grad}] \times \boldsymbol{B}]. \tag{2.70}$$

Using the theorems of the vector analysis we transform the integrand in a following way:

$$[[d\boldsymbol{s}_1 \times \text{grad}] \times \boldsymbol{B}] = -d\boldsymbol{s}_1 \text{div}\boldsymbol{B} + \text{grad}\,(d\boldsymbol{s}_1 \cdot \boldsymbol{B})$$

$$= -d\boldsymbol{s}_1 \text{div}\boldsymbol{B} + [d\boldsymbol{s}_1 \times \text{curl}\boldsymbol{B}] + (d\boldsymbol{s}_1 \cdot \text{grad})\,\boldsymbol{B}. \tag{2.71}$$

The first term in the right-hand side of (2.71) vanishes according to the Maxwell equation (2.2). In the second term $\text{curl}\boldsymbol{B} = 0$ in the space outside the

2.1 Fundamentals of Magnetostatics

volume containing the current distribution. The remaining term yields after the substitution into (2.70):

$$F = I_1 \int_{S_1} (ds_1 \cdot \text{grad}) \, B. \quad (2.72)$$

In the case of the almost uniform magnetic field its derivatives can be taken outside the integral. Then (2.72) takes the form:

$$F = \left(I_1 \left(\int_{S_1} ds_1 \right) \cdot \text{grad} \right) B. \quad (2.73)$$

Comparison of (2.73) and (2.40) shows that the first factor in the brackets is a magnetic moment μ_m of the contour with the current. Then we have:

$$F = (\mu_m \cdot \text{grad}) \, B. \quad (2.74)$$

Since the magnetic moment μ_m is a constant vector, (2.74) can be rewritten:

$$F = \text{grad} \, (\mu_m \cdot B). \quad (2.75)$$

It is known from mechanics that a force F can be presented as a gradient of a potential energy U of some field:

$$F = -\text{grad} U. \quad (2.76)$$

Comparing (2.75) and (2.76) one can define the potential energy of a magnetic dipole in a magnetic field. It has the form:

$$U_M = -(\mu_m \cdot B). \quad (2.77)$$

Finally, consider torques acting on a contour with current in a magnetic field. The total torque acting on the current distribution is, by definition

$$N = \int [r \times [j(r) \times B(r)]] \, dV. \quad (2.78)$$

The derivations similar to the previous ones carried out for the force show that in the case of a linear current and almost uniform magnetic field the torque (2.78) can be also expressed in terms of a magnetic moment of the current distribution. Namely, (2.78) takes the form

$$N = [\mu_m \times B]. \quad (2.79)$$

The torque N determines the temporal evolution of an angular momentum L_A of the system according to the well known equation of mechanics

$$\frac{d\boldsymbol{L}_A}{dt} = \boldsymbol{N} = [\boldsymbol{\mu}_m \times \boldsymbol{B}], \tag{2.80}$$

where the angular momentum $\boldsymbol{L}_A$ is defined as follows

$$\boldsymbol{L}_A = \int [\boldsymbol{r} \times \rho \boldsymbol{v}] \, dV, \tag{2.81}$$

and $\boldsymbol{v}$ is the linear velocity of the mass element. Comparison of (2.40) and (2.81) shows that the magnetic moment and angular momentum of the system can be easily expressed through one another, if the charged particles carrying the current possess the same ratio of the electric charge and the mass e/m. Namely, we write

$$\boldsymbol{\mu}_m = \frac{e}{2m} \boldsymbol{L}_A. \tag{2.82}$$

Substituting (2.82) we get

$$\frac{d\boldsymbol{L}_A}{dt} = -[\boldsymbol{\Omega} \times \boldsymbol{L}_A], \quad \frac{d\boldsymbol{\mu}_m}{dt} = -[\boldsymbol{\Omega} \times \boldsymbol{\mu}_m], \tag{2.83}$$

where $\boldsymbol{\Omega}$ is the so-called Larmor frequency:

$$\boldsymbol{\Omega} = \frac{e}{2m} \boldsymbol{B}. \tag{2.84}$$

Equations (2.83) describe the rotation of the angular momentum and the magnetic moment of the system around the direction of the magnetic field $\boldsymbol{B}$ with the angular velocity (2.84) while the moduli of these vectors as well as the angles between them and the field are conserved. This phenomenon is called the Larmor precession. In general, considering the Larmor precession we remain in the framework of the stationary field theory. It should be emphasized that for a comparatively weak magnetic field the Larmor frequency is small compared with frequencies of internal motion of the system. For this reason, the quantities described by (2.79)–(2.83) are varying slowly with respect to the fast internal processes and have in fact the values that have already been averaged over the short time intervals of the internal motion.

Equations (2.82) and (2.83) show that even without an external magnetic field a uniform rotation causes a magnetization which depends linearly on the angular velocity $\boldsymbol{\Omega}$. This is known as the Barnett effect. The linear relation between the vectors of the magnetic moment $\boldsymbol{\mu}_m$ and the angular velocity $\boldsymbol{\Omega}$ is possible because both these vectors change sign when the sign of time is reversed. The vectors $\boldsymbol{\mu}_m$ and $\boldsymbol{\Omega}$ are axial, and the linear relation between them is valid also in an isotropic body where it reduces to a simple proportionality. Inversely, as it is seen from the second of (2.83), a freely suspended body, when being magnetized, begins to rotate which is called the Einstein–de Haas effect. Both effects in general are known as the gyromagnetic phenomena. We derive the relation between them from the energetic characteristics of the rotating body. From the statistical physics it is known that the angular momentum $\boldsymbol{L}_A$ of the body can be expressed in terms of its free energy $\mathcal{F}$ as follows:

$$\boldsymbol{L}_A = -\frac{\partial \mathcal{F}}{\partial \boldsymbol{\Omega}}. \tag{2.85}$$

The gyromagnetic phenomena are described by additional terms in the free energy $\mathcal{F}$ which are linear in both the magnetization $\boldsymbol{M}$ at each point in the body and the angular frequency $\boldsymbol{\Omega}$:

$$\mathcal{F}_{\text{gyro}} = -\int \lambda_{ik} \Omega_i M_k dV = -\lambda_{ik} \mu_{mk} \Omega_i, \tag{2.86}$$

where λ_{ik} is a constant tensor, in general unsymmetrical.. Substituting (2.86) into (2.85) we find that the angular momentum acquired by the body due to the magnetization has the form:

$$L_i^{\text{gyro}} = \lambda_{ik} \mu_{mk} \tag{2.87}$$

and vice versa, the magnetic moment caused by a rotation of the body is

$$\mu_{mk} = \frac{e}{2m} g_{ik} L_k^{\text{gyro}}, \quad g_{ik} = \frac{2m}{e} \lambda_{ik}^{-1}, \tag{2.88}$$

where g_{ik} are the gyromagnetic coefficients. Comparison of (2.80), the first of (2.83), (2.87) and (2.88) shows that the rotation of the body with the angular velocity $\boldsymbol{\Omega}$ is equivalent to some external magnetic field $\boldsymbol{B}$ such that

$$B_i = \lambda_{ik} \Omega_k, \tag{2.89}$$

or, taking into account relations (2.88)

$$B_i = \frac{2m}{e} g_{ik}^{-1} \Omega_k, \tag{2.90}$$

which permits, in principle, the calculation of the magnetization caused by the rotation. Relations (2.88) and (2.90) describe the Einstein–de Haas and Barnett effects respectively which are determined by the same tensor g_{ik}.

2.2 Magnetic Field of a Given Current Distribution

2.2.1 Magnetic Field of Linear Currents

Consider some typical examples of calculation of the magnetic field caused by wires and loops with a linear current. We start with a simplest case of an infinite thin wire having a negligibly small cross-section. Let the current I flows along the wire. Obviously, the problem possesses the axial symmetry,

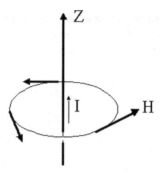

Fig. 2.3. The magnetic field $\boldsymbol{H}$ of the thin infinite conductor with the linear current I

and the magnetic field would depend on a distance R from the wire only. In order to calculate the magnetic field intensity caused by this current we use the Maxwell equation in the integral form (2.9), or the so-called Ampère's law. We introduce the cylindrical system of coordinates (r, φ, z) where the z-axis is chosen to be parallel to the wire with the current I. Taking as an integration contour a circle of a radius r in a plane perpendicular to the wire and assuming that the magnetic field intensity $\boldsymbol{H}$ is tangential with respect to this circle in each point we write the integrand in (2.9):

$$\boldsymbol{H}d\boldsymbol{l} = Hrd\varphi,$$

which yields

$$\oint_L \boldsymbol{H}d\boldsymbol{l} = \int_0^{2\pi} Hrd\varphi = I$$

and

$$H = \frac{I}{2\pi r}. \tag{2.91}$$

The geometry of the problem is shown in Fig. 2.3.

Consider a similar problem but this time we assume that the cylindrical conductor has a finite circular cross-section with a radius R_0. In such a case we should solve the problem separately for the region inside the conductor $0 < r < R_0$ and outside it $r > R_0$. Both solutions must coincide at the surface $r = R_0$ since the tangential component of the magnetic field is continuous at the boundary of two media. The geometry of the problem remains essentially the same. In the inner region the current at the contour of radius r is

$$I(r) = I\left(\frac{r}{R_0}\right)^2. \tag{2.92}$$

2.2 Magnetic Field of a Given Current Distribution   47

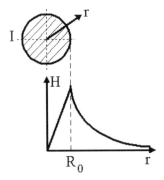

Fig. 2.4. The distribution of the magnetic field H of a cylinder with the current I

Substituting (2.92) into (2.91) we obtain

$$H = \frac{Ir}{2\pi R_0^2}, \quad 0 < r < R_0. \tag{2.93}$$

For the external region $r > R_0$ (2.91) remains valid. Comparison of (2.91) and (2.93) shows that the solution is continuous at $r = R_0$. The solution is presented in Fig. 2.4.

Consider now the coaxial cylinder with the radius of the core R_1 and the internal and external radii of the shell R_2 and R_3, respectively. In the core and in the shell the equal currents I flow in the opposite directions. The field in the core region $0 \le r \le R_1$ and between the core and the shell obeys (2.93) and (2.91), respectively:

$$H = \frac{Ir}{2\pi R_1^2}, \quad 0 \le r \le R_1; \quad H = \frac{I}{2\pi r}, \quad R_1 \le r \le R_2. \tag{2.94}$$

The current distribution in the shell has the form

$$I(r) = I \frac{r^2 - R_2^2}{R_3^2 - R_2^2}, \quad R_2 \le r \le R_3. \tag{2.95}$$

Choosing the current direction in the core as positive we obtain the field in the shell:

$$H = \frac{I}{2\pi r}\left[1 - \frac{r^2 - R_2^2}{R_3^2 - R_2^2}\right] = \frac{I}{2\pi r}\frac{R_3^2 - r^2}{R_3^2 - R_2^2}, \quad R_2 \le r \le R_3. \tag{2.96}$$

Clearly, outside the cylinder the magnetic field vanishes since the currents in the shell and the core compensate each other. In the opposite limiting case of a hollow cylinder of a finite thickness, i.e. $R_1 = 0$, the field inside the cavity $0 \le r \le R_2$ is zero, the field inside the shell has the form

$$H = \frac{I}{2\pi r}\frac{r^2 - R_2^2}{R_3^2 - R_2^2}, \quad R_2 \le r \le R_3 \tag{2.97}$$

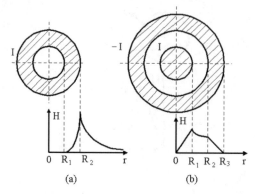

Fig. 2.5. The magnetic field distribution of the hollow cylinder (**a**) and the coaxial cylinders (**b**) with currents

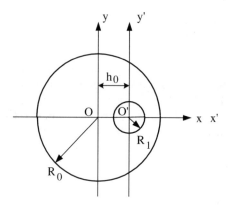

Fig. 2.6. The cross-section of the conducting cylinder with a cylindrical hole

and the field in the external region $r \geq R_3$ is determined by (2.91). These results are shown in Fig. 2.5.

We determine finally the magnetic field in a cylindrical hole in an infinite length cylinder with a current density $\boldsymbol{j}$ which is uniform over its cross section. The cross section of the conducting cylinder is presented in Fig. 2.6.

We use the principle of the superposition of fields. Indeed, if there are no hole, the magnetic field intensity $\boldsymbol{H}_1$ would be according to the first equation (2.94)

$$\boldsymbol{H}_1 = \frac{jr}{2}\boldsymbol{e}_\varphi,$$

where the unit vector $\boldsymbol{e}_\varphi$ is directed tangentially to the contour of the cylinder cross section. The cylindrical hole is equivalent to the current flowing in the opposite direction with respect to the existing one which would correspond to the current density $-\boldsymbol{j}$. The magnetic field $\boldsymbol{H}_2$ of this fictitious current in the hole has the form

$$\boldsymbol{H}_2 = -\frac{jr'}{2}\boldsymbol{e}_\varphi,$$

where r' the radius vector in the coordinates connected with the hole. It is convenient to convert both field intensities to the Cartesian coordinates connected with the origin, i.e. the center of the cylinder cross section. Then they take the form

$$H_{1x} = -\frac{jy}{2},\ H_{1y} = \frac{jx}{2},\ H_{2x} = \frac{jy}{2},\ H_{2y} = -\frac{j(x-h_0)}{2}.$$

The superposition of the two fields shows that the field in the hole is perpendicular to the line connecting the centers of the cylinder and the hole and uniform:

$$H_x = 0,\ H_y = \frac{jh_0}{2} = \frac{Ih_0}{2\pi(R_0^2 - R_1^2)}, \tag{2.98}$$

where I is the total current in the cylinder.

2.2.2 Magnetic Field of a Contour with a Current

Consider the magnetic field of the circular contour with a current I. First of all we calculate the vector potential $\boldsymbol{A}(\boldsymbol{r})$ according to (2.27). For the sake of definiteness we assume that $\mu^r = 1$. Introducing the spherical coordinates (r, θ, φ) we get for the length element $d\boldsymbol{l}'$:

$$d\boldsymbol{l}' = (-a\sin\varphi'\, d\varphi', a\cos\varphi'\, d\varphi', 0), \tag{2.99}$$

where a is the radius of the contour with current. It is easy to see that in the spherical coordinates the element $d\boldsymbol{l}'$ has the only component $dl'_\varphi = a\, d\varphi'$. Similarly we obtain for the distance R :

$$\begin{aligned}|r - r'|^2 &= (x - a\cos\varphi')^2 + (y - a\sin\varphi')^2 + z^2 \\ &= x^2 + y^2 + z^2 - 2ax\cos\varphi' - 2ay\sin\varphi' + a^2 \\ &= r^2 + a^2 - 2ar\sin\theta\cos(\varphi - \varphi').\end{aligned} \tag{2.100}$$

Transformation of the vector-potential to the spherical coordinates yields:

$$\begin{aligned}A_\varphi &= -A_x\sin\varphi + A_y\cos\varphi \\ &= \frac{\mu_0 I a}{4\pi}\int \frac{d\varphi'\cos(\varphi - \varphi')}{\sqrt{r^2 + a^2 - 2ar\sin\theta\cos(\varphi - \varphi')}}.\end{aligned} \tag{2.101}$$

Introduce the new variables

$$\psi = \varphi - \varphi';\ \lambda = \frac{2ar}{r^2 + a^2}\sin\theta. \tag{2.102}$$

Then we have

$$A_\varphi = \frac{\mu_0 I a}{4\pi} \frac{1}{\sqrt{r^2+a^2}} \oint \frac{\cos\psi \, d\psi}{\sqrt{1-\lambda\cos\psi}}. \qquad (2.103)$$

The last integral can be expressed in terms of the complete elliptical integrals of the first and second types $\mathcal{K}(k)$ and $\mathsf{E}(k)$ which are tabulated:

$$\oint \frac{\cos\psi \, d\psi}{\sqrt{1-\lambda\cos\psi}} = 2k\sqrt{\frac{2}{\lambda}}\left[\frac{1}{\lambda}\mathcal{K}(k) - \frac{2}{\lambda^2}\mathsf{E}(k)\right], \qquad (2.104)$$

where

$$\cos\frac{\psi}{2} = \sin\chi; \quad k^2 = \frac{2\lambda}{1+\lambda}; \qquad (2.105)$$

$$\mathcal{K}(k) = \int_0^{\frac{\pi}{2}} \frac{d\theta}{\sqrt{1-k^2\sin^2\theta}}; \quad \mathsf{E}(k) = \int_0^{\frac{\pi}{2}} \sqrt{1-k^2\sin^2\theta}\,d\theta.$$

In order to obtain the approximate results the following useful expansions can be used:

$$\oint \cos^{2n}\psi \, d\psi = 2\pi \frac{(2n)!}{2^{2n}(n!)^2}; \quad \oint \cos^{2n+1}\psi \, d\psi = 0. \qquad (2.106)$$

Substituting these results into the expression for the vector-potential we immediately obtain:

$$\oint \frac{\cos\psi \, d\psi}{\sqrt{1-\lambda\cos\psi}} = \oint d\psi \cos\psi \qquad (2.107)$$

$$\times \left[1 + \frac{1}{2}\lambda\cos\psi + \frac{3}{8}\lambda^2\cos^2\psi + \frac{5}{16}\cos^3\psi + \ldots\right]$$

$$= \frac{\pi}{2}\lambda\left(1 + \frac{15}{32}\lambda^2 + \ldots\right) \qquad (2.108)$$

and

$$A_\varphi = \frac{\mu_0 I a^2}{4} \frac{r\sin\theta}{(r^2+a^2)^{\frac{3}{2}}}\left[1 + \frac{15}{8}\frac{a^2 r^2}{(r^2+a^2)^2}\sin^2\theta + \ldots\right]. \qquad (2.109)$$

It should be noted that this expansion is valid in the region close to the center of the contour where $r \ll a$ and also at large distances from the contour $r \gg a$. Bearing in mind that $\boldsymbol{B}(\boldsymbol{r}) = \mathrm{curl}\boldsymbol{A}(\boldsymbol{r})$ we now can calculate the components of the magnetic field in these zones. Namely, we have:

$$B_r = \frac{1}{r\sin\theta}\frac{d}{d\theta}(A_\varphi \sin\theta) = \frac{\mu_0 I}{4\pi}\pi a^2 \frac{2\cos\theta}{(r^2+a^2)^{\frac{3}{2}}} \qquad (2.110)$$

$$\times \left[1 + \frac{15}{4}\frac{a^2 r^2}{(r^2+a^2)^2}\sin^2\theta + \ldots\right]$$

2.2 Magnetic Field of a Given Current Distribution

$$B_\theta = -\frac{1}{r}\frac{\partial}{\partial r}(rA_\varphi) = \frac{\mu_0 I}{4\pi}\pi a^2 \frac{(r^2 - 2a^2)\sin\theta}{(r^2 + a^2)^{\frac{5}{2}}} \qquad (2.111)$$

$$\times \left[1 + \frac{15}{8}\frac{a^2 r^2}{(r^2+a^2)^2}\frac{3r^2 - 4a^2}{r^2 - 2a^2}\sin^2\theta + ...\right]$$

$$B_\varphi = 0. \qquad (2.112)$$

For small values of λ the expansions have especially compact form:

$$B_r = \frac{\mu_0 I}{4\pi}\pi a^2 \frac{2\cos\theta}{(r^2+a^2)^{\frac{3}{2}}}; \quad B_\theta = \frac{\mu_0 I}{4\pi}\pi a^2 \frac{(r^2 - 2a^2)\sin\theta}{(r^2+a^2)^{\frac{5}{2}}}. \qquad (2.113)$$

In the far zone where $r \gg a$ the magnetic field components take the form:

$$B_r = \frac{\mu_0 I}{4\pi}\pi a^2 \frac{2\cos\theta}{r^3}; \quad B_\theta = \frac{\mu_0 I}{4\pi}\pi a^2 \frac{\sin\theta}{r^3}. \qquad (2.114)$$

In the opposite case, $r \ll a$, close to the center of the circle the magnetic field components are:

$$B_r = \frac{\mu_0 I}{4\pi}\frac{2\pi}{a}\cos\theta; \quad B_\theta = -\frac{\mu_0 I}{4\pi}\frac{2\pi}{a}\sin\theta. \qquad (2.115)$$

In the cylindrical coordinates (ρ, φ, z) the magnetic field components can be represented as follows:

$$B_z = B_r \cos\theta - B_\theta \sin\theta = \frac{\mu_0 I}{4\pi}\frac{2\pi}{a} \qquad (2.116)$$

$$B_\rho = B_r \sin\theta + B_\theta \cos\theta = 0. \qquad (2.117)$$

Along the z-axis we have $\sin\theta = 0$, $\cos\theta = 1$, $B_\theta = 0$; $\lambda = 0$. Then,

$$B_z = B_r \cos\theta = \frac{\mu_0 I}{4\pi}\pi a^2 \frac{2}{(z^2 + a^2)^{\frac{3}{2}}}. \qquad (2.118)$$

It is instructive to calculate the magnetic field of the circular contour of radius a directly using the Biot–Savart law expression (2.25). We write:

$$\mathbf{B} = \frac{\mu_0 I}{4\pi}\oint \frac{[d\mathbf{l}' \times (\mathbf{r} - \mathbf{r}')]}{|\mathbf{r} - \mathbf{r}'|^3}. \qquad (2.119)$$

The vectorial product in the numerator of the integrand has the form:

$$[d\mathbf{l}' \times (\mathbf{r} - \mathbf{r}')] = \begin{vmatrix} ar\cos\theta\cos\varphi' d\varphi' \\ ar\cos\theta\sin\varphi' d\varphi' \\ a[a - r\sin\theta\cos(\varphi - \varphi')]d\varphi' \end{vmatrix}. \qquad (2.120)$$

Then, presenting the field components in the cylindrical coordinates

$$B_\rho = B_x \cos\varphi + B_y \sin\varphi; \quad B_\varphi = -B_x \sin\varphi + B_y \cos\varphi \tag{2.121}$$

and denoting $\psi = \varphi - \varphi'$, $\lambda = 2ar\sin\theta/(r^2 + a^2)$ we find

$$B_\rho = \frac{\mu_0 I}{4\pi} \frac{ar\cos\theta}{(r^2+a^2)^{\frac{3}{2}}} \oint \frac{d\psi\cos\psi}{(1-\lambda\cos\psi)^{\frac{3}{2}}} \tag{2.122}$$

$$B_z = \frac{\mu_0 I}{4\pi} \frac{a}{(r^2+a^2)^{\frac{3}{2}}} \oint d\psi \frac{a - r\sin\theta\cos\psi}{(1-\lambda\cos\psi)^{\frac{3}{2}}}. \tag{2.123}$$

The integrand can be expanded in powers of λ which yields:

$$\mathcal{A}(\lambda) = \oint \frac{d\psi\cos\psi}{(1-\lambda\cos\psi)^{\frac{3}{2}}} = \frac{3\pi}{2}\lambda + \frac{105\pi}{64}\lambda^2 + \ldots \tag{2.124}$$

$$\mathcal{B}(\lambda) = \oint \frac{d\psi}{(1-\lambda\cos\psi)^{\frac{3}{2}}} = 2\pi + \frac{15\pi}{8}\lambda^2 + \ldots. \tag{2.125}$$

Finally, transforming the field components into the spherical coordinates and substituting the integrals $\mathcal{A}(\lambda)$ and $\mathcal{B}(\lambda)$ we obtain the exact formulas:

$$B_r = B_z \cos\theta + B_\rho \sin\theta = \frac{\mu_0 I}{4\pi} \frac{a^2}{(r^2+a^2)^{\frac{3}{2}}} \mathcal{B}(\lambda)\cos\theta \tag{2.126}$$

$$B_\theta = -B_z \sin\theta + B_\rho \cos\theta \tag{2.127}$$

$$= \frac{\mu_0 I}{4\pi} \frac{a^2}{(r^2+a^2)^{\frac{3}{2}}} \left\{ -\mathcal{B}(\lambda)\sin\theta + \frac{r}{a}\mathcal{A}(\lambda) \right\}. \tag{2.128}$$

2.2.3 A Cylinder in an External Magnetic Field

Consider the classical problem of an infinite cylinder having a circular cross-section of a radius R_0 with a current I which is subject to a uniform external magnetic field $\boldsymbol{B}_0$ perpendicular to the axis of the cylinder. The relative magnetic permeabilities of the cylinder and the medium are μ_1^r and μ_2^r respectively. The z-axis coincides with the axis of the cylinder, and the x-axis is parallel to the external magnetic field. The geometry of the problem is shown in Fig. 2.7.

We need to calculate the resulting magnetic field inside and outside the cylinder. For this purpose we first of all calculate the vector potential $\boldsymbol{A}$ which obeys (2.7). Inside the cylinder we have

2.2 Magnetic Field of a Given Current Distribution

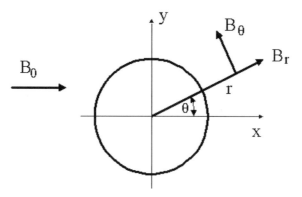

Fig. 2.7. A cylinder in a uniform magnetic field B_0

$$\nabla^2 \mathbf{A} = -\mu_0 \mu_1^r \mathbf{j}, \; r < R_0 \tag{2.129}$$

and outside it

$$\nabla^2 \mathbf{A} = 0, \; r > R_0, \tag{2.130}$$

where the current density has the form

$$\mathbf{j} = \left(0, 0, \frac{I}{\pi R_0^2}\right). \tag{2.131}$$

The vector potential consists of two parts: the potential of the external field $\mathbf{A}_0$ and the secondary potential $\mathbf{A}_1$ caused by the current I and the induced magnetization. Taking into account the direction of the field $\mathbf{B}_0$, equation (2.5) and the gauge $\mathrm{div}\,\mathbf{A} = 0$ we define $\mathbf{A}_0$ as follows:

$$\mathbf{A}_0 = \mathbf{z}_0 B_0 y = \mathbf{z}_0 B_0 r \sin\theta, \tag{2.132}$$

where θ is the angle in the xy plane between the radius vector $\mathbf{r}$ and the x-axis, and $\mathbf{z}_0$ is the unit vector in the z direction. Equations (2.26) and (2.129) show that the secondary vector potential $\mathbf{A}_1$ is parallel to the current density vector $\mathbf{j}$ (2.131):

$$\mathbf{A}_1 = (0, 0, A). \tag{2.133}$$

Substituting (2.131) and (2.133) into (2.129) and (2.130) and taking into account that $\mathbf{A}_1$ does not depend on z due to homogeneity in the axial direction we obtain in the cylindrical coordinates (r, θ, z):

$$\frac{1}{r}\frac{\partial}{\partial r}\left(r\frac{\partial A_i}{\partial r}\right) + \frac{1}{r^2}\frac{\partial^2 A_i}{\partial \theta^2} = -\mu_0 \mu_1^r \frac{I}{\pi R_0^2}, \; r < R_0 \tag{2.134}$$

and

2 Magnetostatics

$$\frac{1}{r}\frac{\partial}{\partial r}\left(r\frac{\partial A_e}{\partial r}\right) + \frac{1}{r^2}\frac{\partial^2 A_e}{\partial \theta^2} = 0, \; r > R_0, \tag{2.135}$$

where A_i and A_e are the values of the vector potential inside and outside the cylinder, respectively. Consider first the solution of (2.134). It consists of the general solution of the corresponding homogeneous equation A_{hi} and of a some particular solution A_{pi} of (2.134). The solution A_{hi} is sought by virtue of the separation of variables:

$$A_{hi} = F_1(r) F_2(\theta). \tag{2.136}$$

Substituting (2.136) into (2.134) we get:

$$A_{hi} = \sum_{n=0}^{\infty}(a_n \cos n\theta + b_n \sin n\theta) r^n + \sum_{n=0}^{\infty}(c_n \cos n\theta + d_n \sin n\theta) r^{-n}. \tag{2.137}$$

The potential (2.137) must be finite in the origin $r = 0$ and unique with respect to the change $2\pi n$ of the angle θ. Consequently, n is integer and $c_n = d_n = 0$. The particular solution A_{pi} does not depend on θ since the right-hand side does not depend on it. It is determined by the equation

$$\frac{\partial}{\partial r}\left(r\frac{\partial A_{pi}}{\partial r}\right) = -\mu_0 \mu_1^r \frac{Ir}{\pi R_0^2}, \tag{2.138}$$

which yields

$$A_{pi} = -\frac{\mu_0 \mu_1^r I}{4\pi}\left(\frac{r}{R_0}\right)^2. \tag{2.139}$$

The general solution A_i inside the cylinder then takes the form:

$$A_i = A_{hi} + A_{pi} = \sum_{n=0}^{\infty}(a_n \cos n\theta + b_n \sin n\theta) r^n - \frac{\mu_0 \mu_1^r I}{4\pi}\left(\frac{r}{R_0}\right)^2, \tag{2.140}$$

$$r < R_0. \tag{2.141}$$

The vector potential in the external region includes the term (2.132), the general solution (2.137) of the homogeneous equation regular at infinity and the particular solution of the equation

$$\frac{\partial}{\partial r}\left(r\frac{\partial A_{pe}}{\partial r}\right) = 0, \tag{2.142}$$

which has the form

$$A_{pe} = q_0 \ln r. \tag{2.143}$$

2.2 Magnetic Field of a Given Current Distribution

Combining all these contributions we obtain the expression of the vector potential A_e outside the cylinder. It has the form

$$A_e = B_0 r \sin\theta + q_0 \ln r + \sum_{n=0}^{\infty} (c_n \cos n\theta + d_n \sin n\theta) r^{-n}, \quad r > R_0. \quad (2.144)$$

The solutions (2.141) and (2.144) must satisfy the boundary conditions at the surface of the cylinder $r = R_0$. The vector potential itself should be continuous at $r = R_0$ as well as the tangential components of the magnetic field intensities H_θ.

$$A_i(R_0) = A_e(R_0), \quad \frac{1}{\mu_2^r}\frac{\partial A_e}{\partial r}(R_0) = \frac{1}{\mu_1^r}\frac{\partial A_i}{\partial r}(R_0). \quad (2.145)$$

The integration constant q_0 can be evaluated separately using the superposition of fields and the Maxwell equation in the integral form (2.9). Indeed, the tangential $H_{\theta 1}$ component of the magnetic field caused by the current I outside the cylinder according to (2.9) is $I/(2\pi r)$. On the other hand, this part of the field is due to the second term in the right-hand side of (2.144):

$$H_{\theta 1} = -\frac{1}{\mu_0 \mu_2^r}\frac{\partial(q_0 \ln r)}{\partial r} = -\frac{q_0}{\mu_0 \mu_2^r r},$$

which yields

$$q_0 = -\frac{\mu_0 \mu_2^r I}{2\pi}$$

and

$$A_e = B_0 r \sin\theta + \frac{\mu_0 \mu_2^r I}{2\pi}\ln\left(\frac{1}{r}\right) + \sum_{n=0}^{\infty}(c_n \cos n\theta + d_n \sin n\theta) r^{-n},$$
$$r > R_0. \quad (2.146)$$

Substitution of (2.141) and (2.146) into (2.145) and comparison of the corresponding terms give:

$$a_0 = 0, \quad c_0 = \frac{\mu_0 I}{4\pi}\left(\mu_2^r \ln R_0^2 - \mu_1^r\right) \quad (2.147)$$

$$b_1 R_0 - d_1 R_0^{-1} = B_0 R_0 \quad (2.148)$$

$$b_1 \frac{1}{\mu_0 \mu_1^r} + d_1 \frac{R_0^2}{\mu_0 \mu_2^r} = \frac{B_0}{\mu_0 \mu_2^r} \quad (2.149)$$

$$a_n, c_n = 0, \; n \geq 1; \; b_n, d_n = 0, \; n > 1. \quad (2.150)$$

The system of (2.148) and (2.149) has a non-trivial solution:

$$b_1 = \frac{2\mu_1^r}{\mu_1^r + \mu_2^r} B_0, \quad d_1 = \frac{\mu_1^r - \mu_2^r}{\mu_1^r + \mu_2^r} B_0 R_0^2. \tag{2.151}$$

Substituting the coefficients (2.147), (2.150) and (2.151) into expressions for the vector potential (2.141) and (2.146) we finally obtain:

$$A_i = -\frac{\mu_0 \mu_1^r I}{4\pi} \left(\frac{r}{R_0}\right)^2 + \frac{2\mu_1^r}{\mu_1^r + \mu_2^r} B_0 r \sin\theta, \quad r < R_0 \tag{2.152}$$

$$A_e = \frac{\mu_0 \mu_2^r I}{2\pi} \ln\left(\frac{R_0}{r}\right) - \frac{\mu_0 \mu_1^r I}{4\pi} + \left(r + \frac{R_0^2}{r} \frac{\mu_1^r - \mu_2^r}{\mu_1^r + \mu_2^r}\right) B_0 \sin\theta,$$
$$r > R_0. \tag{2.153}$$

The components of the magnetic induction $\boldsymbol{B}$ are calculated according to (2.5) in the cylindrical coordinates:

$$B_r = \frac{1}{r}\frac{\partial A}{\partial \theta}, \quad B_\theta = -\frac{\partial A}{\partial r}.$$

Then the induction inside the cylinder is

$$B_{ri} = \frac{2\mu_1^r}{\mu_1^r + \mu_2^r} B_0 \cos\theta, \quad B_{\theta i} = \frac{\mu_0 \mu_1^r I}{2\pi} \frac{r}{R_0^2} - \frac{2\mu_1^r}{\mu_1^r + \mu_2^r} B_0 \sin\theta, \quad r < R_0 \tag{2.154}$$

and outside it

$$B_{re} = \left(1 + \frac{R_0^2}{r^2}\frac{\mu_1^r - \mu_2^r}{\mu_1^r + \mu_2^r}\right) B_0 \cos\theta$$

$$B_{\theta e} = \frac{\mu_0 \mu_2^r I}{2\pi r} - \left(1 - \frac{R_0^2}{r^2}\frac{\mu_1^r - \mu_2^r}{\mu_1^r + \mu_2^r}\right) B_0 \sin\theta, \quad (r > R_0). \tag{2.155}$$

The induced magnetization of the cylinder $\boldsymbol{M}$ can be calculated by using the results (2.154), the material equation (2.3) and the definition (2.4).

$$\boldsymbol{M} = (\mu_1^r - 1) \boldsymbol{B}_i \frac{1}{\mu_0 \mu_1^r}. \tag{2.156}$$

Transforming the induction $\boldsymbol{B}_i$ to the Cartesian coordinates

$$B_{xi} = B_{ri}\cos\theta - B_{\theta i}\sin\theta, \quad B_{yi} = B_{ri}\sin\theta + B_{\theta i}\cos\theta,$$

we obtain

2.2 Magnetic Field of a Given Current Distribution

$$M_x = \frac{(\mu_1^r - 1)}{\mu_0}\left[\frac{2}{\mu_1^r + \mu_2^r}B_0 - \frac{\mu_0 I}{2\pi R_0^2}y\right], \quad M_y = (\mu_1^r - 1)\frac{I}{2\pi R_0^2}x. \quad (2.157)$$

The force exerted by the magnetic field on the cylinder with the current can be evaluated directly from expression (2.66). In calculating the total force we can take the induction $\boldsymbol{B}$ to be the external field $\boldsymbol{B}_e$ (2.155) in which the cylinder with the total current I is placed. The field of the current itself cannot contribute to the total force acting on the conductor because of the law of conservation of momentum. In the Cartesian coordinates the external field has the form

$$B_{xe} = B_0\left[1 + \frac{R_0^2}{r^2}\frac{\mu_1^r - \mu_2^r}{\mu_1^r + \mu_2^r}\cos 2\theta\right] - \frac{\mu_0\mu_2^r I}{2\pi r}\sin\theta$$

$$B_{ye} = \frac{\mu_0\mu_2^r I}{2\pi r}\cos\theta + B_0\frac{R_0^2}{r^2}\frac{\mu_1^r - \mu_2^r}{\mu_1^r + \mu_2^r}\sin 2\theta. \quad (2.158)$$

The vector product $[\boldsymbol{j} \times \boldsymbol{B}_e]$ yields two components of the force (2.66), F_x and F_y:

$$F_x = -\int_V jB_{ye}dV, \quad F_y = \int_V jB_{xe}dV. \quad (2.159)$$

In the cylindrical coordinates (r, θ, z)

$$dV = rdrdzd\theta.$$

We should calculate the force per unit length of the infinite length cylinder. The integration on z can, therefore, be omitted. The integral of the uniform current density j over the cross-section of the cylinder is proportional to the total current.

$$\int_0^{R_0} jrdr = \frac{1}{2}jR_0^2 = \frac{I}{2\pi}. \quad (2.160)$$

The remaining integration over the angle θ is carried out on the surface of the cylinder, i.e. for $r = R_0$. We have finally:

$$F_x = -\frac{I}{2\pi}\int_0^{2\pi}d\theta\left[\frac{\mu_0\mu_2^r I}{2\pi R_0}\cos\theta + B_0\frac{\mu_1^r - \mu_2^r}{\mu_1^r + \mu_2^r}\sin 2\theta\right] = 0 \quad (2.161)$$

$$F_y = \frac{I}{2\pi}\int_0^{2\pi}d\theta\left\{B_0\left[1 + \frac{\mu_1^r - \mu_2^r}{\mu_1^r + \mu_2^r}\cos 2\theta\right] - \frac{\mu_0\mu_2^r I}{2\pi R_0}\sin\theta\right\} = IB_0, \quad (2.162)$$

where the integrals over the sine and cosine functions clearly vanish. The induced magnetic moment μ_m per unit length of the cylinder can be calculated by integration of the magnetization (2.157) over the cylinder volume. However, it does not contribute to the total force due to the uniformity of the field applied which is seen from (2.74).

2.2.4 Magnetic Field of a Solenoid

Consider a magnetic field of a solenoid which represents a closely wound cylindrical coil of a length h. Let the radius of the cylinder, the number of turns per unit length and the current in a single turn of the solenoid be R_0, n and I respectively. The solenoid is shown in Fig. 2.8.

In the case of the sufficiently dense winding of the solenoid $n \gg 1$ the current may be considered as quasicontinuous with the density per unit length

$$dI = Indz, \qquad (2.163)$$

where the z-axis coincides with the axis of the solenoid. The geometry of the problem is presented in Fig. 2.9.

The magnetic field created by the contour with a current has been calculated above. Combining the result (2.118) and (2.163) we obtain the induction created by the elementary contour along the solenoid axis:

$$dB_z = \frac{\mu_0 Indz}{2R_0} \frac{R_0^3}{(z^2 + R_0^2)^{\frac{3}{2}}} = \frac{\mu_0 Indz}{2R_0} \sin^3 \alpha, \qquad (2.164)$$

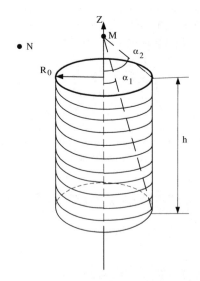

Fig. 2.8. The solenoid of the length h and radius R_0 with nh turns

2.2 Magnetic Field of a Given Current Distribution

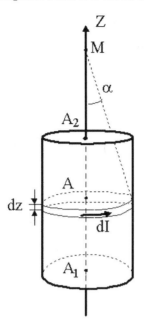

Fig. 2.9. The elementary contour with the current $dI = Indz$

where α is the angle between the z-axis and the vector connecting the observation point M and the elementary contour. The distance z between the observation point M and the plane of the elementary contour can be expressed through α:

$$z = -R_0 \cot\alpha, \quad dz = R_0 \frac{d\alpha}{\sin^2\alpha}. \tag{2.165}$$

Substituting (2.164) into (2.165) we immediately obtain after the integration:

$$B_z = \mu_0 In \frac{\cos\alpha_1 - \cos\alpha_2}{2}, \tag{2.166}$$

where

$$\cos\alpha_1 = \frac{z_M}{\sqrt{R_0^2 + z_M^2}}; \quad \cos\alpha_2 = \frac{z_M + h}{\sqrt{R_0^2 + (z_M + h)^2}} \tag{2.167}$$

and z_M is the distance between the observation point M and the upper base of the solenoid. If the point M is situated inside a very long solenoid, then

$$\alpha_1 \to 0, \quad \alpha_2 \to \pi, \quad B_z \to \mu_0 In. \tag{2.168}$$

Clearly, the induction along the axis of the infinite solenoid is simply $\mu_0 In$. Consider now the field of the infinite solenoid in points which do not belong

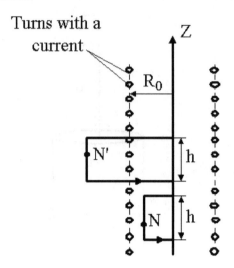

Fig. 2.10. The calculation of the magnetic field inside and outside the infinite solenoid

to the axis. Let it be the point N shown in Fig. 2.8. First of all, we take into account that any plane containing such a point and perpendicular to the axis of the solenoid is the symmetry plane with respect to the magnetic field in this point and to the currents. Consequently the magnetic field must be perpendicular to this plane, i.e. the lines of the parallel to the z-axis:

$$\boldsymbol{B}(N) = B_z \boldsymbol{z}_0, \tag{2.169}$$

where $\boldsymbol{z}_0$ is the unit vector along the z-axis. The infinite solenoid is invariant with respect to the translation along the z-axis as well as with respect to the rotation around it. As a result, in the cylindrical coordinates (r, φ, z) we have

$$\frac{\partial \boldsymbol{B}}{\partial \varphi} = 0, \quad \frac{\partial \boldsymbol{B}}{\partial z} = 0 \tag{2.170}$$

and

$$\boldsymbol{B}(N) = B_z(r) \boldsymbol{z}_0. \tag{2.171}$$

Now we shall try to calculate the field inside the infinite solenoid by virtue of the Ampère's law (2.9). We choose the rectangular contour passing through the point N, as it is shown in Fig. 2.10.

Since this contour does not contain any turn with the current, the magnetic field circulation vanishes:

$$\oint \boldsymbol{B} \cdot d\boldsymbol{r} = 0. \tag{2.172}$$

Taking into account the field property (2.171) one can see that the contributions along the sides of the rectangular contour which are perpendicular to the axis of the solenoid are mutually compensating and yield zero. The contribution due to the integration along the sides parallel to the axis of the solenoid has the form:

$$h\left[B_z\left(r\right) - B_z\left(0\right)\right] = 0 \tag{2.173}$$

and, finally, using the value of the induction along the axis (2.168) we obtain:

$$B_z\left(r\right) = \mu_0 I n, \ r < R_0, \tag{2.174}$$

which means that the magnetic field inside the sufficiently long solenoid is uniform.

Consider now the contour traversing the surface of the solenoid and passing through the point N' outside the solenoid. Such a contour contains the total current nhI. The procedure similar to the previous one yields:

$$B_z\left(r\right) - B_z\left(0\right) = \mu_0 I n. \tag{2.175}$$

Using once more the value of the induction along the solenoid axis $B_z\left(0\right) = \mu_0 I n$ we find that the magnetic field vanishes outside the solenoid:

$$B_z\left(r\right) = 0, \ r > R_0. \tag{2.176}$$

The comparison of results (2.174) and (2.176) shows that the magnetic induction discontinuity occurs at the lateral surface of the solenoid:

$$B_z\left(R_0^+\right) - B_z\left(R_0^-\right) = -\mu_0 I n. \tag{2.177}$$

We have seen previously that according to the boundary conditions (2.14) such a discontinuity is due to the surface current density $\varkappa$ that can be easily identified by using (2.163) and (2.177).

$$\varkappa = \frac{dI}{dz} \boldsymbol{e}_\varphi = In\boldsymbol{e}_\varphi, \tag{2.178}$$

where $\boldsymbol{e}_\varphi$ is the unit vector corresponding to the φ direction in the cylindrical system of coordinate, i.e. it is at any point perpendicular to the unit vectors $\boldsymbol{z}_0$ and $\boldsymbol{r}_0$ in such a way that

$$\left[\boldsymbol{r}_0 \times \boldsymbol{e}_\varphi\right] = \boldsymbol{z}_0.$$

These directions are shown in Fig. 2.11.

Evaluate the vector potential $\boldsymbol{A}$ of the solenoid magnetic field. The symmetry of the problem implies that it takes the form:

$$\boldsymbol{A} = A\left(r\right)\boldsymbol{e}_\varphi. \tag{2.179}$$

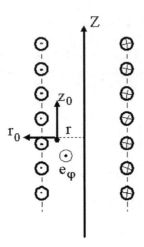

Fig. 2.11. The directions of the magnetic field $\boldsymbol{B}$ and the surface current density $\varkappa$ in the solenoid

Obviously, the vector potential (2.179) satisfies the Coulomb gauge $\text{div}\,\boldsymbol{A} = 0$. Then using (2.5) and the expressions for the magnetic field inside and outside the solenoid (2.174) and (2.176) we have

$$\frac{1}{r}\frac{d}{dr}(rA) = \mu_0 In, \ r < R_0 \tag{2.180}$$

and

$$\frac{1}{r}\frac{d}{dr}(rA) = 0, \ r > R_0. \tag{2.181}$$

The equation (2.180) yields immediately:

$$A = \frac{\mu_0 Inr}{2} + \frac{C_1}{r} = \frac{\mu_0 Inr}{2}, \ r < R_0. \tag{2.182}$$

The integration constant $C_1 = 0$ since the vector potential must be finite at the axis. In the external region we get from (2.181):

$$A = \frac{C_2}{r}, \ r > R_0. \tag{2.183}$$

The integration constant C_2 this time is determined from the continuity condition at the surface $r = R_0$:

$$C_2 = \frac{\mu_0 In R_0^2}{2}, \ A = \frac{\mu_0 In R_0^2}{2r}, \ r > R_0. \tag{2.184}$$

Clearly, the vector potential is due to the surface current density and vanishes at large distances.

2.2.5 Magnetic Field of a Torus

Consider the magnetic field of a torus with N closely wound circular turns of the radius a supporting the current I. The torus which is presented in Fig. 2.12 has the middle radius $R > a$.

From the symmetry considerations, one can see that the magnetic field in some point M shown in Fig. 2.13 has the only component which is tangential with respect to the torus and does not depend on the angle φ in the cylindrical coordinates (r, φ, z).

We therefore have:

$$\boldsymbol{B} = B(r, z)\, \boldsymbol{e}_\varphi. \tag{2.185}$$

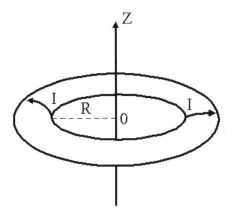

Fig. 2.12. The torus with the electric current I

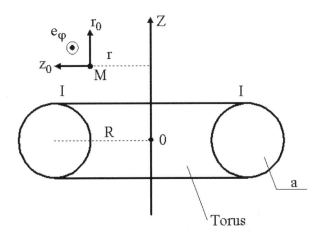

Fig. 2.13. The cross-section of the torus. M is an observation point, $\boldsymbol{r}_0, \boldsymbol{e}_\varphi, \boldsymbol{z}_0$ are the unit vectors of the cylindrical coordinate system (r, φ, z)

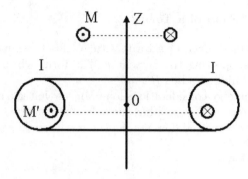

Fig. 2.14. The external and internal circles of the field lines

The lines of the field represent the circles, and we get applying the Ampère's law to one of these circles:

$$\oint \boldsymbol{B} \cdot d\boldsymbol{r} = \int_0^{2\pi} B(r,z) \, r d\varphi = 2\pi r B(r,z). \tag{2.186}$$

The total current for the points M outside the torus is zero since the external circles do not traverse any turns with the current. On the contrary, for the points M' inside the torus the total current is NI since any internal circle encloses all turns with the current. The geometry of the problem is shown in Fig. 2.14.

Consequently we obtain for the fields B_{out} and B_{in} outside and inside the torus respectively:

$$B_{\text{out}} = 0, \ B_{\text{in}} = \frac{\mu_0 NI}{2\pi r}. \tag{2.187}$$

Clearly, the magnetic field reaches its minimal and maximal values at $r = R \pm a$ respectively. At the surface of the torus the magnetic induction has a discontinuity due to the surface current density $\boldsymbol{j}_s$

$$\boldsymbol{B}_{\text{out}} - \boldsymbol{B}_{\text{in}} = \mu_0 \left[\boldsymbol{j}_s \times \boldsymbol{n} \right], \tag{2.188}$$

where $\boldsymbol{n}$ is the unit vector perpendicular to the contour of the torus cross section. The comparison of (2.187) and (2.188) shows that the surface current density $\boldsymbol{j}_s$ has the form:

$$\boldsymbol{j}_s = \frac{NI}{2\pi r} \left[\boldsymbol{n} \times \boldsymbol{e}_\varphi \right]. \tag{2.189}$$

2.2.6 Magnetic Field of a Rotating Charged Sphere

In this section we consider a magnetic field of a hollow metallic sphere of the radius R_0 charged with an electric charge Q uniformly distributed on its

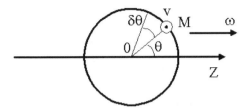

Fig. 2.15. The rotating charged sphere

surface. The sphere is rotating around its axis with the constant frequency ω. The charges are assumed to be rotating along with the sphere. The geometry of the problem is presented in Fig. 2.15. The z-axis is chosen to be the rotation axis. The surface current density $\boldsymbol{j}_s$ in such a case has the form:

$$\boldsymbol{j}_s = \frac{Q\omega \sin\theta}{4\pi R_0} \boldsymbol{e}_\varphi. \tag{2.190}$$

We divide the sphere into the elementary layers limited by the solid angles $(\theta, \theta + d\theta)$. The radius $R(\theta)$ of such a layer is

$$R(\theta) = R_0 \sin\theta.$$

Each elementary layer can be identified as a contour with an elementary current dI which has the form:

$$dI = j_s R_0 d\theta = \frac{Q\omega \sin\theta}{4\pi} d\theta. \tag{2.191}$$

It has been shown above that the magnetic induction $d\boldsymbol{B}$ of an elementary contour with a current is directed along the axis of the contour, and its value on the z-axis is according to (2.164):

$$d\boldsymbol{B} = dB\boldsymbol{z}_0 = \boldsymbol{z}_0 \frac{\mu_0 dI}{2R(\theta)} \sin^3\theta = \boldsymbol{z}_0 \frac{\mu_0 Q\omega}{8\pi R_0} \sin^3\theta d\theta, \tag{2.192}$$

where (2.191) has been used.

For the sake of definiteness we calculate the magnetic induction B in the center O of the sphere. By integrating of (2.192) we obtain:

$$B_z(O) = \frac{\mu_0 Q\omega}{8\pi R_0} \int_0^\pi \sin^3\theta d\theta = \frac{\mu_0 Q\omega}{6\pi R_0}. \tag{2.193}$$

The magnetic moment $d\boldsymbol{\mu}_m$ of the elementary contour is also directed along the z-axis, and it can be easily calculated according to (2.40):

$$d\boldsymbol{\mu}_m = \boldsymbol{z}_0 dI \pi R^2(\theta) = \boldsymbol{z}_0 \frac{Q\omega R_0^2}{4} \sin^3\theta d\theta, \tag{2.194}$$

2 Magnetostatics

which yields for the magnetic moment of the rotating charged sphere

$$\boldsymbol{\mu}_m = \boldsymbol{z}_0 \frac{Q\omega R_0^2}{4} \int_0^\pi \sin^3\theta d\theta = \boldsymbol{z}_0 \frac{Q\omega R_0^2}{3}. \tag{2.195}$$

The vector potential $\boldsymbol{A}(\boldsymbol{r})$ of the rotating charged sphere in an observation point M characterized by the radius vector $\boldsymbol{r}$ can be written according to (2.26) and taking into account (2.190):

$$\boldsymbol{A}(\boldsymbol{r}) = \frac{\mu_0}{4\pi} \iint \frac{\boldsymbol{j}_s dS}{|\boldsymbol{r} - \boldsymbol{R}_0|} = \frac{\mu_0 Q}{16\pi R_0^2} \iint \frac{[\boldsymbol{\omega} \times \boldsymbol{R}_0] dS}{|\boldsymbol{r} - \boldsymbol{R}_0|}, \tag{2.196}$$

where the origin is chosen in the center of the sphere. Introducing in the integrand the vector of the surface element

$$d\boldsymbol{S} = \boldsymbol{r}_0 dS$$

we get

$$\boldsymbol{A}(\boldsymbol{r}) = \frac{\mu_0 Q}{16\pi^2 R_0} \left[\boldsymbol{\omega} \times \iint \frac{d\boldsymbol{S}}{|\boldsymbol{r} - \boldsymbol{R}_0|} \right]. \tag{2.197}$$

The integral over the surface of the sphere in (2.197) can be transformed into the integral over the volume V of the spheroid which gives

$$\boldsymbol{A}(\boldsymbol{r}) = \frac{\mu_0 Q}{16\pi^2 R_0} \left[\boldsymbol{\omega} \times \iiint dV \operatorname{grad}_R \frac{1}{|\boldsymbol{r} - \boldsymbol{R}|} \right], \tag{2.198}$$

where $\boldsymbol{R}$ is the radius vector of any point inside the spheroid: $R \leq R_0$. The integrand in (2.198) has the form

$$\operatorname{grad}_R \frac{1}{|\boldsymbol{r} - \boldsymbol{R}|} = \frac{\boldsymbol{r} - \boldsymbol{R}}{|\boldsymbol{r} - \boldsymbol{R}|^3}$$

and (2.198) can be rewritten as follows:

$$\boldsymbol{A}(\boldsymbol{r}) = \frac{\mu_0 Q}{4\pi R_0} \left[\boldsymbol{\omega} \times \frac{q_{\text{eff}}}{4\pi\varepsilon_0} \iiint dV \frac{\boldsymbol{r} - \boldsymbol{R}}{|\boldsymbol{r} - \boldsymbol{R}|^3} \right], \tag{2.199}$$

where $q_{\text{eff}} = \varepsilon_0$. In this denotion the second factor in brackets in (2.199) can be formally identified with the effective electrostatic field intensity $\boldsymbol{E}$ of the uniformly charged spheroid. Such an electric field has the form according to the Gauss theorem:

$$\boldsymbol{E} = \frac{q_{\text{eff}}}{4\pi\varepsilon_0} \iiint dV \frac{\boldsymbol{r} - \boldsymbol{R}}{|\boldsymbol{r} - \boldsymbol{R}|^3} = E(r) \boldsymbol{r}_0, \tag{2.200}$$

where
$$E(r) = \frac{q(r)}{4\pi\varepsilon_0 r^2}$$
and
$$q(r) = \begin{cases} \frac{4}{3}\pi R_0^3 q_{\text{eff}}, & r > R_0 \\ \frac{4}{3}\pi r^3 q_{\text{eff}}, & r < R_0 \end{cases}.$$

Then the effective electric field takes the form:
$$E(r) = \frac{R_0^3}{3r^2}, \quad r > R_0 \tag{2.201}$$
and
$$E(r) = \frac{r}{3}, \quad r < R_0. \tag{2.202}$$

Substituting (2.200)–(2.202) into (2.199) we obtain for the vector potential
$$A(r) = \frac{\mu_0 Q R_0^2}{12\pi r^2}[\omega \times r_0] = e_\varphi \frac{\mu_0 Q R_0^2}{12\pi r^2} \omega \sin\theta, \quad r > R_0 \tag{2.203}$$
and
$$A(r) = \frac{\mu_0 Q r}{12\pi R_0}[\omega \times r_0] = e_\varphi \frac{\mu_0 Q r}{12\pi R_0} \omega \sin\theta, \quad r < R_0. \tag{2.204}$$

Finally, using (2.5) in spherical coordinates, we find the magnetic induction B
$$B = \frac{1}{r\sin\theta} \frac{\partial}{\partial\theta}[A\sin\theta]\, r_0 - \frac{1}{r}\frac{\partial}{\partial r}[rA]\, e_\theta, \tag{2.205}$$
which yields inside and outside the rotating sphere respectively
$$B = \frac{\mu_0 Q\omega}{6\pi R_0}[r_0 \cos\theta - e_\theta \sin\theta] = \frac{\mu_0 Q\omega}{6\pi R_0} z_0 = z_0 B_z(O), \quad r < R_0 \tag{2.206}$$
and
$$B = \frac{\mu_0 Q\omega R_0^2}{12\pi r^3}[2r_0 \cos\theta + e_\theta \sin\theta] = \frac{\mu_0 \mu_m}{4\pi r^3}[2r_0 \cos\theta + e_\theta \sin\theta], \quad r > R_0. \tag{2.207}$$

Here e_θ is the unit vector which determines the direction of the angular coordinate θ; we used results (2.193) and (2.195). Equation (2.206) shows that the induction inside the rotating charged sphere is uniform and equal to the value in the center of the sphere. The magnetic induction outside the rotating charged sphere is equal to the one created by the magnetic moment μ_m at large distances as it is seen from (2.207). The magnetic field has a discontinuity at the transition through the surface of the sphere:
$$B(R_0^+) - B(R_0^-) = \frac{\mu_0 Q\omega}{4\pi R_0}[e_\varphi \times r_0]\sin\theta = \mu_0[j_s \times r_0], \tag{2.208}$$
which coincides with the boundary conditions (2.14).

3 Quasi-static Electromagnetic Field

3.1 Conductors in a Variable Magnetic Field

3.1.1 Basic Equations

In the previous chapter we have analyzed phenomena in a constant magnetic field. Now we consider the processes in conductors subject to a time dependent external magnetic field. The field and current variations are assumed to be sufficiently slow such that the field distribution at any moment of time can be described as static and created by the currents existing at this moment. This means, in particular, that if the current temporal dependence is determined by a function $f(t)$, then the magnetic field $\boldsymbol{H}(\boldsymbol{r},t)$ can be presented as a product of this function and the spatial distribution $\boldsymbol{H}(\boldsymbol{r})$ of the magnetic field [4]:

$$\boldsymbol{H}(\boldsymbol{r},t) \sim \boldsymbol{H}(\boldsymbol{r}) f(t). \tag{3.1}$$

The validity of this approach is determined by the so-called quasi-static condition which requires that the wavelength λ of the electromagnetic field would be much greater than the dimension l of the conductor [4]:

$$\lambda \gg l. \tag{3.2}$$

Indeed, in the quasi-static approximation at a moment t an electric current should have the same value $I(t)$ at any cross section of the conductor. Suppose, for the sake of definiteness, that the current has a harmonic time dependence

$$I(t) = I_0 \sin \omega t. \tag{3.3}$$

Then, the phase difference $\Delta\varphi$ between the values of the current at two cross sections separated by a distance l is in fact finite due to the finite velocity v of the signal propagation:

$$\Delta\varphi = \frac{l}{v}\omega = \frac{2\pi l}{\lambda}. \tag{3.4}$$

Obviously, the phase difference $\Delta\varphi$ must be negligibly small with respect to the period 2π which immediately yields the condition (3.2).

3.1 Conductors in a Variable Magnetic Field

In the quasi-static approximation the Maxwell equations (2.1)–(2.4) can be applied to the magnetic field description, since in the Maxwell equation (1.19) the so-called displacement current determined by the term $\partial \boldsymbol{D}/\partial t$ can be neglected. Generally,

$$\frac{\partial \boldsymbol{D}}{\partial t} \sim \varepsilon_0 \varepsilon^r \omega \boldsymbol{E}. \tag{3.5}$$

The current density inside a conductor is related to the electric field $\boldsymbol{E}$ by the Ohm's law

$$\boldsymbol{j} = \sigma \boldsymbol{E} \tag{3.6}$$

with a constant conductivity σ. Expression (3.6) remains valid when the field frequencies are small compared to the inverse mean free time of the electrons in the conductor. For typical metals at room temperatures this condition is satisfied up to the infrared region of the spectrum [4]. In good conductors, i.e. metals the conductivity $\sigma \gg \varepsilon_0 \varepsilon^r \omega$ throughout the whole frequency interval where (3.6) holds, and therefore the conduction current density $\boldsymbol{j}$ is much larger than the displacement current $\partial D/\partial t$. In semiconductors both quantities σ, $\varepsilon_0 \varepsilon^r \omega$ are comparable, and this case will be considered separately. Outside the conductor a current is absent and

$$\text{curl} \boldsymbol{H} = 0. \tag{3.7}$$

The time dependent magnetic field creates an electric field $\boldsymbol{E}$ according to the Maxwell equation (1.17)

$$\text{curl} \boldsymbol{E} = -\frac{\partial \boldsymbol{B}}{\partial t}. \tag{3.8}$$

Combining (2.1), (2.3), (3.6) and (3.8) we obtain:

$$\text{curl} \frac{\text{curl} \boldsymbol{H}}{\sigma} = -\mu_0 \mu^r \frac{\partial \boldsymbol{H}}{\partial t}. \tag{3.9}$$

In a homogeneous medium the conductivity σ and the relative permeability μ^r are constant. Then using the Maxwell equation (2.2) we get from (3.9):

$$\nabla^2 \boldsymbol{H} = \mu_0 \mu^r \sigma \frac{\partial \boldsymbol{H}}{\partial t}. \tag{3.10}$$

The boundary conditions on the surface of the conductor are defined by relations (2.13) and (2.17). For non-ferromagnetic materials $\mu^r = 1$ and we can simply conclude that the magnetic field intensity $\boldsymbol{H}$ as well as magnetic induction $\boldsymbol{B}$ are continuous at the boundary:

$$\boldsymbol{H}_1 = \boldsymbol{H}_2. \tag{3.11}$$

The continuity of the tangential field components $\boldsymbol{H}_t$ together with (3.7) results, in particular, in the continuity of the normal component of the rotation $(\operatorname{curl}\boldsymbol{H})_n$. Consequently, the quantity $(\sigma\boldsymbol{E})_n$ is also continuous on the surface of the conductor. Outside the conductor $\sigma = 0$. Hence, the normal component of the electric field E_n^{int} inside the conductor also vanishes at the boundary as well as the current density normal component j_n does:

$$E_n^{\text{int}} = 0, \; j_n = 0 \text{ on the surface.} \quad (3.12)$$

One can conclude that in the quasi-static approximation a variable magnetic field does not create free charges on the surface of the conductor.

Consider a conductor placed in an external magnetic field which is suddenly removed. Then the solution of (3.10) is sought to be:

$$\boldsymbol{H}(x,y,z,t) = \boldsymbol{H}_m(x,y,z) f(t). \quad (3.13)$$

The substitution of (3.13) into (3.10) gives

$$\boldsymbol{H}(x,y,z,t) = \boldsymbol{H}_m(x,y,z) \exp(-\gamma_m t) \quad (3.14)$$

and

$$\nabla^2 \boldsymbol{H}_m = -\mu_0 \sigma \gamma_m \boldsymbol{H}_m, \quad (3.15)$$

where γ_m is a constant. For a conductor of a given shape (3.15) has non-zero solutions satisfying the appropriate boundary conditions only for some certain values, or eigenvalues of γ_m. In order to clarify the type of possible eigenvalues γ_m without specifying the boundary conditions we substitute solution (3.14) into (3.9), multiply both sides by $\boldsymbol{H}_m^*$ and integrate over all space. We obtain:

$$\mu_0 \mu^r \gamma_m \int_V |\boldsymbol{H}_m|^2 \, dV = \int_V \boldsymbol{H}_m^* \cdot \operatorname{curl} \frac{\operatorname{curl} \boldsymbol{H}}{\sigma} \, dV = \frac{1}{\sigma} \int_V |\operatorname{curl} \boldsymbol{H}|^2 \, dV \quad (3.16)$$

and

$$\gamma_m = \frac{1}{\mu_0 \mu^r \sigma} \left[\int_V |\operatorname{curl} \boldsymbol{H}|^2 \, dV \right] \left[\int_V |\boldsymbol{H}_m|^2 \, dV \right]^{-1}. \quad (3.17)$$

It is seen from (3.17) that all eigenvalues γ_m are real and positive. Then the general solution of (3.10) takes the form:

$$\boldsymbol{H}(x,y,z,t) = \sum_m C_m \boldsymbol{H}_m(x,y,z) \exp(-\gamma_m t). \quad (3.18)$$

The eigenfunctions $\boldsymbol{H}_m(x,y,z)$ are the solutions of (3.15) corresponding to eigenvalues γ_m. They form a complete set of orthogonal vector functions. Solution (3.18) should obey the initial condition

$$\boldsymbol{H}(x,y,z,0) = \boldsymbol{H}_0(x,y,z) = \sum_m C_m \boldsymbol{H}_m(x,y,z). \quad (3.19)$$

The coefficients C_m of expansion (3.19) are determined according to a standard procedure

$$C_m = \left[\int_V \boldsymbol{H}_0(x,y,z)\,\boldsymbol{H}_m^*(x,y,z)\,dV\right]\left[\int_V |\boldsymbol{H}_m|^2\,dV\right]^{-1}, \quad (3.20)$$

where the orthogonality of $\boldsymbol{H}_m(x,y,z)$ is taken into account.

Expressions (3.17) and (3.18) show that the magnetic field is decaying with time. The rate of decay is mainly determined by the term in the series with the minimal $\gamma_m = \gamma_m^{\min}$. We define the decay time τ of the magnetic field as follows:

$$\tau = \frac{1}{\gamma_m^{\min}}. \quad (3.21)$$

Evaluate the order of magnitude of the decay time τ. Note that the left-hand side of (3.15) is proportional to $-\boldsymbol{H}/l^2$ where l is the conductor dimension. Then we obtain combining (3.15) and (3.21)

$$\tau \sim \mu_0 \mu^r \sigma l^2. \quad (3.22)$$

3.1.2 Eddy Currents

Consider now a situation where a conductor is subject to an external magnetic field varying in time as $\exp(-i\omega t)$. The variable magnetic field penetrates into the conductor and creates in it a variable electric field according to (3.8). The electric field in its turn gives rise to electric currents (3.6). These currents are called eddy currents [4]. Analysis of (3.10) shows that the magnetic field, the induced electric field and eddy currents penetrate into the conductor to a distance $\delta \sim (\mu_0 \sigma \omega)^{-1/2}$. In a variable field of frequency ω all quantities depend on the time through a factor $\exp(-i\omega t)$. In particular, (3.10) reduces to

$$\nabla^2 \boldsymbol{H} = -i\mu_0 \mu^r \sigma \omega \boldsymbol{H}. \quad (3.23)$$

We consider two limiting cases.

1. The penetration depth δ is large compared with the dimension of the conductor l. This occurs at low frequencies. Then, in the first approximation we can put the right-hand side of (3.23) equal to zero which results in the same magnetic field distribution $\boldsymbol{H}_{st}$ at any instant as it would be in a steady state:

$$\mathrm{curl}\,\boldsymbol{H}_{st} = 0 \quad (3.24)$$

and, consequently, the electric field $\boldsymbol{E}$ appears only in the next approximation according to which

$$\operatorname{curl} \boldsymbol{E} = i\mu_0 \omega \boldsymbol{H}_{\mathrm{st}}. \tag{3.25}$$

Equations (2.1) and (3.6) for a constant σ prove that

$$\operatorname{div} \boldsymbol{E} = 0. \tag{3.26}$$

Equations (3.25) and (3.26) completely determine the electric field distribution. Obviously, the magnetic field amplitude does not depend on frequency. Hence, the electric field amplitude is proportional to the frequency ω.

2. The penetration depth δ is small: $\delta \ll l$. This case corresponds to high frequencies. Note that the macroscopic equations are still applicable if δ is at the same time is large compared to the mean free path of conduction electrons [4]. In such a situation the magnetic field penetrates only into a thin surface layer of the conductor. The field distribution in the surface layer of the conductor can be studied by regarding small regions of the surface as plane. We should, therefore, solve (3.23) for a conducting medium bounded by a plane surface. The magnetic field outside the conductor is parallel to its surface and has the form

$$\boldsymbol{H}_{\mathrm{out}} = \boldsymbol{H}_0 \exp(-i\omega t). \tag{3.27}$$

The magnetic field in the conductor coincides at the surface with (3.27) according to the boundary condition (3.11). We choose the surface of the conductor as the xy plane and the conducting medium in the half-space $z > 0$. The conditions of the problem do not depend on x, y, and the magnetic field is a function of the coordinate z and time t. Then (3.23) takes the form

$$\frac{\partial^2 \boldsymbol{H}}{\partial z^2} + k^2 \boldsymbol{H} = 0, \tag{3.28}$$

where

$$k = \sqrt{i\mu_0 \sigma \omega} = (1+i)\sqrt{\frac{\mu_0 \sigma \omega}{2}} = \frac{1+i}{\delta} \tag{3.29}$$

and $\mu^r = 1$. The solution of (3.28) vanishing at $z \to \infty$ is given by

$$\boldsymbol{H} = \boldsymbol{H}(0) \exp\left(-\frac{z}{\delta}\right) \exp i\left(\frac{z}{\delta} - \omega t\right), \quad \delta = \sqrt{\frac{2}{\mu_0 \sigma \omega}} = c\sqrt{\frac{2\varepsilon_0}{\sigma \omega}}. \tag{3.30}$$

Using boundary condition (3.11) at $z = 0$ for (3.27) and (3.28) we finally get

$$\boldsymbol{H} = \boldsymbol{H}_0 \exp\left(-\frac{z}{\delta}\right) \exp i\left(\frac{z}{\delta} - \omega t\right). \tag{3.31}$$

Equation (2.1) yields immediately the eddy current $\boldsymbol{j}$

$$\boldsymbol{j} = (i-1)\sqrt{\frac{\mu_0 \sigma \omega}{2}}\, [\boldsymbol{z}_0 \times \boldsymbol{H}]. \tag{3.32}$$

The electric field is obtained from the Ohm's law (3.6)

$$\boldsymbol{E} = (i-1)\sqrt{\frac{\mu_0 \omega}{2\sigma}}\, [\boldsymbol{z}_0 \times \boldsymbol{H}], \tag{3.33}$$

where $\boldsymbol{z}_0$ is the unit vector along the z-axis. For the sake of definiteness we choose the magnetic field amplitude $\boldsymbol{H}_0$ to be real and parallel to the y-axis. Then taking the real part of the magnetic and electric fields (3.31) and (3.33) we get

$$H_y = H_0 \exp\left(-\frac{z}{\delta}\right) \cos\left(\frac{z}{\delta} - \omega t\right) \tag{3.34}$$

and

$$E_x = \sqrt{\frac{\mu_0 \omega}{\sigma}}\, H_0 \exp\left(-\frac{z}{\delta}\right) \cos\left(\frac{z}{\delta} - \omega t - \frac{\pi}{4}\right). \tag{3.35}$$

The eddy current is given by

$$j_x = \sqrt{\mu_0 \sigma \omega}\, H_0 \exp\left(-\frac{z}{\delta}\right) \cos\left(\frac{z}{\delta} - \omega t - \frac{\pi}{4}\right). \tag{3.36}$$

The eddy currents result in a dissipation of the field energy in the form of the Joule heat. The time average energy losses per unit volume Q_{loss} have the form

$$Q_{\text{loss}} = \int_V \overline{\boldsymbol{j} \cdot \boldsymbol{E}}\, dV = \int_V \sigma \overline{E^2}\, dV. \tag{3.37}$$

3.1.3 Skin Effect

Consider the distribution of current density over the cross section of a conductor with a non-zero variable total current I. Expressions (3.30) and (3.36) show that, as the frequency increases, the current will concentrate near the surface of the conductor. This phenomenon is known as the skin effect [4]. In general, the exact solution of the problem of the skin effect depends both on the shape of the conductor and on the nature of the variable external magnetic field which gives rise to the current. We limit our consideration with the practically important particular case where the current flows in a wire of thickness small compared with its length. In such a case the current distribution is independent of the manner of excitation [4]. We assume that the wire is straight, with a circular cross section of a radius R, the electric field $\boldsymbol{E}$ is parallel to the axis of the wire, and the magnetic field $\boldsymbol{H}$ is in a plane perpendicular to the axis. From the symmetry considerations, $\boldsymbol{E}$ is uniform over the surface of the wire varying only with time. With this boundary condition the only solution of (3.26) and

$$\operatorname{curl} \boldsymbol{E} = 0 \qquad (3.38)$$

outside the wire is

$$\boldsymbol{E} = \operatorname{const}. \qquad (3.39)$$

The magnetic field outside the wire must be the same as it would be outside a wire carrying a constant current equal to the instantaneous value of the variable current. In order to evaluate the electric field inside the wire we apply the curl operation to (3.8) and obtain by using (2.3), (3.6) and (3.26) the same equation as (3.10) for the magnetic field:

$$\nabla^2 \boldsymbol{E} = \mu_0 \sigma \frac{\partial \boldsymbol{E}}{\partial t}. \qquad (3.40)$$

Introduce the cylindrical coordinates, with the z-axis along the axis of the wire. The only component of the electric field is E_z which depends on radial coordinate r due to the symmetry of the problem. Then for a periodic field $E_z \sim \exp(-i\omega t)$ we get

$$\frac{1}{r} \frac{\partial}{\partial r} \left(r \frac{\partial E_z}{\partial r} \right) + k^2 E_z = 0, \qquad (3.41)$$

where k is determined by (3.29). The solution of (3.41) finite at $r = 0$ is expressed in terms of the Bessel function of the zeroth order $J_0(kr)$:

$$E_z = E_0 J_0(kr) \exp(-i\omega t). \qquad (3.42)$$

Obviously, the eddy current has the same distribution. The magnetic field is calculated by virtue of (3.8). It has only the azimuthal component H_φ:

$$H_\varphi = \frac{1}{i\omega\mu_0} (\operatorname{curl} \boldsymbol{E})_\varphi = -\frac{1}{i\omega\mu_0} \frac{\partial E_z}{\partial r} = -i\sqrt{\frac{i\sigma}{\omega\mu_0}} E_0 J_1(kr) \exp(-i\omega t), \qquad (3.43)$$

where the well known identity has been used:

$$\frac{dJ_0(x)}{dx} = -J_1(x)$$

and $J_1(kr)$ is the first order Bessel function.

Similarly to the previous problem we consider the limiting cases of low frequencies $\delta \ll R$ and high frequencies $\delta \gg R$ respectively. The quantity δ is defined by (3.30).

1. In the low frequency limit when $\delta \gg R$ we take the first terms of the Bessel functions $J_0(kr)$ and $J_1(kr)$ expansions. Then (3.42) and (3.43) take the form:

$$E_z = E_0 \left[1 - \frac{1}{2}i\left(\frac{r}{\delta}\right)^2 - \frac{1}{16}\left(\frac{r}{\delta}\right)^4\right] \exp(-i\omega t) \qquad (3.44)$$

and

$$H_\varphi = E_0 \frac{\sigma}{2} r \left[1 - \frac{1}{4}i\left(\frac{r}{\delta}\right)^2 - \frac{1}{48}\left(\frac{r}{\delta}\right)^4\right] \exp(-i\omega t). \qquad (3.45)$$

Equation (3.44) shows that the amplitudes of the electric field and the current density increase as $[1 + (r/2\delta)^4]$ with the increase of the distance r from the axis.

2. In the opposite limiting case of high frequencies when $\delta \ll R$ we use the asymptotic formula for the Bessel function

$$J_0\left(x\sqrt{2i}\right) \simeq x^{-1/2} \exp\left[(1-i)x\right],$$

which is valid for large values of argument over most of the cross section [4]. Retaining the rapidly varying exponential factor we find

$$E_z = E_0 \exp\left[-\frac{(R-r)}{\delta}\right] \exp i\left[\frac{(R-r)}{\delta} - \omega t\right] \qquad (3.46)$$

and

$$H_\varphi = E_0(1+i)\sqrt{\frac{\sigma}{2\mu_0\omega}} \exp\left[-\frac{(R-r)}{\delta}\right] \exp i\left[\frac{(R-r)}{\delta} - \omega t\right]. \qquad (3.47)$$

These expressions are similar to (3.31) and (3.33), which are valid near the surface of a conductor of any shape when the skin effect is strong [4].

3.2 Quasi-steady Current Circuit

Consider a linear circuit characterized by resistance, self-inductance and capacitance and containing a source of a variable electromotive force (e.m.f.) $\mathcal{V}(t)$. The circuit is shown in Fig. 3.1. By definition, an e.m.f. is the work done by external forces on the unit positive charge transfer through a closed circuit. It is equal to the total potential difference in the circuit. The quasi-static approximation means, in particular, that the electric current is at any moment of time the same at any cross section of the circuit as it was mentioned above. Then, the work $\mathcal{W}$ done per unit time by the electric field on the charges in the circuit is [4]:

$$\mathcal{W} = \mathcal{V}(t) I(t), \qquad (3.48)$$

where $I(t)$ is the instantaneous current in the circuit. This work in general case goes partly into Joule heat W^J, partly into the energy per unit time W^M of the magnetic field of the current, and partly into the energy per unit time W^E of the electric field of the capacitor since a variable current $I(t)$, unlike a constant one, can flow in an open circuit. The contributions into the energy per unit time have the form respectively [4]:

$$W^J = RI^2, \quad W^M = \frac{d}{dt}\left(\frac{1}{2}\mathcal{L}I^2\right), \quad W^E = \frac{d}{dt}\left(\frac{q^2}{2C}\right), \qquad (3.49)$$

where R is the resistance, and $\mathcal{L}$ is the self-inductance of the conductor, C is the capacitance of the capacitor, and q is the charge on its plates. Then the energy balance is given by

$$\mathcal{V}I = RI^2 + \frac{d}{dt}\left(\frac{1}{2}\mathcal{L}I^2\right) + \frac{d}{dt}\left(\frac{q^2}{2C}\right). \qquad (3.50)$$

The continuity of the current in the circuit requires that

$$I = \frac{dq}{dt}. \qquad (3.51)$$

As a result, (3.50) reduces to the differential equation of the balance of potentials

$$\mathcal{V} = R\frac{dq}{dt} + \mathcal{L}\frac{d^2q}{dt^2} + \frac{q}{C}. \qquad (3.52)$$

The right-hand side of (3.52) can be also expressed in terms of the current I:

$$\mathcal{V} = RI + \mathcal{L}\frac{dI}{dt} + \frac{1}{C}\int_0^t I(t')\,dt'. \qquad (3.53)$$

Consider the case of the periodic e.m.f. and current. Equations (3.52) and (3.53) are linear, and can use complex quantities

$$\mathcal{V} = \mathcal{V}_0 \exp(-i\omega t), \quad I = I_0 \exp(-i\omega t). \qquad (3.54)$$

Substituting (3.54) into any of (3.52), (3.53) and taking into account that

$$q = i\frac{I}{\omega}, \qquad (3.55)$$

we obtain:

$$\mathcal{V} = I\left(R - i\omega\mathcal{L} + i\frac{1}{\omega C}\right) = I\mathcal{Z}(\omega), \qquad (3.56)$$

3.2 Quasi-steady Current Circuit

where $\mathcal{Z}(\omega)$ is the complex resistance or impedance of the circuit. The real part R of $\mathcal{Z}(\omega)$ determines the energy dissipation in the circuit. Equation (3.56) can be also written in a real form.

$$I(t)\sqrt{R^2 + \left(\frac{1}{\omega\mathcal{C}} - \omega\mathcal{L}\right)^2} = \mathcal{V}_0 \cos(\omega t - \phi), \quad \tan\phi = \frac{1}{R}\left(\frac{1}{\omega\mathcal{C}} - \omega\mathcal{L}\right). \tag{3.57}$$

It is seen from (3.56) that in the absence of the e.m.f. ($\mathcal{V}=0$) the current in the circuit occurs in the form of free electric oscillations with the complex frequency ω_0 that is given by the condition

$$\mathcal{Z}(\omega_0) = 0, \tag{3.58}$$

which yields

$$\omega_0 = -i\frac{R}{2\mathcal{L}} \pm \sqrt{\frac{1}{\mathcal{L}\mathcal{C}} - \left(\frac{R}{2\mathcal{L}}\right)^2}. \tag{3.59}$$

If $1/(\mathcal{L}\mathcal{C}) > (R/2\mathcal{L})^2$, then the periodic oscillations damped with the decrement $R/(2\mathcal{L})$ would exist. In the opposite case an aperiodic regime of discharge occurs. In the limit of resistanceless circuit $R \to 0$ undamped oscillations take place with the frequency given by the Thomson formula [4]:

$$\omega_0 = \frac{1}{\sqrt{\mathcal{L}\mathcal{C}}}. \tag{3.60}$$

Equation (3.52) can be generalized to a system of several inductively coupled circuits containing capacitors. The current $I_j(t)$ in the j-th circuit is related to the charges $\pm q_j$ on the corresponding capacitor by (3.51). Equation (3.52) should be replaced by the set of equations

$$\sum_j \mathcal{M}_{ij}\frac{d^2 q_j}{dt^2} + R_i\frac{dq_i}{dt} + \frac{q_i}{\mathcal{C}_i} = \mathcal{V}_i, \tag{3.61}$$

where $\mathcal{M}_{ij}$ is the mutual inductance of the i-th and j-th circuits. For periodic monochromatic currents (3.54), (3.61) transform into the system of linear algebraic equations:

$$\sum_j Z_{ij}(\omega) I_j = \mathcal{V}_i, \tag{3.62}$$

where the matrix elements $Z_{ij}(\omega)$ have the form

$$Z_{ij}(\omega) = \delta_{ij}\left(R_i + \frac{i}{\omega \mathcal{C}_i}\right) - i\omega \mathcal{M}_{ij}. \tag{3.63}$$

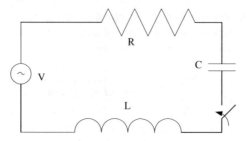

Fig. 3.1. A linear circuit containing a resistance R, a self-inductance L, a capacitor C, and an e.m.f. V

The eigenfrequencies of the current system are determined by condition that the determinant of system (3.62) should vanish. If the resistances R_i are not zero, all eigenfrequencies have a non-zero imaginary part, and the electric oscillations are damped.

Equations (3.61) are formally identical with the mechanical equations of motion of a system with several degrees of freedom which executes small damped oscillations [4]. The generalized coordinates can be identified with the charges q_i and generalized velocities with the currents $I_i = dq_i/dt$. The Lagrangian of such a system is given by

$$\mathcal{L} = \sum_{ij} \frac{1}{2} \mathcal{M}_{ij} \frac{dq_i}{dt} \frac{dq_j}{dt} - \sum_i \frac{q_i^2}{2\mathcal{C}_i} + \sum_i q_i \mathcal{V}_i. \tag{3.64}$$

Here the first sum containing the magnetic energy of the current system corresponds to the kinetic energy of the mechanical system, the second sum describes the energy of the electric field and is similar to the potential energy of the mechanical system, and the e.m.f.'s $\mathcal{V}_i$ can be identified with the externally applied forces which cause the forced oscillations of the system; finally, the resistances R_i compose the dissipative function $\mathcal{R}$ [4].

$$\mathcal{R} = \sum_i \frac{1}{2} R_i \left(\frac{dq_i}{dt}\right)^2. \tag{3.65}$$

Comparison of (5.1), (3.64) and (3.65) shows that (3.61) can be formulated in the Lagrangian terms as follows [4]

$$\frac{d}{dt} \frac{\partial \mathcal{L}}{\partial (dq_i/dt)} - \frac{\partial \mathcal{L}}{\partial q_i} = -\frac{\partial \mathcal{R}}{\partial (dq_i/dt)}. \tag{3.66}$$

3.3 Electromagnetic Induction Law

Consider a circuit in an external variable magnetic field $\boldsymbol{H}$. In the absence of conductors this magnetic field would induce the electric field $\boldsymbol{E}$. Both fields vary slightly over the thickness of a thin wire. Calculate the circulation of the electric field $\boldsymbol{E}$ round the current circuit using the Maxwell equation (3.8). It has the form

$$\oint_C \boldsymbol{E} \cdot d\boldsymbol{l} = \int_S \mathrm{curl}\boldsymbol{E} \cdot d\boldsymbol{s} = -\frac{\partial}{\partial t}\int_S \mu_0 \boldsymbol{H} \cdot d\boldsymbol{s} = -\frac{\partial \Phi}{\partial t}, \qquad (3.67)$$

where Φ is the flux of the external magnetic field through the circuit. The electric field circulation is the work done by the externally induced force on the unit positive charge transfer through a closed circuit which coincides by definition with the e.m.f. $\mathcal{V}$ [4]. Then we can write

$$\mathcal{V} = -\frac{\partial \Phi}{\partial t}. \qquad (3.68)$$

Equations (3.67) and (3.68) express the Faraday's law of electromagnetic induction which claims that the e.m.f. in the closed circuit can be induced by the change of the external magnetic flux traversing the circuit [4]. Substituting (3.50) and (3.51) into (3.68) we obtain for the circuit consisting of the resistance and self-inductance with a current in an external magnetic field:

$$RI + \mathfrak{L}\frac{dI}{dt} = -\frac{\partial \Phi}{\partial t}. \qquad (3.69)$$

According to the result (2.60) of the previous chapter, the second term in the left-hand side of (3.69) is a time derivative of the intrinsic magnetic flux Φ_i which is due to the current in the circuit. Then (3.69) takes the form

$$RI = -\frac{d}{dt}(\Phi + \Phi_i) = -\frac{d\Phi^{\mathrm{tot}}}{dt}, \qquad (3.70)$$

where Φ^{tot} is the total magnetic flux from the external magnetic field and the magnetic field of the current. In this form (3.70) gives the Ohm's law for the whole circuit, i.e. the equality of RI to the total e.m.f. in the circuit [4]. Equation (3.70) can be generalized to the case where the shape of the circuit varies with time. Then the self-inductance $\mathfrak{L}$ depends on time, and (3.70) becomes

$$RI = -\frac{d}{dt}(\mathfrak{L}I) - \frac{\partial \Phi}{\partial t}. \qquad (3.71)$$

Until now we assumed implicitly that a conductor in an electromagnetic field is at rest in the frame of reference K in which $\boldsymbol{H}$, $\boldsymbol{E}$ and the Ohm's law (3.6) are defined. In the frame of reference K' connected with the conductor moving with the non-relativistic velocity $v \ll c$ the electric field $\boldsymbol{E}'$ has the form according to (1.101), (1.102):

$$\boldsymbol{E}' = \boldsymbol{E} + [\boldsymbol{v} \times \boldsymbol{B}]. \qquad (3.72)$$

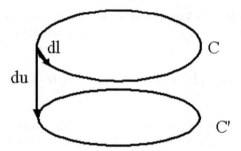

Fig. 3.2. The motion of the circuit C

Substituting the effective electric field (3.72) into (3.67) we get for the e.m.f. of the moving conductor

$$\mathcal{V}^{tot} = \oint_C \{E + [v \times B]\} \cdot dl. \tag{3.73}$$

The first term in the integrand yields the already known contribution (3.68) due to the magnetic flux change for the circuit at rest $(\partial \Phi / \partial t)_{v=0}$. The second term can be transformed as follows

$$\oint_C [v \times B] \cdot dl = \oint_C \left[\frac{du}{dt} \times B\right] \cdot dl = -\frac{\partial}{\partial t} \oint_S B \cdot ds = -\left(\frac{\partial \Phi}{\partial t}\right)_{B=\text{const}}. \tag{3.74}$$

Here du is the current circuit displacement during the infinitesimal interval of time dt, and $ds = [du \times dl]$ is an element of area on the side surface S between two infinitely close positions C and C' of the current circuit, which it occupies at the time moments t and $t + dt$. The evolution of the circuit is shown in Fig. 3.2.

The total magnetic flux through any closed surface is zero. Therefore, the magnetic flux through the side surface S is equal to the difference of fluxes through surfaces spanning contours C and C'. Finally the total e.m.f. induced by the change of the magnetic flux takes the form :

$$\mathcal{V}^{tot} = -\left(\frac{\partial \Phi}{\partial t}\right)_{v=0} - \left(\frac{\partial \Phi}{\partial t}\right)_{B=\text{const}}. \tag{3.75}$$

It is seen from (3.75) that the e.m.f. in a closed circuit can be excited by the external magnetic field by the change of the field, by the motion of the circuit, or by a combination of both mechanisms [4]. It should be noted that if the circuit moves in such a way that each point of it remains parallel to the field lines, then the flux through the circuit does not vary since the field component B_n normal to the side surface equals zero. The side surface itself

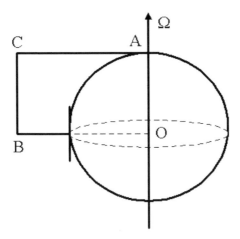

Fig. 3.3. The rotating magnet with a wire ACB

reduces to zero if the circuit displacement $d\boldsymbol{u}$ remains parallel to the circuit element $d\boldsymbol{l}$. Consequently, in order to induce the e.m.f., the conductor must in its motion cross lines of the magnetic field.

The Maxwell equations (3.8) and (2.2) for a moving conductor remain unchanged. On the contrary, (2.1) must be modified taking into account (3.72):

$$\mathrm{curl}\boldsymbol{H} = \boldsymbol{j} = \sigma\left(\boldsymbol{E} + [\boldsymbol{v}\times\boldsymbol{B}]\right). \tag{3.76}$$

Then (3.10) for a homogeneous conductor with a constant conductivity σ and relative magnetic permeability μ^r can be generalized to the case of a moving conductor:

$$\frac{1}{\sigma\mu_0\mu^r}\nabla^2\boldsymbol{H} = \frac{\partial\boldsymbol{H}}{\partial t} - \mathrm{curl}\left[\boldsymbol{v}\times\boldsymbol{H}\right]. \tag{3.77}$$

Investigate a practically important case of a unipolar induction which occurs when a magnetized conductor rotates with an angular velocity Ω. A stationary wire ACB connected to the rotating magnet by means of two sliding contacts A and B is shown in Fig. 3.3. A current flows in the wire.

Evaluate the e.m.f. which produces a current in the circuit. We choose the coordinate system rotating with the magnet. In such a system the magnet is at rest while the wire rotates with the angular velocity $-\Omega$. As a result, the conductor moves in a constant magnetic field $\boldsymbol{B}$ caused by the magnet. The e.m.f. is due to the second term in (3.75):

$$\mathcal{V} = \oint_C [\boldsymbol{v}\times\boldsymbol{B}]\cdot d\boldsymbol{l} = -\int_{ACB}[\boldsymbol{B}\times[\boldsymbol{r}\times\boldsymbol{\Omega}]]\cdot d\boldsymbol{l} \tag{3.78}$$

where the integration is carried out along the wire.

4 Electromagnetic Waves

4.1 Electromagnetic Waves in a Dielectric

Consider a variable electromagnetic field in an isotropic homogeneous dielectric without absorption. We limit our analysis in this section with the frequencies small compared to the eigenfrequencies of the molecular and electronic oscillations. Then the material equations, i.e. the relations between the vectors D and E, and B and H remain the same as in the case of the constant fields. The relative permittivity ε^r and permeability μ^r of the isotropic homogeneous dielectric without absorption in such case reduce to real scalars, and we can use the material equations (1.20), (1.21):

$$\boldsymbol{D} = \varepsilon_0 \varepsilon^r \boldsymbol{E}, \quad \boldsymbol{B} = \mu_0 \mu^r \boldsymbol{H}.$$

In the dielectric considered we may set $\rho = 0$, $\boldsymbol{j} = 0$. Then the electromagnetic field is determined by the Maxwell equations in the form

$$a) \ \mathrm{div}\,\boldsymbol{B} = 0; \quad b) \ \mathrm{curl}\,\boldsymbol{E} = -\frac{\partial \boldsymbol{B}}{\partial t} \quad (4.1)$$

$$c) \ \mathrm{div}\,\boldsymbol{D} = 0; \quad d) \ \mathrm{curl}\,\boldsymbol{H} = \frac{\partial \boldsymbol{D}}{\partial t}.$$

The electromagnetic potentials in Lorentz gauge (1.46)

$$\mathrm{div}\,\boldsymbol{A} + \mu_0 \varepsilon_0 \varepsilon^r \mu^r \frac{\partial \varphi}{\partial t} = 0$$

are determined by (1.47), (1.48) with $\rho = 0$, $\boldsymbol{j} = 0$:

$$a) \ \Box \boldsymbol{A} \equiv \Delta \boldsymbol{A} - \mu_0 \varepsilon_0 \varepsilon^r \mu^r \frac{\partial^2 \boldsymbol{A}}{\partial t^2} = 0 \quad (4.2)$$

$$b) \ \Box \varphi \equiv \Delta \varphi - \mu_0 \varepsilon_0 \varepsilon^r \mu^r \frac{\partial^2 \varphi}{\partial t^2} = 0$$

and the fields are related to the potentials according to (1.39), (1.41)

$$\boldsymbol{B} = \mathrm{curl}\,\boldsymbol{A}; \quad \boldsymbol{E} = -\mathrm{grad}\,\varphi - \frac{\partial \boldsymbol{A}}{\partial t}. \quad (4.3)$$

4.1 Electromagnetic Waves in a Dielectric

It appears that (4.1) has non-zero solutions; thus, the electromagnetic field can exist even in the absence of any charges. This field is, however, necessarily, time-dependent. Indeed, if

$$\frac{\partial \boldsymbol{D}}{\partial t} = \frac{\partial \boldsymbol{B}}{\partial t} = 0,$$

then both div and curl of $\boldsymbol{E}$ and $\boldsymbol{B}$ are equal to zero in the whole space, and the only solution is a trivial one: $\boldsymbol{E} = \boldsymbol{B} = 0$.

First of all we point out that (4.1) and (4.2) are satisfied if the scalar potential is zero

$$\varphi = 0. \tag{4.4}$$

In this case the Lorentz condition reduces to the Coulomb gauge $\operatorname{div} \boldsymbol{A} = 0$, and the direct substitution of $\boldsymbol{B} = \operatorname{curl} \boldsymbol{A}$ and $\boldsymbol{E} = -\partial \boldsymbol{A}/\partial t$ into (4.1) shows that the first three of them are trivially satisfied, while the fourth is identical to (4.2a). Applying the curl operation to (4.1b) we obtain

$$\operatorname{curl}\operatorname{curl}\boldsymbol{E} \equiv \operatorname{grad}\operatorname{div}\boldsymbol{E} - \Delta \boldsymbol{E} = -\frac{\partial}{\partial t}\operatorname{curl}\boldsymbol{B}.$$

After substitutions

$$\operatorname{div}\boldsymbol{E} = 0; \quad \operatorname{curl}\boldsymbol{H} = \varepsilon_0 \varepsilon^r \frac{\partial \boldsymbol{E}}{\partial t}; \quad \boldsymbol{B} = \mu_0 \mu^r \boldsymbol{H}$$

from (4.1c,d) this equation takes form

$$\Box \boldsymbol{E} \equiv \Delta \boldsymbol{E} - \frac{1}{c_m^2}\frac{\partial^2 \boldsymbol{E}}{\partial t^2} = 0,$$

where the light velocity c_m in the medium is introduced as follows

$$c_m^2 = \frac{1}{\mu_0 \varepsilon_0 \varepsilon^r \mu^r},$$

or, taking into account that the vacuum light velocity c is given by

$$c^2 = \frac{1}{\mu_0 \varepsilon_0}$$

we write

$$c_m = \frac{c}{\sqrt{\varepsilon^r \mu^r}}. \tag{4.5}$$

Usually in non-magnetic media $\mu^r = 1$. The quantity

$$\sqrt{\varepsilon^r} = n \tag{4.6}$$

is called a refraction coefficient of a medium. By definition, $n > 1$, and the light velocity in a medium is always less than c

$$c_m < c. \tag{4.7}$$

Similar procedure with (4.1d) results in

$$\Box \boldsymbol{B} \equiv \Delta \boldsymbol{B} - \frac{1}{c^2} \frac{\partial^2 \boldsymbol{B}}{\partial t^2} = 0.$$

Thus, the electric $\boldsymbol{E}$ and magnetic $\boldsymbol{B}$ fields, as well as the vector-potential $\boldsymbol{A}$ satisfy the same – d'Alambert or wave – equation

$$\Box f = 0, \tag{4.8}$$

where f is any one of the components of $\boldsymbol{E}, \boldsymbol{B}$, or $\boldsymbol{A}$. The solutions of this equation for $\boldsymbol{E}$ and $\boldsymbol{B}$ are called electromagnetic waves.

Consider the energy characteristics of the electromagnetic waves. We start with multiplying (4.1b) by $\boldsymbol{H}$, (4.1d) by $\boldsymbol{E}$, and subtracting the former from the latter we obtain:

$$\boldsymbol{E} \frac{\partial \boldsymbol{D}}{\partial t} + \boldsymbol{H} \frac{\partial \boldsymbol{B}}{\partial t} = -\left(\boldsymbol{H} \operatorname{curl} \boldsymbol{E} - \boldsymbol{E} \operatorname{curl} \boldsymbol{H}\right). \tag{4.9}$$

Using the vector calculus identity

$$\operatorname{div}[\boldsymbol{a} \times \boldsymbol{b}] = \boldsymbol{b} \operatorname{curl} \boldsymbol{a} - \boldsymbol{a} \operatorname{curl} \boldsymbol{b},$$

we rewrite the last equation in the form

$$\frac{1}{2} \frac{\partial}{\partial t} \left(\varepsilon_0 \varepsilon^r E^2 + \mu_0 \mu^r H^2 \right) = -\operatorname{div}[\boldsymbol{E} \times \boldsymbol{H}],$$

or

$$\frac{\partial}{\partial t} \left(\frac{\varepsilon_0 \varepsilon^r E^2 + \mu_0 \mu^r H^2}{2} \right) = -\operatorname{div} \boldsymbol{S}, \tag{4.10}$$

where

$$\boldsymbol{S} = [\boldsymbol{E} \times \boldsymbol{H}] \tag{4.11}$$

is called the Poynting vector. The quantity in brackets in the left hand side of (4.10)

$$W = \frac{\varepsilon_0 \varepsilon^r E^2 + \mu_0 \mu^r H^2}{2} \tag{4.12}$$

should be seen as the energy density of an electromagnetic field. Equation (4.10) can be interpreted as an "energy continuity equation," in close analogy with the density continuity equation

$$\frac{\partial \rho}{\partial t} + \operatorname{div} \boldsymbol{j} = 0. \tag{4.13}$$

Thus, the Poynting vector $\boldsymbol{S}$ (4.11) is an energy current, i.e. an amount of energy which flows through the unit area in the unit of time; this quantity is called the energy flux of an electromagnetic field.

4.2 Plane Waves

In the special case of electromagnetic waves depending only on one space coordinate (say x) and time the equation for field becomes

$$\frac{\partial^2 f}{\partial t^2} - c_m^2 \frac{\partial^2 f}{\partial x^2} = 0 \tag{4.14}$$

and its solutions are called plane waves. Without a loss of generality, we will carry out the further analysis for the case of vacuum where

$$\varepsilon^r = 1, \ \mu^r = 1, \ c_m = c = \frac{1}{\sqrt{\mu_0 \varepsilon_0}}. \tag{4.15}$$

The general solution of (4.14) can be obtained in the following way. First, we rewrite it in the form

$$\left(\frac{\partial}{\partial t} - c\frac{\partial}{\partial x}\right)\left(\frac{\partial}{\partial t} + c\frac{\partial}{\partial x}\right) f = 0$$

and introduce new variables

$$\zeta = t - \frac{x}{c}; \ \eta = t + \frac{x}{c}.$$

Now

$$\frac{\partial}{\partial \zeta} = \left(\frac{\partial}{\partial t} - c\frac{\partial}{\partial x}\right); \ \frac{\partial}{\partial \eta} = \left(\frac{\partial}{\partial t} + c\frac{\partial}{\partial x}\right)$$

and the wave equation takes the form

$$\frac{\partial^2 f}{\partial \zeta \partial \eta} = 0.$$

Integration of this equation with respect to ζ gives

$$\frac{\partial f}{\partial \eta} = F(\eta),$$

where $F(\eta)$ is an arbitrary function. Integrating now with respect to η we get

$$f = f_1(\zeta) + f_2(\eta),$$

where f_1 and f_2 are arbitrary functions ($f_2(\eta) = \int^\eta F(\eta') d\eta'$ and is arbitrary since $F(\eta)$ is arbitrary). Therefore

$$f = f_1\left(t - \frac{x}{c}\right) + f_2\left(t + \frac{x}{c}\right). \tag{4.16}$$

Let us clarify the meaning of this solution separately for each of two terms. Thus, first we set $f_2 = 0$ and

$$f = f_1\left(t - \frac{x}{c}\right).$$

In the given plane $x = \text{const}$ the field changes with time; at each given moment t the field has different values for different x. If the field has some definite value f_1 at some time t_1 at the point x_1, it will have the same value at the point $x_1 + \Delta x$ at the moment $t_1 + \Delta t$, providing that $\Delta x = c\Delta t$. In other words, all the values of the electromagnetic field are propagating in space along the x-axis with a constant velocity c (the velocity of light). Hence, $f_1(t - x/c)$ represents a plane wave moving in the positive x direction; similar analysis shows that $f_2(t + x/c)$ represents a wave moving in the opposite, negative direction along the x-axis.

Let us now establish the relations between the directions of the field vectors $\boldsymbol{E}$ and $\boldsymbol{B}$ and the direction of the wave propagation. Since all the quantities in plane wave $f_1(t - x/c)$ depend only on the coordinate x, the gauge condition $\text{div}\,\boldsymbol{A} = 0$ reduces to

$$\frac{\partial A_x}{\partial x} = 0$$

in the whole space. Thus

$$\frac{\partial^2 A_x}{\partial x^2} = 0$$

and according to (4.14)

$$\frac{\partial^2 A_x}{\partial t^2} = 0,$$

that is

$$\frac{\partial A_x}{\partial t} = \text{const}.$$

But $\partial A_x / \partial t = E_x$, i.e. the non-zero A_x represents the existence of a constant longitudinal electric field. As has been noted in the beginning of the chapter, time-independent field do not satisfy (4.1). Thus, $A_x = 0$, which means that the vector-potential of the plane wave is perpendicular to the x-axis, i.e. to the direction $\boldsymbol{n}$ of the propagation of this wave, $\boldsymbol{A} \perp \boldsymbol{n}$. Since

$$\boldsymbol{E} = -\frac{\partial \boldsymbol{A}}{\partial t}, \quad \boldsymbol{B} = \text{curl}\boldsymbol{A}$$

and for $\boldsymbol{A} = \boldsymbol{A}(\zeta) = \boldsymbol{A}(t - x/c)$,

$$\frac{\partial \boldsymbol{A}}{\partial t} = \frac{\partial \boldsymbol{A}}{\partial \zeta}, \quad \frac{\partial \boldsymbol{A}}{\partial x} = -\frac{1}{c}\frac{\partial \boldsymbol{A}}{\partial \zeta}$$

we obtain

$$\boldsymbol{E} = -\{0, A'_y, A'_z\}, \quad \boldsymbol{B} = \frac{1}{c}\{0, A'_z, -A'_y\}, \tag{4.17}$$

where the prime denotes the derivative with respect to ζ. Thus, both $\boldsymbol{E}$ and $\boldsymbol{B}$ are perpendicular with respect to the direction of propagation of the plane wave, i.e. the electromagnetic wavesin vacuum are transverse. Introducing the unit vector $\boldsymbol{n}$ in the direction of propagation ($\boldsymbol{n} = (1,0,0)$) we can verify that (4.17) implies

$$\boldsymbol{B} = \frac{1}{c}[\boldsymbol{n} \times \boldsymbol{E}]. \tag{4.18}$$

From (4.18) it is clear that the electric and magnetic fields of the plane wave are perpendicular to each other. The energy flux in the plane wave (its Poynting vector) is

$$\boldsymbol{S} = [\boldsymbol{E} \times \boldsymbol{H}] = \frac{1}{\mu_0 c}[\boldsymbol{E} \times [\boldsymbol{n} \times \boldsymbol{E}]] = \sqrt{\frac{\varepsilon_0}{\mu_0}}\,[\boldsymbol{n}E^2 - \boldsymbol{E}(\boldsymbol{n}\cdot\boldsymbol{E})] = \sqrt{\frac{\varepsilon_0}{\mu_0}}\,E^2\boldsymbol{n}$$

since $(\boldsymbol{n}\cdot\boldsymbol{E}) = 0$. Similarly, the Poynting vector can be expressed in terms of the magnetic field H:

$$\boldsymbol{S} = \sqrt{\frac{\mu_0}{\varepsilon_0}}\,H^2\boldsymbol{n}.$$

The energy density W is:

$$W = \frac{1}{2}\left(\varepsilon_0 E^2 + \mu_0 H^2\right)$$

and it is easy to see that

$$\boldsymbol{S} = cW\boldsymbol{n} \tag{4.19}$$

in accordance with the fact that fields propagate with the velocity of light.

4.3 Monochromatic Plane Waves

A very important special case of electromagnetic waves is a wave, in which the field is a simple periodic function of time of the type $\cos(\omega t + \alpha)$, where $\omega \equiv 2\pi/T$, T being the period of the wave and ω – its cyclic frequency, or simply frequency. In a plane wave propagating along the z-axis the field is a function of $t - z/c$; thus, in a monochromatic plane wave the field dependence on z and t has the form

$$\boldsymbol{C} = \boldsymbol{c}_0 \cos\left[\omega\left(t - z/c\right) + \alpha\right],$$

where c_0 is a real amplitude of the field. It is more convenient to present this form as a real part of a complex expression

$$\boldsymbol{C} = \mathrm{Re}\left\{\boldsymbol{C}_0 \exp\left[-i\omega\left(t - z/c\right)\right]\right\}, \quad (4.20)$$

where $\boldsymbol{C}_0$ is a constant complex vector

$$\boldsymbol{C}_0 = \boldsymbol{c}_0 \exp\left(-i\varphi\right). \quad (4.21)$$

The quantity

$$\lambda = \frac{2\pi c}{\omega} \quad (4.22)$$

determines the period of the space variation of the field at a fixed time t; it is called the wavelength. If the wave is propagating in the direction determined by the unit vector $\boldsymbol{n}$, we can introduce the wave vector $\boldsymbol{k}$

$$\boldsymbol{k} = \frac{\omega}{c}\boldsymbol{n} \quad (4.23)$$

in term of which the field of the wave can be written as

$$\boldsymbol{C} = \mathrm{Re}\left\{\boldsymbol{C}_0 \exp i\left(\boldsymbol{k}\boldsymbol{r} - \omega t\right)\right\}. \quad (4.24)$$

The main advantage of the exponential expressions of the type (4.24) lies in the fact that so long as we apply only linear operators on the field vectors (like curl, div, $\partial/\partial t$, etc.) we can omit the symbol Re and operate with complex exponential expressions as such, while the real part can be taken from the final result only. Thus, writing instead of (4.24)

$$\boldsymbol{C} = \boldsymbol{C}_0 \exp i\left(\boldsymbol{k}\boldsymbol{r} - \omega t\right)$$

we have for the differential operations entering the Maxwell equations

$$\mathrm{div}\boldsymbol{C} = i\left(\boldsymbol{k}\cdot\boldsymbol{C}\right);\ \mathrm{curl}\boldsymbol{C} = [\boldsymbol{k}\times\boldsymbol{C}];\ \frac{\partial\boldsymbol{C}}{\partial t} = -i\omega\boldsymbol{C}. \quad (4.25)$$

Writing the expressions for the electric $\boldsymbol{E}(\boldsymbol{r},t)$ and the magnetic $\boldsymbol{B}(\boldsymbol{r},t)$ fields of the monochromatic wave in the form $\boldsymbol{E}(\boldsymbol{r},t)$

$$\boldsymbol{E}(\boldsymbol{r},t) = \boldsymbol{e}_E E_0 \exp i\left(\boldsymbol{k}\boldsymbol{r} - \omega t\right) \quad (4.26)$$
$$\boldsymbol{B}(\boldsymbol{r},t) = \boldsymbol{e}_B B_0 \exp i\left(\boldsymbol{k}\boldsymbol{r} - \omega t\right),$$

where $\boldsymbol{e}_E, \boldsymbol{e}_B$ are two constant unit vectors, and E_0, B_0 are the constant complex amplitudes, and substituting them into the Maxwell equations (4.1), we have with the help of (4.25)

$$(\boldsymbol{e}_E \cdot \boldsymbol{k}) = (\boldsymbol{e}_B \cdot \boldsymbol{k}) = 0 \quad (4.27)$$
$$E_0\left[\boldsymbol{n}\times\boldsymbol{e}_E\right] = B_0\boldsymbol{e}_B;\ B_0\left[\boldsymbol{n}\times\boldsymbol{e}_B\right] = -\frac{1}{c}E_0\boldsymbol{e}_E.$$

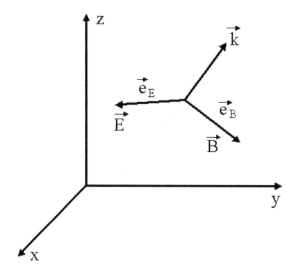

Fig. 4.1. A wave vector $\boldsymbol{k}$ and field vectors $\boldsymbol{E}$ and $\boldsymbol{B}$ of the transverse electromagnetic wave. The unit vectors $\boldsymbol{e}_E$ and $\boldsymbol{e}_B$ define the direction of the corresponding field vectors

The first pair of equations (4.27) implies that both the electric and the magnetic field of the wave are perpendicular to the direction of its propagation (the wave is transverse), while the second pair shows that these fields are orthogonal to one another and their complex amplitudes are related as follows $E_0 = cB_0$; $\boldsymbol{e}_E = -[\boldsymbol{n} \times \boldsymbol{e}_B]$; $\boldsymbol{e}_B = [\boldsymbol{n} \times \boldsymbol{e}_E]$. The mutual orientation of the vectors $\boldsymbol{k}, \boldsymbol{E}$ and $\boldsymbol{B}$ is shown in Fig. 4.1.

All these statements are in complete agreement with the general properties of the plane electromagnetic waves established in the last section.

4.4 Polarization of a Monochromatic Plane Wave

Consider now an important specific property of a monochromatic electromagnetic wave called polarization which is defined by the orientation of the electric vector $\boldsymbol{E}(\boldsymbol{r},t)$ in the plane perpendicular to the propagation direction. Let the plane monochromatic wave is propagating along the z-axis. We decompose $\boldsymbol{E}(\boldsymbol{r},t)$ into two mutually orthogonal components $E_{x,y}$ choosing their directions as the x- and y-axes respectively. In general case both components $E_{x,y}$ can possess different phases and amplitudes. In a real form they can be written as follows

$$E_x = a\cos(kz - \omega t + \varphi_1), \quad E_y = b\cos(kz - \omega t + \varphi_2), \qquad (4.28)$$

where $a, b, \varphi_{1,2}$ are real numbers. In order to exclude the time and coordinate dependent factor we transform components (4.28):

90   4 Electromagnetic Waves

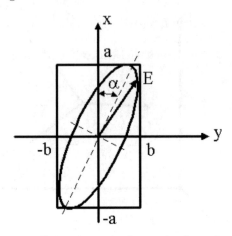

Fig. 4.2. Polarization ellipse of a monochromatic plane electromagnetic wave

$$\frac{E_x}{a} = \cos\left(kz - \omega t + \varphi_1\right) \tag{4.29}$$

$$\frac{E_y}{b} = \cos\left(kz - \omega t + \varphi_1\right)\cos\delta - \sin\left(kz - \omega t + \varphi_1\right)\sin\delta \tag{4.30}$$

and

$$\left(\frac{E_x}{a}\right)^2 - 2\frac{E_x E_y}{ab}\cos\delta + \left(\frac{E_y}{b}\right)^2 = \sin^2\delta, \tag{4.31}$$

where

$$\delta = \varphi_2 - \varphi_1. \tag{4.32}$$

The discriminant $\mathcal{D}$ of the quadratic form (4.31) is negative

$$\mathcal{D} = \frac{4\cos^2\delta}{(ab)^2} - \frac{4}{(ab)^2} = -\frac{4\sin^2\delta}{(ab)^2} \leq 0, \tag{4.33}$$

which means that the electric vector $\boldsymbol{E}\left(\boldsymbol{r}, t\right)$ rotates in the xy plane along the ellipse defined by (4.31). In such a case the wave is called elliptically polarized. The principal axes of the ellipse are turned around the z-axis by the angle α with respect to the x-axis:

$$\tan 2\alpha = \frac{2ab}{a^2 - b^2}\cos\delta. \tag{4.34}$$

The polarization ellipse is shown in Fig. 4.2.

4.4 Polarization of a Monochromatic Plane Wave

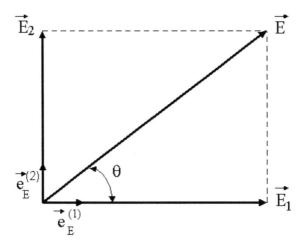

Fig. 4.3. The electric field E of a linearly polarized electromagnetic wave. The vectors $e_E^{(1)}$ and $e_E^{(2)}$ are the unit polarization vectors of the wave components

Address now special cases.

1. E_x and E_y have the phases differing by $\delta = m\pi$, $m = 0, 1, 2, ...$:

$$E_x = a \exp i\varphi, \quad E_y = (-1)^m b \exp i\varphi.$$

Then

$$\boldsymbol{E}(\boldsymbol{r}, t) = (\boldsymbol{x}_0 a + (-1)^m \boldsymbol{y}_0 b) \exp i (\boldsymbol{kr} - \omega t + \varphi),$$

while the physically meaningful electric field is given by

$$\mathrm{Re}\boldsymbol{E}(\boldsymbol{r}, t) = (\boldsymbol{x}_0 a + (-1)^m \boldsymbol{y}_0 b) \cos (\boldsymbol{kr} - \omega t + \varphi),$$

where $\boldsymbol{x}_0$ and $\boldsymbol{y}_0$ are the unit vectors parallel to the x- and y-axes, respectively. Then ellipse reduces to a direct line making the angle θ with the direction $\boldsymbol{x}_0$ (see Fig. 4.3) determined by

$$\tan \theta = \frac{E_y}{E_x} = (-1)^m \frac{b}{a}$$

and the electric vector $\boldsymbol{E}(\boldsymbol{r}, t)$ oscillating with frequency ω between the values $\pm\sqrt{a^2 + b^2}$. Such an electromagnetic wave is called linearly polarized.

2. E_x and E_y have the phase difference $\delta = \pm m\pi/2$, $m = 1, 3, ...$ and $a = b$:

$$E_x = a \exp i\varphi_1, \quad E_y = a \exp i\left(\varphi_1 \pm \frac{\pi}{2}\right).$$

In this case $\boldsymbol{E}(\boldsymbol{r}, t)$ takes the form

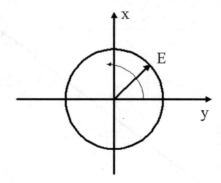

Fig. 4.4. The electric field $\boldsymbol{E}$ of a circularly polarized wave

$$\boldsymbol{E}(\boldsymbol{r},t) = a(\boldsymbol{x}_0 \pm i\boldsymbol{y}_0)\exp i(\boldsymbol{kr} - \omega t + \varphi_1). \tag{4.35}$$

Then the components of the real electric field obtained by taking a real part of (4.35) are

$$E_x(\boldsymbol{r},t) = a\cos(kz - \omega t + \varphi_1) \tag{4.36}$$
$$E_y(\boldsymbol{r},t) = \pm a\sin(kz - \omega t + \varphi_1).$$

The the electric vector $\boldsymbol{E}(\boldsymbol{r},t)$ rotates with a frequency ω around the circle of a radius a

$$E_x^2 + E_y^2 = a^2$$

with the axes along the x and y directions. For the upper sign in (4.35) the rotation is counterclockwise when the observer is facing into oncoming wave; this wave is called left polarized, or having positive helicity. For the lower sign in (4.35) the rotation is clockwise and the wave is called right polarized or having negative helicity. The wave in such a case is called circularly polarized (see Fig. 4.4).

4.5 Reflection and Refraction of Electromagnetic Waves

Until now we studied the electromagnetic wave propagation in a free space. However, in practice the phenomena at a boundary between vacuum and a condensed medium, or between two condensed media with different characteristics play an important role in electrodynamics. Below we consider the reflection and refraction of a monochromatic plane wave at a plane boundary between two homogeneous non-magnetic media ($\mu_{1,2}^r = 1$). The x- and z-axes are chosen to be parallel and perpendicular to the boundary, respectively. We assume that the plane monochromatic linearly polarized wave is propagated

4.5 Reflection and Refraction of Electromagnetic Waves 93

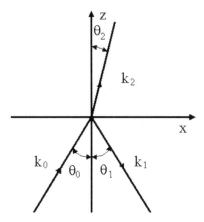

Fig. 4.5. The reflection and refraction of electromagnetic waves at the boundary between two media. The vectors $k_{0,1,2}$ define the propagation direction of the incident, reflected and refracted waves respectively. $\theta_{0,1,2}$ are the angles of incidence, reflection and refraction respectively

from the medium 1 with the relative dielectric permittivity ε_1^r to the medium 2 with the relative dielectric permittivity ε_2^r. The medium 1 is assumed to be transparent while the medium 2 can be in general case absorbing. The geometry of the problem is shown in Fig. 4.5. As we have seen in the previous section, the electric and magnetic fields $\boldsymbol{E}_i$ and $\boldsymbol{H}_i$ in such a wave have a form for the incident wave

$$\boldsymbol{E}_i = \boldsymbol{E}_0 \exp i\,(\boldsymbol{k}_0 \boldsymbol{r} - \omega t);\ \boldsymbol{H}_i = \boldsymbol{H}_0 \exp i\,(\boldsymbol{k}_0 \boldsymbol{r} - \omega t) \qquad (4.37)$$

and for the reflected and refracted waves the fields $\boldsymbol{E}_{e,r}$ and $\boldsymbol{H}_{e,r}$ are given by, respectively

$$\boldsymbol{E}_{e,r} = \boldsymbol{E}_{1,2} \exp i\,(\boldsymbol{k}_{1,2}\boldsymbol{r} - \omega t);\ \boldsymbol{H}_{e,r} = \boldsymbol{H}_{1,2} \exp i\,(\boldsymbol{k}_{1,2}\boldsymbol{r} - \omega t).$$

The complete homogeneity in the xy plane results in the same dependence of the incident, reflected and refracted waves on x, y. Hence, $k_{x,y}$ components must be the same for all three waves, and their directions of propagation lie in the same plane xz. Then we have

$$k_{0x} = k_{1x} = k_{2x} \qquad (4.38)$$

$$k_{1z} = -k_{0z} = -\frac{\omega}{c}\sqrt{\varepsilon_1^r}\cos\theta_0 \qquad (4.39)$$

$$k_{2z} = \sqrt{\left(\frac{\omega^2}{c^2}\varepsilon_2^r - k_{0x}^2\right)} = \frac{\omega}{c}\sqrt{\varepsilon_2^r - \varepsilon_1^r \sin^2\theta_0}. \qquad (4.40)$$

94   4 Electromagnetic Waves

The quantity k_2 can be complex in an absorbing medium. Therefore, the sign of the root k_{2z} must be chosen in such a way that $\mathrm{Im}\,k_{2z} > 0$, and the refracted wave would be damped towards the interior of medium 2. For the case when both media are transparent, (4.38)–(4.40) yield the well-known Snellius laws of reflection and refraction of electromagnetic waves which read [4].

$$\theta_1 = \theta_0, \quad \frac{\sin\theta_2}{\sin\theta_0} = \sqrt{\frac{\varepsilon_1^r}{\varepsilon_2^r}} = \frac{n_1}{n_2} = \frac{\lambda_2}{\lambda_1} = \frac{c_2}{c_1}. \tag{4.41}$$

1. The incidence angle and the reflection angle are equal.
2. The ratio of the sines of the refraction and incidence angles is equal to the ratio of the refraction indices $n_{1,2}$ of these media, and it is inversely proportional to the ratio of wavelengths $\lambda_{1,2}$, or the light phase velocities $c_{1,2}$ in the corresponding media.

In order to determine the amplitudes of the reflected and refracted waves, we need to use the boundary conditions for the electric and magnetic fields at the surface $z = 0$. The continuity of the tangential components of both fields yields the necessary equations. We should account for two possibilities: when the wave is polarized in the plane of incidence, and perpendicular to it. In general case the electric field of any incident wave can be decomposed into two these components. Consider first the case when the electric field of the incident wave $\boldsymbol{E}_i$ is perpendicular to the xz plane. Then we have from the Maxwell equations the following expression for the amplitude of the magnetic field tangential component H_x in each wave [4]:

$$H_{0x} = -\frac{k_{0z}}{\omega\mu_0}E_0, \quad H_{1x} = -\frac{k_{1z}}{\omega\mu_0}E_1, \quad H_{2x} = -\frac{k_{2z}}{\omega\mu_0}E_2 \tag{4.42}$$

The field in the medium 1 represents the sum of the incident and reflected waves which yields

$$E_0 + E_1 = E_2, \quad k_{0z}E_0 + k_{1z}E_1 = k_{2z}E_2 \tag{4.43}$$

Taking into account (4.39) and (4.40) we obtain the so-called Fresnel's formulae [4]:

$$E_1 = \frac{k_{0z} - k_{2z}}{k_{0z} + k_{2z}}E_0 = \frac{\sqrt{\varepsilon_1^r}\cos\theta_0 - \sqrt{\varepsilon_2^r - \varepsilon_1^r\sin^2\theta_0}}{\sqrt{\varepsilon_1^r}\cos\theta_0 + \sqrt{\varepsilon_2^r - \varepsilon_1^r\sin^2\theta_0}}E_0 \tag{4.44}$$

$$E_2 = \frac{2k_{0z}}{k_{0z} + k_{2z}}E_0 = \frac{2\sqrt{\varepsilon_1^r}\cos\theta_0}{\sqrt{\varepsilon_1^r}\cos\theta_0 + \sqrt{\varepsilon_2^r - \varepsilon_1^r\sin^2\theta_0}}E_0. \tag{4.45}$$

If both media are transparent, then, substituting the second equation (4.41) into (4.44) and (4.45) we find

4.5 Reflection and Refraction of Electromagnetic Waves

$$E_1 = \frac{\sin(\theta_2 - \theta_0)}{\sin(\theta_2 + \theta_0)} E_0, \quad E_2 = \frac{2\cos\theta_0 \sin\theta_2}{\sin(\theta_2 + \theta_0)} E_0. \tag{4.46}$$

In the case when the electric field $\boldsymbol{E}_i$ lies in the incidence plane it is more convenient to use the magnetic field $\boldsymbol{H}_i$ which this time is parallel to the y-axis. The direct calculation similar to the previous one gives [4]

$$H_1 = \frac{\varepsilon_2^r k_{0z} - \varepsilon_1^r k_{2z}}{\varepsilon_2^r k_{0z} + \varepsilon_1^r k_{2z}} H_0 = \frac{\varepsilon_2^r \cos\theta_0 - \sqrt{\varepsilon_1^r \varepsilon_2^r - \varepsilon_1^{r2}\sin^2\theta_0}}{\varepsilon_2^r \cos\theta_0 + \sqrt{\varepsilon_1^r \varepsilon_2^r - \varepsilon_1^{r2}\sin^2\theta_0}} H_0 \tag{4.47}$$

$$H_2 = \frac{2\varepsilon_2^r k_{0z}}{\varepsilon_2^r k_{0z} + \varepsilon_1^r k_{2z}} H_0 = \frac{2\varepsilon_2^r \cos\theta_0}{\varepsilon_2^r \cos\theta_0 + \sqrt{\varepsilon_1^r \varepsilon_2^r - \varepsilon_1^{r2}\sin^2\theta_0}} H_0. \tag{4.48}$$

For the transparent media equations (4.47) and (4.48) take the form

$$H_1 = \frac{\tan(\theta_2 - \theta_0)}{\tan(\theta_2 + \theta_0)} H_0, \quad H_2 = \frac{\sin 2\theta_0}{\sin(\theta_2 + \theta_0)\sin(\theta_2 - \theta_0)} H_0. \tag{4.49}$$

The reflection coefficient R can be defined as the ratio of the time average energy flux reflected from the surface to the incident flux. Each of these fluxes is given by the averaged z component of the Poynting vector for the corresponding wave. Combining the expression for the Poynting vector and (4.39) and (4.44) we get [4]

$$R = \frac{|E_1|^2}{|E_0|^2} = \left| \frac{\sqrt{\varepsilon_1^r}\cos\theta_0 - \sqrt{\varepsilon_2^r - \varepsilon_1^r \sin^2\theta_0}}{\sqrt{\varepsilon_1^r}\cos\theta_0 + \sqrt{\varepsilon_2^r - \varepsilon_1^r \sin^2\theta_0}} \right|^2. \tag{4.50}$$

This expression (4.50) simplifies for the case of the normal incidence when $\theta = 0$:

$$R = \left| \frac{\sqrt{\varepsilon_1^r} - \sqrt{\varepsilon_2^r}}{\sqrt{\varepsilon_1^r} + \sqrt{\varepsilon_2^r}} \right|^2. \tag{4.51}$$

Generally, in the transparent media the reflection coefficients $R_\perp$ and $R_\parallel$ for the electric field perpendicular and parallel to the plane of incidence, respectively, have the form

$$R_\perp = \frac{\sin^2(\theta_2 - \theta_0)}{\sin^2(\theta_2 + \theta_0)} \tag{4.52}$$

$$R_\parallel = \frac{\tan^2(\theta_2 - \theta_0)}{\tan^2(\theta_2 + \theta_0)}. \tag{4.53}$$

Suppose now, that the second medium is the absorbing one. In such a case, the refraction index is complex:

$$n_2 = \text{Re} n_2 + i \text{Im} n_2.$$

Then, if the first medium is vacuum with $\varepsilon_1^r = 1$, the reflection coefficient for the normal incidence case is

$$R = \frac{(\text{Re} n_2 - 1)^2 + (\text{Im} n_2)^2}{(\text{Re} n_2 + 1)^2 + (\text{Im} n_2)^2}. \tag{4.54}$$

In the case of the reflection from a transparent medium the coefficients of proportionality between E_0 and $E_{1,2}$ are real. Consequently, the phase of the wave either remains the same, or changes by π which depends on the sign of the coefficients. The phases of the incident and the refracted waves always coincide since the coefficient connecting these waves is positive definite. The factor connecting E_0 and E_1 may be negative, for example, in the case of normal incidence when $\varepsilon_1^r < \varepsilon_2^r$. Then the directions of the vectors E_0 and E_1 are opposite, which corresponds to the phase change by π [4].

Equation (4.53) shows that in the case when the reflected and refracted rays are perpendicular, the reflection coefficient $R_\parallel$ for the wave polarized in the incidence plane vanishes:

$$\theta_2 + \theta_0 = \frac{\pi}{2}, \quad \tan(\theta_2 + \theta_0) \to \infty, \quad R_\parallel \to 0. \tag{4.55}$$

The second equation of the Snellius laws (4.41) yields

$$\tan \theta_0 = \tan \theta_b = \sqrt{\frac{\varepsilon_1^r}{\varepsilon_2^r}}. \tag{4.56}$$

The angle of incidence θ_b is called the angle of total polarization or the Brewster angle. The total polarization effect means that for any polarization direction of light incident at the angle θ_b the reflected light will be polarized in such a way that the electric field is perpendicular to the plane of incidence. It should be noted that a natural light can be totally polarized by reflection, but this effect cannot be produced by refraction.

The reflection and refraction of linearly polarized light always results in linearly polarized light, but the direction of polarization is in general not the same as in the incident light. Let γ_0 be the angle between the direction of polarization E_0 and the plane of incidence, and $\gamma_{1,2}$ the corresponding angles for the reflected and refracted waves. Then using expressions (4.46) and (4.49) we obtain [4]:

$$\tan \gamma_1 = -\frac{\cos(\theta_0 - \theta_2)}{\cos(\theta_0 + \theta_2)} \tan \gamma_0, \quad \tan \gamma_2 = \cos(\theta_0 - \theta_2) \tan \gamma_0. \tag{4.57}$$

The angles $\gamma_{0,1,2}$ are equal for all angles of incidence under following conditions.

4.5 Reflection and Refraction of Electromagnetic Waves

1. $\gamma_0 = 0$ and $\gamma_0 = \pi/2$;
2. the case of the normal incidence $\theta_0 = \theta_2 = 0$;
3. the case of grazing incidence $\theta_0 = \pi/2$ and the refracted wave is absent.

It is seen from relation (4.57) that for $\theta_0 > 0$, $\theta_2 < \pi/2$, $0 < \gamma_0 < \pi/2$, $\gamma_1 > 0$, $\gamma_2 < \pi$ the following inequalities are valid:

$$\gamma_1 > \gamma_0, \quad \gamma_2 < \gamma_0.$$

These inequalities show that the direction of $\boldsymbol{E}$ is turned away from the plane of incidence on reflection, but towards it on refraction. A comparison of relations (4.52) and (4.53) shows that at all angles of incidence, except $\theta_0 = 0$ and $\theta_0 = \pi/2$, $R_\parallel < R_\perp$. Hence, when the incident light is natural the reflected is partly polarized, and the predominant direction of the electric field is perpendicular to the plane of incidence. The refracted light is partly polarized, with the predominant direction of $\boldsymbol{E}$ lying in the plane of incidence.

Consider the dependence of the reflection coefficients $R_\parallel$ and $R_\perp$ on the angle of incidence θ_0. The coefficient $R_\perp$ increases monotonically with the angle θ_0 from the value (4.51) for $\theta_0 = 0$. The coefficient $R_\parallel$ has the same value for $\theta_0 = 0$, but as θ_0 increases $R_\parallel$ decreases to zero at $\theta_0 = \theta_b$ before monotonically increasing. If the reflection is from the optically more dense medium, i.e. $\varepsilon_2^r > \varepsilon_1^r$, then $R_\parallel$ and $R_\perp$ increase to the common value of unity at $\theta_0 = \pi/2$ (grazing incidence). If, on the other hand, the reflecting medium is optically less dense, i.e. $\varepsilon_2^r < \varepsilon_1^r$, both coefficients become equal to unity for $\theta_0 = \theta_r$,

$$\sin\theta_r = \sqrt{\frac{\varepsilon_2^r}{\varepsilon_1^r}} = \frac{n_2}{n_1}, \qquad (4.58)$$

where θ_r is called the angle of total reflection. When $\theta_0 = \theta_r$ the angle of refraction $\theta_2 = \pi/2$, i.e. the refracted wave is propagated along the surface separating the media. Reflection from an optically less dense medium at angles $\theta_0 > \theta_r$ requires special consideration. In this case, as it is seen from (4.40), k_{2z} is purely imaginary, and the field is damped in medium 2. The damping of the wave without true absorption, i.e. dissipation of energy signifies that the average energy flux from medium 1 to medium 2 is zero. That means that all the energy incident on the boundary is reflected back into medium 1, so that the reflection coefficients are $R_\parallel = R_\perp = 1$. This phenomenon is called total reflection [4].

For $\theta_0 > \theta_r$ the proportionality coefficients between $\boldsymbol{E}_1$ and $\boldsymbol{E}_0$ become complex quantities:

$$E_1 = \frac{\sqrt{\varepsilon_1^r}\cos\theta_0 - i\sqrt{\varepsilon_1^r \sin^2\theta_0 - \varepsilon_2^r}}{\sqrt{\varepsilon_1^r}\cos\theta_0 + i\sqrt{\varepsilon_1^r \sin^2\theta_0 - \varepsilon_2^r}} E_0$$

$$H_{0e} = \frac{\varepsilon_2^r \cos\theta_0 - i\sqrt{\varepsilon_1^{r2}\sin^2\theta_0 - \varepsilon_1^r\varepsilon_2^r}}{\varepsilon_2^r \cos\theta_0 + i\sqrt{\varepsilon_1^{r2}\sin^2\theta_0 - \varepsilon_1^r\varepsilon_2^r}} H_{0i}.$$

The quantities $R_\parallel$ and $R_\perp$ are given by the squared moduli of these coefficients, which are equal to unity. In order to calculate the differences in the phases of the waves we write:

$$E_{1\perp} = E_{0\perp}\exp(-i\delta_\perp), \quad E_{1\parallel} = E_{0\parallel}\exp(-i\delta_\parallel).$$

Then it is easy to see that [4]

$$\tan\frac{\delta_\perp}{2} = \frac{\sqrt{\varepsilon_1^r \sin^2\theta_0 - \varepsilon_2^r}}{\sqrt{\varepsilon_1^r}\cos\theta_0} \tag{4.59}$$

$$\tan\frac{\delta_\parallel}{2} = \frac{\sqrt{\varepsilon_1^{r2} \sin^2\theta_0 - \varepsilon_1^r\varepsilon_2^r}}{\varepsilon_2^r \cos\theta_0}. \tag{4.60}$$

The total reflection involves a change in the wave phase which is in general different for the field components parallel and perpendicular to the plane of incidence. Hence, on reflection of a wave polarized in a plane inclined to the plane of incidence, the reflected wave will be elliptically polarized. The phase difference $\delta = \delta_\perp - \delta_\parallel$ is found to be

$$\tan\frac{\delta}{2} = \frac{\cos\theta_0\sqrt{\varepsilon_1^r \sin^2\theta_0 - \varepsilon_2^r}}{\sqrt{\varepsilon_1^r}\sin^2\theta_0}. \tag{4.61}$$

The difference is zero only for $\theta_0 = \theta_r$ or $\theta_0 = \pi/2$.

4.6 Electromagnetic Waves in Anisotropic Media

4.6.1 Plane Wave in an Anisotropic Medium

The properties of an anisotropic medium with respect to electromagnetic waves are defined by the tensors of dielectric and magnetic permeability $\varepsilon_{ik}(\omega)$ and $\mu_{ik}(\omega)$, respectively. They relate the inductions and the fields [4]:

$$D_i = \varepsilon_{ik}(\omega) E_k, \quad B_i = \mu_{ik}(\omega) H_k. \tag{4.62}$$

Here and elsewhere in this section we apply the Einstein rule of the summation over the repeated indices. In the MKS system of units which is used elsewhere in this book both tensors have their dimensionality, farad/m and henry/m, respectively, and incorporate the vacuum factors ε_0 and μ_0. The tensors of the relative dielectric constant (permittivity) $\varepsilon_{ik}^r(\omega)$ and magnetic permeability $\mu_{ik}^r(\omega)$ are dimensionless quantities, they completely characterize the electrodynamic properties of a medium and have the form

$$\varepsilon_{ik}^r(\omega) = \frac{1}{\varepsilon_0}\varepsilon_{ik}(\omega), \quad \mu_{ik}^r(\omega) = \frac{1}{\mu_0}\mu_{ik}(\omega). \tag{4.63}$$

In this section we consider non-magnetic media in the absence of an external magnetic field. Therefore we neglect the relative magnetic permeability tensor limiting with μ_0 and concentrate on the dielectric constant tensor $\varepsilon_{ik}(\omega)$. The medium is also supposed to be transparent in a given range of frequencies, such that losses can be ignored. Using the generalized principle of the symmetry of the kinetic coefficients it can be shown that in the absence of an external magnetic field the tensor $\varepsilon_{ik}(\omega)$ is symmetrical:

$$\varepsilon_{ik}(\omega) = \varepsilon_{ki}(\omega). \tag{4.64}$$

The components of $\varepsilon_{ik}(\omega)$ in a medium without losses are all real, and its principal values are positive.

Consider the field of the monochromatic wave with the frequency ω. In such a wave the temporal dependence of all vectors is described by the exponential phase factor $\exp(\pm i\omega t)$, as it was mentioned above. For the sake of definiteness we choose the sign minus. Substituting this factor into the Maxwell equations we obtain

$$\text{curl}\boldsymbol{E} = i\omega\boldsymbol{B}, \quad \text{curl}\boldsymbol{H} = -i\omega\boldsymbol{D}. \tag{4.65}$$

It is seen from (4.65) that the material equations are essential for the further analysis. In a non-magnetic medium the relation between the magnetic field $\boldsymbol{H}$ and the magnetic induction $\boldsymbol{B}$ remains simple

$$\boldsymbol{B} = \mu_0 \boldsymbol{H}. \tag{4.66}$$

Being multiplied by a scalar the vector of the magnetic field intensity $\boldsymbol{H}$ is parallel to the vector of the magnetic induction $\boldsymbol{B}$: $\boldsymbol{B} \parallel \boldsymbol{H}$. The vector of the electric field $\boldsymbol{E}$ and the vector of the electric induction $\boldsymbol{D}$ in an anisotropic medium have, on the contrary, essentially different directions since they are related by the second rank tensor according to the first equation of (4.62). In a plane wave propagated in a transparent medium all quantities are proportional to $\exp i\boldsymbol{kr}$, with a real vector $\boldsymbol{k}$. Effecting the differentiation with respect to the coordinates, we obtain

$$[\boldsymbol{k} \times \boldsymbol{E}] = \omega\mu_0\boldsymbol{H}, \quad -[\boldsymbol{k} \times \boldsymbol{H}] = \omega\boldsymbol{D}. \tag{4.67}$$

Hence we see, that the vectors $\boldsymbol{k}, \boldsymbol{D}, \boldsymbol{H}$ are mutually perpendicular. The vector $\boldsymbol{H}$ is perpendicular to $\boldsymbol{E}$ and, consequently, the three vectors $\boldsymbol{D}, \boldsymbol{E}, \boldsymbol{k}$, being all perpendicular to $\boldsymbol{H}$, must be coplanar.

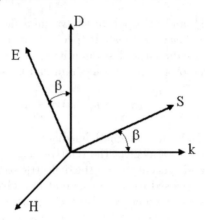

Fig. 4.6. The relative position of the vectors E, D, H, k, S for a plane wave propagated in an anisotropic medium

The relative position of all these vectors is shown in Fig. 4.6. It should be noted that, in contrast to isotropic media, the vectors D and H are transverse with respect to the wave vector k, but the vector E is not. Figure 4.6 also shows the direction of energy flux S in the wave. It is given by the vector product $[E \times H]$, i.e. it is perpendicular to both E and H. The direction of S is not the same as that of k, unlike what happens for an isotropic medium. Clearly the vector S is coplanar with E, D and k, and the angle between S and k is equal to that between E and D. For the further analysis it is useful to define a dimensionless vector n by

$$k = \frac{\omega}{c} n \qquad (4.68)$$

The magnitude of this vector in an anisotropic medium depends on its direction, while in an isotropic medium $n = \sqrt{\varepsilon^r}$ depends only on the frequency due to the dispersion of dielectric constant. Substituting expression (4.68) into (4.67) and taking into account that the vacuum light velocity $c = (\varepsilon_0 \mu_0)^{-1/2}$ we obtain

$$\sqrt{\frac{\mu_0}{\varepsilon_0}} H = [n \times E], \quad -[n \times H] = \frac{1}{\sqrt{\varepsilon_0 \mu_0}} D \qquad (4.69)$$

The energy flux vector in a plane wave is

$$S = [E \times H] = \sqrt{\frac{\varepsilon_0}{\mu_0}} \left[nE^2 - E(n \cdot E) \right] \qquad (4.70)$$

Unlike the situation in an isotropic medium described in the previous section, the scalar product $(n \cdot E)$ does not vanish, since in anisotropic medium n is not perpendicular to E.

4.6 Electromagnetic Waves in Anisotropic Media

Combining (4.69) and the first of Eqs. (4.62) and (4.63) we get three linear homogeneous equations for the three components of $\boldsymbol{E}$ [4]:

$$n^2 E_i - n_i n_k E_k = \varepsilon^r_{ik} E_k, \quad i,k = 1,2,3. \tag{4.71}$$

Here the summation is carried out over the repeated indices. The compatibility condition for these equations is that the determinant of their coefficients should vanish:

$$\left| n^2 \delta_{ik} - n_i n_k - \varepsilon^r_{ik} \right| = 0. \tag{4.72}$$

The determinant (4.72) can be conveniently evaluated by choosing as the axes of x, y, z the principal axes of the tensor ε^r_{ik}, or the so-called the principal dielectric axes. Let the principal values of the tensor be $\varepsilon^r_{xx}, \varepsilon^r_{yy}, \varepsilon^r_{zz}$. Then (4.72) yields

$$n^2 \left(\varepsilon^r_{xx} n_x^2 + \varepsilon^r_{yy} n_y^2 + \varepsilon^r_{zz} n_z^2 \right) - \varepsilon^r_{xx} n_x^2 \left(\varepsilon^r_{yy} + \varepsilon^r_{zz} \right) -$$

$$- \varepsilon^r_{yy} n_y^2 \left(\varepsilon^r_{xx} + \varepsilon^r_{zz} \right) - \varepsilon^r_{zz} n_z^2 \left(\varepsilon^r_{xx} + \varepsilon^r_{yy} \right) + \varepsilon^r_{xx} \varepsilon^r_{yy} \varepsilon^r_{zz} = 0. \tag{4.73}$$

Equation (4.73) is called the Fresnel's equation [4]. It is one of the fundamental equations of crystal optics. It determines implicitly the dispersion relation, i.e. the frequency as a function of the wave vector. In general case, the principal values ε^r_{ii} are themselves the functions of frequency. For waves of a single frequency, however, ω and therefore all the ε^r_{ii}, are usually given constants, and (4.73) gives the magnitude of wave vector as a function of its direction. When the direction of $\boldsymbol{n}$ is given, (4.73) is a quadratic equation, for n^2, with real coefficients. Hence two different magnitudes of the wave vector correspond, in general, to each direction of $\boldsymbol{n}$. Equation (4.73) defines in the coordinates n_x, n_y, n_z the wave-vector surface. In general this is the surface of the fourth order, whose properties for some practically important particular cases will be discussed below.

Consider some general properties of the wave-vector surface. We introduce another quantity characterizing the propagation of light in an anisotropic medium. The direction of the light rays in geometrical optics is given by the group velocity vector $\partial \omega / \partial \boldsymbol{k}$. In an isotropic medium, the direction of this vector is always the same as that of the wave vector, but in an anisotropic medium the two do not coincide in general case. The rays can be characterized by a vector $\boldsymbol{s}$, whose direction is that of the group velocity, while its magnitude is given by

$$\boldsymbol{n} \cdot \boldsymbol{s} = 1. \tag{4.74}$$

The vector $\boldsymbol{s}$ is called the ray vector. In order to clarify its significance consider a beam of rays of a single frequency propagated in all directions. The value of the wave phase at any point is given by the integral $\int \boldsymbol{n} \cdot d\boldsymbol{l}$ taken along the ray. Using the vector s which determines the direction of the ray, we can put

$$\psi = \int \boldsymbol{n} \cdot d\boldsymbol{l} = \int \frac{(\boldsymbol{n} \cdot \boldsymbol{s})}{s} dl = \int \frac{dl}{s}.$$

In a homogeneous medium, s is constant along the ray, so that $\psi = L/s$, where L is the length of the ray segment concerned. We see that, if a segment equal or proportional to s is taken along each ray from the centre, the resulting surface is such that the phase of the rays is the same at every point. This is called the ray surface. The wave-vector surface and the ray surface are in a certain dual relationship. Let the equation of the wave-vector surface be written

$$f(k_x, k_y, k_z, \omega) = 0.$$

Then the components of the group velocity vector are

$$\frac{\partial \omega}{\partial k_i} = -\frac{\partial f}{\partial k_i} \left(\frac{\partial f}{\partial \omega}\right)^{-1}, \tag{4.75}$$

i.e. they are proportional to the derivatives $\partial f/\partial k_i$, or, what is the same thing (since the derivatives are taken for constant ω), to the derivatives $\partial f/\partial n_i$. The components of the ray vector, therefore, are also proportional to $\partial f/\partial n_i$:

$$s_i = \gamma \frac{\partial f}{\partial n_i}, \tag{4.76}$$

where γ is a proportionality factor. Differentiating the left-hand side of (4.73) with respect to n_i and combining the results with (4.74) and (4.76) we obtain:

$$s_x = \frac{n_x \left[\varepsilon^r_{xx}\left(\varepsilon^r_{yy} + \varepsilon^r_{zz}\right) - 2\varepsilon^r_{xx}n_x^2 - n_y^2\left(\varepsilon^r_{xx} + \varepsilon^r_{zz}\right) - n_z^2\left(\varepsilon^r_{xx} + \varepsilon^r_{yy}\right)\right]}{2\varepsilon^r_{xx}\varepsilon^r_{yy}\varepsilon^r_{zz} - \varepsilon^r_{xx}n_x^2\left(\varepsilon^r_{yy} + \varepsilon^r_{zz}\right) - \varepsilon^r_{yy}n_y^2\left(\varepsilon^r_{xx} + \varepsilon^r_{zz}\right) - \varepsilon^r_{zz}n_z^2\left(\varepsilon^r_{xx} + \varepsilon^r_{yy}\right)} \tag{4.77}$$

and similarly for s_y, s_z. The vector $\partial f/\partial \boldsymbol{n}$ is normal to the surface $f = 0$. Thus we conclude that the direction of the ray vector of a wave with given $\boldsymbol{n}$ is that of the normal at the corresponding point of the wave-vector surface. It can be shown that the reverse is also true: the normal to the ray surface gives the direction of the corresponding wave vectors. This relation between the surfaces of $\boldsymbol{n}$ and $\boldsymbol{s}$ can be made more precise. Let $\boldsymbol{n}_0$ be the radius vector of a point on the wave-vector surface, and $\boldsymbol{s}_0$ the corresponding ray vector. The equation in coordinates n_x, n_y, n_z of the tangent plane at this point is

$$\boldsymbol{s}_0 \cdot (\boldsymbol{n} - \boldsymbol{n}_0) = 0,$$

which states that $\boldsymbol{s}_0$ is perpendicular to any vector $\boldsymbol{n} - \boldsymbol{n}_0$ in the plane. Since $\boldsymbol{s}_0 \cdot \boldsymbol{n}_0 = 1$, we can write the equation as

4.6 Electromagnetic Waves in Anisotropic Media

$$s_0 \cdot n = 1.$$

Hence it follows that s_0^{-1} is the length of the perpendicular from the origin to the tangent plane to the wave-vector surface at the point n_0. Conversely, the length of the perpendicular from the origin to the tangent plane to the ray surface at the point s_0 is $1/n_0$.

Consider the location of the ray vector relative to the field vectors in the wave. The group velocity is always in the same direction as the time average energy flux vector. When the wave packet, occupying a small region of space, moves, the energy concentrated in it move must move with it, and the direction of the energy flux is therefore the same as the direction of the velocity of the packet, i.e. the group velocity. Since the Poynting vector is perpendicular to H and E, the same is true of s:

$$s \cdot H = 0, \quad s \cdot E = 0. \tag{4.78}$$

A direct calculation, using expressions (4.69), (4.74) and (4.78) gives

$$H = \frac{1}{\sqrt{\varepsilon_0 \mu_0}} [s \times D], \quad -E = \sqrt{\frac{\mu_0}{\varepsilon_0}} [s \times H]. \tag{4.79}$$

Comparison of (4.69) and (4.79) shows that they differ by the interchange

$$\sqrt{\varepsilon_0} E \text{ and } D/\sqrt{\varepsilon_0}, \; n \text{ and } s, \; \varepsilon^r_{ik} \text{ and } (\varepsilon^r_{ik})^{-1}. \tag{4.80}$$

Obviously, the last of these pairs must be included in order that the first relation (4.62) should remain valid. Thus the following rule may be formulated: an equation valid for one set of quantities can be converted into one valid for another set by means of the interchanges (4.80). The application of this rule to the Fresnel's equation (4.73) gives an analogous equation for s [4]

$$s^2 \left(\varepsilon^r_{yy} \varepsilon^r_{zz} s_x^2 + \varepsilon^r_{xx} \varepsilon^r_{zz} s_y^2 + \varepsilon^r_{xx} \varepsilon^r_{yy} s_z^2 \right) - s_x^2 \left(\varepsilon^r_{yy} + \varepsilon^r_{zz} \right) - s_y^2 \left(\varepsilon^r_{xx} + \varepsilon^r_{zz} \right)$$

$$- s_z^2 \left(\varepsilon^r_{xx} + \varepsilon^r_{yy} \right) + 1 = 0. \tag{4.81}$$

This equation yields the form of the ray surface. When the direction of s is given, (4.81) is a quadratic equation for s^2, which in general has two different real roots. Thus two rays with different wave vectors can be propagated in any direction in the crystal.

Consider now the polarization of waves propagated in an anisotropic medium [4]. Taking into account that the induction D is transverse to the given n in the wave, we use a new coordinate system with one axis in the direction of the wave vector, and denote the transverse axes by α, β which take the values 1,2. Substituting the first equation (4.69) into the second one we obtain

$$\boldsymbol{D} = \varepsilon_0 \left[n^2 \boldsymbol{E} - \boldsymbol{n}\left(\boldsymbol{n}\cdot\boldsymbol{E}\right)\right]. \tag{4.82}$$

The transverse components of $\boldsymbol{D}$ in this new coordinate system are

$$D_\alpha = \varepsilon_0 n^2 E_\alpha.$$

Substituting into the last equation

$$E_\alpha = \frac{1}{\varepsilon_0}\left(\varepsilon^r_{\alpha\beta}\right)^{-1} D_\beta,$$

where $\left(\varepsilon^r_{\alpha\beta}\right)^{-1}$ is a component of the tensor inverse to $\varepsilon^r_{\alpha\beta}$, we have

$$D_\alpha - n^2 \left(\varepsilon^r_{\alpha\beta}\right)^{-1} D_\beta = 0,$$

or

$$\left(\frac{1}{n^2}\delta_{\alpha\beta} - \left(\varepsilon^r_{\alpha\beta}\right)^{-1}\right) D_\beta = 0. \tag{4.83}$$

The condition for the two equations ($\alpha, \beta = 1, 2$) in the two unknowns $D_{1,2}$ to be compatible is that their determinant should be zero. This condition results in the Fresnel's equation mentioned above. However, the vectors $\boldsymbol{D}$ corresponding to the two values of n are along the principal axes of the symmetrical two-dimensional tensor of the second rank $(\varepsilon^r_{\alpha\beta})^{-1}$. It can be shown that these two vectors are perpendicular. Thus, in the two waves with the wave vector in the same direction, the electric induction vectors are linearly polarized in two perpendicular planes. Equations (4.83) have a simple geometrical interpretation. The tensor $(\varepsilon^r_{\alpha\beta})^{-1}$ corresponds to the tensor ellipsoid in the principal dielectric axes which is shown in Fig. 4.7

It is described by the equation

$$\left(\varepsilon^r_{ik}\right)^{-1} x_i x_k = \frac{x^2}{\varepsilon^r_{xx}} + \frac{y^2}{\varepsilon^r_{yy}} + \frac{z^2}{\varepsilon^r_{zz}} = 1. \tag{4.84}$$

Let this ellipsoid be cut by a plane through its centre perpendicular to the given direction $\boldsymbol{n}$. The section is in general an ellipse. The lengths of its axes determine the values of n, and their directions determine the directions of the oscillations, i.e. the vectors $\boldsymbol{D}$. If the vector $\boldsymbol{n}$ lies in one of the coordinate planes, e.g. the xy plane, one of the directions of polarization is also in that plane, and the other is in the z direction.

The polarizations of two waves with the ray vector in the same direction have entirely similar properties. Instead of the directions of the induction $\boldsymbol{D}$, we must now consider those of the vector $\boldsymbol{E}$, which is transverse to the ray vector $\boldsymbol{s}$, and (4.83) should be replaced by the analogous equations

$$\left(\frac{1}{s^2}\delta_{\alpha\beta} - \varepsilon^r_{\alpha\beta}\right) E_\beta = 0. \tag{4.85}$$

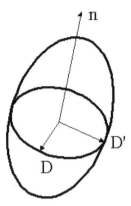

Fig. 4.7. The tensor ellipsoid corresponding to the tensor $(\varepsilon^r_{ik})^{-1}$; n is the wave vector, D, D' are the electric induction vectors

The geometrical construction is here based on the tensor ellipsoid called the Fresnel ellipsoid:

$$\varepsilon^r_{ik} x_i x_k = \varepsilon^r_{xx} x^2 + \varepsilon^r_{yy} y^2 + \varepsilon^r_{zz} z^2 = 1.$$

It should be emphasized that plane waves propagated in anisotropic medium are completely linearly polarized in certain planes. In this respect the optical properties of anisotropic media are essentially different from those of isotropic media. A plane wave propagated in an isotropic medium is in general elliptically polarized, and is linearly polarized only in particularly cases. This important difference arises because the case of complete isotropy of the medium is in a sense one of degeneracy, in which a single wave vector corresponds to two directions of polarization, whereas in an anisotropic media there are in general two different wave vectors in the same direction. The two linearly polarized waves propagated with the same value of n combine to form one elliptically polarized wave.

4.6.2 Optical Properties of Uniaxial Crystals

The optical properties of a crystal depend primarily on the symmetry of its dielectric tensor ε^r_{ik}. In this respect all crystals are divided into three types: cubic, uniaxial and biaxial. In a crystal of the cubic system $\varepsilon^r_{ik} = \varepsilon \delta_{ik}$, i.e. the three principal values of the tensor are equal, and the directions of the principal axes are arbitrary. The optical properties of cubic crystals are the same as those of isotropic media.

Consider the optical properties of uniaxial crystals. Here one of the principal axes of the tensor ε^r_{ik} coincides with the axis of symmetry of the third, fourth or sixth order for the crystals of rhombohedral, tetragonal or hexagonal system, respectively. This axis is called the optical axis of the crystal.

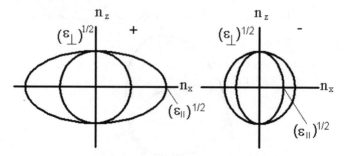

Fig. 4.8. The cross-section of the wave vector surfaces in the uniaxial crystal. The positive uniaxial crystal corresponds to $\varepsilon_\perp < \varepsilon_\parallel$ (the sphere lies inside the ellipsoid). The negative uniaxial crystal corresponds to $\varepsilon_\perp > \varepsilon_\parallel$ (the sphere lies outside the ellipsoid)

In the further analysis we will be chosen to be the z-axis. The corresponding value of ε_{ik}^r will be denoted by $\varepsilon_\parallel$. The directions of the other two principal axes, in a plane perpendicular to the optical axis, are arbitrary, and the corresponding principal values which are denoted by $\varepsilon_\perp$, are equal. Substituting in the Fresnel's equation (4.73) the principal values $\varepsilon_{xx}^r = \varepsilon_{yy}^r = \varepsilon_\perp$, $\varepsilon_{zz}^r = \varepsilon_\parallel$ we obtain

$$\left(n^2 - \varepsilon_\perp\right)\left[\varepsilon_\parallel n_z^2 + \varepsilon_\perp\left(n_x^2 + n_y^2\right) - \varepsilon_\perp \varepsilon_\parallel\right] = 0. \tag{4.86}$$

The quartic equation gives the two quadratic equations

$$n^2 = \varepsilon_\perp \tag{4.87}$$

and

$$\frac{n_z^2}{\varepsilon_\perp} + \frac{n_x^2 + n_y^2}{\varepsilon_\parallel} = 1. \tag{4.88}$$

Geometrically, this signifies that the wave-vector surface, which is in general of the fourth order, becomes two separate surfaces, a sphere and an ellipsoid.

Figure 4.8 shows a cross-section of these surfaces. Two cases are possible: if $\varepsilon_\perp > \varepsilon_\parallel$, the sphere lies outside the ellipsoid, and if $\varepsilon_\perp < \varepsilon_\parallel$, it lies inside. In the first case we speak of a negative uniaxial crystal, and in the second case of a positive one. The two surfaces touch at opposite poles on the n_z-axis. The direction of the optical axis therefore corresponds to only one value of the wave vector.

The ray surface is exactly similar in form. Its equation can be obtained from (4.87) and (4.88) by replacing $\boldsymbol{n}$ by $\boldsymbol{s}$ and ε by $1/\varepsilon$:

$$s^2 = \frac{1}{\varepsilon_\perp}, \quad \varepsilon_\perp s_z^2 + \varepsilon_\parallel\left(s_x^2 + s_y^2\right) = 1. \tag{4.89}$$

We see that two types of wave can be propagated in a uniaxial crystal. With respect to one type, called ordinary waves, the crystal behaves like an isotropic medium of a refractive index $n = \sqrt{\varepsilon_\perp}$. The magnitude of the wave vector is $\omega n/c$ whatever its direction, and the direction of the ray vector is that of $\boldsymbol{n}$. In waves of the second type, called extraordinary waves, the magnitude of the wave vector depends on the angle θ which it makes with the optical axis:

$$\frac{1}{n^2} = \frac{\sin^2\theta}{\varepsilon_\parallel} + \frac{\cos^2\theta}{\varepsilon_\perp}. \tag{4.90}$$

The direction of the ray vector in an extraordinary wave does not coincide with that of the wave vector, but the ray vector is coplanar with the wave vector and the optical axis, their common plane being called the principal section for the given $\boldsymbol{n}$. Let this be the xz plane; then it can be shown that the angle θ' between the ray vector and the optical axis and the angle θ satisfy the simple relation

$$\tan\theta' = \frac{\varepsilon_\perp}{\varepsilon_\parallel}\tan\theta. \tag{4.91}$$

The directions of $\boldsymbol{n}$ and $\boldsymbol{s}$ are the same only for waves propagated along or perpendicular to the optical axis. In the extraordinary wave $\boldsymbol{n}$ and $\boldsymbol{s}$ are not in the same direction, but lie in the same principal section. This wave is therefore polarized so that the vectors $\boldsymbol{E}$ and $\boldsymbol{D}$ lie in the same principal section as $\boldsymbol{n}$ and $\boldsymbol{s}$. The vectors $\boldsymbol{D}$ in the ordinary and extraordinary waves with the same direction of $\boldsymbol{n}$ are perpendicular. Hence the polarization of ordinary wave is such that $\boldsymbol{E}$ and $\boldsymbol{D}$ lie in a plane perpendicular to the principal section. An exception is formed by waves propagated in the direction of the optical axis. In this direction there is no difference between the ordinary and the extraordinary wave, and so their polarizations combine to give a wave which is, in general, elliptically polarized. The refraction of a plane wave incident on the surface of a crystal is different from refraction at a boundary between two isotropic media. The laws of refraction and reflection are again obtained from the continuity of the component n_t tangential to the plane of separation. The wave vectors of the refracted and reflected waves lie in the plane of incidence. In a crystal, however, two different refracted waves are formed, a phenomenon called double refraction. They correspond to the two possible values of the normal component n_n which satisfy the Fresnel's equation for a given tangential component n_t. It should be emphasized that the observed direction of propagation of the rays is determined not by the wave vector, but by the ray vector $\boldsymbol{s}$, whose direction is different from that of $\boldsymbol{n}$ and in general does not lie in the plane of incidence.

4.6.3 Birefringence in an Electric Field

An isotropic body becomes optically anisotropic when placed in a constant electric field. This anisotropy may be regarded as the result of a change in the dielectric constant due to the constant external field. This change usually is very small, but it leads to a qualitative change in the optical properties of bodies. Denote by $\boldsymbol{E}_0$ the constant electric field in the body and expand the dielectric tensor ε^r_{ik} in powers of $\boldsymbol{E}_0$. In an isotropic body in the zero-order approximation, we have

$$\varepsilon^r_{ik} = \varepsilon^r_d \delta_{ik}.$$

There can be no terms in ε^r_{ik} which are of the first order in the field, since in an isotropic body there is no constant vector with which a second rank tensor linear in $\boldsymbol{E}_0$ could be constructed. The next terms in the expansion of ε^r_{ik} must therefore be quadratic in the field. From the components of the vector $\boldsymbol{E}_0$ we can form two symmetrical tensors of rank two, $E_0^2 \delta_{ik}$ and $E_{0i} E_{0k}$. The former does not alter the symmetry of the tensor $\varepsilon^r_d \delta_{ik}$, and the addition of it amounts to a small correction in the scalar constant ε^r_d, which does not result in optical anisotropy and is therefore of no interest. Consequently, the dielectric constant tensor as a function of the field has the form

$$\varepsilon^r_{ik} = \varepsilon^r_d \delta_{ik} + \alpha E_{0i} E_{0k}, \tag{4.92}$$

where α is a scalar constant. One of the principal axes of this tensor coincides with the direction of the electric field, and the corresponding principal value is

$$\varepsilon_\| = \varepsilon^r_d + \alpha E_0^2. \tag{4.93}$$

The other two principal values are both equal to

$$\varepsilon_\perp = \varepsilon^r_d \tag{4.94}$$

and the position of the corresponding principal axes in a plane perpendicular to the field is arbitrary. As a result, an isotropic body in an electric field behaves optically as a uniaxial crystal. This phenomenon is known as the Kerr effect.

The change in optical symmetry in an electric field may occur also in a crystal. An optically uniaxial crystal may become biaxial, and a cubic crystal may cease to be optically isotropic. Here the effect may be of the first order in the field. This linear effect corresponds to a dielectric tensor of the form

$$\varepsilon^r_{ik} = \varepsilon^{(0)}_{ik} + \alpha_{ikl} E_{0l}, \tag{4.95}$$

where the coefficients α_{ikl} form a tensor of rank three symmetrical in the indices i, k ($\alpha_{ikl} = \alpha_{kil}$). The symmetry of this tensor is the same as that of the piezoelectric tensor. The effect in question therefore occurs in the twenty crystal classes which admit piezoelectricity.

4.6.4 Magnetic-Optical Effects

In the presence of a constant magnetic field $\boldsymbol{H}$ the tensor ε_{ik}^r is no longer symmetrical. The generalized principle of symmetry of the kinetic coefficients requires that

$$\varepsilon_{ik}^r(\boldsymbol{H}) = \varepsilon_{ki}^r(-\boldsymbol{H}). \tag{4.96}$$

The condition that absorption is absent requires that the tensor should be Hermitian, but not that it should be real:

$$\varepsilon_{ik}^r = \varepsilon_{ki}^{r*}, \quad \varepsilon_{ik}^r = \varepsilon_{ik}^{r'} + i\varepsilon_{ik}^{r''}, \tag{4.97}$$

where the asterisk stands for the operation of complex conjugation, $\varepsilon_{ik}^{r'}$ and $\varepsilon_{ik}^{r''}$ are the real and imaginary parts of the dielectric constant, respectively. Equations (4.97) show that the real and imaginary parts of ε_{ik}^r must be respectively symmetrical and antisymmetric:

$$\varepsilon_{ik}^{r'} = \varepsilon_{ki}^{r'}, \quad \varepsilon_{ik}^{r''} = -\varepsilon_{ki}^{r''}. \tag{4.98}$$

Using (4.96), we have

$$\varepsilon_{ik}^{r'}(\boldsymbol{H}) = \varepsilon_{ki}^{r'}(\boldsymbol{H}) = \varepsilon_{ik}^{r'}(-\boldsymbol{H})$$

$$\varepsilon_{ik}^{r''}(\boldsymbol{H}) = -\varepsilon_{ki}^{r''}(\boldsymbol{H}) = -\varepsilon_{ik}^{r''}(-\boldsymbol{H}), \tag{4.99}$$

i.e. in a non-absorbing medium $\varepsilon_{ik}^{r'}$ is an even function of $\boldsymbol{H}$, and $\varepsilon_{ik}^{r''}$ is an odd function. The inverse tensor $(\varepsilon_{ik}^r)^{-1}$ has the same symmetry properties, and is more convenient for use in following calculations. To simplify the notation we shall write

$$(\varepsilon_{ik}^r)^{-1} = \eta_{ik} = \eta_{ik}' + i\eta_{ik}''. \tag{4.100}$$

Any antisymmetric tensor of rank two is equivalent to some axial vector; let the vector corresponding to the tensor η_{ik}'' be $\boldsymbol{G}$. Then we can write

$$\eta_{ik}'' = e_{ikl}G_l, \tag{4.101}$$

where e_{ikl} is the antisymmetric unit tensor. In components equation (4.101) has the form

$$\eta_{xy}'' = G_z, \quad \eta_{zx}'' = G_y, \quad \eta_{yz}'' = G_x. \tag{4.102}$$

The relation between the electric field and induction then becomes

$$E_i = \frac{1}{\varepsilon_0}(\eta_{ik}' + ie_{ikl}G_l)D_k = \frac{1}{\varepsilon_0}\{\eta_{ik}'D_k + i[\boldsymbol{D}\times\boldsymbol{G}]_i\}. \tag{4.103}$$

A medium in which the relation between $\boldsymbol{E}$ and $\boldsymbol{D}$ has the form (4.103) is called gyrotropic.

Consider the propagation of waves in an arbitrary anisotropic gyrotropic medium, with no restriction on the magnitude of the magnetic field. We choose the direction of the wave vector as the z-axis. Then (4.83) become

$$\left(\eta_{\alpha\beta} - \frac{1}{n^2}\delta_{\alpha\beta}\right) D_\beta = \left(\eta'_{\alpha\beta} + i\eta''_{\alpha\beta} - \frac{1}{n^2}\delta_{\alpha\beta}\right) D_\beta = 0, \quad (4.104)$$

where the indices α, β take the values x, y. The directions of the x- and y-axes are chosen along the principal axes of the two-dimensional tensor $\eta'_{\alpha\beta}$; we denote the corresponding principal values of this tensor by $1/n_{01}^2$ and $1/n_{02}^2$. Then the Eqs. (4.104) become

$$\left(\frac{1}{n_{01}^2} - \frac{1}{n^2}\right) D_x + iG_z D_y = 0$$

$$-iG_z D_x + \left(\frac{1}{n_{02}^2} - \frac{1}{n^2}\right) D_y = 0. \quad (4.105)$$

The condition that the determinant of (4.105) vanishes yields an equation quadratic in n^2:

$$\left(\frac{1}{n_{01}^2} - \frac{1}{n^2}\right)\left(\frac{1}{n_{02}^2} - \frac{1}{n^2}\right) = G_z^2. \quad (4.106)$$

The roots of (4.106) give two values of n for a given direction of $\boldsymbol{n}$:

$$\frac{1}{n^2} = \frac{1}{2}\left(\frac{1}{n_{01}^2} + \frac{1}{n_{02}^2}\right) \pm \sqrt{\frac{1}{4}\left(\frac{1}{n_{01}^2} - \frac{1}{n_{02}^2}\right)^2 + G_z^2}. \quad (4.107)$$

Substituting the values (4.107) in (4.105), we find the corresponding ratios D_y/D_x:

$$\frac{D_y}{D_x} = \frac{i}{G_z}\left\{\frac{1}{2}\left(\frac{1}{n_{01}^2} - \frac{1}{n_{02}^2}\right) \mp \sqrt{\frac{1}{4}\left(\frac{1}{n_{01}^2} - \frac{1}{n_{02}^2}\right)^2 + G_z^2}\right\}. \quad (4.108)$$

The purely imaginary value of the ratio D_y/D_x means that the waves are elliptically polarized, and the principal axes of the ellipses are the x and y. The product of the two values of the ratio is easily seen to be unity. If in one wave $D_y = i\rho D_x$, where the real quantity ρ is the ratio of the axes of the polarization ellipse, then in the other wave, then in the other wave $D_y = -iD_x/\rho$. This means that the polarization ellipses of the two waves have the same axis ratio, but are 90° apart, and the directions of rotation are opposite which is shown in Fig. 4.9.

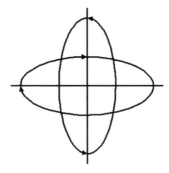

Fig. 4.9. The polarization ellipses of the two waves with the same axis ratio, perpendicular axes and opposite rotation directions

The components G_i and η'_{ik} are functions of the magnetic field. If the magnetic field is weak, we can expand in powers of the field. The vector $\boldsymbol{G}$ is zero in the absence of the field, and so for a weak field we can put

$$G_i = f_{ik} H_k, \qquad (4.109)$$

where f_{ik} is a tensor of rank two, in general not symmetrical. This dependence is in accordance with the general rule (4.99) whereby, in a transparent medium, the components of the antisymmetric tensor η''_{ik} (and $\varepsilon^{r''}_{ik}$) must be odd functions of $\boldsymbol{H}$. The symmetrical components η'_{ik} are even functions of the magnetic field. The first correction terms in η'_{ik} are therefore quadratic in the field.

In the general case of an arbitrary directed wave vector, the magnetic field has little effect on the propagation of light in the crystal, causing only a slight ellipticity of the oscillations, with a small axis ratio of the polarization ellipse (of the first order with respect to the field). The directions of the optical axes and neighboring directions form an exception. The two values of n are equal in the absence of the field when the wave vector is along one of these axes. The roots of (4.106) then differ from these values by first-order quantities, and the resulting effects are analogous to those in isotropic bodies, which we shall now consider.

The magnetic-optical effect in isotropic bodies and in crystals of the cubic system is of particular interest on account of its nature and its comparatively large magnitude. Neglecting second-order quantities, we have

$$\eta'_{ik} = (\varepsilon^r)^{-1} \delta_{ik},$$

where ε^r is the dielectric constant of the isotropic medium in the absence of the magnetic field. The relation between $\boldsymbol{D}$ and $\boldsymbol{E}$ is

$$\boldsymbol{E} = \frac{1}{\varepsilon_0} \left(\frac{1}{\varepsilon^r} \boldsymbol{D} + i\, [\boldsymbol{D} \times \boldsymbol{G}] \right), \qquad (4.110)$$

or

$$\boldsymbol{D} = \varepsilon_0 \left(\varepsilon^r \boldsymbol{E} + i \left[\boldsymbol{E} \times \boldsymbol{g} \right] \right), \tag{4.111}$$

where

$$\boldsymbol{G} = -\frac{\boldsymbol{g}}{(\varepsilon^r)^2}.$$

The dependence of $\boldsymbol{g}$ or $\boldsymbol{G}$ on the external field reduces in an isotropic medium to simple proportionality:

$$\boldsymbol{g} = f\boldsymbol{H},$$

in which the scalar constant f may be either positive or negative. In (4.106) we now have

$$n_{01} = n_{02} = n_0 = \sqrt{\varepsilon^r}.$$

Hence

$$\frac{1}{n^2} = \mp G_z + \frac{1}{n_0^2},$$

or, to the same accuracy,

$$n_\mp^2 = n_0^2 \pm n_0^4 G_z = n_0^2 \mp g_z. \tag{4.112}$$

Since the z-axis is in the direction of $\boldsymbol{n}$, we can write this formula in the vector form

$$\left(\boldsymbol{n} \pm \frac{1}{2n_0} \boldsymbol{g} \right)^2 = n_0^2. \tag{4.113}$$

Hence we see that the wave-vector surface in this case consists of two spheres of radius n_0, whose centres are at distances $\pm g/2n_0$ from the origin in the direction of $\boldsymbol{G}$. A different polarization of the wave corresponds to each of the two values of n. We have

$$D_x = \mp i D_y, \tag{4.114}$$

where the signs correspond to those in (4.112). The equality of the magnitudes of D_x and D_y and their phase difference of $\mp \pi/2$ signify a circular polarization of the wave, with the rotation direction of the vector $\boldsymbol{D}$ respectively anticlockwise and clockwise looking along the wave vector, or, to use the customary expressions, with right-hand and left-hand polarization, respectively. The difference between the refractive indices in the right-hand and left-hand polarized waves has the result that two circularly polarized refracted waves are formed at the surface of a gyrotropic body. This phenomenon is called double circular refraction.

4.6 Electromagnetic Waves in Anisotropic Media

Let a linearly polarized plane wave be incident normally on a slab of thickness l. We take the direction of incidence as the z-axis, and that of the vector $\boldsymbol{E}$ ($\boldsymbol{D}$) in the incident wave as the x-axis. The linear oscillation can be represented as the sum of two circular oscillations with opposite directions of rotation, which are then propagated through the slab with different wave vectors

$$k_\pm = \frac{\omega}{c} n_\pm.$$

Arbitrarily taking the wave amplitude as unity, we have

$$D_x = \frac{1}{2}\left[\exp\left(ik_+z\right) + \exp\left(ik_-z\right)\right], \ D_y = \frac{1}{2}i\left[-\exp\left(ik_+z\right) + \exp\left(ik_-z\right)\right].$$

Introducing the quantities

$$k = \frac{1}{2}(k_+ + k_-), \ \kappa = \frac{1}{2}(k_+ - k_-),$$

we obtain

$$D_x = \frac{1}{2}\exp ikz\left[\exp i\kappa z + \exp\left(-i\kappa z\right)\right] = \exp ikz \cos \kappa z$$

$$D_y = \frac{1}{2}i\exp ikz\left[-\exp i\kappa z + \exp\left(-i\kappa z\right)\right] = \exp ikz \sin \kappa z.$$

When the wave leaves the slab we have

$$\frac{D_y}{D_x} = \tan \kappa l = \tan \frac{l\omega g}{2cn_0}.$$

Since this ratio is real, the wave remains linearly polarized, but the direction of polarization is changed. This is known as the Faraday effect. The angle through which the plane of polarization is rotated is proportional to the path traversed by the wave; the angle per unit length in the direction of the wave vector is $(\omega g/2cn_0)\cos\theta$, where θ is the angle between $\boldsymbol{n}$ and $\boldsymbol{g}$.

It should be noted that, when the direction of the magnetic field is given, the direction of rotation of the plane of polarization with respect to the direction of $\boldsymbol{n}$ is reversed (left-hand becoming right-hand, and vice versa) when the sign of $\boldsymbol{n}$ is changed. If the ray traverses the same path twice, the total rotation of the plane of polarization is therefore double the value resulting from a single traversal.

For $\theta = \pi/2$ which corresponds to the wave vector perpendicular to the magnetic field, the effect linear in the field given by expression (4.112) disappears, according to the general rule that only the component of $\boldsymbol{g}$ in the direction of $\boldsymbol{n}$ affects the propagation of light. For angles θ close to $\pi/2$ we must therefore take account of the terms proportional to the square of the magnetic field, and in particular these terms must be included in the tensor η'_{ik}. By virtue of the axial symmetry about the direction of the field, two principal values of the symmetrical tensor η'_{ik} are equal, as for a uniaxial crystal. We shall take the x-axis in the direction of the field, and denote by $\eta_\parallel$ and $\eta_\perp$ the principal values of η'_{ik} in the directions parallel and perpendicular to the magnetic field. The difference $(\eta_\parallel - \eta_\perp)$ is proportional to H^2.

Let us consider the purely quadratic effect called the Cotton–Mouton effect which occurs when $\boldsymbol{n}$ and $\boldsymbol{g}$ are perpendicular. In (4.105) and (4.106) we have $G_z = 0$, and $1/n_{01}^2, 1/n_{02}^2$ are $\eta_\parallel$ and $\eta_\perp$, respectively. Thus in one wave we have $1/n^2 = \eta_\parallel, D_y = 0$; this wave is linearly polarized, and the vector $\boldsymbol{D}$ is parallel to the x-axis. In the other wave $1/n^2 = \eta_\perp, D_x = 0$, i.e. $\boldsymbol{D}$ is parallel to the y-axis. Let linearly polarized light be incident normally on a slab in a magnetic field parallel to its surface. The two components in the slab with vectors $\boldsymbol{D}$ in the xz and yz planes are propagated with different values of n. Consequently the light leaving the slab is elliptically polarized.

4.7 Electromagnetic Waves in Dispersive Media

4.7.1 Dispersion of the Dielectric Permittivity

The material equations

$$\boldsymbol{D} = \varepsilon_0 \varepsilon^r \boldsymbol{E}, \quad \boldsymbol{B} = \mu_0 \mu^r \boldsymbol{H} \tag{4.115}$$

using the simple proportionality between the field intensities and inductions are not valid at frequencies comparable with eigenfrequencies of the molecular or electronic vibrations which lead to the electric or magnetic polarization of matter. In such a case ε^r and μ^r exhibit dispersion, i.e. the frequency dependence. In general case there also exists the wave vector dependence, or a spatial dispersion. However, it is usually small, being of the order of magnitude of a/λ, or even $(a/\lambda)^2$ for centrosymmetric media, and it manifests in some specific cases that are not considered in this book. Here a is a crystal lattice constant, and λ is a light wavelength. The order of magnitude of frequencies typical for a dispersion appearance depends on the substance concerned and varies widely. It may be entirely different for electric and magnetic phenomena.

Consider rapidly varying electromagnetic fields whose frequencies are not small in comparison with the frequencies which characterize the establishment of the electric and magnetic polarization of the substances concerned. An electromagnetic field variable in time must be variable in space also. For a frequency ω the spatial periodicity is characterized by a wavelength $\lambda \sim c/\omega$. With the increase of a frequency λ may become comparable with the atomic dimensions a. The macroscopic description of the field is thereafter invalid. The frequency range where dispersion phenomena are important and, on the

4.7 Electromagnetic Waves in Dispersive Media

other hand, the macroscopic formulation is still valid can be defined as follows. The most rapid mechanism of establishment of the electric or magnetic polarization in matter is the electronic one. Its relaxation time is of the order of magnitude of the atomic time a/v, where v is the velocity of the electrons in the atom. Since $v \ll c$, even the wavelength $\lambda \sim ac/v$ corresponding to these times is large compared with a. Below we assume that the condition $\lambda \gg a$ is met. However, this condition may not be sufficient: for metals at low temperatures there is a range of frequencies where an electromagnetic wave penetration depth becomes of the same order of magnitude as the mean free path of the conduction electrons, and the macroscopic theory is inapplicable, although the inequality $c/\omega \gg a$ is satisfied.

The formal theory given below is equally applicable to metals and dielectrics. At frequencies corresponding to the motion of electrons within the atoms, or the optical frequencies and at higher frequencies, there is not difference in the properties of metals and dielectrics [4]. The Maxwell's equations remain formally the same in arbitrary variable electromagnetic field. However, these equations need to be completed with the relations between the vectors $\boldsymbol{D}, \boldsymbol{B}$ and $\boldsymbol{E}, \boldsymbol{H}$, respectively. At the high frequencies these relation are totally different from (4.115). The principal property of these relations, i.e. the dependence of $\boldsymbol{D}, \boldsymbol{B}$ only on the values of $\boldsymbol{E}, \boldsymbol{H}$ at the instant considered is no longer valid. They depend in general on the values of $\boldsymbol{E}(t), \boldsymbol{H}(t)$ at every previous moment. This expresses the fact that the establishment of the electric or magnetic polarization of the matter cannot keep up with the change in the electromagnetic field. The frequencies at which dispersion phenomena appear may be different for the electric and magnetic properties of the substance.

We first investigate the dependence $\boldsymbol{D}$ on $\boldsymbol{E}$. We limit our study with the case of small field intensity when the relation between $\boldsymbol{D}$ and $\boldsymbol{E}$ is linear. The linear dependence of $\boldsymbol{D}$ on $\boldsymbol{H}$ can be neglected here as small effect of the order a/λ. The most general linear relation between $\boldsymbol{D}(t)$ and the values of $\boldsymbol{E}(t)$ at all previous instants has the integral form:

$$\boldsymbol{D}(t) = \varepsilon_0 \left[\boldsymbol{E}(t) + \int_0^\infty f(\tau) \boldsymbol{E}(t-\tau) d\tau \right]. \tag{4.116}$$

By analogy with the formula (4.115) we can write the relation (4.116) in the symbolic form

$$\boldsymbol{D} = \varepsilon_0 \bar{\varepsilon}^r \boldsymbol{E},$$

where $\bar{\varepsilon}$ is a linear integral operator transforming $\boldsymbol{E}$ according to (4.116). Any variable field can be expanded into a Fourier series of monochromatic components, in which all quantities depend on time through the factor $\exp(-i\omega t)$. For such fields the relation (4.116) takes the form

$$\boldsymbol{D} = \varepsilon_0 \varepsilon^r(\omega) \boldsymbol{E}, \tag{4.117}$$

where

$$\varepsilon^r(\omega) = 1 + \int_0^\infty f(\tau) \exp i\omega\tau \, d\tau. \tag{4.118}$$

For periodic fields, the dielectric permittivity is a function of the frequency and the properties of the medium. The dependence of ε^r on the frequency is called its dispersion law. The function $\varepsilon^r(\omega)$ is in general complex:

$$\varepsilon^r(\omega) = \varepsilon^{r'}(\omega) + i\varepsilon^{r''}(\omega), \tag{4.119}$$

where $\varepsilon^{r'}$ and $\varepsilon^{r''}$ are its real and imaginary parts, respectively. Relation (4.118) shows that for a real $f(\tau)$

$$\varepsilon^r(-\omega) = \varepsilon^{r*}(\omega), \tag{4.120}$$

where the asterisk means the complex conjugation operation. Separating the real and imaginary parts we find

$$\varepsilon^{r'}(-\omega) = \varepsilon^{r'}(\omega), \ \varepsilon^{r''}(-\omega) = -\varepsilon^{r''}(\omega). \tag{4.121}$$

The real part of the dielectric permittivity is an even function of frequency, while the imaginary part is an odd function of it. For frequencies which are small compared with those at which the dispersion is large the function $\varepsilon^r(\omega)$ can be expanded in powers of ω. The expansion of $\varepsilon^{r'}(\omega)$ includes only even powers, and that of $\varepsilon^{r''}(\omega)$ includes only odd powers. In the limit $\omega \to 0$ the function $\varepsilon^r(\omega)$ in dielectrics tends to the electrostatic dielectric constant $\varepsilon^r(0)$. Therefore, in dielectrics the expansion of $\varepsilon^{r'}(\omega)$ begins with the constant term $\varepsilon^r(0)$, while that of $\varepsilon^{r''}(\omega)$ begins with the term in ω. For metals the function $\varepsilon^r(\omega)$ at low frequencies can be defined in a following way. The Maxwell equation

$$\text{curl}\boldsymbol{H} - \frac{\partial \boldsymbol{D}}{\partial t} = \boldsymbol{j} \tag{4.122}$$

should become in the limit $\omega \to 0$

$$\text{curl}\boldsymbol{H} = \boldsymbol{j} = \sigma \boldsymbol{E}, \tag{4.123}$$

where σ is a conductivity. At high frequencies, on the contrary, only the displacement current $\partial \boldsymbol{D}/\partial t$ is essential. Comparing (4.122) and (4.123) we see that for $\omega \to 0$

$$\frac{\partial \boldsymbol{D}}{\partial t} \to \sigma \boldsymbol{E}. \tag{4.124}$$

In a periodic field

$$\frac{\partial \boldsymbol{D}}{\partial t} = -i\omega\varepsilon_0 \varepsilon^r(\omega) \boldsymbol{E}. \tag{4.125}$$

4.7 Electromagnetic Waves in Dispersive Media

Comparing (4.123) and (4.124) we obtain that at low frequencies in metals the dielectric permittivity has the form:

$$\varepsilon^r(\omega) = i\frac{\sigma}{\omega \varepsilon_0}. \tag{4.126}$$

Equation (4.126) shows that the expansion of $\varepsilon^r(\omega)$ in conductors begins with an imaginary term in ω^{-1} which is expressed in terms of the conductivity for constant currents. The next term is a real constant which for metals does not coincide with the same electrostatic value as it does for dielectrics.

In the opposite limit $\omega \to \infty$, the function $\varepsilon^r(\omega)$ tends to unity, because when the field changes sufficiently rapidly, the polarization processes responsible for the difference between the field $\boldsymbol{E}$ and the induction $\boldsymbol{D}$ cannot occur. We now establish the limiting high frequency form of the function $\varepsilon^r(\omega)$ which is valid for both metals and dielectrics. The field frequency is assumed to be large compared with the frequencies of the motion of electrons in the atoms forming the body. Then we can calculate the polarization of the substance by regarding the electrons as free and neglecting their interaction with one another and with the nuclei of the atoms. The velocities v of the motion of the electrons in the atoms are small compared with the velocity of light. Hence the distances v/ω which they traverse during one period of the electromagnetic wave are small compared with the wavelength $\lambda = c/\omega$, and we can assume the wave field uniform in determining the velocity acquired by an electron in that field. The equation of motion of electron has the form

$$m\frac{d\boldsymbol{v}'}{dt} = e\boldsymbol{E}_0 \exp(-i\omega t), \tag{4.127}$$

where $\boldsymbol{v}'$ is the additional velocity acquired by the electron in the wave field, m, e are the electron mass and charge, respectively. Equation (4.127) yields

$$\boldsymbol{v}' = \frac{ie}{m\omega}\boldsymbol{E}. \tag{4.128}$$

The displacement $\boldsymbol{r}$ of the electron is given by

$$\frac{d\boldsymbol{r}}{dt} = \boldsymbol{v}'$$

and

$$\boldsymbol{r} = -\frac{e}{m\omega^2}\boldsymbol{E}. \tag{4.129}$$

The polarization $\boldsymbol{P}$ of the body is the dipole moment per unit volume:

$$\boldsymbol{P} = \sum e\boldsymbol{r} = -\frac{e^2 N}{m\omega^2}\boldsymbol{E}, \tag{4.130}$$

where N is the number of electrons in all the atoms in unit volume of the substance. Combining the definition of the electric induction and (4.130) we obtain

$$\boldsymbol{D} = \varepsilon_0 \varepsilon^r(\omega) \boldsymbol{E} = \varepsilon_0 \boldsymbol{E} + \boldsymbol{P} \tag{4.131}$$

and finally we have

$$\varepsilon^r(\omega) = 1 - \frac{e^2 N}{\varepsilon_0 m \omega^2} = 1 - \frac{\omega_p^2}{\omega^2}, \tag{4.132}$$

where ω_p is the so-called plasma frequency:

$$\omega_p = \sqrt{\frac{e^2 N}{\varepsilon_0 m}}. \tag{4.133}$$

It should be noted that, unlike the dielectric polarizability, the magnetic susceptibility ceases to have any physical meaning at relatively low frequencies. To take into account the deviation of $\mu^r(\omega)$ from unity would be an unwarranted refinement [4]. It has been shown that the concept of magnetic susceptibility χ can be meaningful only if two inequalities are compatible [4]:

$$l^2 \ll \chi \frac{c^2}{\omega^2}, \ l \gg a, \tag{4.134}$$

where l is the dimension of the sample considered. These inequalities require the dimensions l of the body which are macroscopic. The first condition (4.134) is not fulfilled for the optical frequency range. For such frequencies, the magnetic susceptibility $\chi \sim v^2/c^2$; the optical frequencies themselves are $\omega \sim v/a$. Therefore, the right-hand side of the first inequality (4.134) is $\sim a^2$, and it is incompatible with the second inequality. Thus there is no meaning in using the magnetic susceptibility from optical frequencies onward, and for such phenomena we must put

$$\mu^r = 1, \ \boldsymbol{B} = \mu_0 \boldsymbol{H}.$$

Actually, the same is true for many phenomena even at frequencies well below the optical range.

4.7.2 Field Energy in Dispersive Media

The expression for energy flux density, or the Poynting vector

$$\boldsymbol{S} = [\boldsymbol{E} \times \boldsymbol{H}] \tag{4.135}$$

obtained in previous sections remains valid in variable electromagnetic fields, even if dispersion is present. Indeed, the normal component of the Poynting vector $\boldsymbol{S}$ is continuous at the boundary of a medium and the vacuum outside it due to the continuity of the tangential components of the field intensities $\boldsymbol{E}$ and $\boldsymbol{H}$, and formula (4.135) is valid for vacuum. In a dielectric medium

4.7 Electromagnetic Waves in Dispersive Media

without dispersion, when ε^r and μ^r are real constants, the rate of change of the energy in unit volume of the body is

$$\mathrm{div}\,\boldsymbol{S} = -\left(\boldsymbol{E}\frac{\partial \boldsymbol{D}}{\partial t} + \boldsymbol{H}\frac{\partial \boldsymbol{B}}{\partial t}\right) \tag{4.136}$$

and it can be regarded as the rate of change of the electromagnetic energy

$$W = \frac{1}{2}\left(\varepsilon_0 \varepsilon^r E^2 + \mu_0 \mu^r H^2\right), \tag{4.137}$$

which has an exact thermodynamic significance: it is a difference between the internal energy per unit volume with and without the field, the density and entropy remaining constant. In the general case of arbitrary dispersion the electromagnetic energy cannot be rationally defined as a thermodynamic quantity because a presence of dispersion in general signifies a dissipation of energy, i.e. a dispersive medium is also an absorbing medium. To determine the dissipation we first consider an electromagnetic field of a single frequency. By averaging with respect to time (4.136), we find the steady rate of change of the energy, and this is the mean quantity Q of heat evolved per unit time and volume. It should be noted that the expression (4.136) is quadratic in the fields, and therefore all quantities must be written in real form. We represent real parts of all complex vectors $\boldsymbol{F}$ as

$$\frac{1}{2}\left(\boldsymbol{F} + \boldsymbol{F}^*\right),$$

where the asterisk stands for the complex conjugation operation. On averaging with respect to time the products $\boldsymbol{E}\cdot\boldsymbol{E}, \boldsymbol{E}^*\cdot\boldsymbol{E}^*, \boldsymbol{H}\cdot\boldsymbol{H}, \boldsymbol{H}^*\cdot\boldsymbol{H}^*$, which contain factors $\exp(\mp 2i\omega t)$, give zero, leaving

$$Q = \frac{i\omega}{4}\left[\varepsilon_0\left(\varepsilon^{r*} - \varepsilon^r\right)\boldsymbol{E}\cdot\boldsymbol{E}^* + \mu_0\left(\mu^{r*} - \mu^r\right)\boldsymbol{H}\cdot\boldsymbol{H}^*\right]$$

$$= \frac{\omega}{2}\left[\varepsilon_0\varepsilon^{r''}|\boldsymbol{E}|^2 + \mu_0\mu^{r''}|\boldsymbol{H}|^2\right]. \tag{4.138}$$

The expression (4.138) shows that the absorption (dissipation) of energy is determined by the imaginary parts of ε^r and μ^r. The two terms are called the electric and magnetic losses, respectively. According to the law of increase of entropy the sign of these losses is determinate: the dissipation of energy is accompanied by the evolution of heat, i.e. $Q > 0$. Therefore, the imaginary parts of ε^r and μ^r are always positive:

$$\varepsilon^{r''} > 0, \ \mu^{r''} > 0 \tag{4.139}$$

for all substances and all frequencies. The signs of the real parts of ε^r and μ^r for $\omega \neq 0$ are subject to no physical restrictions.

120 4 Electromagnetic Waves

Any non-steady process in an actual body is to some extent thermodynamically irreversible. The electric and magnetic losses in a variable electromagnetic field always occur to some extent, however slight, and the it has been shown that the functions $\varepsilon^{r''}(\omega)$ and $\mu^{r''}(\omega)$ are not exactly zero for any non-zero frequency [4]. The ranges, in which $\varepsilon^{r''}(\omega)$ and $\mu^{r''}(\omega)$ are very small in comparison with the real parts $\varepsilon^{r'}$ and $\mu^{r'}$, are called the transparency ranges. It is possible to neglect the absorption in these ranges and to introduce the concept of the internal energy of the body in the same sense as in a constant field. To determine this quantity, it is necessary to consider a field whose components have frequencies in a narrow range about some mean value ω_0 since the strict periodicity of a monochromatic wave results in no steady accumulation of electromagnetic energy. The field intensities can be written

$$\boldsymbol{E} = \boldsymbol{E}_0(t)\exp(-i\omega_0 t), \quad \boldsymbol{H} = \boldsymbol{H}_0(t)\exp(-i\omega_0 t), \quad (4.140)$$

where $\boldsymbol{E}_0(t)$ and $\boldsymbol{H}_0(t)$ are functions of time which vary slowly in comparison with the factor $\exp(-i\omega_0 t)$. The real parts of these expressions are to be substituted on the right-hand side of (4.136) and then averaged with respect to time over the period $2\pi/\omega_0$, which is small compared to the time of variation of the functions $\boldsymbol{E}_0(t)$ and $\boldsymbol{H}_0(t)$. The products $\boldsymbol{E} \cdot \partial \boldsymbol{D}/\partial t, \boldsymbol{E}^* \cdot \partial \boldsymbol{D}^*/\partial t, \boldsymbol{H} \cdot \partial \boldsymbol{B}/\partial t$ and $\boldsymbol{H}^* \cdot \partial \boldsymbol{B}^*/\partial t$ vanish when averaged over time, and can be ignored. Then we obtain

$$\frac{1}{4}\left(\boldsymbol{E} \cdot \frac{\partial \boldsymbol{D}^*}{\partial t} + \boldsymbol{E}^* \cdot \frac{\partial \boldsymbol{D}}{\partial t}\right) \quad (4.141)$$

and the similar terms for the magnetic field.

We write the derivative $\partial \boldsymbol{D}/\partial t$ as $\widehat{f}\boldsymbol{E}$, where $\widehat{f}$ is the operator

$$\widehat{f} = \varepsilon_0 \frac{\partial}{\partial t}\widehat{\varepsilon^r}.$$

If $\boldsymbol{E}_0$ were a constant, we should have

$$\widehat{f}\boldsymbol{E} = f(\omega)\boldsymbol{E} = -i\omega\varepsilon_0\varepsilon^r(\omega)\boldsymbol{E}.$$

We expand the function $\boldsymbol{E}_0(t)$ as a series of Fourier components $\boldsymbol{E}_{0\alpha}\exp(-i\alpha t)$, with constant $\boldsymbol{E}_{0\alpha}$. Since $\boldsymbol{E}_0(t)$ varies slowly, the series will include only components with $\alpha \ll \omega_0$. Then we obtain

$$\widehat{f}\boldsymbol{E}_{0\alpha}\exp(-i\alpha t - i\omega_0 t) = f(\alpha + \omega_0)\boldsymbol{E}_{0\alpha}\exp[-i(\alpha + \omega_0)t]$$

$$\approx \left[f(\omega_0) + \alpha\frac{df(\omega_0)}{d\omega_0}\right]\boldsymbol{E}_{0\alpha}\exp[-i(\alpha + \omega_0)t]. \quad (4.142)$$

Summing the Fourier components, we get

$$\widehat{f}\boldsymbol{E}_0(t)\exp(-i\omega_0 t) = \left[f(\omega_0)\boldsymbol{E}_0(t) + i\frac{df(\omega_0)}{d\omega_0}\frac{\partial \boldsymbol{E}_0(t)}{\partial t}\right]\exp(-i\omega_0 t).$$

Omitting the suffix 0 to ω we finally obtain

$$\frac{\partial \boldsymbol{D}}{\partial t} = \varepsilon_0\left[-i\omega\varepsilon^r(\omega)\boldsymbol{E}_0(t) + \frac{d(\omega\varepsilon^r(\omega))}{d\omega}\frac{\partial \boldsymbol{E}_0(t)}{\partial t}\right]\exp(-i\omega t). \qquad (4.143)$$

Substituting the result (4.143) into (4.141), neglecting the imaginary part of $\varepsilon^r(\omega)$ and adding a similar expression involving the magnetic field, we conclude that the steady rate of change of the energy in unit volume has the form

$$\frac{d\overline{U}}{dt} = \frac{d}{dt}\frac{1}{4}\left[\varepsilon_0\frac{d(\omega\varepsilon^r(\omega))}{d\omega}|\boldsymbol{E}|^2 + \mu_0\frac{d(\omega\mu^r(\omega))}{d\omega}|\boldsymbol{H}|^2\right]. \qquad (4.144)$$

In terms of real fields $\mathrm{Re}\,\boldsymbol{E} = (\boldsymbol{E}+\boldsymbol{E}^*)/2$ and $\mathrm{Re}\,\boldsymbol{H} = (\boldsymbol{H}+\boldsymbol{H}^*)/2$ the energy $\overline{U}$ can be written as follows

$$\overline{U} = \frac{1}{2}\left[\varepsilon_0\frac{d(\omega\varepsilon^r(\omega))}{d\omega}\overline{(\mathrm{Re}\,\boldsymbol{E})^2} + \mu_0\frac{d(\omega\mu^r(\omega))}{d\omega}\overline{(\mathrm{Re}\,\boldsymbol{H})^2}\right]. \qquad (4.145)$$

If there is no dispersion, ε^r and μ^r are constants, and expression (4.145) coincides with the mean value of the energy (4.137), as it should.

If the external supply of electromagnetic energy to the body is cut off, the absorption which is always present (even though very small) ultimately converts the energy $\overline{U}$ entirely into heat. Since, by the law of entropy increase, there must be evolution and absorption of heat, we must have $\overline{U} > 0$. It therefore follows that the inequalities

$$\frac{d(\omega\varepsilon^r(\omega))}{d\omega} > 0, \quad \frac{d(\omega\mu^r(\omega))}{d\omega} > 0$$

must hold. In reality, these conditions are necessarily fulfilled, by virtue of more stringent inequalities always satisfied by the functions $\varepsilon^r(\omega)$ and $\mu^r(\omega)$.

4.7.3 The Kramers and Kronig's Relations

The function $f(\tau)$ in (4.116) is finite for all values of τ including zero. For dielectrics it to zero as $\tau \to \infty$. This expresses the fact that the value of $\boldsymbol{D}(t)$ at any moment cannot be appreciably affected by the values of $\boldsymbol{E}(t)$ at remote instants. The physical mechanism underlying the integral relation (4.116) consists in the processes of the establishment of the electric polarization. Hence the range of values in which the function $f(\tau)$ differs considerably from zero is of the order of magnitude of the relaxation time characterizing these processes. The same is also true for metals, the only difference being that the function $f(\tau)-\sigma/\varepsilon_0$, rather than $f(\tau)$ itself, tends to zero as $\tau \to \infty$.

This difference arises because the passage of a steady conduction current leads formally to the presence of an induction

$$D(t) = \sigma \int_0^\infty E(t-\tau)\,d\tau.$$

Previously we defined the function $\varepsilon^r(\omega)$ by (4.118). It is possible to derive some very general relations concerning this function by using the methods of the theory of functions of a complex variable. We regard ω as a complex variable

$$\omega = \omega' + i\omega''$$

and ascertain the properties of the function $\varepsilon^r(\omega)$ in the upper half of the ω-plane. From the definition (4.118) and the properties of $f(\tau)$ mentioned above, it follows that $\varepsilon^r(\omega)$ is a one-valued regular function everywhere in the upper half-plane. When $\omega'' > 0$, the integrand in (4.118) includes the exponentially decreasing factor $\exp(-\omega''\tau)$ and since the function $f(\tau)$ is finite throughout the region of integration, the integral converges. The function $\varepsilon^r(\omega)$ has no singularity on the real axis, except possibly at the origin where for metals it has a simple pole. The conclusion that $\varepsilon^r(\omega)$ is regular in the upper half-plane is a consequence of the causality principle. The integration in expression (4.116) is, on account of this principle, taken only over times previous to t, and the region of integration in expression (4.118) therefore extends from 0 to ∞, rather than from $-\infty$ to ∞. It follows from expression 4.118 that

$$\varepsilon^r(-\omega^*) = \varepsilon^{r*}(\omega). \tag{4.146}$$

In particular, for purely imaginary ω we have

$$\varepsilon^r(i\omega'') = \varepsilon^{r*}(i\omega''), \tag{4.147}$$

i.e. the function $\varepsilon^r(\omega)$ is real on imaginary axis:

$$\mathrm{Im}\,\varepsilon^r(\omega) = 0 \text{ for } \omega = i\omega''. \tag{4.148}$$

It should be emphasized that the property (4.146) expresses the fact that the operator relation

$$D = \hat{\varepsilon} E$$

must give real values of D for real E. If the function $E(t)$ is given by the real expression

$$E(t) = E_0 \exp(-i\omega t) + E_0^* \exp(i\omega^* t) \tag{4.149}$$

4.7 Electromagnetic Waves in Dispersive Media   123

then, applying the operator $\hat{\varepsilon}$ to each term, we have

$$\boldsymbol{D} = \varepsilon_0 \left[\varepsilon^r(\omega) \boldsymbol{E}_0 \exp(-i\omega t) + \varepsilon^r(-\omega^*) \boldsymbol{E}_0^* \exp(i\omega^* t) \right] \quad (4.150)$$

The condition for $\boldsymbol{D}$ to be real immediately yields relation (4.146).

It has been shown above that the imaginary part of $\varepsilon^r(\omega)$ is positive for positive real $\omega = \omega'$, i.e. on the right-hand half of the real axis. Since, according to (4.146),

$$\mathrm{Im}\,\varepsilon^r(-\omega') = -\mathrm{Im}\,\varepsilon^r(\omega')$$

the imaginary part of $\varepsilon^r(\omega)$ is negative on the left-hand half of this axis. Thus

$$\mathrm{Im}\,\varepsilon^r(\omega) \gtrless 0 \text{ for } \omega = \omega' \gtrless 0. \quad (4.151)$$

At $\omega = 0$ $\mathrm{Im}\,\varepsilon^r(\omega)$ changes sign, passing through zero for dielectrics and through infinity for metals. This is the only point on the real axis for which $\mathrm{Im}\,\varepsilon^r(\omega)$ can vanish.

When ω tends to infinity in any manner in the upper half-plane, $\varepsilon^r(\omega)$ tends to unity. This has been already shown for the case where ω tends to infinity along the real axis. The general result is seen from relation (4.118). If $\omega \to \infty$ in such a way that $\omega'' \to \infty$, the integral in (4.118) vanishes because of the factor $\exp(-\omega'' \tau)$ in the integrand, while if ω'' remains finite but $|\omega'| \to \infty$ the integral vanishes because of the rapidly oscillating factor $\exp(i\omega' \tau)$. The above properties of the function $\varepsilon^r(\omega)$ are sufficient to prove the following theorem: the function $\varepsilon^r(\omega)$ does not take real values at any finite point in the upper half-plane except on the imaginary axis, where it decreases monotonically from $\varepsilon^r(0) > 1$ for dielectrics or from ∞ for metals at $\omega = i0$ to 1 at $\omega = i\infty$. In particular, it follows that the function $\varepsilon^r(\omega)$ has no zeroes in the upper half-plane.

The function $\varepsilon^r(\omega)$ satisfies the general relations between the real and imaginary parts of the generalized susceptibility. We derive these relations emphasizing the differences between dielectrics and metals. Let us take some real value ω_0 of ω and integrate the expression

$$(\varepsilon^r - 1)/(\omega - \omega_0)$$

round the contour C shown in Fig. 4.10.

This contour includes the whole of the real axis, indented upwards at the point $\omega = \omega_0 > 0$, and also at the point $\omega = 0$ if the latter is, as in metals, a pole of the function $\varepsilon^r(\omega)$, and is completed by a semicircle of infinite radius. At infinity $\varepsilon^r \to 1$, and the function $(\varepsilon^r - 1)/(\omega - \omega_0)$ therefore tends to zero more rapidly than $1/\omega$. The integral

$$I = \int_C \frac{(\varepsilon^r - 1)}{(\omega - \omega_0)} d\omega \quad (4.152)$$

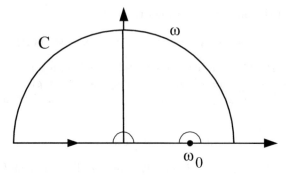

Fig. 4.10. The contour C of integration of the expression $(\varepsilon^r - 1)/(\omega - \omega_0)$

consequently converges. The function $\varepsilon^r(\omega)$ is regular in the upper half-plane, as it has been mentioned above, and the point $\omega = \omega_0$ has been excluded from the region of integration. As a result, the function $(\varepsilon^r - 1)/(\omega - \omega_0)$ is analytic everywhere inside the contour C, and the integral (4.152) is zero. The integral along the semicircle at infinity is also zero. We pass round the point ω_0 along a semicircle whose radius $\rho \to 0$. The direction of integration is clockwise, and the contribution to the integral (4.152) is $-i\pi[\varepsilon^r(\omega_0) - 1]$. If the function $\varepsilon^r(\omega)$ pertains to a dielectric, the indentation at the origin is unnecessary, and we therefore have

$$\lim_{\rho \to 0} \left\{ \int_{-\infty}^{-\rho+\omega_0} \frac{(\varepsilon^r - 1)}{(\omega - \omega_0)} d\omega + \int_{\rho+\omega_0}^{\infty} \frac{(\varepsilon^r - 1)}{(\omega - \omega_0)} d\omega \right\} - i\pi[\varepsilon^r(\omega_0) - 1] = 0. \tag{4.153}$$

The expression in the brackets is the integral from $-\infty$ to ∞, taken as a principal value. Thus we have

$$\text{P} \int_{-\infty}^{\infty} \frac{(\varepsilon^r - 1)}{(\omega - \omega_0)} d\omega - i\pi[\varepsilon^r(\omega_0) - 1] = 0. \tag{4.154}$$

Here the variable of integration ω takes only real values, and P stands for the principal value. We replace the variable ω by x, ω_0 by ω, and write the function of the real variable ω in the form (4.119) $\varepsilon^r(\omega) = \varepsilon^{r'}(\omega) + i\varepsilon^{r''}(\omega)$. Taking the real and imaginary parts of (4.154), we obtain

$$\varepsilon^{r'}(\omega) - 1 = \frac{1}{\pi} \text{P} \int_{-\infty}^{\infty} \frac{\varepsilon^{r''}(x)}{x - \omega} dx \tag{4.155}$$

$$\varepsilon^{r''}(\omega) = -\frac{1}{\pi} \text{P} \int_{-\infty}^{\infty} \frac{\varepsilon^{r'}(x) - 1}{x - \omega} dx \tag{4.156}$$

4.7 Electromagnetic Waves in Dispersive Media

first derived by H.A. Kramers and R. de L. Kronig (1927). We used in the proof the only important property of the function $\varepsilon^r(\omega)$ that it is regular in the upper half-plane. We can say that Kramers and Kronig's relations, like this property of the function $\varepsilon^r(\omega)$, are a direct consequence of the causality principle. Taking into account that $\varepsilon^{r''}(x)$ is an odd function, we can rewrite (4.155)

$$\varepsilon^{r'}(\omega) - 1 = \frac{1}{\pi}P\int_0^\infty \frac{\varepsilon^{r''}(x)}{x-\omega}dx + \frac{1}{\pi}P\int_0^\infty \frac{\varepsilon^{r''}(x)}{x+\omega}dx$$

$$= \frac{2}{\pi}P\int_0^\infty \frac{x\varepsilon^{r''}(x)}{x^2-\omega^2}dx. \tag{4.157}$$

If a metal is concerned, the function $\varepsilon^r(\omega)$ has a pole at the point $\omega = 0$, near which $\varepsilon^r(\omega) = i\sigma/\varepsilon_0\omega$. The passage along a semicircle round this point yields a further real term $-\pi\sigma/\varepsilon_0\omega_0$ which must be added to the left-hand side of (4.154). Then (4.156) takes the form

$$\varepsilon^{r''}(\omega) = -\frac{1}{\pi}P\int_{-\infty}^\infty \frac{\varepsilon^{r'}(x)-1}{x-\omega}dx + \frac{\sigma}{\varepsilon_0\omega}. \tag{4.158}$$

Equations (4.155) and (4.157) remain unchanged. Equation (4.157) is especially important, since it permits the calculation of the function $\varepsilon^{r'}(\omega)$ if the function $\varepsilon^{r''}(\omega)$ is known even approximately (for example, empirically) for a given body. For any function $\varepsilon^{r''}(\omega)$ satisfying the physically necessary condition $\varepsilon^{r''}(\omega) > 0$ for $\omega > 0$, expression (4.157) gives a function $\varepsilon^{r'}(\omega)$ consistent with all physical requirements, i.e. one which is in principle possible. This make it possible to use expression (4.157) even when the function $\varepsilon^{r'}(\omega)$ is approximate. Expression (4.156), on the contrary, does not give a physically possible function $\varepsilon^{r''}(\omega)$ for any arbitrary choice of the function $\varepsilon^{r'}(\omega)$, since the condition $\varepsilon^{r''}(\omega) > 0$ is not necessarily fulfilled.

All above results are applicable to the magnetic permeability $\mu^r(\omega)$. The differences are due to the fact that the function $\mu^r(\omega)$ ceases to be physically meaningful at relatively low frequencies. For example, Kramers and Kronig's relations must be applied to $\mu^r(\omega)$ as follows. We consider a finite range of ω (from 0 to ω_1), which extends only to frequencies where $\mu^r(\omega)$ is still meaningful but no longer variable, so that its imaginary part may be taken as zero. We denote the real quantity $\mu^r(\omega_1)$ by μ_1^r. Then relation (4.157) must be applied to $\mu^r(\omega)$ as follows:

$$\mu^{r'}(\omega) - \mu_1^r = \frac{2}{\pi}P\int_0^{\omega_1} \frac{x\mu^{r''}(x)}{x^2-\omega^2}dx \tag{4.159}$$

Unlike $\varepsilon^r(0)$, the value of $\mu^r(0)$ may be either less than or greater than unity. The variation of $\mu^r(\omega)$ along the imaginary axis is again a monotonic decrease, from $\mu^r(0)$ to $\mu_1^r < \mu^r(0)$.

Let us apply the general expressions derived above to media which absorb slightly in a given range of frequencies, assuming that for these frequencies the imaginary part of the dielectric permittivity can be neglected. In such a case there is no need to take the principal value in expression (4.157), since the point $x = \omega$ does not in practice lie in the region of integration. The integral in expression (4.157) can be differentiated with respect to the parameter ω, giving

$$\frac{d\varepsilon^r(\omega)}{d\omega} = \frac{4\omega}{\pi} \int_0^\infty \frac{x \varepsilon^{r''}(x)}{(x^2 - \omega^2)^2} dx.$$

The integrand is positive throughout the region of integration, and therefore

$$\frac{d\varepsilon^r(\omega)}{d\omega} > 0, \qquad (4.160)$$

i.e. if absorption is absent the dielectric permittivity is a monotonically increasing function of the frequency. In the same frequency range we obtain another inequality

$$\frac{d}{d\omega}\left[\omega^2 \left(\varepsilon^r(\omega) - 1\right)\right] = \frac{4\omega}{\pi} \int_0^\infty \frac{x^3 \varepsilon^{r''}(x)}{(x^2 - \omega^2)^2} dx > 0,$$

or

$$\frac{d\varepsilon^r(\omega)}{d\omega} > \frac{2\left(1 - \varepsilon^r(\omega)\right)}{\omega}. \qquad (4.161)$$

It may be noted that the inequalities (4.160) and (4.161) together with the corresponding ones for $\mu^r(\omega)$ ensure that the inequality $u < c$ is satisfied by the velocity u of propagation of waves. If $\mu^r = 1$ we have $n = \sqrt{\varepsilon^r}$ and replacing $\varepsilon^r(\omega)$ by n^2 in inequalities (4.160) and (4.161) we obtain

$$\frac{d(n\omega)}{d\omega} > n, \quad \frac{d(n\omega)}{d\omega} > \frac{1}{n}. \qquad (4.162)$$

Introducing the group velocity

$$u = \frac{d\omega}{dk} = c\left[\frac{d(n\omega)}{d\omega}\right]^{-1} \qquad (4.163)$$

we obtain two inequalities for it:

$$u < cn \text{ and } u < \frac{c}{n},$$

which means that $u < c$ whether $n < 1$ or $n > 1$. These inequalities show that $u > 0$, i.e. the group velocity is in the same direction as the wave vector.

Suppose that the weak absorption extends over a wide range of frequencies, from ω_1 to $\omega_2 \gg \omega_1$ and consider frequencies ω such that $\omega_1 \ll \omega \ll \omega_2$. Then the region of integration in relation (4.157) divides into two parts: $x < \omega_1$ and $x > \omega_2$. In the former region we can neglect x in comparison with ω, and in the latter region ω in comparison with x. Then we get

$$\varepsilon^r(\omega) = 1 + \frac{2}{\pi} \int_{\omega_2}^{\infty} \varepsilon^{r''}(x) \frac{dx}{x} - \frac{2}{\pi\omega^2} \int_0^{\omega_1} x \varepsilon^{r''}(x)\, dx, \qquad (4.164)$$

i.e. the function $\varepsilon^r(\omega)$ in this range has the form $a - b/\omega^2$, where a and b are positive constants. The constant b can be expressed in terms of the number of dispersion electrons N responsible for the absorption in the range from 0 to ω_1:

$$\varepsilon^r(\omega) = a - \frac{Ne^2}{\varepsilon_0 m \omega^2}. \qquad (4.165)$$

From this expression it follows that when the region of weak absorption is sufficiently wide, the dielectric permittivity in general passes through zero. In this connection it should be recalled that a literally transparent medium is one in which $\varepsilon^r(\omega)$ is not only real but also positive. If $\varepsilon^r(\omega)$ is negative, the wave is damped inside the medium, even though no true dissipation of energy occurs.

For the frequency at which $\varepsilon^r(\omega) = 0$ the induction $\boldsymbol{D} \equiv 0$, and the Maxwell's equations admit a variable electric field satisfying the single equation

$$\mathrm{curl} \boldsymbol{E} = 0,$$

with zero magnetic field, i.e. longitudinal electric waves can occur. To determine their velocity of propagation, we must take into account the dispersion of the dielectric permittivity not only in frequency, but also with respect to the wave vector. In the isotropic medium the frequency for which $\varepsilon^r[\omega(\boldsymbol{k})] = 0$ has the form

$$\omega = \omega_0 + \frac{1}{2}\alpha k^2.$$

Hence the velocity of a longitudinal electric wave propagation is

$$\boldsymbol{u} = \frac{\partial \omega}{\partial \boldsymbol{k}} = \alpha \boldsymbol{k}$$

and is proportional to the wave vector.

4.7.4 Dispersion Relation for a Monochromatic Plane Wave

We have already studied the propagation of a plane monochromatic wave in vacuum and in an anisotropic dielectric medium. Now we consider the dispersion relation of such a wave in a dispersive isotropic medium with absorption. The Maxwell equations remain the same. For a monochromatic wave we have

$$\mathrm{curl}\boldsymbol{E} = i\omega\mu_0\mu^r(\omega)\boldsymbol{H}, \quad i\omega\varepsilon_0\varepsilon^r(\omega)\boldsymbol{E} = -\mathrm{curl}\boldsymbol{H}. \qquad (4.166)$$

Assuming that the medium is homogeneous, eliminating $\boldsymbol{H}$ from (4.166) and using the vector identity

$$\mathrm{curl}\,\mathrm{curl}\boldsymbol{E} = (\mathrm{grad}\,\mathrm{div} - \Delta)\boldsymbol{E}$$

we obtain the second order equation for $\boldsymbol{E}$

$$\Delta\boldsymbol{E} + \varepsilon^r(\omega)\mu^r(\omega)\frac{\omega^2}{c^2}\boldsymbol{E} = 0. \qquad (4.167)$$

Here we have taken into account that in an isotropic homogeneous medium the equation $\mathrm{div}\boldsymbol{D} = 0$ results in $\mathrm{div}\boldsymbol{E} = 0$. Elimination of $\boldsymbol{E}$ gives a similar equation for $\boldsymbol{H}$. In a plane wave in vacuum the space dependence of the field is given by a factor $\exp i\boldsymbol{k}\boldsymbol{r}$ with a real wave vector $\boldsymbol{k}$, as we have shown in the previous section. In considering wave propagation in matter it is in general necessary to take $\boldsymbol{k}$ complex

$$\boldsymbol{k} = \boldsymbol{k}' + i\boldsymbol{k}''. \qquad (4.168)$$

Substituting the factor $\exp i\boldsymbol{k}\boldsymbol{r}$ with the wave vector (4.168) into (4.167) we obtain:

$$k^2 \equiv k'^2 - k''^2 + 2i\boldsymbol{k}'\cdot\boldsymbol{k}'' = \varepsilon^r(\omega)\mu^r(\omega)\frac{\omega^2}{c^2}. \qquad (4.169)$$

The wave vector $\boldsymbol{k}$ can be real only if $\varepsilon^r(\omega)$ and $\mu^r(\omega)$ are real and positive, except that the case when $\boldsymbol{k}'\cdot\boldsymbol{k}'' = 0$. In the situation when the wave vector $\boldsymbol{k}$ is complex, the term "plane wave" is purely conventional. Putting

$$\exp i\boldsymbol{k}\boldsymbol{r} = (\exp i\boldsymbol{k}'\boldsymbol{r})\exp(-\boldsymbol{k}''\boldsymbol{r}) \qquad (4.170)$$

we see that the planes perpendicular to the vector $\boldsymbol{k}'$ are planes of constant phase. The planes of constant amplitude are those perpendicular to the vector $\boldsymbol{k}''$, the direction in which the wave is damped. The surfaces on which the field itself is constant are in general not planes at all. Such waves are called inhomogeneous plane waves.

Consider now the propagation of an electromagnetic wave in an absorbing medium, where the wave vector have a definite direction, i.e. $\boldsymbol{k}'$ and $\boldsymbol{k}''$ are

4.7 Electromagnetic Waves in Dispersive Media

parallel. Then the wave is literally plane, since the surfaces of constant field in it are planes perpendicular to the direction of propagation (a homogeneous plane wave). In this case we introduce the "length" k of the wave vector given by $\boldsymbol{k} = k\boldsymbol{l}$, where $\boldsymbol{l}$ is a unit vector in the direction of $\boldsymbol{k'}$ and $\boldsymbol{k''}$, and from (4.169) we have

$$k = \frac{\omega}{c}\sqrt{\varepsilon^r(\omega)\mu^r(\omega)}. \qquad (4.171)$$

The complex quantity $\sqrt{\varepsilon^r(\omega)\mu^r(\omega)}$ is usually written in the form $n + in'$, with real n and n', so that

$$k = \frac{\omega}{c}\sqrt{\varepsilon^r(\omega)\mu^r(\omega)} = \frac{\omega}{c}(n + in'). \qquad (4.172)$$

The quantity n is the refractive index of the medium, and n' is the absorption coefficient; the latter gives the rate of damping of the wave during its propagation. It should be emphasized that the damping of the wave need not be due to true absorption. Dissipation of energy occurs only when $\varepsilon^r(\omega)$ and $\mu^r(\omega)$ are complex, but n' is different from zero if $\varepsilon^r(\omega)$ and $\mu^r(\omega)$ are real and have opposite sign. Let us express n and n' in terms of the real and imaginary parts of $\varepsilon^r(\omega)$ taking $\mu^r(\omega) = 1$. Then, from (4.172) we get

$$n^2 - (n')^2 + 2inn' = \varepsilon^{r'} + i\varepsilon^{r''}. \qquad (4.173)$$

Equating the real and imaginary parts in both sides of relation (4.173) and solving the corresponding equations we have

$$n = \sqrt{\frac{1}{2}\left\{\varepsilon^{r'} + \sqrt{(\varepsilon^{r'})^2 + (\varepsilon^{r''})^2}\right\}} \qquad (4.174)$$

and

$$n' = \sqrt{\frac{1}{2}\left\{\sqrt{(\varepsilon^{r'})^2 + (\varepsilon^{r''})^2} - \varepsilon^{r'}\right\}}. \qquad (4.175)$$

For metals and in frequency range where formula (4.126) is valid, the imaginary part of $\varepsilon^r(\omega)$ is large compared with its real part, and is related to the conductivity by $\varepsilon^{r''} = \sigma/\varepsilon_0\omega$. Then neglecting $\varepsilon^{r'}$ we find from (4.174) and (4.175) that n and n' are equal:

$$n = n' = \sqrt{\frac{\sigma}{2\varepsilon_0\omega}}. \qquad (4.176)$$

The relation between the fields $\boldsymbol{E}$ and $\boldsymbol{H}$ in this homogeneous plane wave can be established combining relations (4.166), (4.170) and (4.171):

$$\boldsymbol{H} = \sqrt{\frac{\varepsilon_0\varepsilon^r(\omega)}{\mu_0\mu^r(\omega)}}\,[\boldsymbol{l}\times\boldsymbol{E}]. \qquad (4.177)$$

But this time $\varepsilon^r(\omega)$ and $\mu^r(\omega)$ are complex. Equation (4.177) again shows that two fields and the propagation direction are mutually perpendicular.

4.8 Electromagnetic Wave Propagation in Guiding Systems

4.8.1 Hollow Wave-Guides

A waveguide is a hollow or filled with a dielectric material pipe of infinite length. The characteristic oscillations in a waveguide are stationary in the transverse directions, i.e., in a cross-section of the waveguide; the waves travelling along the waveguide can be propagated under some specific conditions [4, 6–8].

Consider a straight waveguide of any simply-connected cross-section uniform along its length and having the perfectly conducting walls. The z-axis is chosen to be along the waveguide. We start the analysis of the electromagnetic wave propagation in the waveguide with the wave equation for any component E_i of the electric field $\boldsymbol{E}$ [4]

$$\Delta E_i + k^2 E_i = 0, \qquad (4.178)$$

where Δ is the Laplacian, $k = \omega/v$, $v = c/\sqrt{\varepsilon^r \mu^r}$ is the light velocity in the medium filling the waveguide, c is the light velocity in vacuum, the temporal dependence of the field is taken in the form $\exp(i\omega t)$, and this factor is omitted in the further calculations. The similar equation is valid for the magnetic field components H_i. Taking, for example, the longitudinal component E_z, we find the solution by virtue of its factorization into two parts depending only on the transverse coordinates and z, respectively. In the Cartesian system of coordinates we have [4, 6–8]

$$E_z(x, y, z) = F(x, y) f(z). \qquad (4.179)$$

Substituting expression (4.179) into (4.178) we obtain:

$$f\left(\nabla_\perp^2 + k^2\right) F + \frac{\partial^2 f}{\partial z^2} F = 0, \qquad (4.180)$$

where $\nabla_\perp^2$ is the transverse part of the Laplacian. The form of the solution (4.179) permits the separation of the functions $F(x, y)$ and $f(z)$. Equation (4.180) takes the form

$$\frac{\left(\nabla_\perp^2 + k^2\right) F}{F} = -\frac{1}{f} \frac{\partial^2 f}{\partial z^2} = \Gamma^2 = \mathrm{const}. \qquad (4.181)$$

Obviously, both sides of (4.181) depend on different variables, and consequently they are constant. Consider first $f(z)$. It is determined by the following equation

$$\frac{\partial^2 f}{\partial z^2} + \Gamma^2 f = 0, \qquad (4.182)$$

4.8 Electromagnetic Wave Propagation in Guiding Systems

which yields the solution in the form of the running waves that is appropriate for the waveguide:

$$f = A_1 \exp(-i\Gamma z) + A_2 \exp i\Gamma z. \tag{4.183}$$

For the sake of definiteness, we choose $A_2 = 0$. Clearly, all field components have the same structure, and we write [4,6,7]

$$\boldsymbol{E} = \boldsymbol{E}_0(x,y)\exp(-i\Gamma z), \ \boldsymbol{H} = \boldsymbol{H}_0(x,y)\exp(-i\Gamma z). \tag{4.184}$$

The distribution of the complex amplitudes $\boldsymbol{E}_0(x,y)$ and $\boldsymbol{H}_0(x,y)$ in the cross-section $z = 0$ remains constant along the waveguide, while the factor $\exp(-i\Gamma z)$ with the real propagation constant Γ corresponds to an undamped wave which is propagated in the positive z direction. Independently of the explicit form of the factor $f(z)$ (4.183) equations for the amplitudes $\boldsymbol{E}_0(x,y)$ and $\boldsymbol{H}_0(x,y)$ have the form:

$$(\nabla_\perp^2 + \kappa^2)\boldsymbol{E}_0(x,y) = 0, \ (\nabla_\perp^2 + \kappa^2)\boldsymbol{H}_0(x,y) = 0, \tag{4.185}$$

where the so-called transverse wave number κ has the form:

$$\kappa^2 = k^2 - \Gamma^2 \tag{4.186}$$

The propagation constant Γ remains constant in different media inside and outside of the waveguide volume due to the boundary conditions. Consequently the transverse wave number κ depends on the properties of a medium.

Now, having clarified the general structure of the field in the waveguide, we return to the Maxwell equations in order to express the transverse components of the electric and magnetic field in terms of the longitudinal ones. The electromagnetic waves possible in the waveguide are divided into two types [4,6,7].

1. The magnetic-type waves (H-waves), or transverse electric (TE) waves where the longitudinal component of the electric field is absent: $E_z = 0$.
2. The electric-type waves (E-waves), or transverse magnetic (TM) waves where the longitudinal component of the magnetic field is absent: $H_z = 0$.

Using the Maxwell equations and the dependence (4.184) we obtain for the TE waves [4, 6–8]:

$$E_x = -\frac{i\omega\mu_0\mu^r}{\kappa^2}\frac{\partial H_z}{\partial y}, \ E_y = \frac{i\omega\mu_0\mu^r}{\kappa^2}\frac{\partial H_z}{\partial x} \tag{4.187}$$

$$H_x = -\frac{i\Gamma}{\kappa^2}\frac{\partial H_z}{\partial x}, \ H_y = -\frac{i\Gamma}{\kappa^2}\frac{\partial H_z}{\partial y} \tag{4.188}$$

and for the TM waves:

$$E_x = -\frac{i\Gamma}{\kappa^2}\frac{\partial E_z}{\partial x}, \quad E_y = -\frac{i\Gamma}{\kappa^2}\frac{\partial E_z}{\partial y} \tag{4.189}$$

$$H_x = \frac{i\omega\varepsilon_0\varepsilon^r}{\kappa^2}\frac{\partial E_z}{\partial y}, \quad H_y = -\frac{i\omega\varepsilon_0\varepsilon^r}{\kappa^2}\frac{\partial E_z}{\partial x}. \tag{4.190}$$

It should be noted that the purely transverse electromagnetic (TEM) wave, where $H_z = 0$ and $E_z = 0$ belongs to the particular class of the plane waves, and it cannot exist for $\kappa^2 \neq 0$. In such a case all components of both electric and magnetic field vanish, as it is seen from (4.187)–(4.189). When $\kappa^2 = 0$, (4.178) reduces to the Laplace equation describing an electrostatic field that cannot exist inside a conductor. Consequently, the existence of TEM waves in hollow waveguides with the conducting walls is impossible.

Generally, the problem of calculation of the TE or the TM waves consists of two stages [4, 8].

1. First, it is necessary to find the longitudinal components of the field, i.e., to solve in the case of TE waves the equation

$$\nabla_\perp^2 H_z + \kappa^2 H_z = 0 \tag{4.191}$$

with the boundary condition on the circumference of the cross-section $\partial H/\partial n = 0$, and in the case of TM waves the equation

$$\nabla_\perp^2 E_z + \kappa^2 E_z = 0 \tag{4.192}$$

with the boundary condition on the circumference of the cross-section $E_z = 0$.

2. Secondly, one calculates the transverse components of the TE waves according to (4.187) and (4.188) using the result of (4.191), and the transverse components of the TM waves according to (4.189) and (4.190) using the result of (4.192).

For a given cross-section, solutions of (4.191) and (4.192) exist only for certain definite eigenvalues of the parameter κ^2. For each eigenvalue κ^2 we have the relation between the frequency ω and the propagation constant Γ

$$\omega^2 = v^2\left(\Gamma^2 + \kappa^2\right). \tag{4.193}$$

The group velocity of propagation of the wave along the waveguide is given by the following expression:

$$v_{gr} = \frac{\partial\omega}{\partial\Gamma} = \frac{v\Gamma}{\sqrt{\Gamma^2+\kappa^2}} = \frac{v^2\Gamma}{\omega}. \tag{4.194}$$

4.8 Electromagnetic Wave Propagation in Guiding Systems

It is seen from (4.194) that for a given value of κ the group velocity v_{gr} varies from 0 to v as Γ varies from 0 to ∞. The phase velocity v_f has the form:

$$v_f = \frac{\omega}{\Gamma} = \frac{v}{\sqrt{1 - \kappa^2/k^2}} = \frac{v}{\sqrt{1 - \omega_{min}^2/\omega^2}}. \tag{4.195}$$

Comparison of (4.194) and (4.195) shows that, if the waveguide is filled with the dielectric, $v_f v_{gr} = c^2/(\varepsilon^r \mu^r)$, or $v_f v_{gr} = c^2$ in the hollow waveguide. It is seen from (4.193) and (4.195) that for the smallest eigenvalue κ_{min} which is not zero there exist the minimal frequency $\omega_{min} = v\kappa_{min}$ such that for $\omega > \omega_{min}$ the TE or TM wave propagates along the waveguide without damping. At frequencies lower than ω_{min} the propagation of the waves of each type is impossible. The order of magnitude of ω_{min} is v/a where a is the transverse dimension of the waveguide. For frequencies $\omega < \omega_{min}$ the propagation constant Γ becomes purely imaginary, the field components are proportional to the factor $\exp(-|\Gamma|z)$ being attenuated with the waveguide length, and the energy transfer through the waveguide is stopped. Such a regime is called a cut-off one.

Consider the energetic characteristics of the wave propagation in a waveguide [4, 6, 7]. The time average energy flux density along the waveguide is given by the z component of the Poynting vector. For the magnetic-type (TE) and electric-type (TM) waves we find using (4.187), (4.188) and (4.189), (4.190), respectively:

$$S_{zTE} = \frac{\mu_0 \mu^r \omega \Gamma}{2\kappa^4} |\text{grad}_\perp H_z|^2, \quad S_{zTM} = \frac{\varepsilon_0 \varepsilon^r \omega \Gamma}{2\kappa^4} |\text{grad}_\perp E_z|^2. \tag{4.196}$$

The total energy flux Q is obtained by integrating S_{zTE} or S_{zTM} over the cross-section s of the waveguide. For TE wave we have

$$Q_{TE} = \frac{\mu_0 \mu^r \omega \Gamma}{2\kappa^4} \int_s |\text{grad}_\perp H_z|^2 \, ds = \frac{\mu_0 \mu^r \omega \Gamma}{2\kappa^4} \left\{ \oint H_z^* \frac{\partial H_z}{\partial n} dl - \int_s H_z^* \nabla_\perp^2 H_z \, ds \right\}. \tag{4.197}$$

The first integral in the right-hand side of (4.197) vanishes due to the zero boundary conditions for $\partial H_z/\partial n$ on the circumference of the cross-section. In the integrand of the second integral we change $\nabla_\perp^2 H_z$ with $-\kappa^2 H_z$ according to (4.191). Then we get

$$Q_{TE} = \frac{\mu_0 \mu^r \omega \Gamma}{2\kappa^2} \int_s |H_z|^2 \, ds. \tag{4.198}$$

Similar calculations yield for the energy flux of the TM wave the following expression:

$$Q_{TM} = \frac{\varepsilon_0 \varepsilon^r \omega \Gamma}{2\kappa^2} \int_s |E_z|^2 \, ds. \tag{4.199}$$

The electromagnetic energy density per unit length of the waveguide

$$W = Q/v_{\text{gr}}.$$

Combining (4.198), (4.199) and (4.194) we immediately get for both types of wave:

$$W_{\text{TE}} = \frac{\mu_0 \mu^r \omega^2}{2\kappa^2 v^2} \int_s |H_z|^2 \, ds, \quad W_{\text{TM}} = \frac{\varepsilon_0 \varepsilon^r \omega^2}{2\kappa^2 v^2} \int_s |E_z|^2 \, ds. \tag{4.200}$$

4.8.2 Rectangular Waveguides

Until now we have not used any additional assumptions on form of the waveguide cross-section and a medium that filling it. For the sake of definiteness we consider the practically important case of a hollow waveguide with a rectangular cross-section and ideally conducting walls [4, 8].

The geometry of the problem is shown in Fig. 4.11. We start the analysis with the case of the electric, or TM waves. For this purpose we solve (4.192) with the transverse Laplacian

$$\nabla_\perp^2 = \frac{\partial^2}{\partial x^2} + \frac{\partial^2}{\partial y^2} \tag{4.201}$$

in the area $0 < x < a$, $0 < y < b$ and with the following boundary conditions:

$$E_z = 0: \quad x = 0, \ x = a, \ y = 0, \ y = b. \tag{4.202}$$

The well-known solution of this problem has the form:

$$E_{zmn} = E_0 \sin \frac{m\pi x}{a} \sin \frac{n\pi y}{b}, \quad m, n = 1, 2, \ldots. \tag{4.203}$$

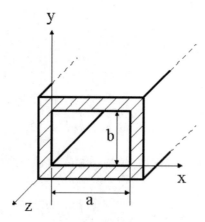

Fig. 4.11. A cross-section of a hollow rectangular waveguide

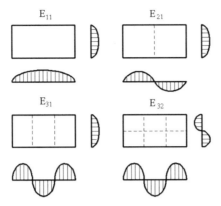

Fig. 4.12. The distribution of the longitudinal component of the TM wave electric field in the cross-section of the rectangular hollow waveguide for the waves $E_{11}, E_{21}, E_{31}, E_{32}$

These eigenfunctions of the problem have the eigenvalues κ_{mn} that obey the condition:

$$\kappa_{mn}^2 = \left(\frac{m\pi}{a}\right)^2 + \left(\frac{n\pi}{b}\right)^2. \tag{4.204}$$

The field distributions in the waveguide cross-section for the waves E_{11}, E_{21}, E_{31} and E_{32} are presented in Fig. 4.12.

Each solution (4.203) is defined with the accuracy to the complex factor E_0. It provides the distribution of the longitudinal component of the vector $\mathbf{E}$ in the waveguide cross-section. The solution of the Maxwell equations corresponding to the eigenvalues m, n is called the wave of the type E_{mn}. The complete field of the running wave of the type E_{mn} is determined by (4.189), (4.190). We write the complete solution explicitly [6–8]:

$$\boldsymbol{E}_{mn} = -E_0 \frac{i\Gamma}{\kappa^2} \{\boldsymbol{x}_0 \kappa_x \cos \kappa_x x \sin \kappa_y y + \boldsymbol{y}_0 \kappa_y \sin \kappa_x x \cos \kappa_y y\} \exp(-i\Gamma z)$$

$$+ E_0 \boldsymbol{z}_0 \sin \kappa_x x \sin \kappa_y y \exp(-i\Gamma z) \tag{4.205}$$

$$\boldsymbol{H}_{mn} = \frac{iE_0 \Gamma}{\kappa^2 w^E} \{\boldsymbol{x}_0 \kappa_y \sin \kappa_x x \cos \kappa_y y - \boldsymbol{y}_0 \kappa_x \cos \kappa_x x \sin \kappa_y y\} \exp(-i\Gamma z), \tag{4.206}$$

where $\boldsymbol{x}_0, \boldsymbol{y}_0$ and $\boldsymbol{z}_0$ are the unit vectors of the corresponding axes,

$$w^E = \frac{\Gamma}{\omega \varepsilon_0}, \quad \kappa_x = \frac{m\pi}{a}, \quad \kappa_y = \frac{n\pi}{b}, \quad \kappa^2 = \kappa_x^2 + \kappa_y^2$$

$$\Gamma = \sqrt{k^2 - \kappa^2} = \sqrt{\frac{\omega^2}{c^2} - \left(\frac{m\pi}{a}\right)^2 - \left(\frac{n\pi}{b}\right)^2}.$$

The period of the field in the z direction Λ is defined as follows

$$\Lambda = \frac{\lambda}{\sqrt{1 - \omega_{\min}^2/\omega^2}}, \qquad (4.207)$$

where $\lambda = 2\pi c/\omega$ is the wavelength of a TEM wave in vacuum.

The indices m and n determine the number of the semi-periods of the structure along the axes x and y, respectively, in the intervals $(0, a)$ and $(0, b)$. The simplest type of the field is the E_{11}. Its structure consists of one set of the closed lines of the vector $\boldsymbol{H}$ that are symmetric with respect to the middle point of the cross-section where the maximum of E_z is situated. More complicated structures are built of these elementary units.

Consider the second type of waves, namely, the magnetic-type, or TE waves. They possess the longitudinal component of the field H_z that obeys (4.191) with the Laplacian (4.201) and the following boundary conditions

$$\frac{\partial H_z}{\partial x} = 0: \; x = 0, \; x = a; \quad \frac{\partial H_z}{\partial y} = 0: \; y = 0, \; y = b. \qquad (4.208)$$

The solution has the form

$$H_{zmn} = H_0 \cos\frac{m\pi x}{a} \cos\frac{n\pi y}{b}, \quad m, n = 0, 1, 2, \dots . \qquad (4.209)$$

Unlike TM waves, the eigenvalues of the eigenfunctions (4.209) can be equal to zero, although not simultaneously. Each function (4.209) gives the distribution of the longitudinal component H_z in the cross-section of the waveguide. Some of these distributions are shown in Fig. 4.13.

The total field of the magnetic-type running wave is calculated by virtue of (4.187) and (4.188), and it has the form [4, 8]

$$\boldsymbol{E}_{mn} = \frac{w^H i H_0 \Gamma}{\kappa^2} \{\boldsymbol{x}_0 \kappa_y \cos\kappa_x x \sin\kappa_y y - \boldsymbol{y}_0 \kappa_x \sin\kappa_x x \cos\kappa_y y\} \exp(-i\Gamma z) \qquad (4.210)$$

$$\boldsymbol{H}_{mn} = \frac{i H_0 \Gamma}{\kappa^2} [\boldsymbol{x}_0 \kappa_x \sin\kappa_x x \cos\kappa_y y + \boldsymbol{y}_0 \kappa_y \cos\kappa_x x \sin\kappa_y y] \exp(-i\Gamma z)$$

$$+ H_0 \boldsymbol{z}_0 \cos\kappa_x x \cos\kappa_y y \exp(-i\Gamma z), \qquad (4.211)$$

where $w^H = \omega\mu_0/\Gamma$.

The minimal frequency and the corresponding maximal wavelength in the case considered have the form:

4.8 Electromagnetic Wave Propagation in Guiding Systems

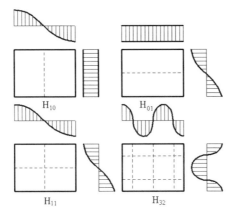

Fig. 4.13. The distribution of the TE wave longitudinal magnetic field in the cross-section of the rectangular hollow waveguide for the waves $H_{z10}, H_{z01}, H_{z11}$ and H_{z32}

$$\omega_{\min}^{mn} = \pi c \sqrt{\frac{m^2}{a^2} + \frac{n^2}{b^2}}, \quad \lambda_{\mathrm{cr}}^{mn} = 2\left(\sqrt{\frac{m^2}{a^2} + \frac{n^2}{b^2}}\right)^{-1}. \quad (4.212)$$

Any field of the type E_{mn} and H_{mn} cannot propagate along the waveguide, if

$$\omega < \omega_{\min}^{mn}, \text{ or } \lambda > \lambda_{\mathrm{cr}}^{mn}, \quad (4.213)$$

since in such a case the propagation constant Γ is a purely imaginary quantity. The energy transfer is forbidden, if the inequality (4.213) is valid for all types of the field. With the increase of ω this inequality first breaks for only one type of the wave: $\omega > \min(\omega_{\min}^{mn})$. The only propagated wave in such a case is called a principal wave. Let for the sake of definiteness $a > b$. Then the principal wave is the eigenfunction H_{10}, and

$$\min(\omega_{\min}^{mn}) = \omega_{\min}^{10} = \frac{c\pi}{a}, \quad \max(\lambda_{\mathrm{cr}}^{mn}) = 2a. \quad (4.214)$$

It should be noted that the number of propagated waves remains finite, since $\omega_{\min}^{mn}$ increases infinitely with the increase of m, n.

The fields of the principal wave have the form

$$\boldsymbol{E}_{10} = -\boldsymbol{y}_0 \frac{iH_0 w_{10}^H \Gamma_{10}}{\kappa_{10}} \sin\frac{\pi x}{a} \exp(-i\Gamma_{10} z) \quad (4.215)$$

$$\boldsymbol{H}_{10} = \frac{iH_0 \Gamma_{10}}{\kappa_{10}} \left[\boldsymbol{x}_0 \sin\frac{\pi x}{a} - \boldsymbol{z}_0 \frac{i\pi}{\Gamma_{10} a} \cos\frac{\pi x}{a}\right] \exp(-i\Gamma_{10} z), \quad (4.216)$$

where

$$\Gamma_{10} = k\sqrt{1 - \left(\frac{\lambda}{2a}\right)^2}.$$

The quantities Γ, v_f, v_{gr} and Λ appeared to be the same for any b since the critical wavelength $\lambda_{cr}^{10} = 2a$ and does not depend on b. The energy flux of the principal wave is

$$\mathcal{Q}_{TE10} = \frac{\mu_0 |H_0|^2 \Gamma_{10}\omega}{2\kappa_{10}^2} \int_0^a \int_0^b \sin^2 \frac{\pi x}{a} dx dy = \frac{ab\mu_0 |H_0|^2 \Gamma_{10}\omega}{4\kappa_{10}^2}. \qquad (4.217)$$

It is seen from (4.217) that the energy flux is determined by the amplitude of the transverse magnetic field component or, which is the same, by the amplitude of the electric field at the crest point $x = a/2$. The expressions (4.216) and (4.217) show that at the critical frequency ω_{min}^{10} magnetic field becomes purely longitudinal, while the longitudinal component of the Poynting vector vanishes.

4.8.3 Energy Absorption in Waveguides

Consider briefly the problem of energy absorption in waveguides [4, 6, 7]. The different mechanisms of losses are possible: the finite absorption of a dielectric material in a dielectric filled waveguide, small but finite impedance of the walls, etc. In any case the propagation constant of the waveguide with absorption is a complex quantity:

$$\Gamma = \Gamma' - i\Gamma'', \qquad (4.218)$$

where Γ' and Γ'' are real numbers. The analytical expression of the imaginary part Γ'' in many practically important cases in unknown. It is typical for hollow waveguides when we account for the conducting walls impedance. Comparison of (4.184), (4.198), (4.199) and (4.218) shows that the energy flux $\mathcal{Q}$ is changing along the z direction according to the following expression:

$$\mathcal{Q} = \mathcal{Q}(0) \exp\left(-2\Gamma''z\right). \qquad (4.219)$$

The energy flux decrease $\Delta \mathcal{Q}$ after the traversing of the small distance along the waveguide Δz is therefore

$$\Delta \mathcal{Q} = -\frac{d\mathcal{Q}}{dz}\Delta z + ... = 2\Gamma''\mathcal{Q}\Delta z + ..., \qquad (4.220)$$

where the small terms of higher order are neglected. Similarly, the losses $\mathcal{Q}_{loss}$ can be written as follows:

$$\Delta \mathcal{Q}_{loss} = \frac{d\mathcal{Q}_{loss}}{dz}\Delta z + ... \qquad (4.221)$$

4.8 Electromagnetic Wave Propagation in Guiding Systems

Relations (4.220) and (4.221) become exact when $\Delta z \to 0$. According to the law of the conservation of energy $\Delta Q = \Delta Q_{\text{loss}}$. Then, we get

$$\Gamma'' = \frac{1}{2Q} \frac{dQ_{\text{loss}}}{dz}. \tag{4.222}$$

The denominator of the right-hand side of (4.222) is calculated by virtue of expressions (4.198) or (4.199) depending on the type of waves. The numerator is expressed in terms of the imaginary parts of dielectric constant and magnetic permeability:

$$\frac{dQ_{\text{loss}}}{dz} = \frac{\omega}{2} \int_s \left[\varepsilon_0 \varepsilon^{r''} |\boldsymbol{E}|^2 + \mu_0 \mu^{r''} |\boldsymbol{H}|^2 \right] ds, \tag{4.223}$$

where s is the cross-section area of a waveguide. Usually the losses in the dielectric and conducting walls are evaluated separately:

$$\Gamma'' = \Gamma''_d + \Gamma''_w. \tag{4.224}$$

The latter term can be expressed in terms of the walls conductivity σ and a penetration depth d_0, i.e. the distance in an absorbing medium at which the wave amplitude is decreased by e.

$$\Gamma''_w = \frac{1}{4\sigma d_0 Q} \int_{L_\perp} |\boldsymbol{H}|^2 \, dl, \tag{4.225}$$

where $L_\perp$ is the contour of the waveguide cross-section.

We evaluate, for example, the attenuation of the H_{10} in the hollow waveguide with the rectangular cross-section due to the finite conductivity of the walls σ. Using (4.216) we first find the numerator of relation (4.225):

$$\int_{L_\perp} |\boldsymbol{H}|^2 \, dl = 2 \int_0^b (H_{10z}^2)_{x=0} \, dy + 2 \int_0^a (H_{10z}^2 + H_{10x}^2)_{y=0} \, dx$$

$$= 2 \frac{|H_0|^2 \Gamma_{10}^2}{\kappa_{10}^2 \left[1 - (\lambda/2a)^2\right]} \left[\left(\frac{\lambda}{2a}\right)^2 b + \frac{a}{2} \right]. \tag{4.226}$$

Substituting expressions (4.226) and (4.217) into (4.225) we finally obtain:

$$\Gamma''_w = \frac{1}{120\pi\sigma d_0 b \sqrt{1 - (\lambda/2a)^2}} \left[\left(\frac{\lambda}{2a}\right)^2 \frac{2b}{a} + 1 \right]. \tag{4.227}$$

The analysis of expression (4.227) shows that at $\omega \to \omega_{\min}^{10}$ it becomes meaningless since in fact the attenuation constant Γ''_w remains finite. It has a minimum. The further increase of Γ_w''' is due to the decrease of the penetration depth

$$d_0 = \sqrt{2/(\omega\mu_0\sigma)}. \tag{4.228}$$

It should be also noted that the attenuation constant Γ_w''' is approximately proportional to the inverse dimension of the waveguide cross-section $1/b$.

4.8.4 Optical Dielectric Waveguides

The analysis of the electromagnetic wave propagation in the guiding structures carried out in the previous section has shown that the wavelength of the propagated wave should be of the order of magnitude of the dimension of a waveguide cross-section. Clearly, this condition cannot be met for the radiation of millimeter, sub-millimeter, infrared and visible range with a characteristic wavelength $\lambda \sim \left(10^{-3} \div 1\right)$ mm. In such a case so-called dielectric waveguides are used which, in general, represent dielectric slab or film with thickness comparable to the radiation wavelength and contained between the conducting films or between dielectric slabs with different refraction indices. Usually, the thickness of a dielectric waveguide is much less than its width, and it can be considered as a planar one. For the sake of definiteness, we limit ourselves with the analysis of light wave propagation in optical dielectric waveguides. Note that materials considered are assumed to be non-magnetic, and $\mu^r = 1$.

Consider the problems concerning the TE and TM mode propagation in slab dielectric waveguides ([9]). Such a waveguide is shown in Fig. 4.14. The basic features of the behavior of a dielectric waveguide can be extracted from a planar model in which no variation exists in one dimension [9]. In our case this is the y direction, and therefore we set $\partial/\partial y = 0$ in the further study. Channel waveguides, in which the waveguide dimensions are finite in both transverse directions, approach the behavior of the planar waveguide when

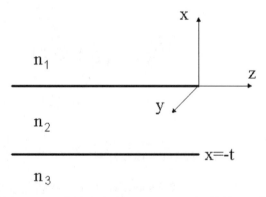

Fig. 4.14. A slab dielectric waveguide $(\partial/\partial y = 0)$

4.8 Electromagnetic Wave Propagation in Guiding Systems

one dimension is considerably larger than the other. A mode of a dielectric waveguide is the solution of (4.178). However, unlike the situation considered in the previous section, the waves are propagated in all three media, i.e., in the waveguide ($-t < x < 0$) and out of it ($x < -t, x > 0$). The boundary conditions at the upper ($x = 0$) and lower ($x = -t$) surfaces of the waveguide are no longer homogeneous. Instead, the tangential components of the fields $\boldsymbol{E}$ and $\boldsymbol{H}$ must be continuous at these dielectric surfaces. We find the solution in the form (4.184), except that the wave amplitude do not depend on y. We write (4.178) separately for regions $j = 1, 2, 3$ taking into account (4.184).

$$\frac{\partial^2}{\partial x^2} E_i + \left(k_j^2 - \Gamma^2\right) E_i = 0, \tag{4.229}$$

where $k_j^2 = k_0^2 n_j^2$, $k_0 = \omega/c$, and the refraction indices for each region are

$$n_j = \sqrt{\varepsilon_j^r}, \tag{4.230}$$

where ε_j^r is the relative dielectric constant for each region, respectively. We first investigate the behavior of possible solution of (4.229) qualitatively [9]. Let us assume that $n_2 > n_3 > n_1$. Then there exist the following possible solutions [9].

1. For $\Gamma > k_0 n_2$ it is seen from (4.229) that

$$\frac{1}{E_i} \frac{\partial^2}{\partial x^2} E_i > 0$$

everywhere, and $E_i(x)$ is exponential in all three regions 1, 2, 3 of the slab. The field distribution is due to the continuity of $E_{y,z}(x)$ and $H_{y,z}(x)$ at the two interfaces $x = -t, x = 0$. The field infinitely increases away from the waveguide so that the solution is not physically realizable and does not correspond to any real wave.

2. For $k_0 n_3 < \Gamma < k_0 n_2$, (4.229) shows that in the region 2

$$\frac{1}{E_i} \frac{\partial^2}{\partial x^2} E_i < 0,$$

which makes it possible to have a sinusoidal solution for $-t < x < 0$. In the external regions 1, 3

$$\frac{1}{E_i} \frac{\partial^2}{\partial x^2} E_i > 0$$

and the exponentially decaying solutions can exist which also satisfy the boundary conditions. The energy carried by these modes is confined to the vicinity of the guiding layer 2, and these modes are referred to as confined, or guided modes. A necessary condition for their existence is that $k_0 n_1, k_0 n_3 < \Gamma < k_0 n_2$ so that confined modes are possible only when $n_2 > n_3, n_1$, i.e., the inner region possesses the highest index of refraction.

3. For $k_0 n_1 < \Gamma < k_0 n_3$ solutions correspond to exponential behavior to exponential behavior in region 1 and to sinusoidal behavior in regions 2 and 3. These modes are referred to as substrate radiation modes.
4. For $0 < \Gamma < k_0 n_1$ (e) the solution for $E_i(x)$ becomes sinusoidal in all three regions. These are so-called radiation modes of the waveguides.

A useful point of view is one of considering the wave propagation in the region 2 as that of a plane wave propagating at some angle θ to the horizontal axis and undergoing a series of total internal reflections at the interface 2–1 and 2–3 [9]. Assuming $E \sim \sin(\kappa x + \alpha) \exp(-i\Gamma z)$ we obtain

$$\Gamma^2 + \kappa^2 = k_0^2 n_2^2. \tag{4.231}$$

Consider now in details the analytical solutions of (4.229) with the appropriate boundary conditions for the dielectric waveguide of thickness t with an index of refraction n_2 sandwiched between media with refraction indices n_1 and n_3. We start with the TE waves known from the previous section. Unlike (4.187) and (4.188), in a planar waveguide a TE mode consists of the components E_y, H_x and H_z due to the absence of dependence on y. We take E_y in the form

$$E_y(x, z, t) = E_0(x) \exp i(\omega t - \Gamma z). \tag{4.232}$$

The transverse function $E_0(x)$ is taken in different regions as [9]

$$E_0(x) = C \exp(-qx) \quad 0 \le x < \infty \tag{4.233}$$

$$E_0(x) = C\left[\cos \kappa x - \frac{q}{\kappa} \sin \kappa x\right] \quad -t \le x \le 0 \tag{4.234}$$

$$E_0(x) = C\left[\cos \kappa t + \frac{q}{\kappa} \sin \kappa t\right] \exp p(x+t), \quad -\infty < x \le -t. \tag{4.235}$$

The dispersion relations in regions 1,2,3 yield, respectively:

$$q = \sqrt{\Gamma^2 - k_0^2 n_1^2}, \quad \kappa = \sqrt{k_0^2 n_2^2 - \Gamma^2}, \quad p = \sqrt{\Gamma^2 - k_0^2 n_3^2}. \tag{4.236}$$

It is seen that only the second relation coincides with the corresponding equation (4.186) for the ordinary waveguide. The acceptable solutions for E_y and $H_z = (i/\omega\mu_0)\partial E_y/\partial x$ are continuous at $x = 0$ and $x = -t$. The coefficients in expressions (4.233)–(4.235) are defined in such a way that these solutions satisfy the continuity conditions of E_y at $x = 0$ and $x = -t$, and $\partial E_y/\partial x$ at $x = 0$. By imposing the continuity condition on $\partial E_y/\partial x$ at $x = -t$ we get from (4.234) and (4.235):

$$\kappa \sin \kappa t - q \cos \kappa t = p\left[\cos \kappa t + \frac{q}{\kappa} \sin \kappa t\right] \tag{4.237}$$

or

4.8 Electromagnetic Wave Propagation in Guiding Systems

$$\tan \kappa t = \frac{q+p}{\kappa}\left(1 - \frac{pq}{h^2}\right)^{-1}. \tag{4.238}$$

In the symmetric case $n_1 = n_3$ the field expressions (4.233)–(4.235) must possess even or odd symmetry with respect to $x = -t/2$. This condition is satisfied by (4.233)–(4.235) if

$$pt = \kappa t \tan \frac{\kappa t}{2}$$

for even symmetry and

$$pt = -\kappa t \cot \frac{\kappa t}{2}$$

for odd symmetry. In general case $n_1 \neq n_3$ relation (4.238) is used together with expressions (4.236) in order to obtain the eigenvalues Γ for the confined TE modes. Equation (4.238) shows that the values of allowed Γ in the propagation regime $k_0 n_3 < \Gamma < k_0 n_2$ are discrete. The number of modes depends on the width t, the frequency and the indices of refraction n_1, n_2, n_3. At a given wavelength the number of guided modes increases from 0 with increasing t. At some t the mode TE_0 becomes confined. Further increases in t will allow TE_1 to exist as well, and so on.

The constant C appearing in expressions (4.233)–(4.235) is arbitrary yet, for many applications, it is advantageous to define C in such a way that it is related to total power in the mode. We choose C so that the field $E_0(x)$ in expressions (4.233)–(4.235) corresponds to the power flow of one watt per unit width in y direction in the mode. A mode for which $E_y = AE_0(x)$ will correspond to a power flow of $|A|^2$ watts/m. the normalization condition becomes [9]

$$-\frac{1}{2}\int_{-\infty}^{\infty} E_y H_x^* dx = \frac{\Gamma}{2\omega\mu_0}\int_{-\infty}^{\infty} [E_{0m}(x)]^2 \, dx = 1, \tag{4.239}$$

where the number m denotes the mth confined TE mode corresponding to mth eigenvalue of (4.238) and

$$H_x = -\frac{i}{\omega\mu_0}\frac{\partial E_y}{\partial z}.$$

Substituting explicit expressions of E_y (4.233)–(4.235) into (4.239) we get

$$C_m = 2\kappa_m \sqrt{\frac{\omega\mu_0}{|\Gamma_m|\left(t + \frac{1}{q_m} + \frac{1}{p_m}\right)(\kappa_m^2 + q_m^2)}}. \tag{4.240}$$

It can be shown that the eigenmodes of the waveguide $E_{0m}(x)$ are orthogonal which yield the additional condition

$$\int_{-\infty}^{\infty} E_{0l}(x) E_{0m}(x)\,dx = \frac{2\omega\mu_0}{\Gamma_m}\delta_{lm}. \tag{4.241}$$

Finally, we briefly discuss the structure and behavior of the TM modes. The field components are [9]

$$H_y(x,z,t) = H_0(x)\exp i(\omega t - \Gamma z) \tag{4.242}$$

$$E_x(x,z,t) = \frac{i}{\omega\varepsilon_0\varepsilon^r}\frac{\partial H_y}{\partial z} = \frac{\Gamma}{\omega\varepsilon_0\varepsilon^r}H_0(x)\exp i(\omega t - \Gamma z) \tag{4.243}$$

$$E_z(x,z,t) = -\frac{i}{\omega\varepsilon_0\varepsilon^r}\frac{\partial H_y}{\partial x}. \tag{4.244}$$

The transverse function $H_0(x)$ is taken as

$$H_0(x) = -C_1\frac{\kappa}{q_1}\exp(-qx),\quad \infty > x \geq 0 \tag{4.245}$$

$$H_0(x) = C_1\left[-\frac{\kappa}{q_1}\cos\kappa x + \sin\kappa x\right],\quad -t < x < 0 \tag{4.246}$$

$$H_0(x) = -C_1\left[\frac{\kappa}{q_1}\cos\kappa t + \sin\kappa t\right]\exp p(x+t),\quad -\infty < x \leq -t. \tag{4.247}$$

The continuity of H_y and E_z at the two interfaces requires that the various propagation constants obey the eigenvalue equation

$$\tan\kappa t = \frac{\kappa(p_1 + q_1)}{\kappa^2 - p_1 q_1}, \tag{4.248}$$

where

$$p_1 \equiv p\frac{n_2^2}{n_3^2},\quad q_1 \equiv q\frac{n_2^2}{n_1^2}.$$

The normalization constant, C_1, is chosen similarly to the case of the TE modes, so that the field determined by (4.242)–(4.247) carries one watt per unit length in the y direction:

$$\frac{1}{2}\int_{-\infty}^{\infty} H_y E_x^* dx = \frac{\Gamma}{2\omega\varepsilon_0}\int_{-\infty}^{\infty}\frac{H_0^2(x)}{n^2(x)}dx = 1,$$

or

4.8 Electromagnetic Wave Propagation in Guiding Systems 145

$$\int_{-\infty}^{\infty} \frac{H_{0m}^2(x)}{n^2(x)} dx = \frac{2\omega\varepsilon_0}{\Gamma_m}. \tag{4.249}$$

Carrying out the integration and using (4.245)–(4.247) we get:

$$C_{1m} = 2\sqrt{\frac{\omega\varepsilon_0}{\Gamma_m t_{\text{eff}}}}, \tag{4.250}$$

where

$$t_{\text{eff}} \equiv \frac{q_1^2 + \kappa^2}{q_1^2} \left[\frac{t}{n_2^2} + \left(\frac{q^2 + \kappa^2}{q_1^2 + \kappa^2}\right) \frac{1}{n_1^2 q} + \left(\frac{p^2 + \kappa^2}{p_1^2 + \kappa^2}\right) \frac{1}{n_3^2 p} \right].$$

The dispersion curves of the TE and TM modes can be found in [9]. In general a mode becomes confined above a certain cutoff value of t/λ which is similar to the case of an ordinary waveguide mentioned above. However, the presence of dielectric media instead of conducting walls results in some specific features of dielectric waveguides. At the cutoff value $p = 0$ and the mode extends to $x \to -\infty$. For increasing values of t/λ $p > 0$, and the mode becomes increasingly confined to layer 2. In a symmetric waveguide where $n_1 = n_3$ the lowest order modes TE_0 and TM_0 have no cutoff and are confined for all values of t/λ.

4.8.5 Cavity Resonators

In guiding systems studied above the wave is propagated freely in one direction while its evolution in two other directions is limited by totally or partially reflecting surfaces. As a result, an electromagnetic energy transfer occurred in the direction of the wave propagation. Introducing additionally the ideally conducting surfaces in the propagation direction, we provide the conditions for the conservation of energy of the electromagnetic field inside the screened volume, or a so-called cavity resonator where free oscillations of the electromagnetic field are possible [4]. When the shape and size of the cavity are known the certain values of a frequency w, or eigenfrequences of the electromagnetic oscillations of the resonator can be calculated which correspond to the possible modes of the electromagnetic field, or the eigenmodes of the resonator. The eigenmodes of the resonator are standing waves oscillating in time, unlike the guiding systems which possess the running waves as their eigenmodes. In the case of zero losses the electromagnetic field does not penetrate into the metallic walls, and these standing waves are undamped. The corresponding eigenfrequences are real. In the further analysis it will be shown that the number of eigenmodes is infinite, and the order of magnitude of the lowest eigenfrequency $\omega_{\min}$ is c/l where l the linear dimension of the cavity. The high eigenfrequencies $\omega \gg c/l$ lie very close together forming a quasi-continuous spectrum, and the number of them per unit range of ω

is $V\omega^2/2\pi^2 c^3$, which depends on the volume V of the resonator but on its shape [4].

Consider first the case of a hollow evacuated resonator of an arbitrary form, i.e. a totally screened volume V limited with a perfectly conducting surface s [4]. In order to determine the electromagnetic field inside the resonator we must solve in V either (4.178) for E_i with the boundary condition providing that the tangential component E_t of the electric field would vanish on the surface s

$$E_t = 0 \text{ on } s, \qquad (4.251)$$

or the similar equation for the magnetic field $\boldsymbol{H}$

$$\Delta H_i + k^2 H_i = 0 \qquad (4.252)$$

with the boundary condition providing that the normal component H_n of the magnetic field would vanish on the surface s

$$H_n = 0 \text{ on } s. \qquad (4.253)$$

The fields $\boldsymbol{E}$ and $\boldsymbol{H}$ in resonator must also obey the conditions

$$\operatorname{div} \boldsymbol{E} = 0, \; \operatorname{div} \boldsymbol{H} = 0. \qquad (4.254)$$

The calculation of the eigenvalues k of these problems yields [4]

$$k^2 = \frac{\int_V |\operatorname{curl} \boldsymbol{E}_m|^2 \, dV}{\int_V |\boldsymbol{E}_m|^2 \, dV} = \frac{\int_V |\operatorname{curl} \boldsymbol{H}_m|^2 \, dV}{\int_V |\boldsymbol{H}_m|^2 \, dV}. \qquad (4.255)$$

It can also be shown that in a resonator the energy of the electric and the magnetic fields are equal [4]:

$$\int_V |\boldsymbol{E}_m|^2 \, dV = \int_V |\boldsymbol{H}_m|^2 \, dV \qquad (4.256)$$

It is seen from (4.255) that the eigenvalues of k are real for even for the resonator with dissipation. In the latter case the eigenfrequencies appeared to be complex:

$$\omega = \omega' + i\omega''. \qquad (4.257)$$

The complex eigenfrequency results in the decaying oscillations of a standing wave in the cavity with a period $T' = 2\pi/\omega'$ and a decay time constant $T'' = 1/\omega''$. The time factor in such a case has the form

$$\operatorname{Re} \exp i\omega t = \exp(-\omega'' t) \cos \omega' t. \qquad (4.258)$$

4.8 Electromagnetic Wave Propagation in Guiding Systems

Generally speaking, any kind of losses in resonators that can be due to absorption in dielectric, finite conduction of cavity walls, or radiation from a resonator result in the exponential attenuation of free oscillations. Typically, cavity resonators possess very low losses, and the imaginary part of the frequency is small compared to the real part: $\omega'' \ll \omega'$. Hence, the decay time T'' is much larger than the oscillation period T', and the time averaging of energetic characteristics of oscillations can be carried out similarly to the purely harmonic case. The time average energetic quantities quadratic in field behave as $\exp(-2\omega''t)$ only for time intervals $t \gg T'$. The time average energy $\overline{W(t)}$ and dissipated power $\overline{P(t)}$ of the electromagnetic field in the cavity resonator are, respectively [4]:

$$\overline{W(t)} = W(0)\exp(-2\omega''t), \quad \overline{P(t)} = -\frac{d\overline{W(t)}}{dt} = 2\omega''\overline{W(t)}. \quad (4.259)$$

Using relations (4.257)–(4.259) we introduce the quality factor Q_p of a cavity resonator as follows

$$Q_p = \frac{\omega'\overline{W(t)}}{\overline{P(t)}} = \frac{\omega'}{2\omega''}. \quad (4.260)$$

Then expression (4.257) takes the form

$$\omega = \omega'\left(1 + \frac{i}{2Q_p}\right). \quad (4.261)$$

In order to evaluate the role of different losses mechanisms in the quality factor Q_p we write the time average dissipated power $\overline{P(t)}$ as a sum of the losses in dielectric $\overline{P_D}$, in conducting walls $\overline{P_C}$ and the radiation losses $\overline{P_R}$

$$\overline{P(t)} = \overline{P_D} + \overline{P_C} + \overline{P_R}. \quad (4.262)$$

Substituting the sum (4.262) into relation (4.260) we get

$$\frac{1}{Q_p} = \frac{1}{Q_D} + \frac{1}{Q_C} + \frac{1}{Q_R}, \quad (4.263)$$

where $Q_D = \omega'\overline{W(t)}/\overline{P_D}$, $Q_C = \omega'\overline{W(t)}/\overline{P_C}$ and $Q_R = \omega'\overline{W(t)}/\overline{P_R}$. In particular, the quality factor Q_D due to the losses in a dielectric can be calculated as a ratio of the real and imaginary parts of the dielectric constant

$$Q_D = \frac{\omega\varepsilon_0\varepsilon^{r'}}{\sigma}, \quad (4.264)$$

where σ is the conductivity of the medium filling the cavity. The quality factor Q_C is calculated using the expression of energy (4.256) and the penetration depth d_0 (4.228):

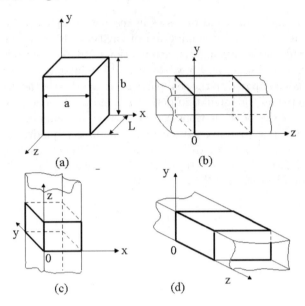

Fig. 4.15. The rectangular cavity resonator (**a**). The longitudinal z-axis can be chosen in any of three positive directions (**b**), (**c**), (**d**). The resulting fields have different structures but the same frequency being triply degenerate

$$Q_C = \frac{2\int_V |\boldsymbol{H}_m|^2 \, dV}{d_0 \oint_s |\boldsymbol{H}_m|^2 \, ds}, \qquad (4.265)$$

where V is the volume of the cavity, s is its conducting surface, and $\boldsymbol{H}_m$ is a magnetic field of the m-th mode.

As an example, we consider the structure of the field in the rectangular resonator with the length L in the z direction and the cross-section dimensions $x = a$, $y = b$ filled with a dielectric material characterized by the relative dielectric constant ε^r and relative magnetic permeability μ^r. The cavity resonators of such a type are shown in Fig. 4.15.

The field in the resonator has the form of standing waves, where the dependence on the longitudinal coordinate z is described by sinusoidal functions instead of exponential one. For the sake of definiteness we start with the TE waves. The longitudinal magnetic field component H_z has the form

$$H_z = H_0(x, y)(A \sin \varGamma z + B \cos \varGamma z). \qquad (4.266)$$

The condition (4.252) on the totally reflecting walls $z = 0$ and $z = L$ yields $B = 0$ and the discrete spectrum of values for $\varGamma$

$$\varGamma = \frac{p\pi}{L}, \quad p = 0, 1, 2..., \qquad (4.267)$$

and

4.8 Electromagnetic Wave Propagation in Guiding Systems

$$k^2 = \kappa^2 + \left(\frac{p\pi}{L}\right)^2, \quad \omega = \frac{c}{\sqrt{\varepsilon^r \mu^r}} \sqrt{\kappa^2 + \left(\frac{p\pi}{L}\right)^2}. \tag{4.268}$$

The amplitude $H_0(x,y)$ of the field (4.266) is given by expression (4.209) since the boundary conditions (4.252) coincide with the ones (4.208) for the waveguide. The transverse components of the magnetic and electric field in the cavity resonator are calculated according to the following formulae [4]:

$$H_x = \frac{1}{\kappa^2} \frac{\partial^2 H_z}{\partial z \partial x}, \quad H_y = \frac{1}{\kappa^2} \frac{\partial^2 H_z}{\partial z \partial y} \tag{4.269}$$

$$E_x = -\frac{i\omega\mu_0\mu^r}{\kappa^2} \frac{\partial H_z}{\partial y}, \quad E_y = \frac{i\omega\mu_0\mu^r}{\kappa^2} \frac{\partial H_z}{\partial y}. \tag{4.270}$$

We finally obtain [4]:

$$\boldsymbol{H}_{mnp} = \boldsymbol{z}_0 H_0 \cos\frac{m\pi x}{a} \cos\frac{n\pi y}{b} \sin\frac{p\pi z}{L}$$

$$-\frac{p\pi}{\kappa^2 L} H_0 \left[\boldsymbol{x}_0 \frac{m\pi}{a} \sin\frac{m\pi x}{a} \cos\frac{n\pi y}{b} + \boldsymbol{y}_0 \frac{n\pi}{b} \cos\frac{m\pi x}{a} \sin\frac{n\pi y}{b} \right] \cos\frac{p\pi z}{L} \tag{4.271}$$

$$\boldsymbol{E}_{mnp} = -\frac{i\omega\mu_0\mu^r H_0}{\kappa^2} \left[-\boldsymbol{x}_0 \frac{n\pi}{b} \cos\frac{m\pi x}{a} \sin\frac{n\pi y}{b} + \boldsymbol{y}_0 \frac{m\pi}{a} \sin\frac{m\pi x}{a} \cos\frac{n\pi y}{b} \right]$$

$$\times \sin\frac{p\pi z}{L} \tag{4.272}$$

$$\kappa^2 = \left(\frac{m\pi}{a}\right)^2 + \left(\frac{n\pi}{b}\right)^2, \quad \omega_{mnp} = \frac{c}{\sqrt{\varepsilon^r \mu^r}} \sqrt{\left(\frac{m\pi}{a}\right)^2 + \left(\frac{n\pi}{b}\right)^2 + \left(\frac{p\pi}{L}\right)^2} \tag{4.273}$$

where $m = 0, 1, 2, ...$, $n = 0, 1, 2, ...$, m and n should not be zero simultaneously.

The field components of the TM waves can be expressed in terms of the longitudinal electric field E_z:

$$E_x = \frac{1}{\kappa^2} \frac{\partial^2 E_z}{\partial z \partial x}, \quad E_y = \frac{1}{\kappa^2} \frac{\partial^2 E_z}{\partial z \partial y} \tag{4.274}$$

$$H_x = \frac{i\omega\varepsilon_0\varepsilon^r}{\kappa^2} \frac{\partial E_z}{\partial y}, \quad H_y = -\frac{i\omega\varepsilon_0\varepsilon^r}{\kappa^2} \frac{\partial E_z}{\partial x}. \tag{4.275}$$

4 Electromagnetic Waves

Taking

$$E_z = E_0(x,y)(C \sin \Gamma z + D \cos \Gamma z)$$

and using the boundary conditions (4.251) at $z=0$, $z=L$ for E_x, at $x=0$, $x=a$ for E_y and at $y=0$, $y=b$ for E_z we find [4]

$$C = 0, \quad \Gamma = \frac{p\pi}{L}, \quad p = 0, 1, 2..., \tag{4.276}$$

$$\boldsymbol{E}_{mnp} = \boldsymbol{z}_0 E_0 \sin\frac{m\pi x}{a} \sin\frac{n\pi y}{b} \cos\frac{p\pi z}{L}$$

$$-\frac{p\pi}{\kappa^2 L} E_0 \left[\boldsymbol{x}_0 \frac{m\pi}{a} \cos\frac{m\pi x}{a} \sin\frac{n\pi y}{b} + \boldsymbol{y}_0 \frac{n\pi}{b} \sin\frac{m\pi x}{a} \cos\frac{n\pi y}{b} \right] \sin\frac{p\pi z}{L} \tag{4.277}$$

$$\boldsymbol{H}_{mnp} = \frac{i\omega\varepsilon_0\varepsilon^r E_0}{\kappa^2} \left[\boldsymbol{x}_0 \frac{n\pi}{b} \sin\frac{m\pi x}{a} \cos\frac{n\pi y}{b} - \boldsymbol{y}_0 \frac{m\pi}{a} \cos\frac{m\pi x}{a} \sin\frac{n\pi y}{b} \right]$$

$$\times \cos\frac{p\pi z}{L}, \tag{4.278}$$

where $m, n = 1, 2, ...$, and relations (4.273) remain valid. The expressions (4.268)–(4.278) show that any combination of three numbers m, n, p determines the eigenmode of oscillations of the cavity resonator with the eigenfrequency ω_{mnp} and the fields $\boldsymbol{E}_{mnp}$ and $\boldsymbol{H}_{mnp}$. The type of oscillations with the minimal eigenvalue of the wave number $k_{mnp} = \omega_{mnp}\sqrt{\varepsilon^r\mu^r}/c$ is called the principal oscillation mode. It corresponds to the zero value of one of the numbers m, n, p and the unity value of two others. Clearly, we should choose zero number for the minimal length of a, b, L. Let, for instance, $L < a, L < b$. Then we have

$$(k_{mnp})_{\min} = k_{110} = \pi\sqrt{\frac{1}{a^2} + \frac{1}{b^2}} \tag{4.279}$$

and the principal oscillation mode is E_{110}. Changing the axes notations one may identify the same mode as H_{011} or H_{101}.

Estimate the quality factor Q_p for the principal oscillation mode H_{101}. Assuming that the losses are due to the finite conduction of the walls of the empty resonator and substituting

$$|H_{101}|^2 = H_0^2 \left[\cos^2\frac{\pi x}{a} \sin^2\frac{\pi z}{L} + \frac{a^2}{L^2} \sin^2\frac{\pi x}{a} \cos^2\frac{\pi z}{L} \right]$$

into relation (4.265) we find [4]

4.8 Electromagnetic Wave Propagation in Guiding Systems

$$\int_V |\mathbf{H}_{101}|^2 \, dV = \int_0^a \int_0^b \int_0^L |H_{101}|^2 \, dxdydz = H_0^2 \frac{abL}{4}\left(1 + \frac{a^2}{L^2}\right)$$

$$\oint_S |\mathbf{H}_m|^2 \, ds = 2\int_0^L \int_0^a |H_{101}|^2_{y=0} \, dzdx + 2\int_0^a \int_0^b |H_{101}|^2_{z=0} \, dxdy$$

$$+ 2\int_0^b \int_0^L |H_{101}|^2_{x=0} \, dydz = H_0^2 \left[\frac{aL}{2}\left(1 + \frac{a^2}{L^2}\right) + \frac{a^3 b}{L^2} + bL\right]$$

and, finally, the quality factor has the form

$$Q_p = \frac{1}{d_0} \frac{abL\left(L^2 + a^2\right)}{\left[aL\left(L^2 + a^2\right) + 2b\left(L^3 + a^3\right)\right]}. \tag{4.280}$$

In the particular cases of the cubic cavity $a = b = L$ and the "plane" cavity $a \ll L, b \ll L$ expression (4.280) yields, respectively

$$Q_{pcub} = \frac{1}{d_0}\frac{a}{3}, \quad Q_{pplane} \approx \frac{1}{d_0}b. \tag{4.281}$$

The penetration length d_0 is proportional to the square root of the cavity linear dimension according to relations (4.227) and (4.273):

$$d_0 \sim \frac{1}{\sqrt{\omega}} \sim \sqrt{a}.$$

Then, clearly, the quality factor is also proportional to the square root of the cavity linear dimension as it is seen from expressions (4.281).

5 Charged Particle in Electromagnetic Field

5.1 Equations of Motion

In this section we derive the equations of motion of a single particle in an electromagnetic field, using the expressions for Lagrangians and actions obtained above [15]. The particle equations of motion are the Lagrange equations

$$\frac{d}{dt}\frac{\partial \mathcal{L}}{\partial \boldsymbol{v}} = \frac{\partial \mathcal{L}}{\partial \boldsymbol{r}}, \tag{5.1}$$

where Lagrangian is given by (1.122). The derivative $\partial \mathcal{L}/\partial \boldsymbol{v}$ is a generalized momentum of the particle (1.123), while the space derivative is given by

$$\frac{\partial \mathcal{L}}{\partial \boldsymbol{r}} \equiv \operatorname{grad} \mathcal{L} = e \operatorname{grad}(\boldsymbol{A}\boldsymbol{v}) - e \operatorname{grad}\varphi.$$

Using the known formula of vector calculus

$$\operatorname{grad}(\boldsymbol{a} \cdot \boldsymbol{b}) = (\boldsymbol{a} \cdot \nabla)\boldsymbol{b} + (\boldsymbol{b} \cdot \nabla)\boldsymbol{a} + [\boldsymbol{b} \times \operatorname{curl}\boldsymbol{a}] + [\boldsymbol{a} \times \operatorname{curl}\boldsymbol{b}]$$

and taking into account that $\boldsymbol{r}$ and $\boldsymbol{v}$ are independent variables in Lagrange's equations (the derivative with respect to $\boldsymbol{r}$ is evaluated while keeping $\boldsymbol{v}$ constant), we obtain

$$\frac{\partial \mathcal{L}}{\partial \boldsymbol{r}} = e(\boldsymbol{v} \cdot \nabla)\boldsymbol{A} + e[\boldsymbol{v} \times \operatorname{curl}\boldsymbol{A}] - e \operatorname{grad}\varphi.$$

Thus, the Lagrange's equations take the form

$$\frac{d}{dt}(\boldsymbol{p} + e\boldsymbol{A}) = e(\boldsymbol{v} \cdot \nabla)\boldsymbol{A} + e[\boldsymbol{v} \times \operatorname{curl}\boldsymbol{A}] - e \operatorname{grad}\varphi.$$

The complete differential $(d\boldsymbol{A}/dt)\,dt$ consists of two parts:

1. The change $(\partial \boldsymbol{A}/\partial t)\,dt$ of the potential on the particle at rest due to the explicit time dependence of the potential.
2. The change of the potential due to the motion of the particle: during the time dt it moves across the distance $d\boldsymbol{r}$ and the potential on the particle changes by

$$(\partial \boldsymbol{r} \cdot \nabla)\boldsymbol{A} \equiv \left(\frac{\partial \boldsymbol{r}}{dt} \cdot \nabla\right)\boldsymbol{A}\,dt = (\boldsymbol{v} \cdot \nabla)\boldsymbol{A}\,dt.$$

Then, the complete time derivative $d\boldsymbol{A}/dt$ is given by[1]

$$\frac{d\boldsymbol{A}}{dt} = \frac{\partial \boldsymbol{A}}{\partial t} + (\boldsymbol{v} \cdot \nabla) \boldsymbol{A}. \tag{5.2}$$

Substituting (5.2) into (5.1) we obtain

$$\frac{d\boldsymbol{p}}{dt} = -e\frac{\partial \boldsymbol{A}}{\partial t} - e\mathrm{grad}\varphi + e\left[\boldsymbol{v} \times \mathrm{curl}\boldsymbol{A}\right], \tag{5.3}$$

or, since

$$-\frac{\partial \boldsymbol{A}}{\partial t} - \mathrm{grad}\varphi = \boldsymbol{E}$$

and

$$\mathrm{curl}\boldsymbol{A} = \boldsymbol{B}$$

(Eqs. (1.39), (1.41)),

$$\frac{d\boldsymbol{p}}{dt} = e\boldsymbol{E} + e\left[\boldsymbol{v} \times \boldsymbol{B}\right]. \tag{5.4}$$

In the classical (non-relativistic) limit ($\boldsymbol{p} = m\boldsymbol{v}$), (5.4) takes the standard form of Newton equation with Lorentz force

$$\boldsymbol{F} = e\boldsymbol{E} + e\left[\boldsymbol{v} \times \boldsymbol{B}\right]$$

$$m\frac{d\boldsymbol{v}}{dt} = e\boldsymbol{E} + e\left[\boldsymbol{v} \times \boldsymbol{B}\right]. \tag{5.5}$$

5.2 Energy of Particles and Fields

As we have seen, the charged particles and the electromagnetic field interact with each other and thus exchange energy. If the entire system, particles and field, is isolated, its total energy should be conserved, whereas in the opposite case some energy flow from the system into the outside world or vice versa ought to take place. The basic laws governing these processes of energy exchange will be outlined in this section [2].

---

[1] Expression 5.2 is of a very general validity; it is not restricted to vector potential, but rather holds for any quantity which is related to a particle (or group of particles) rather than to a specific point in space. The differential operator on the right-hand side of (5.2) is called the *convective derivative* and it enters the equations of hydrodynamics, magnetohydrodynamics, the Boltzmann kinetic equation etc.

Let us start with particles, and, for the sake of generality, consider the relativistic case. As has been shown in the previous section, the kinetic energy of a relativistic particle is

$$\mathcal{E}_{\text{kin}} = \frac{mc^2}{\sqrt{1 - v^2/c^2}},$$

while its momentum is given by

$$\boldsymbol{p} = \frac{m\boldsymbol{v}}{\sqrt{1 - v^2/c^2}}.$$

Differentiating $\mathcal{E}_{\text{kin}}$ and $\boldsymbol{p}$ with respect to time we obtain

$$\frac{d\mathcal{E}_{\text{kin}}}{dt} = m\boldsymbol{v}\frac{d\boldsymbol{v}}{dt}\frac{1}{(1 - v^2/c^2)^{\frac{3}{2}}}$$

and

$$\frac{d\boldsymbol{p}}{dt} = \frac{m}{(1 - v^2/c^2)^{\frac{3}{2}}}\left[\frac{d\boldsymbol{v}}{dt} - \frac{v^2}{c^2}\left(\frac{d\boldsymbol{v}}{dt}\right)_{\perp}\right],$$

where

$$\left(\frac{d\boldsymbol{v}}{dt}\right)_{\perp} \equiv \frac{d\boldsymbol{v}}{dt} - \frac{\boldsymbol{v}}{v^2}\left(\boldsymbol{v}\frac{d\boldsymbol{v}}{dt}\right)$$

is the component of $d\boldsymbol{v}/dt$ perpendicular to $\boldsymbol{v}$,

$$\boldsymbol{v}\left(\frac{d\boldsymbol{v}}{dt}\right)_{\perp} = 0.$$

Thus,

$$\boldsymbol{v}\frac{d\boldsymbol{p}}{dt} = \frac{m\boldsymbol{v}}{(1 - v^2/c^2)^{\frac{3}{2}}}\frac{d\boldsymbol{v}}{dt}$$

and

$$\frac{d\mathcal{E}_{\text{kin}}}{dt} = \boldsymbol{v}\frac{d\boldsymbol{p}}{dt}. \tag{5.6}$$

Substituting in (5.6) $d\boldsymbol{p}/dt$ from the equation of motion, (5.5), and taking into account that

$$\boldsymbol{v}\left[\boldsymbol{v} \times \boldsymbol{B}\right] = 0,$$

we obtain

$$\frac{d\mathcal{E}_{\text{kin}}}{dt} = e\boldsymbol{E}\boldsymbol{v}. \tag{5.7}$$

If, instead of one particle, we consider the particle distribution with charge density $\rho(\boldsymbol{r})$, the time variation of the total kinetic energy $\mathcal{E}^p_{\text{kin}}$ of the particles is given, according to (5.7) by

$$\frac{d\mathcal{E}^p_{\text{kin}}}{dt} = \int \boldsymbol{j}\boldsymbol{E}dV. \tag{5.8}$$

Equations (5.7), (5.8) complete our consideration of the energy variation of particles; (5.7) simply states that the change of the particle kinetic energy is equal to the work done by the electromagnetic force $e\boldsymbol{E}$ (or, more precisely, by its component in $\boldsymbol{v}$ direction) acting on the particle. Concerning the electromagnetic field itself, we use the derivation of the Poynting vector $\boldsymbol{S}$ mentioned above, this time for the system with a finite current density. Equations (4.9) and (4.10) should be modified to become

$$\boldsymbol{E}\frac{\partial \boldsymbol{D}}{\partial t} + \boldsymbol{H}\frac{\partial \boldsymbol{B}}{\partial t} = -\boldsymbol{j}\boldsymbol{E} - (\boldsymbol{H}\text{curl}\boldsymbol{E} - \boldsymbol{E}\text{curl}\boldsymbol{H}) \tag{5.9}$$

and

$$\frac{\partial}{\partial t}\left(\frac{\varepsilon_0 \varepsilon^r E^2 + \mu_0 \mu^r H^2}{2}\right) = -\boldsymbol{j}\boldsymbol{E} - \text{div}\boldsymbol{S}. \tag{5.10}$$

Integrating (5.10) over a certain volume V and applying the Gauss theorem to the second term on the right-hand side, we arrive at

$$\frac{\partial}{\partial t}\iiint_V \left(\frac{\varepsilon_0 \varepsilon^r E^2 + \mu_0 \mu^r H^2}{2}\right) dV = -\iiint_V \boldsymbol{j}\boldsymbol{E}dV - \oint_A \boldsymbol{S}d\boldsymbol{a}, \tag{5.11}$$

where A is the surface, enclosing the volume V of the integration, and $d\boldsymbol{a}$ is the surface element. Taking into account (5.8) we turn (5.11) into the form

$$\frac{\partial}{\partial t}\left\{\iiint \left(\frac{\varepsilon_0 \varepsilon^r E^2 + \mu_0 \mu^r H^2}{2}\right) dV + \mathcal{E}^p_{\text{kin}}\right\} = -\oint_A \boldsymbol{S}d\boldsymbol{a}. \tag{5.12}$$

If the integration in (5.12) is expanded to infinity, the integral on the right-hand side vanishes since the fields at infinity tend to zero; the system of fields and particles becomes an isolated system, satisfying

$$\frac{\partial}{\partial t}\left\{\iiint \left(\frac{\varepsilon_0 \varepsilon^r E^2 + \mu_0 \mu^r H^2}{2}\right) dV + \mathcal{E}^p_{\text{kin}}\right\} = 0. \tag{5.13}$$

Equation (5.13) has the form of conservation law. The conserved quantity (the expression in brackets) contains two terms, the first being a characteristic of an electromagnetic field and the second that of particles. This second term is just the particles' energy, which makes it more than plausible that the first term stands for the energy of the electromagnetic field, and that the conserved quantity under consideration is the total energy of the fields and the particles in the entire space where the quantity in brackets in the integrand (5.13)

$$W = \frac{\varepsilon_0 \varepsilon^r E^2 + \mu_0 \mu^r H^2}{2} \tag{5.14}$$

is the energy density of an electromagnetic field (4.12). Returning now to (5.12) in the finite volume or to its differential form

$$\frac{\partial}{\partial t}\left[\frac{\varepsilon_0 \varepsilon^r E^2 + \mu_0 \mu^r H^2}{2} + \overline{\mathcal{E}^p_{\text{kin}}}\right] = -\text{div}\boldsymbol{S},$$

where $\overline{\mathcal{E}^p_{\text{kin}}}$ is the energy of the particles in the unit volume, we interpret it as a total energy continuity equation, in close analogy with the density continuity equation (4.13).

5.3 A Charged Particle Motion in Static Uniform Fields

5.3.1 Static Uniform Magnetic Field

In the current and the following sections we consider a number of concrete problems concerning the behavior of the combined system, consisting of charged particles and an electromagnetic field [2]. Two common features distinguish these problems. The first may be thought of as methodical; in all particular cases we decouple the system of equations determining the motion of the particles and the form of the fields and consider either the motion of the particles in a given external field, or the fields created by a given distribution of charges and currents. It appears that this approach is a good approximation to the final self-consistent treatment in some cases, and serves as at least a starting point for qualitative understanding in others. The second common feature relates to the choice of problems, dictated by the general plan of the series. Thus, the relatively simple problem of the classical motion of a charged particle in external fields provides the necessary basis for the understanding of different kinds of quantum oscillations in two- and three-dimensional solids, as well as of special kinds of electromagnetic excitations in plasma. Consider the situation where

$$\boldsymbol{E} = 0; \quad \boldsymbol{B}(\boldsymbol{r}, t) = \text{const.}$$

We start with a few remarks concerning the motion of a charge in a constant in time electromagnetic field. The potentials of such a field can always be chosen to be time-independent. It follows from (1.39), (1.41) that in this case $\boldsymbol{E} = -\text{grad}\varphi$ and $\boldsymbol{B} = \text{curl}\boldsymbol{A}$; thus a constant in time electric field is

5.3 A Charged Particle Motion in Static Uniform Fields

determined only by the scalar potential and a constant in time magnetic field is determined only by the vector potential, irrespective of their variation in space. If the potentials are time-independent, the Lagrangian of the charged particle does not depend explicitly on time as well, and its energy is conserved, being equal to the Hamiltonian (see (1.125))

$$\mathcal{E} = \mathcal{H} = \frac{mc^2}{\sqrt{1-v^2/c^2}} + e\varphi.$$

The vector potential does not enter the expression for the energy, which means that the presence of a constant in time magnetic field does not affect the energy of a particle. In the absence of the electric field ($\varphi = 0$) the total energy of the particle is reduced to its kinetic energy, so that we may say that the kinetic energy of the particle remains constant in the course of its motion in any time-independent magnetic field. This, of course, is a direct consequence of the fact that the magnetic part of Lorentz force, being perpendicular to the particle's velocity, does not produce a work on the particle.

We turn now to the special case of a constant in time and uniform in space magnetic field $\mathbf{B}$. Choosing the direction of the field as a z-axis, we adopt the non-relativistic approximation (the non-relativistic limit is sufficient for most applications); and we rewrite the equations of motion in components

$$\frac{dv_x}{dt} = \omega_c v_y; \quad \frac{dv_y}{dt} = -\omega_c v_x; \quad \frac{dv_z}{dt} = 0. \tag{5.15}$$

Here the notation

$$\omega_c = \frac{eB}{m} \tag{5.16}$$

is introduced. Equations (5.15) show that the motion along the magnetic field direction is not affected by the field and is decoupled from the motion in the perpendicular plane. This implies that not only is the total kinetic energy of the particle conserved, but the kinetic energy of the longitudinal motion

$$\mathcal{E}_\| = \frac{mv_\|^2}{2}; \quad v_\| \equiv v_z$$

and the kinetic energy of the transverse motion

$$\mathcal{E}_\perp = \frac{mv_\perp^2}{2}; \quad \mathbf{v}_\perp = \{v_x, v_y\}$$

are separately conserved too. For a free motion in the z (

$$v_\| = v_{\|0} = \text{const}; \quad z = z_0 + v_{\|0} t.$$

The motion in the xy plane is governed by the first tw this is a system of two homogeneous linear ordinary which can be easily solved; we offer an outline of thre of solution, to demonstrate more clearly its different

1. Taking the time derivatives of both equations and substituting for dv_y/dt and dv_x/dt their values from 5.15, we arrive at

$$\frac{d^2 v_x}{dt^2} + \omega_c^2 v_x = 0; \quad \frac{d^2 v_y}{dt^2} + \omega_c^2 v_y = 0. \quad (5.18)$$

Equations (5.18) imply that the particle performs two oscillatory motions in perpendicular directions. However, it is not clear enough that these oscillations are strongly correlated.

2. Looking for a solution in the form

$$v_x = v_{x0} \exp i\omega_c t, \quad v_y = v_{y0} \exp i\omega_c t$$

we arrive at the system of two linear homogeneous equations for v_{x0}, v_{y0}, whose solution determines the ratio v_{x0}/v_{y0}; thus, the correlated character of the motion in the x and y directions is evident, while the independent oscillators are somewhat hidden.

3. Multiplying the second equation of (5.15) by i and adding it to the first equation we obtain

$$\frac{d}{dt}(v_x + iv_y) = -i\omega_c(v_x + iv_y). \quad (5.19)$$

The solution of (5.19) can be written in the form

$$v_x + iv_y = V_{\perp 0} \exp[-i(\omega_c t + \alpha)], \quad (5.20)$$

where $V_{\perp 0}$ is real. Separating real and imaginary parts in (5.20), we get

$$v_x = V_{\perp 0} \cos(\omega_c t + \alpha); \quad v_y = -V_{\perp 0} \sin(\omega_c t + \alpha). \quad (5.21)$$

It is clear from (5.21) that

$$v_x^2 + v_y^2 = V_{\perp 0}^2 = \text{const}, \quad (5.22)$$

which means that $V_{\perp 0}$ is a (constant) velocity in the xy-plane. Its value, as well as a phase α are determined by initial conditions; (5.22) reflects the conservation of the transverse kinetic energy, discussed above.

Once again integrated, (5.21) results in

$$x = x_0 + R_c \sin(\omega_c t + \alpha); \quad y = y_0 + R_c \cos(\omega_c t + \alpha), \quad (5.23)$$

where

$$R_c = \frac{V_{\perp 0}}{\omega_c}; \quad |R_c| = \frac{|V_{\perp 0}|}{\omega_c} = \frac{\sqrt{m}}{|e|}\sqrt{\frac{mV_{\perp 0}^2}{|B|}}\frac{1}{\sqrt{|B|}}.$$

The reason for writing $|R_c|$ in this form will become apparent later. Equations (5.21)–(5.23) describe the rotational motion of a charged particle along a circle of a radius $|R_c|$ with angular frequency of ω_c in the xy-plane. $|R_c|$ and ω_c are called the cyclotron radius and the cyclotron frequency, respectively. The coordinates x_0, y_0 of the circle's center and the linear velocity of rotation $V_{\perp 0}$ are determined by the initial conditions. The direction of rotation is determined by the sign of the product eB_z, or, alternatively, by the sign of ω_c: for positive ω_c the rotation is clockwise, while for negative ω_c it is counter-clockwise. For future reference, let us mention that in both cases ($eB_z \lessgtr 0$) the vector product

$$[\boldsymbol{r}_\perp \times \boldsymbol{v}_\perp] = -\left\{0, 0, \frac{V_{\perp 0}^2}{\omega_c}\right\}$$

and thus has the opposite sign of ω_c. If convenient, this transverse motion may be viewed as two strongly correlated oscillatory motions in x and y directions with a constant phase shift equal to $\pi/2$. The overall motion (including the z direction) is performed along the helix of radius R_c, having its axis along the direction of the magnetic field.

5.3.2 Adiabatic Invariance

Since the transverse motion of the charge particle in a constant magnetic field is periodic, the associated action integral

$$J_\perp = \frac{1}{2\pi} \oint \boldsymbol{P}_{g\perp} d\boldsymbol{l} \qquad (5.24)$$

(where $\boldsymbol{P}_g$ is the canonical momentum (1.123), $d\boldsymbol{l}$ a directed line element along the circular path and integration is over a complete circle) is a constant of motion and, most importantly, an adiabatic invariant, that remains constant under slow variation of external parameters [15, 21], such as the magnetic field. All the physical quantities related to the transverse motion (kinetic energy, angular momentum etc.) are functions of two parameters, $\boldsymbol{v}_\perp$ and $\boldsymbol{B}$. An evaluation of the integral in (5.24) results in the relation

$$J_\perp(\boldsymbol{v}_\perp, \boldsymbol{B}) \equiv J_\perp = \text{const}$$

and allows a division of all these quantities into two groups: the one comprised of quantities dependent only on $J_\perp$, which are therefore themselves adiabatic invariants [15]; and the other comprised of quantities whose explicit dependence on slow varying $\boldsymbol{B}$ is completely established. In order to evaluate $J_\perp$ we substitute $\boldsymbol{P}_{g\perp}$ from (1.123) into (5.24). Then we obtain

$$J_\perp = \frac{1}{2\pi}\left[\oint m\boldsymbol{v}_\perp d\boldsymbol{l} + e \oint \boldsymbol{A}_\perp d\boldsymbol{l}\right]. \qquad (5.25)$$

Since $\boldsymbol{v}_\perp$ is parallel to $d\boldsymbol{l}$ and constant in magnitude, the first term in (5.25) is easily integrated to give $m|V_{\perp 0} R_c|$; the second term after the substitution can be treated as follows

$$\mathbf{A}_\perp = \frac{1}{2}[\mathbf{B} \times \mathbf{r}]_\perp$$

$$\frac{1}{2\pi} e \frac{1}{2} \oint [\mathbf{B} \times \mathbf{r}]_\perp \, dl = \frac{eB}{4\pi} \oint [\mathbf{r} \times d\mathbf{l}]_z = \frac{eB}{4\pi} \oint \left[\mathbf{r} \times \frac{\mathbf{v}_\perp}{|\mathbf{V}_{\perp 0}|} |R_c| \, d\varphi \right]_z$$

$$= \frac{eB}{4\pi} \oint \left[-\frac{|\mathbf{V}_{\perp 0}|}{\omega_c} |R_c| \right] d\varphi = -\frac{m|\mathbf{V}_{\perp 0} R_c|}{2},$$

and, finally

$$J_\perp = \frac{m|\mathbf{V}_{\perp 0} R_c|}{2} = \frac{m|\omega_c| R_c^2}{2} = \frac{mV_{\perp 0}^2}{2|\omega_c|} = \text{const.} \quad (5.26)$$

Consider some essential physical quantities from the point of view of an adiabatic invariance.

1. **The Angular Momentum L_z.** According to the definition

$$L_z = [\mathbf{r} \times \mathbf{P}_g]_z = [\mathbf{r} \times (m\mathbf{v} + e\mathbf{A})]_z = m[\mathbf{r}_\perp \times \mathbf{v}]_z + e[\mathbf{r} \times \mathbf{A}]_z$$

$$= -\frac{mV_{\perp 0}^2}{\omega_c} + \frac{e}{2}[\mathbf{r} \times [\mathbf{B} \times \mathbf{r}]]_z = -\frac{mV_{\perp 0}^2}{\omega_c} + \frac{eBR_c^2}{2} =$$

$$= -\frac{mV_{\perp 0}^2}{\omega_c} + \frac{m\omega_c}{2} \frac{V_{\perp 0}^2}{\omega_c^2} = -\frac{mV_{\perp 0}^2}{2\omega_c}.$$

Thus

$$L_z = -\frac{mV_{\perp 0}^2}{2\omega_c} = -J_\perp \operatorname{sign}\omega_c \quad (5.27)$$

is a constant of motion and an adiabatic invariant. In the most important case of electron motion in a constant magnetic field aligned along the $+z$ direction, $\omega_c < 0$ and

$$L_z = \frac{mV_{\perp 0}^2}{2|\omega_c|} = J_\perp.$$

2. **The Orbital Magnetic moment M_z.** As will be shown in the next section, the moving charged particle creates a magnetic field, equal to the one produced by a magnetic field dipole with a magnetic moment

$$\boldsymbol{\mu} = \frac{e}{2}[\mathbf{r} \times \mathbf{v}], \quad (5.28)$$

where $\mathbf{r}$ and $\mathbf{v}$ are the radius-vector and the velocity of the particle. Since for the transverse motion

$$[\mathbf{r} \times \mathbf{v}] = \left\{ 0, 0, -\frac{V_{\perp 0}^2}{\omega_c} \right\},$$

we find

$$\mu_z = \left(-\frac{e}{m}\right)\frac{mV_{\perp 0}^2}{2\omega_c} = -\frac{e}{m}J_\perp \text{sign}\omega_c = \frac{e}{m}L_z. \quad (5.29)$$

Thus, the magnetic moment is a constant of motion and an adiabatic invariant. The magnetic moment of an electron, moving in a constant $\boldsymbol{B} \parallel Z$ ($\omega_c > 0$) is equal to

$$\mu_{z(el)} = -\frac{|e|}{m}L_z = -\frac{mV_{\perp 0}^2}{2|B|}. \quad (5.30)$$

It is worth noting that the gyromagnetic ratio (the ratio of the magnetic moment to the mechanical angular momentum) is equal to e/m (5.29) and differs by factor 2 from the standard value $e/2m$.

3. **Magnetic Flux.** The magnetic flux through the area bounded by a circular orbit of radius R_c is

$$\Phi = \pi R_c^2 B = \frac{\pi m V_{\perp 0}^2}{e|\omega_c|} = \frac{2\pi}{e}J_\perp. \quad (5.31)$$

Thus, the magnetic flux is an adiabatic invariant.

4. **Kinetic Energy.** It follows from (5.26) that the kinetic energy $\mathcal{E}_\perp$ of the transverse motion is given by

$$\mathcal{E}_\perp = \frac{mV_{\perp 0}^2}{2} = J_\perp|\omega_c| = -\mu_z B. \quad (5.32)$$

Thus, transverse kinetic energy, being a constant of motion is not an adiabatic invariant; in slow varying magnetic fields, it changes proportionately to B.

5. **The Cyclotron Radius** R_c. Using (5.26), we obtain

$$|R_c| = \sqrt{\frac{2J_\perp}{|e|}}\frac{1}{\sqrt{B}}. \quad (5.33)$$

The cyclotron radius is not an adiabatic invariant; it changes in slow varying fields in inverse proportion to the square root B.

5.3.3 Static Uniform Electric Field

Let us consider the motion of a charge e in a uniform constant in time electric field $\boldsymbol{E}$. Taking the direction of the field as the y-axis, the plane, containing the vectors of the field $\boldsymbol{E}$ and the initial velocity $\boldsymbol{V}_0$, as the xy plane, and the initial coordinates as the origin of the coordinate system, we write the equations of motion (1.123) in a non-relativistic approximation:

$$m\frac{dv_x}{dt} = 0; \quad m\frac{dv_y}{dt} = eE.$$

The solution

$$v_x = V_{x0}; \quad v_y = V_{y0} + \frac{eE}{m}t$$

and

$$x = V_{x0}t; \quad y = V_{y0}t + \frac{eE}{2m}t^2$$

determines the trajectory

$$y = \frac{eE}{2mV_{x0}^2}x^é + \frac{V_{y0}}{V_{x0}}x. \tag{5.34}$$

According to (5.34), the charge moves along a parabola. The motion direction is determined by the signs of the initial velocity components V_{x0} and V_{y0}. If both velocity components are positive $V_{x0}, V_{y0} > 0$, then the charge starting from the origin $x = 0, y = 0$ moves along the positive branch of the parabola where $x, y > 0$. In the opposite case $V_{x0}, V_{y0} < 0$ the charge first moves along the negative branch of the parabola $x, y < 0$, passes through the minimum point $\left(-m\left|V_{x0}V_{y0}\right|/eE; -mV_{y0}^2/2eE\right)$ where it changes the direction of motion along the y-axis, and finally, for $x < -2m\left|V_{x0}V_{y0}\right|/eE$, the charge continues the motion in the second quarter $x < 0, y > 0$. Consider now the charge trajectories when the initial velocity components possess the different signs. If $V_{x0} < 0, V_{y0} > 0$ the charge moves monotonically in the second quarter $x < 0, y > 0$. If $V_{x0} > 0, V_{y0} < 0$ then the charge first moves in the fourth quarter $x > 0, y < 0$ down to the minimum point $\left(m\left|V_{x0}V_{y0}\right|/eE; -mV_{y0}^2/2eE\right)$ then changes the direction of motion along the y-axis and for $x > 2m\left|V_{x0}V_{y0}\right|/eE$ the charge moves along the positive branch of the parabola $x, y > 0$.

5.3.4 Combined Static Uniform Electric and Magnetic Fields

Choosing the direction of $\boldsymbol{B}$ as z-axis, we decouple the equation of motion

$$m\frac{d\boldsymbol{v}}{dt} = e\boldsymbol{E} + e\left[\boldsymbol{v} \times \boldsymbol{B}\right]$$

into an equation of motion along the magnetic field

$$m\frac{dv_z}{dt} = eE_z \tag{5.35}$$

and in a perpendicular direction

$$m\frac{d\boldsymbol{v}_\perp}{dt} = e\boldsymbol{E}_\perp + e\left[\boldsymbol{v}_\perp \times \boldsymbol{B}\right]. \tag{5.36}$$

5.3 A Charged Particle Motion in Static Uniform Fields

Equation (5.35) describes the motion of the charge with constant acceleration in z direction:

$$z = \frac{eE_z}{2m}t^2 + V_{z0}t. \tag{5.37}$$

Concerning the transverse motion, we note that (5.36) is a system of two in homogeneous linear differential equations with respect to $v_{x,y}$; its general solution is equal to the sum of the general solution of the corresponding homogeneous equation

$$m\frac{d\boldsymbol{v}_\perp}{dt} = e\left[\boldsymbol{v}_\perp \times \boldsymbol{B}\right] \tag{5.38}$$

and the particular solution of the inhomogeneous (5.36). The general solution of (5.38) has been found above; it is given by (5.21)–(5.23), which describe the rotational motion of a charge in the xy-plane with angular frequency ω_c and radius R_c. The particular solution of (5.36) can be obtained as follows. Assuming in (5.36) $\boldsymbol{v}_\perp = \text{const}$, we have

$$\boldsymbol{E}_\perp = -\left[\boldsymbol{v}_{\perp d} \times \boldsymbol{B}\right]. \tag{5.39}$$

Multiplying (5.39) vectorially from the left by $\boldsymbol{B}$, and using the identity

$$[\boldsymbol{a} \times [\boldsymbol{b} \times \boldsymbol{c}]] = \boldsymbol{b}\left(\boldsymbol{a}\cdot\boldsymbol{c}\right) - \boldsymbol{c}\left(\boldsymbol{a}\cdot\boldsymbol{b}\right),$$

we arrive at

$$[\boldsymbol{E}_\perp \times \boldsymbol{B}] = \left[\boldsymbol{v}_{\perp d}B^2 - (\boldsymbol{v}_{\perp d}\cdot\boldsymbol{B})\boldsymbol{B}\right]. \tag{5.40}$$

Since $(\boldsymbol{v}_{\perp d}\cdot\boldsymbol{B}) = 0$, (5.40) results in

$$\boldsymbol{v}_{\perp d} = \frac{[\boldsymbol{E}_\perp \times \boldsymbol{B}]}{B^2}. \tag{5.41}$$

Equation (5.41) describes motion with a constant velocity equal to $E_\perp/B$ in a direction perpendicular to both $\boldsymbol{B}$ and $\boldsymbol{E}$; thus, the complete transverse motion can be viewed as a cyclotron rotation with a simultaneous drift of the circle-center, with a constant velocity v_d given by (5.41). It is important to note that the drift velocity does not depend either on the parameters of the particle (e, m) or on the initial conditions of its motion (like $\boldsymbol{V}_{\perp 0}$). Let us mention also that the velocity components of the rotational motion are periodic functions of time (see (5.21)); thus, if the time-average is taken over the time $t \gg \omega_c^{-1}$, their contribution vanishes and the average velocity is simply equal to the drift velocity.[2]

---

[2] This is precisely the situation when the current density $\boldsymbol{j} = \sum_i \overline{e_i\delta\left(\boldsymbol{r} - \boldsymbol{r}_i\right)\boldsymbol{v}_i} = \rho\boldsymbol{v}$ is calculated. Experimentally, the drift of the charge carriers gives rise to the Hall effect – the appearance of the potential difference in the direction perpendicular to the applied electric and magnetic fields.

In the particular case of $E_\perp = E_y$ (the so-called Hall geometry) and for initial conditions $x = y = 0$ at $t = 0$, the parametric equations of the projection of the trajectory on the xy plane are given by

$$x = R_c \sin \omega_c t + \frac{E_y}{B} t; \quad y = R_c \cos \omega_c t - 1, \tag{5.42}$$

which are equations of trochoid.

5.4 Hall Effect in Semiconductors

Consider the influence of the magnetic field on a current flowing in a semiconductor. Assume that the electric field $\boldsymbol{E}$ is parallel to the x-axis and the uniform magnetic field $\boldsymbol{B}$ is parallel to the y-axis:

$$\boldsymbol{E} = (E, 0, 0), \quad \boldsymbol{B} = (0, B, 0). \tag{5.43}$$

The electric current density caused by the electric field in the absence of the magnetic field has the form

$$\boldsymbol{j} = \sigma \boldsymbol{E} \tag{5.44}$$

and the charge carriers possess the drift velocity $\boldsymbol{v}_\mathrm{d}$ which is given by expression (6.42). We have seen previously that the charged particle in the magnetic field is affected by the Lorentz force. The component $\boldsymbol{F}$ of this force caused by the magnetic field is

$$\boldsymbol{F} = e \left[\boldsymbol{v}_\mathrm{d} \times \boldsymbol{B} \right]. \tag{5.45}$$

Substituting expressions (6.42) and (6.43) into (5.45) we get

$$\boldsymbol{F} = \frac{e^2 \tau}{m^*} \left[\boldsymbol{E} \times \boldsymbol{B} \right]. \tag{5.46}$$

Suppose first that the semiconductor sample is infinite in the z direction. Expression (5.46) shows clearly that the Lorentz force direction in the case of charge carrier drift is independent of its sign and is determined by the direction of fields $\boldsymbol{E}$ and $\boldsymbol{B}$, or by the direction of the current density $\boldsymbol{j}$ and the magnetic field $\boldsymbol{B}$. Hence, in the case of the drift motion both electrons and holes deviate in the same direction under the influence of the Lorentz force (5.46). In general, due to the collisions, electrons and holes subjected to the the fields (5.43) are moving along the trajectory which represents a straight line resulting from the averaging of elementary cycloids. The current density $\boldsymbol{j}$ in the magnetic field $\boldsymbol{B}$ rotates by the so-called Hall angle φ with respect to the electric field applied. The rotation direction depends on the type of charge carriers due to the fact that both types are deviated by the magnetic field in the same direction while they drift in the opposite directions.

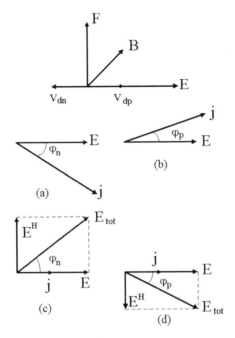

Fig. 5.1. The Hall angle φ_n and φ_p for electrons and holes respectively, for the case of an infinite sample (**a**), (**b**) and a sample of a finite thickness d (**c**), (**d**). The drift velocities of electrons and holes are v_{dn} and v_{dp}

Consider now the sample with the one type of charge carriers and with the finite dimensions in the z direction. Then, in the chosen geometry, the upper and the lower surfaces of the sample become negatively and positively charged, respectively, due to the transverse component of the current density $j_z \neq 0$. As a result, an electric field $\boldsymbol{E}^H$ occurs transverse with respect to the Lorentz force (5.46) which is called the Hall field. The emergence of the transverse electric field due to the magnetic field applied is known as the Hall effect [4]. Obviously, the direction of the Hall field $\boldsymbol{E}^H$ is defined by the types of the charge carriers. The transverse electric field increases until it compensates the Lorentz force (5.46). Then, the deviation of charge carriers is stopped and they are moving parallel to the external electric field (5.43). The total electric field $\boldsymbol{E}_{tot}$ consists of the external field $\boldsymbol{E}$ (5.43) and the Hall field $\boldsymbol{E}^H$

$$\boldsymbol{E}_{tot} = \boldsymbol{E} + \boldsymbol{E}^H \tag{5.47}$$

It rotates by the Hall angle φ with respect to the initial current density (5.44). The Hall angles φ_n for electrons and φ_p for holes for the infinite sample and for the sample of a finite thickness d are shown in Fig. 5.1.

The Hall voltage V^H occurs between the surfaces of the sample with the thickness d in the z direction

$$V^H = dE^H. \tag{5.48}$$

The four quantities $\boldsymbol{E}^H, \boldsymbol{j}, \boldsymbol{B}$ and the so-called Hall constant C^H are related as follows

$$\boldsymbol{E}^H = C^H \left[\boldsymbol{B} \times \boldsymbol{j}\right]. \tag{5.49}$$

The Hall constant C^H can be calculated from the condition of the equilibrium between the Lorentz force (5.46) and the electric force due to the Hall field $\boldsymbol{E}^H$. We have

$$e\boldsymbol{E}^H + \boldsymbol{F} = 0. \tag{5.50}$$

Combining (5.45) and (6.42) we obtain

$$\boldsymbol{E}^H = -\mu \left[\boldsymbol{E} \times \boldsymbol{B}\right]. \tag{5.51}$$

Comparing (5.49) and (5.51) and taking into account (5.44) we immediately get

$$C^H \sigma = \mu \tag{5.52}$$

and

$$C^H = \frac{\mu}{\sigma} = \frac{1}{en}, \tag{5.53}$$

where n is the concentration of the charge carriers, either electrons, or holes. It is seen from (5.53) that the sign of the Hall constant C^H coincides with the sign of the electric charge of a carrier. Consequently, one can determine the type of conduction in semiconductor once the Hall constant C^H is known. The sign of the Hall constant C^H itself is identified by using the Hall potential (5.48). The Hall angle φ is given by

$$\tan \varphi = \frac{E^H}{E} = -\frac{C^H B \sigma E}{E} = -\mu B. \tag{5.54}$$

It should be noted that at given values of the fields $\boldsymbol{E}$ and $\boldsymbol{B}$ (5.43) the Hall field $\boldsymbol{E}^H$ is determined only by the charge carriers mobility μ. The process of the Hall buildup in a semiconductor with an electron and a hole type of conduction is illustrated in Fig. 5.2.

In a sample with both electrons and holes contributing to the conduction the Hall field vanishes, if the mobilities of two types of charge carriers coincide. For a typical value of concentration $n \sim 10^{16}$ cm^{-3} we find that $C^H \approx 6 \times 10^{-4}$ m^3/C. Experimentally, the Hall potential V^H (5.48) and the total current

$$I = jad \tag{5.55}$$

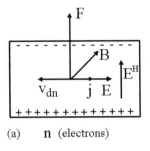

(a) n (electrons)

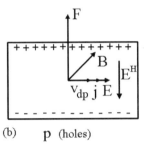

(b) p (holes)

Fig. 5.2. The Hall field E^H buildup in a semiconductor sample with an electron type of conduction (**a**) and in a semiconductor sample with a hole type of conduction (**b**). v_{dn} and v_{dp} are the drift velocities of electrons and holes respectively

through the sample are measured where a is the sample dimension in the y direction parallel to the magnetic field. Then the Hall constant C^H takes the form

$$C^H = \frac{V^H a}{IB}. \tag{5.56}$$

Consider now the Hall effect in an anisotropic material [4]. This time the current density j and the electric field E are related by the conductivity tensor σ_{ik}

$$j_i = \sigma_{ik} E_k. \tag{5.57}$$

In the presence of the magnetic field B the components of the conductivity tensor σ_{ik} obey the conditions

$$\sigma_{ik}(B) = \sigma_{ki}(-B). \tag{5.58}$$

Any tensor of the rank two can be divided into symmetrical and anti-symmetrical parts s_{ik} and a_{ik} respectively.

$$\sigma_{ik} = s_{ik} + a_{ik}. \tag{5.59}$$

By definition

$$s_{ik}(\boldsymbol{B}) = s_{ki}(\boldsymbol{B}), \quad a_{ik}(\boldsymbol{B}) = -a_{ki}(\boldsymbol{B}). \tag{5.60}$$

Combining relations (5.58) and (5.60) we find that the components of the tensor s_{ik} are even functions of the magnetic field and those of a_{ik} are odd functions:

$$s_{ik}(\boldsymbol{B}) = s_{ki}(-\boldsymbol{B}) = s_{ik}(-\boldsymbol{B}), \quad a_{ik}(\boldsymbol{B}) = a_{ki}(-\boldsymbol{B}) = -a_{ik}(-\boldsymbol{B}). \tag{5.61}$$

It is known that any antisymmetrical tensor of rank two corresponds to some axial vector. Then we have for a_{ik}

$$a_{yz} = a_x, \quad -a_{xz} = a_y, \quad a_{xy} = a_z. \tag{5.62}$$

In a comparatively weak external magnetic field the conductivity tensor σ_{ik} can be expanded in powers of the magnetic field $\boldsymbol{B}$. The expansion of tensors a_{ik} and s_{ik} would contain the odd and even powers of $\boldsymbol{B}$, respectively, according to (5.61). It is known that any antisymmetrical tensor of rank two corresponds to some axial vector. Then we have for a_{ik} and s_{ik} taking into account (5.62) and keeping the terms up to the second power in $\boldsymbol{B}$:

$$a_i = A_{ik}B_k, \quad s_{ik} = \sigma_{0ik} + \gamma_{iklm}B_l B_m, \tag{5.63}$$

where σ_{0ik} is conductivity tensor in the absence of the magnetic field, A_{ik} is a polar tensor of the rank two, and γ_{iklm} is the tensor of the rank four which is symmetrical with respect to the indices i, k and l, m. Substituting relations (5.59), (5.62) and (5.63) into (5.57) we obtain

$$j_i = \sigma_{0ik}E_k + [\boldsymbol{E} \times \boldsymbol{a}]_i + \gamma_{iklm}B_l B_m E_k. \tag{5.64}$$

Comparison of the (5.64) and the first equation of (5.63) shows that the second term in the right-hand side of (5.64) linear in the magnetic field is responsible for the Hall effect since it creates the component of the current density proportional to the magnetic field and perpendicular to the electric field applied. It should be noted that the third term in the right-hand side of (5.64) also contains the current density component perpendicular to the electric field but it is quadratic in the magnetic field and consequently it is small compared to the principal term.

Finally, we should emphasize that the Hall term does not contribute to the Joule heat generated by the passage of the current. Indeed, the Joule heat P is given by a scalar product of the current density and electric field:

$$P = \boldsymbol{j} \cdot \boldsymbol{E}. \tag{5.65}$$

The Hall term is perpendicular to $\boldsymbol{E}$ and their product is zero. Then, substituting (5.63) and (5.64) into (5.65) we get

$$P = s_{ik}E_i E_k. \tag{5.66}$$

The Joule heat P is determined only by the symmetrical part of the conductivity tensor.

It can be shown that the Hall effect in an anisotropic medium can be described in another form by expressing the electric field $\boldsymbol{E}$ in terms of the current density $\boldsymbol{j}$

$$E_i = \rho_{ik} j_k + [\boldsymbol{j} \times \boldsymbol{b}]_i . \tag{5.67}$$

Here the tensor ρ_{ik} and the vector $\boldsymbol{b}$ have the same properties as s_{ik} and $\boldsymbol{a}$. The vector $\boldsymbol{b}$ is linear in the magnetic field, and the Hall effect is represented by the term $[\boldsymbol{j} \times \boldsymbol{b}]$. Then the Hall electric field is perpendicular to the current and proportional to the magnetic and electric fields applied similarly to (5.51).

5.5 A Charged Particle Motion in Static Non-uniform Magnetic Fields

Studying the motion of a charged particle in an uniform static electric and magnetic fields, we succeeded in finding exact solutions. Unfortunately, this can not be done in the general case of fields varying in space and time. However, if the variations are slow enough, a perturbation solution to the motion is an adequate approximation. Slow enough generally means that the distance (or time) over which the field appreciably changes in magnitude or direction is large compared to the cyclotron radium R_c or the period $T_c = 2\pi/\omega_c$ of rotation. More precisely, these conditions may be formulated as

$$R_c |(\text{grad} B)_\perp| \ll B \tag{5.68}$$

$$T_c v_\parallel \left|(\text{grad} B)_\parallel\right| \ll B \tag{5.69}$$

$$T_c \frac{\partial B}{\partial t} \ll B. \tag{5.70}$$

If the conditions (5.68)–(5.70) are fulfilled, the overall motion of the particle is effectively decoupled into two almost independent motions [21]:

1. A fast rotation (spiralling) around the lines of a field at a frequency $\omega_c(B)$ and with a radius $R_c(B(\boldsymbol{r}))$ given by the local value of the field.
2. Slow changes which can be described as a drift of the center of the orbit (a drifting of the guiding center), and which 'feel' the fast motion only through its averaged characteristics.

5.5.1 Gradient Drift

The first type of the spatial variation of the field to be considered is a field B gradient is perpendicular to the direction of B: $\nabla |B| \perp B$. Let the gradient at the point of interest (the center of the orbit) be in the direction of the unit vector $n \perp B_0$ (B_0 being the field of this point). Then, to the first order, the magnetic field of the point r_c at the orbit can be written as

$$B(r_c) = B_0 + \left(\frac{\partial |B|}{\partial \zeta}\right)_0 (n \cdot r_c) \frac{B_0}{|B_0|}, \qquad (5.71)$$

where $(\partial |B|/\partial \zeta)_0$ is the derivative in the direction n, and the last factor $B_0/|B_0|$ reflects the fact that the direction of B remains unchanged. Thus, the non-uniformity considered does not affect the longitudinal motion which remains a uniform translation, and we are left with the problem of transverse motion only. Writing $v_\perp = v_{\perp 0} + v_1$, where $v_{\perp 0}$ is the transverse velocity in the uniform field and v_1 is a small correction term, and substituting (5.71) into the transverse equation of motion (5.38), we obtain for the first-order terms

$$m \frac{dv_1}{dt} = e[v_1 \times B_0] + e[v_{\perp 0} \times B_0] \left(\frac{\partial |B|}{\partial \zeta}\right)_0 (n \cdot r_c) \frac{1}{|B_0|}. \qquad (5.72)$$

The general solution of the homogeneous equation, related to (5.72), is identical to the one for $v_{\perp 0}$ and is included in it. Thus, we are interested only in the particular solution to (5.72). Using (5.21) and (5.23) we can easily show that

$$v_{\perp 0} = -\frac{e}{m}[B_0 \times r_c] \qquad (5.73)$$

and

$$[v_{\perp 0} \times B_0] = -\frac{eB_0^2}{m} r_c. \qquad (5.74)$$

Substituting (5.73) into (5.72) we obtain

$$m \frac{dv_1}{dt} = e \left[\left\{v_1 - \frac{e}{m|B_0|} \left(\frac{\partial |B|}{\partial \zeta}\right)_0 [B_0 \times r_c(n \cdot r_c)]\right\} \times B_0\right]. \qquad (5.75)$$

At this point we approximate the rapidly oscillating with frequency $2\,\omega_c$ term $r_c(n \cdot r_c)$ by its average value, assuming that v_1 is small enough, so that during the period of rotation the particle will not leave the region of $B \approx B_0$. Writing

$$\overline{r_c(n \cdot r_c)} = \overline{(x_c i + y_c j)(x_c n_x + y_c n_y)}$$

5.5 A Charged Particle Motion in Static Non-uniform Magnetic Fields

(where i and j are unit vectors in the x and y directions, respectively) and substituting $x_c(t)$ and $y_c(t)$ from (5.23) we find that (since $\overline{x_c y_c} = 0$; $\overline{x_c^2} = \overline{y_c^2} = R_c^2/2$)

$$\overline{r_c(n \cdot r_c)} = \frac{R_c^2}{2}(n_x i + n_y j) = \frac{R_c^2}{2} n. \tag{5.76}$$

The substitution of (5.76) into (5.75) results in

$$m\frac{dv_1}{dt} \approx e\left[\left\{v_1 - \frac{e}{2m|B_0|}\left(\frac{\partial |B|}{\partial \zeta}\right)_0 R_c^2 [B_0 \times n]\right\} \times B_0\right]. \tag{5.77}$$

The particular solution of this equation has the form

$$v_1 = \text{const} = \frac{eR_c^2}{2m|B_0|}\left(\frac{\partial |B|}{\partial \zeta}\right)_0 [B_0 \times n],$$

or, since

$$\frac{e}{m} = \frac{\omega_c}{|B_0|}$$

and

$$\left(\frac{\partial |B|}{\partial \zeta}\right)_0 n \equiv \nabla |B|,$$

we find

$$v_1 \equiv v_{\nabla |B|} = \frac{R_c^2 \omega_c}{2B^2}[B \times \nabla |B|] = \frac{v_{\perp 0} R_c}{2B^2}[B \times \nabla |B|]. \tag{5.78}$$

Equation (5.78) shows that, if the variation of the magnetic field on the length scale of the radius of the orbit ($R_c \cdot \nabla |B|$) is small in comparison with B itself, the drift velocity v_1 is small in comparison with the orbital velocity $v_\perp = \omega_c R_c$. The particle spirals rapidly, while its center of rotation moves slowly in a direction perpendicular to both B and $\nabla |B|$; positive and negative particles move in opposite directions since ω_c has different signs for both types of carriers. The qualitative features of gradient drift can be understood from the consideration of the variation of the orbit as the particle moves in the regions where the field differs from the average. This qualitative behavior for both signs of the charge is shown in the following Fig. 5.3.

5.5.2 Centrifugal Drift

Consider now the magnetic field whose force lines are curved with a local radius of curvature R; the magnitude of magnetic field $|B|$ is decreasing with a radius ($\nabla |B| \parallel R$) (see Fig. 5.4) [21]:

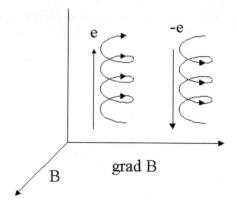

Fig. 5.3. Drift of charged particles due to the transverse gradient of the magnetic field

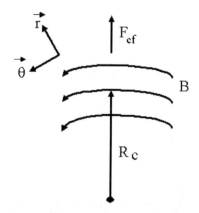

Fig. 5.4. The magnetic field B with curved intensity lines

In this case, the gradient drift (for which the explicit expression of velocity will be derived later) is accompanied by a new kind of drift, which can be understood as follows. The particle rotates around a field line with an orbital velocity $v_\perp$ and moves along this line with a longitudinal velocity $v_\parallel$. But the field lines curve off; as far as the motion of the guiding center (the center of the orbit) is concerned, this gives rise to a centrifugal acceleration

$$a = \frac{v_\parallel^2}{R}\frac{\boldsymbol{R}}{R}$$

which can be viewed as arising from a centrifugal force

$$\boldsymbol{F}_{cf} = m\frac{v_\parallel^2}{R}\frac{\boldsymbol{R}}{R} \tag{5.79}$$

This force causes a curvature drift velocity

5.5 A Charged Particle Motion in Static Non-uniform Magnetic Fields

$$v_R = \frac{1}{e}\frac{[F_{cf} \times B]}{B^2} = \frac{mv_\parallel^2}{eB^2}\frac{[R \times B]}{R^2} \tag{5.80}$$

The direction of the drift is determined by the vector product, in which R is the radius-vector pointed from the center of the curvature to the moving charged particle; for negative and positive particles, the directions of drift are opposite, owing to the different signs of e in (5.80).

Let us now evaluate the velocity of the gradient drift, which accompanies the centrifugal drift. The static magnetic field in vacuum ($\rho = 0, \boldsymbol{j} = 0$) obeys, according to (1.19), the equation

$$\mathrm{curl}\boldsymbol{H} = 0$$

and the material equation $\boldsymbol{B} = \mu_0 \boldsymbol{H}$, i.e. $\boldsymbol{B} \parallel \boldsymbol{H}$ and we can use the equation

$$\mathrm{curl}\boldsymbol{B} = 0.$$

Using the cylindrical coordinates, one immediately concludes that $\boldsymbol{B}$ has only the θ– component, $\nabla |\boldsymbol{B}|$ only the r-component, while $\mathrm{curl}\boldsymbol{B}$ has only the z-component. We obtain in this case

$$(\mathrm{curl}\boldsymbol{B})_z = \frac{1}{r}\frac{\partial}{\partial r}(rB_\theta) = 0; \quad B_\theta = |\boldsymbol{B}| = \frac{\mathrm{const}}{r}.$$

Thus, at a particular point with a radius-vector $\boldsymbol{R}$,

$$|\boldsymbol{B}| = \frac{\mathrm{const}}{|\boldsymbol{R}|}, \quad \frac{\nabla |\boldsymbol{B}|}{|\boldsymbol{B}|} = -\frac{\boldsymbol{R}}{R^2}. \tag{5.81}$$

Substituting (5.81) into (5.78) one obtains

$$v_{\nabla|B|} = -\frac{v_\perp R_c}{2B^2}\left[\boldsymbol{B} \times \frac{|\boldsymbol{B}|\boldsymbol{R}}{R^2}\right] = \frac{v_\perp^2}{2\omega_c}\frac{[\boldsymbol{R} \times \boldsymbol{B}]}{R^2|\boldsymbol{B}|} = \frac{mv_\perp^2}{2eB^2}\frac{[\boldsymbol{R} \times \boldsymbol{B}]}{R^2}. \tag{5.82}$$

Combining (5.80) and (5.82), one finds the full drift velocity v_d:

$$v_d = v_R + v_{\nabla|B|} = \frac{m}{e}\frac{[\boldsymbol{R} \times \boldsymbol{B}]}{B^2 R^2}\left(v_\parallel^2 + \frac{1}{2}v_\perp^2\right). \tag{5.83}$$

For singly charged nonrelativistic particles in thermal equilibrium

$$mv_\parallel^2 = \frac{1}{2}mv_\perp^2 = kT$$

the magnitude of the drift velocity is

$$v_d\left(\frac{\mathrm{cm}}{\mathrm{s}}\right) = \frac{172T\,(^\circ \mathrm{K})}{R\,(\mathrm{m})\,B\,(\mathrm{gauss})}.$$

This means, for example, that in a toroidal tube of $R = 1\mathrm{m}$, placed in a field of $B = 10^3$ gauss, particles of a moderately hot plasma (energy ~ 1 ev, $T^\circ \mathrm{K} \simeq 10^4 \mathrm{K}$) will have drift velocities $v_d \sim 10^3$ cm/s and will drift out to the walls in a small fraction of a second. For hotter plasmas, needed for thermo-nuclear fusion, the drift rate is correspondingly greater – causing one on the troubles in attempts to perform thermo-nuclear synthesis through the confinement of plasma by magnetic fields.

5.5.3 Magnetic Mirrors

Let us now consider a situation in which a static magnetic field $\boldsymbol{B}(z)$ acts mainly in the z direction, with a small gradient in the same direction, i.e. $\nabla B \parallel \boldsymbol{B}$. The general behavior of the lines of force in this case is shown in Fig. 5.5.

The motion of the charged particle in such a field is governed by the law of conservation of energy and by the adiabatic invariance of its orbital magnetic moment $\boldsymbol{\mu}$. Since the energy of the particle's transverse motion is

$$\mathcal{E}_\perp = |\boldsymbol{\mu}| B,$$

$|\boldsymbol{\mu}|$ being independent of $\boldsymbol{B}$ for slow varying fields, and since the total energy of the particle is conserved, it is clear that during its motion in the varying in magnitude magnetic field $B_z(z)$, energy transfer between the transverse and the longitudinal motion should take place. Supposing that a particle is spiralling around the z-axis in an orbit with a radius R_c with a transverse velocity $\boldsymbol{V}_{\perp 0}$, and that the longitudinal velocity $\boldsymbol{V}_{\parallel 0}$ at the point $z = 0$, where the magnetic field $\boldsymbol{B} = \boldsymbol{B}_0$ is minimal in the region considered. The total energy $\mathcal{E}$ of the particle at this point is

$$\mathcal{E} = \frac{mv_0^2}{2} \equiv \mathcal{E}_\perp + \mathcal{E}_\parallel = |\boldsymbol{\mu}| B_0 + \frac{p_{\parallel 0}^2}{2m} \equiv |\boldsymbol{\mu}| B_0 + \frac{mv_{\parallel 0}^2}{2}, \tag{5.84}$$

where $p_{\parallel 0} = mv_{\parallel 0}$. When the particle reaches some point s on the line of force, where the field has a magnitude $B(s)$, its total conserving energy is given by

$$\mathcal{E} = |\boldsymbol{\mu}| B(s) + \frac{p_\parallel^2(s)}{2m} \tag{5.85}$$

with an unchanged magnetic moment $\boldsymbol{\mu}$, due to adiabatic invariance. Thus,

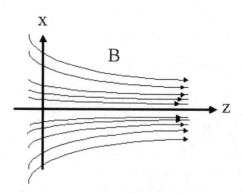

Fig. 5.5. The general behavior of the magnetic field lines of force for the case of a small field gradient in the z direction

5.5 A Charged Particle Motion in Static Non-uniform Magnetic Fields

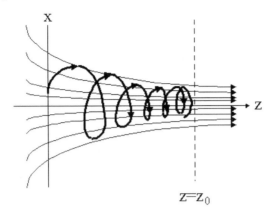

Fig. 5.6. Reflection of a charged particle out the high magnetic field region

$$\frac{p_\|^2(s) - p_{\|0}^2}{2m} = -|\boldsymbol{\mu}|\,(B(s) - B_0) \tag{5.86}$$

Equation (5.85) is actually the first integral of the equation of motion in the longitudinal direction. This equation of motion can be obtained by taking the time derivative of (5.85) and making use of the fact that

$$\frac{dB(s)}{dt} = \frac{\partial B}{\partial s}\frac{ds}{dt} = \frac{\partial B}{\partial s}v_\| = \frac{\partial B}{\partial s}\frac{p_\|}{m}.$$

Then we have

$$\frac{dp_\|}{dt} = -|\boldsymbol{\mu}|\frac{\partial B}{\partial s}. \tag{5.87}$$

Equation (5.87) describes a one-dimensional motion of the particle in the effective potential $U_{eff}(s) = |\boldsymbol{\mu}|\,B(s)$; if, as we have supposed, $B(s)$ has a minimum at $s = z = 0$, the implication is that the particle moves in the potential well. Then, if its kinetic energy $mv_0^2/2$ at the minimum of the well (which at this point is equal to the total energy) is lower than the height of the well (or, equivalently, the magnetic field is strong enough), the particle will be reflected by the potential barrier at the turning point s_1, determined by the relation

$$\frac{mv_0^2}{2} = |\boldsymbol{\mu}|\,B(s_1)$$

(see the following Fig. 5.6). Writing

$$\frac{mv_{\perp 0}^2}{2} = |\boldsymbol{\mu}|\,B_0 = \frac{mv_0^2}{2}\sin^2\alpha$$

we use (5.84) to evaluate $v_\|$ as a function of v_0, B and angle α between its velocity and the magnetic field:

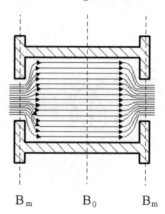

Fig. 5.7. A schematic diagram of "mirror" machine for the containment of a hot plasma

$$\frac{mv_\parallel^2}{2} = \frac{mv_0^2}{2} - |\boldsymbol{\mu}|\, B = \frac{mv_0^2}{2} - \frac{B}{B_0}\frac{mv_0^2}{2}\sin^2\alpha$$

and, finally

$$v_\parallel = v_0\sqrt{1 - \frac{B(s)}{B_0}\sin^2\alpha}. \tag{5.88}$$

Thus, the turning points s_t, if they exist, are given by the equation

$$B(s_t) = \frac{B_0}{\sin^2\alpha}, \tag{5.89}$$

which implies that all the particles moving at point $z = 0$ in the directions given by $\sqrt{B_0/B_{\max}} < \sin\alpha$, will be reflected by the well, and forced to move backwards to the region of the weaker field. They will be reflected and not simply stopped since, according to (5.87), at the turning points the particle, moving in the direction ∇B, will experience a force equal to $-|\boldsymbol{\mu}|\,(\partial B/\partial s)_{s_t}$ which is opposite in sign to the gradient of B. This is the reason why the region of a strong magnetic field is called the magnetic mirror [21]. Let us suppose that there are two such mirrors; then the particle, for which the angle α exceeds some critical value, will be reflected by both mirrors and will be captured in the space between them.

This principle can be applied to the containment of a hot plasma for thermonuclear energy production. A magnetic bottle can be constructed with an axial field produced by solenoidal winding over some region of space, and additional coils at each end to provide a much higher field towards the edges (see Fig. 5.7).Particles created or injected into the field in the central region ($B = B_0$) will spiral along the axis, but will be reflected by the magnetic mirrors of each end. If the ratio of maximum field B_m in the mirror to the

5.5 A Charged Particle Motion in Static Non-uniform Magnetic Fields

field B_0 in the central region is very large, only particles with a very large component of velocity parallel to the axis can penetrate through the ends. The criterion of reflection, mentioned above, may be rewritten as the criterion for trapping:

$$\left|\frac{v_{\|0}}{v_{\perp 0}}\right| < \sqrt{\frac{B_m}{B_0} - 1}. \tag{5.90}$$

If the particles are injected into the apparatus, it is easy to satisfy requirement (5.90). Then the escape of particles is governed by the rate at which they are scattered by the residual atoms etc., in such a way that their velocity component violates (5.90). Provided that the particle's initial velocity satisfies criterion (5.90), its motion in the space between two magnetic mirrors will be periodic. Thus, the action integral $J_\|$ of the longitudinal motion will be a constant of motion and an adiabatic invariant:

$$J_\| = \frac{1}{2\pi} \oint \boldsymbol{P}_{g\|} d\boldsymbol{l} = \text{const}. \tag{5.91}$$

(Here the integral is taken over the period of longitudinal motion, or, equivalently, along the path between the two turning points – forward and back.) Taking into account that

$$\boldsymbol{P}_{g\|} = m\boldsymbol{v}_\| + e\boldsymbol{A}$$

and that

$$\oint \boldsymbol{A}_\| d\boldsymbol{l} = 0$$

($\boldsymbol{A}_\|$ is single-valued), we obtain

$$J_\|^* = \oint \boldsymbol{v}_\| d\boldsymbol{l} = \text{const}. \tag{5.92}$$

The condition for the adiabatic invariance of $J_\|^*$ are satisfied if, on the one hand, the period of longitudinal vibrations is large compared to the period of cyclotron rotation, and, on the other hand, the transverse drift is small at the period of longitudinal motion.

The concept of the adiabatic invariance of $J_\|^*$ is a powerful tool in understanding the Fermi mechanism of the acceleration of cosmic-rays particles. As a preliminary step, we consider the following purely mechanical problem. Let us assume that the particle moves with a velocity $v_\|$ along the z-axis, while at the points $z = \pm L/2$ there are two ideally reflecting planes, perpendicular to the z-axis and moving with velocities $\pm u$ ($u \ll v_\|$). If the collisions of the particle with the planes are totally elastic, every collision gives rise to the change of particle momentum by

$$\Delta P_{g\|} = 2mu.$$

Since the delay between the consequent collisions in $\Delta t = L/v_\|$, the rate of the change in the particle's momentum is

$$\frac{dP_{g\|}}{dt} = \frac{\Delta P_{g\|}}{\Delta t} = \frac{2muv_\|}{L}.$$

But $2u = -dL/dt$ and $mv_\| = P_{g\|}$; thus

$$\frac{dP_{g\|}}{P_{g\|}} + \frac{dL}{L} = 0$$

and

$$P_{g\|} L = \text{const}. \tag{5.93}$$

Of course the calculation leading to (5.93) is very simple; still, even this calculation may be avoided, if the adiabatic invariance if the action is integral (5.92) is taken into account. Indeed, (5.93) is an immediate consequence of (5.92) for $v_\| = \text{const}$.

Consider now the analogous problem for a charged particle in a magnetic field. Let us assume that the field is uniform ($B = B_0$) in the interval $-z_0 < z < z_0$, while for $z > z_0$ and $z < -z_0$ the field is equal to

$$B = B_0 + b|z - z_0|$$

and $bz_0 \gg B_0$. The last inequality implies that the particle moves between two magnetic mirrors, located very close to the points $\pm z_0$. Neglecting the small contributions to the integral in (5.92) from the regions $z > z_0$ and $z < -z_0$, and taking advantage of the fact that, according to (5.88), along the remaining path of integration, $v_\| = \text{const}$ (since $B(s) = B_0$), we obtain for the adiabatic invariant $J_\|^*$:

$$J_\|^* = 2z_0 v_\| = \text{const} \tag{5.94}$$

in complete analogy with (5.93).

Let us assume now that the mirrors are slowly moving one towards the other; then, after some time, they will be located at the points $\pm z_0'$ ($z_0' < z_0$). Since $J_\|^* = \text{const}$, (5.94) results in

$$v_\|' = v_\| \frac{z_0}{z_0'} > v_\|, \tag{5.95}$$

which means that the particle is accelerated in the longitudinal direction. The adiabatic invariant $J_\perp \sim v_\perp^2/B$ (see (5.26)) remains constant too; therefore,

$$v_\perp' = v_\perp. \tag{5.96}$$

5.5 A Charged Particle Motion in Static Non-uniform Magnetic Fields

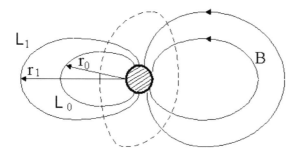

Fig. 5.8. The magnetic field distribution of an asymmetric dipole

The angle between the trajectory and the z-axis is changing according to

$$\tan \alpha' = \frac{v'_\perp}{v'_\parallel} = \frac{z'_0}{z_0} \frac{P_{g\perp}}{P_{g\parallel}} = \frac{z'_0}{z_0} \tan \alpha < \tan \alpha \qquad (5.97)$$

and eventually becomes small enough to allow for the particle's escape. The kinetic energy of the particle increases from

$$\mathcal{E}_{\mathrm{kin}} = \frac{m}{2}\left(v_\parallel^2 + v_\perp^2\right)$$

to

$$\mathcal{E}'_{\mathrm{kin}} = \frac{m}{2}\left(v'^2_\parallel + v'^2_\perp\right) = \frac{m}{2}\left[\left(\frac{z_0}{z'_0}\right)^2 v_\parallel^2 + v_\perp^2\right]. \qquad (5.98)$$

Equations (5.95)–(5.98) describe the main features of the Fermi mechanism of charged particles' acceleration in cosmic rays. These features have been elucidated using only the adiabatic invariance of longitudinal motion, without solving the equations of motion of the particle.

If the configuration of the magnetic field implies that the oscillatory longitudinal motion of the particle is accompanied by its slow transverse drift from one line of force to another, it may appear that this drift motion is periodic, too; then, its action integral will be an adiabatic invariant. Let us consider the special case of the field created by a slightly asymmetric dipole (Fig. 5.8). The particle oscillates along the line of force L_0, with a velocity determined by (5.88), between the turning points S_t, given by (5.89). If in the course of its drift around the dipole the particle will cross from the line of force L_0 to L_1, then, since $B(r_1) < B(r_0)$, it follows from (5.88) that $v_\parallel(r_1) > v_\parallel(r_0)$. Taking into account that the length of the oscillatory longitudinal path is also increasing upon the transition from L_0 to L_1, we come to the conclusion that the integral in (5.92) becomes larger under this transition too. However, since $J^*_\parallel$ is an adiabatic invariant, this cannot happen; thus, the transition from L_0 to L_1 is 'forbidden,' and the drift motion around the dipole is periodic. The action integral of the (third) drift periodic motion J_d is a constant of motion and an adiabatic invariant:

$$J_d = \oint \boldsymbol{P}_{gd} dl = m \oint \boldsymbol{u}_\perp dl + e \oint \boldsymbol{A}_\perp dl$$

$$= m \oint \boldsymbol{u}_\perp dl + e \int_S \mathrm{curl} \boldsymbol{A}_\perp d\boldsymbol{S} = m \oint \boldsymbol{u}_\perp dl + e \int_S \boldsymbol{B} d\boldsymbol{S} = \mathrm{const.} \quad (5.99)$$

Here $\boldsymbol{u}_\perp$ is the velocity of the transverse drift motion, S is the surface bound by the drift orbit and the cyclotron and longitudinal motions are assumed to be averaged. The second term in (5.99) is much larger than the first; indeed, the first term may be estimated as $\sim 2\pi R m \bar{u}_\perp$, while the second may be estimated as $\sim eB\pi R^2$ (R is the radius of the drift orbit). Therefore, the ratio of the two terms is

$$\frac{1}{2} \frac{eBR}{m u_\perp} = \frac{1}{2} \frac{\omega_c}{\omega_d} \gg 1$$

ω_d being the drift frequency. Thus, the adiabatic invariance of J_d (5.99) may be formulated as the invariance of the magnetic flux through the area bounded by the drift orbit

$$\int_S \boldsymbol{B} d\boldsymbol{S} = \mathrm{const.} \quad (5.100)$$

5.6 A Charged Particle Motion in Static Non-uniform Electric Field

The methods developed in the last section allow for investigating the motion of a charged particle in the crossed uniform magnetic field $\boldsymbol{B} = \mathrm{const}$ and in slowly varying in space electric field $\boldsymbol{E}(\boldsymbol{r})$, almost without additional calculations. Indeed, taking advantage of the small variation of the electric field on the length scale of a cyclotron radius, we use the adiabatic principle and treat the orbital motion of the particle as taking place in the quasi-uniform electric field, taken at the points of the cyclotron orbit ($\boldsymbol{E} = \boldsymbol{E}(\boldsymbol{r}_c)$). Therefore, instead of (5.41), we obtain for the Hall drift velocity the following expression

$$\boldsymbol{v}_{\perp d} = \frac{[\boldsymbol{E}(\boldsymbol{r}_c(t)) \times \boldsymbol{B}]}{B^2}. \quad (5.101)$$

In the case of its slow variation, the electric field at the points of the orbit $\boldsymbol{E}(\boldsymbol{r}_c(t))$ can be expressed using the Taylor expansion, in terms of the field at the center of the orbit ($\boldsymbol{r}_c = 0$), as follows:

5.6 A Charged Particle Motion in Static Non-uniform Electric Field

$$\boldsymbol{E}\left(\boldsymbol{r}_{c}(t)\right)=\boldsymbol{E}(0)+\left(\boldsymbol{r}_{c}(t)\nabla\right)\boldsymbol{E}\mid_{0}+\frac{1}{2}\sum_{i,j}r_{ci}(t)\,r_{cj}(t)\,\frac{\partial^{2}\boldsymbol{E}}{\partial r_{ci}\partial r_{cj}}\mid_{0}+... \tag{5.102}$$

where $i(j)$ are x and y. Then, according to the same adiabatic principle, the slow drift motion feels only the averaged fast orbital motion $\boldsymbol{r}_c(t)$; therefore, (5.102) should be averaged over the period of the orbital motion. Since, according to (5.23), $x_c \sim \sin(\omega_c t + \alpha)$ and $y_c \sim \cos(\omega_c t + \alpha)$, the second term in the RHS of (5.102) vanishes by averaging; so does also the mixed term in the last sum of this expression, while $x_c^2(t) = y_c^2(t) = R_c^2/2$. Hence

$$\overline{\boldsymbol{E}\left(\boldsymbol{r}_{c}(t)\right)} \simeq \left[1+\frac{1}{4}R_c^2\nabla^2\right]\boldsymbol{E}. \tag{5.103}$$

Substituting (5.103) in (5.101), we obtain for the Hall drift velocity in the slightly non-uniform electric field

$$\boldsymbol{v}_{\perp d} = \left(1+\frac{1}{4}R_c^2\nabla^2\right)\frac{[\boldsymbol{E}\times\boldsymbol{B}]}{B^2}. \tag{5.104}$$

The correction term in (5.104), being dependent on R_c^2, ascribes different values to different types of particles. For instance, it is much larger for ions than for electrons, since

$$R_c^2 \simeq \frac{v_c^2}{\omega_c^2} = \left(\frac{1}{eB}\right)^2 m^2 v_c^2$$

and in the thermal equilibrium, when

$$\overline{v_c^2} = \frac{kT}{m}$$

R_c^2 is proportional to m. Hence, in the presence of the gradient of the electric field, ions and electrons move with different drift velocities. The electrical gradient may appear in plasma spontaneously, due to the density fluctuation; then, the separation between ions and due to their different drift velocities will create an additional electric field. If this secondary field enhances the primary one, the total electric field can grow to infinity, giving rise to the instability of plasma called drift instability. In principle, the same effect can occur owing to the gradient drift; however, in the latter case the difference in the velocity of the different particles is proportional to the gradient of the (magnetic) field (see (5.78)), and not to its square, as in the case of the Hall drift. Thus, the non-uniformity of the electric field becomes more important for the fluctuations on the small length scale, giving rise to the so-called micro-instability. There is another important consequence to the appearance of the gradient terms in the expressions for drift velocities (5.78) and (5.104).

If the electric and magnetic fields acting on electrons are due to the electromagnetic wave with the components $\boldsymbol{E}, \boldsymbol{B} \sim \exp i(\boldsymbol{k}\boldsymbol{r} - \omega t)$, propagating in the media, the gradient terms give rise to the dependence of the drift velocities on the wave vector $\boldsymbol{k}$ (or the wave length $\lambda = 2\pi/k$), which in turn leads to the spatial dispersion – that is, the dependence of the conductivity and the dielectric constant of the system on the wave vector. Spatial dispersion, though usually small in effect, is responsible for a number of qualitatively new phenomena, such as the propagation of plasma waves in gas plasma and in metals, and the existence of stable non-linear space-confined solutions of Maxwell equations, in the form of standing or propagating solitary waves.

5.7 A Charged Particle Motion in Time Dependent Fields

5.7.1 Time-Dependent Uniform Magnetic Field

If the electromagnetic field varies with time, then the Lagrangian of a charged particle (1.122) is time dependent (through vector $\boldsymbol{A}$ and scalar φ potentials) and the energy $\mathcal{E}$ of the particle is not conserved. On the other hand, since the magnetic part of the Lorentz force is perpendicular to the velocity of the particle, the magnetic field $\boldsymbol{B}(t)$, even if it varies in time, cannot change the energy of the particle directly. Indeed, (5.105) and (5.106)

$$\mathcal{E} = \mathcal{E}_{\text{kin}} + e\varphi \tag{5.105}$$

$$\frac{d\mathcal{E}_{\text{kin}}}{dt} = e\boldsymbol{E} \cdot \boldsymbol{v} \tag{5.106}$$

imply that the energy and the rate of its change depend only on the scalar potential and the electric field, while the magnetic field is completely determined by the vector potential ($\boldsymbol{B}(t) = \operatorname{curl}\boldsymbol{A}(t)$). However, the time-varying magnetic field does affect the energy of the particle indirectly, by creating the electric field in accord with Maxwell's second equation

$$\operatorname{curl}\overrightarrow{E} = -\frac{\partial \boldsymbol{B}}{\partial t}.$$

Thus, in the general case of an arbitrary fast-varying magnetic field, it is necessary to determine first the electric field produced by a given magnetic field (this electric field will be ultimately non-uniform) and then to consider the motion of the charged particle in the resultant time- and space- dependent electromagnetic field. If, however, the time variation of the magnetic field is slow enough

$$\frac{\partial B}{\partial t}\frac{1}{\omega_c} \ll B,$$

5.7 A Charged Particle Motion in Time Dependent Fields

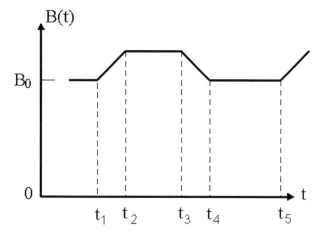

Fig. 5.9. The magnetic field variation $B(t)$

the transverse motion of the particle is completely determined by the adiabatic invariance of the particle's magnetic moment μ. The transverse kinetic energy of the particle is

$$\mathcal{E}_\perp = \frac{mv_\perp^2}{2} = |\mu| B(t)$$

and the magnetic moment is constant

$$\mu = \text{const}.$$

We immediately conclude according to equation (5.32) that the energy simply follows the variation of the field and the transverse velocity is proportional to the $\sqrt{B(t)}$, while the cyclotron radius R_c, according to equation (5.33), is proportional to the $1/\sqrt{B(t)}$. Thus, it is clear that the varying-in-time magnetic field can be used for the heating of charged particles. Let us consider some examples.

One of the very interesting applications of adiabatic invariance of the orbital magnetic moment of an electron in a magnetic field is the Alfven mechanism of the energy transfer from varying in time external magnetic field to ensemble of electrons. This mechanism was studied in connection with acceleration of cosmic particles (magnetic pumping), plasma heating (the gyrorelaxation effect) and in solid state (magnetoviscous damping of helicons and spin waves). We shall illustrate here the main ideas of magnetic pumping mechanism using a schematic construction in the momentum space of electrons under periodically varying classically strong ($\omega_c \tau \gg 1$) magnetic field where τ is a relaxation time.

Consider the following form of a magnetic field variation (Fig. 5.9).

$$B(t) = B_0[1 + b(t)], \quad b(t) \ll 1 \tag{5.107}$$

with a period defined by

$$\frac{db}{dt} = \begin{cases} \alpha & \text{if } t_1 < t < t_2 \\ 0 & \text{if } t_2 < t < t_3 \\ -\alpha & \text{if } t_3 < t < t_4 \\ 0 & \text{if } t_4 < t < t_5 \end{cases}, \qquad (5.108)$$

here $\alpha \equiv db/dt$. As we have seen in the previous sections, an electron in a homogeneous magnetic field possesses an adiabatic invariant:

$$I = \frac{\mathcal{E}_\perp(t)}{B(t)} \simeq \text{const}(t), \qquad (5.109)$$

where $\mathcal{E}_\perp$ is the kinetic energy of the motion in xy-plane. It follows from (5.109) that $\mathcal{E}_\perp(t)$ is proportional to $B(t)$, i.e. a periodic variation in $B(t)$ yields a periodic variation in $\mathcal{E}_\perp(t)$ provided the frequency ω of the applied field is smaller than the cyclotron frequency. Since the kinetic energy along the field $\mathcal{E}_z$ is field independent, the total energy variation of a *single electron* over a period of the field variation average to zero

$$\langle \Delta \mathcal{E}_\perp(t) \rangle_T = 0. \qquad (5.110)$$

In an *electron ensemble*, however, due to the equipartition (the energies, stored in perpendicular and parallel, to the field, electron motion, are periodically restored due to collisions) there is a net energy transfer from the varying in time magnetic field to the electrons. This can be demonstrated in a following way. Let us denote:

$$\delta \equiv t_2 - t_1 = t_4 - t_3 \qquad (5.111)$$

and

$$\Delta \equiv t_3 - t_2 = t_5 - t_4. \qquad (5.112)$$

We assume that:

$$\frac{2\pi}{\omega_c} \ll \delta \ll \tau \ll \Delta, \qquad (5.113)$$

where τ is the *equipartition time* between $\mathcal{E}_\perp$ and $\mathcal{E}_z$.

The inequalities, (5.113) and Fig. 5.10, have the following physical meaning [10]. The field is varying in time fast enough: $\delta \ll \tau$, so that collisions are not efficient in this time interval. Therefore the translation (along the z-axis) motion is decoupled from the oscillatory (in xy-plane) degrees of freedom during the time intervals $t_1 \to t_2$ and $t_3 \to t_4$. The inequality $2\pi/\omega_c \ll \delta$ assures the adiabatic invariance of the orbital moment. Therefore during these time intervals $\mathcal{E}_\perp(t)$ is varying proportionally to the field variations $H(t)$ while $\mathcal{E}_z$ is constant. The inequality $\Delta \gg \tau$ guarantees that the collisions will be

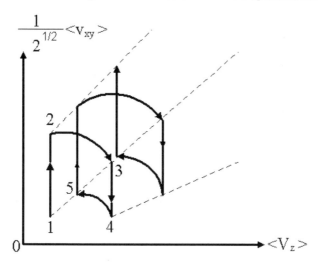

Fig. 5.10. During the time interval $t_1 \Rightarrow t_2$ the field pumps the energy into $\mathcal{E}_\perp$ faster than the collisions couple the oscillatory and translation degrees of freedom (the vertical line). During the time interval $t_2 \Rightarrow t_3$ the collisions restore the equipartition, i.e. some part of the energy is transferred to $\mathcal{E}_z$, and the system returns to equilibrium, point t_3. During the time interval $t_3 \Rightarrow t_4$ only part of the energy, stored in $\mathcal{E}_\perp$ while the field was growing, is returned to the field

effective and the equipartition in the system will be restored during the time intervals $t_2 \rightarrow t_3$ and $t_4 \rightarrow t_5$.

Figure 5.10 shows the irreversible nature of such a process in the momentum space of the electron ensemble. The initial energy gain, 'hidden' in the parallel motion, will be redistributed, finally, between $\mathcal{E}_\perp$ and $\mathcal{E}_z$ during the time interval $t_4 \rightarrow t_5$. Therefore, the temperature of the electron ensemble will grow: $T(t_5) > T(t_1)$. This is a typical irreversible process, resulting in the magnetoviscosity of the three-dimensional electron gas under a strong magnetic fields.

Now the role of the third (along the magnetic field) dimension in the dissipativity of the electron gas under a classically strong: $\omega_c \tau \ll 1$ magnetic field can be understood: the two-dimensional limit corresponds to the trajectory $1 \rightarrow 2 \rightarrow 1$, and the energy of a two-dimensional electron gas (2DEG) at the end of the cycle equals its energy at the beginning.

We have elucidated here the crossover from the three-dimensional, dissipative, electron gas under a strong magnetic field to its nondissipative, two-dimensional limit by designing a billiard of interacting particles with oscillatory and translation degrees of freedom which models the electron gas in a varying in time magnetic field. A transparent graphical presentation in the phase space has shown that the system is dissipative in three spatial dimensions. In $2+1$ dimensional limit, however, our model exhibits an anomaly: it turns to be reversible under a periodic external force. Thus the nondissipa-

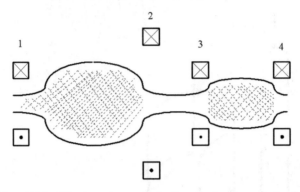

Fig. 5.11. The two-stage adiabatic compression of plasma

tivity of a two-dimensional electron gas in a magnetic field, a widely studied quantum phenomenon could be traced to the classical physics. The central role in this Gedanken Experiment plays the adiabatic invariance of the classical angular momentum and the equipartition between the oscillatory and translation degrees of freedom.

Consider another example of charged particles heating by varying in time magnetic field. Here the directly increasing transverse energy can be transferred into the energy of the longitudinal motion, using the appropriate configuration of a non-uniform in the z direction magnetic field. Figure 5.11 presents schematically the method of plasma heating known as adiabatic compression.

The plasma is injected in the region between coils 1 and 2, and the current pulse, passing through these coils, in used to increase the magnetic field and, accordingly, the transverse energy $\mathcal{E}_\perp$ of the particles in this region. Then the next pulse is applied in coil 1; this creates a larger gradient of the magnetic field in its vicinity and increases the returning force of this magnetic mirror, allowing plasma to pass through the magnetic mirror in point 2 to the region 3-4, where the process can be repeated.

5.7.2 Static Magnetic and Uniform Time-Dependent Electric Field

We consider in this section not only the behavior of a single particle, but also some basic features of the response of a neutral system, consisting of a large number of non-interacting charged particles of opposite signs (electrons and ions), performing the cyclotron motion determined by $\boldsymbol{B}_0$ to the varying in time (especially periodic) and uniform in space electric field: $\boldsymbol{B}(\boldsymbol{r},t) = \boldsymbol{B}_0$; $\boldsymbol{E}(\boldsymbol{r},t) = \boldsymbol{E}(t)$. The existence of adiabatically slow variation will not be assumed in this section. Let us start with the longitudinal motion along the magnetic field $\boldsymbol{B}_0 \parallel z$. The equation of motion in the z direction, unaffected by the presence of a magnetic field, has a simple form

5.7 A Charged Particle Motion in Time Dependent Fields

$$m\frac{dv_z}{dt} = eE_z(t). \quad (5.114)$$

Applying the Fourier transformation to (5.114), we obtain

$$v_z(\omega) = \frac{e}{i\omega m} E_z(\omega) \quad (5.115)$$

and

$$z(\omega) = -\frac{e}{m\omega^2} E_z(\omega). \quad (5.116)$$

In the most interesting case of an oscillating field, (5.115), (5.116) describe the response of the charge to this field – the oscillatory motion of the charge with the amplitude of oscillation and the magnitude of the velocity proportional to the applied field. The proportionality coefficients depend on the frequency of oscillations in a non-resonant manner (if $\omega = 0$, we have motion with a constant acceleration instead of oscillations); they are different for carriers of different masses (simply reflecting the larger inertia of a heavier particle); and they have opposite signs for the negative and positive carriers. Thus, if the system contains both types of carriers (say, electrons and single-charged ions), they will move in opposite directions, creating the oscillatory polarization

$$P_z(\omega) = n|e|[z_i(\omega) - z_e(\omega)] = -\frac{ne^2}{\omega^2}\left(\frac{1}{M_i} + \frac{1}{m_e}\right)E_z(\omega) \quad (5.117)$$

and the oscillatory current density

$$j_z(\omega) = ne(v_{iz} - v_{ez}) = \frac{ne^2}{i\omega}\left(\frac{1}{M_i} + \frac{1}{m_e}\right)E_z(\omega), \quad (5.118)$$

where n is the particle density and M_i and m_e are the ion and the electron mass, respectively. Obviously, due to the large difference in the ion and the electron masses, the longitudinal polarization and the related current are completely determined by the electron motion and can be rewritten with high accuracy as

$$P_z(\omega) = -\frac{ne^2}{m_e\omega^2} E_z(\omega) \quad (5.119)$$

$$j_z(\omega) = -i\frac{ne^2}{m_e\omega} E_z(\omega). \quad (5.120)$$

Turning now to the transverse motion, we write the equations of motion in the xy-plane, in the form (see (5.36)):

$$\frac{dv_x}{dt} = \omega_c v_y + \frac{e}{m} E_x(t) \quad (5.121)$$

$$\frac{dv_y}{dt} = -\omega_c v_x + \frac{e}{m} E_y(t).$$

Taking the time derivatives of (5.121) and substituting (5.121) in these new equations, we obtain

$$\frac{d^2 v_x}{dt^2} + \omega_c^2 v_x = \frac{e}{m}\left(\omega_c E_y(t) + \frac{dE_x(t)}{dt}\right) \quad (5.122)$$

$$\frac{d^2 v_y}{dt^2} + \omega_c^2 v_y = \frac{e}{m}\left(-\omega_c E_x(t) + \frac{dE_y(t)}{dt}\right).$$

The general solution of the homogeneous equations, related to (5.122), describes the cyclotron rotation (see (5.19)–(5.23). On the other hand, the equations with the rhs are formally identical to the equations of the motion of two independent driven harmonic oscillators (or, alternatively, of a two-dimensional harmonic oscillator) – with the sole difference that the oscillating quantity is the velocity and not the coordinate. The simplest way to solve these equations is to use the Fourier transformations. Thus, the particular solution of (5.122) for the Fourier components of the drift velocity $v_{x,y}(\omega)$ in terms of $E_{x,y}(\omega)$ has the form:

$$v_x(\omega) = \frac{e}{m}\frac{i\omega}{\omega_c^2 - \omega^2} E_x(\omega) + \frac{e}{m}\frac{\omega_c}{\omega_c^2 - \omega^2} E_y(\omega) \quad (5.123)$$

$$v_y(\omega) = -\frac{e}{m}\frac{\omega_c}{\omega_c^2 - \omega^2} E_x(\omega) + \frac{e}{m}\frac{i\omega}{\omega_c^2 - \omega^2} E_y(\omega),$$

while the Fourier transformations of coordinates are

$$x(\omega) = \frac{e}{m}\frac{1}{\omega_c^2 - \omega^2} E_x(\omega) - i\frac{e}{m\omega}\frac{\omega_c}{\omega_c^2 - \omega^2} E_y(\omega) \quad (5.124)$$

$$y(\omega) = i\frac{e}{m\omega}\frac{\omega_c}{\omega_c^2 - \omega^2} E_x(\omega) + \frac{e}{m}\frac{1}{\omega_c^2 - \omega^2} E_y(\omega).$$

The 'off-diagonal' terms in (5.123) – those with real coefficients – describe the time-dependent generalization of the effect, – the perpendicular, or the Hall drift of the particle in the combined perpendicular electric and magnetic fields. Indeed, it follows from (5.123) that in this motion $\boldsymbol{vE} = 0$ and in the static case

$$\omega \to 0, \quad |v_i| = \frac{1}{B_0} E_j, \quad (i, j = x, y; i \ne j)$$

in agreement with (5.41). As in the static case, this Hall velocity is the same for negative and positive carriers (the mentioned terms are proportional to e^2, since $\omega_c \sim e$), but its magnitude depends on their mass for $\omega \ne 0$; this dependence is most pronounced when $\omega \longrightarrow \omega_c$. Therefore, in gaseous plasma with two types of carriers (electrons and ions, with particle density n, masses m_e and M_i and cyclotron frequencies ω_{ce} and ω_{ci}) the oscillatory Hall drift will be accompanied by the oscillatory current $j_\perp^{(H)}(\omega)$

5.7 A Charged Particle Motion in Time Dependent Fields

$$j^{(H)}_{\perp y,x}(\omega) = \pm \frac{n\,|e|^3\,B_0}{m_e^2}\,\frac{\omega^2}{(\omega_{ci}^2 - \omega^2)(\omega_{ce}^2 - \omega^2)}\,E_{x,y}(\omega). \tag{5.125}$$

The latter may be viewed as produced by the oscillatory polarization

$$P^{(H)}_{y,x}(\omega) = \mp i\,\frac{n\,|e|^3\,B_0}{m_e^2}\,\frac{\omega}{(\omega_{ci}^2 - \omega^2)(\omega_{ce}^2 - \omega^2)}\,E_{x,y}(\omega). \tag{5.126}$$

Here the upper sign corresponds to the first, and the lower to the second pair of indices. Both the polarization and the current vanish for a static field ($E_{x,y}(\omega) = E_{x,y}(0)\,\delta(\omega)$), as it should be. The Hall current, related to the perpendicular drift of one type of carriers (electrons), is obtained from (5.125) by assuming first $\omega_{ci} \longrightarrow 0$ (i.e., $M_i \longrightarrow \infty$ and the ions are at rest) and then $\omega \longrightarrow 0$

$$j^{(H)}_{y,x}(0) = \mp n\,|e|\,\frac{1}{B_0}\,E_{x,y}(0).$$

The resonant structure of the denominators in (5.125), (5.126) will be discussed later, together with a similar structure of the diagonal terms (terms with imaginary coefficients) in (5.123). These last terms, which differ from zero only for $\omega \neq 0$, e.g., for the varying-in-time electric field, describe a phenomenon that is new with respect to the static case: oscillatory drift in the direction of the electric field, creating an oscillatory polarization and current. It follows from (5.124), (5.123) that the polarization is given by

$$\boldsymbol{P}_{\perp}(\omega) = ne^2\left[\frac{1}{m_e\,(\omega_{ce}^2 - \omega^2)} + \frac{1}{M_i\,(\omega_{ci}^2 - \omega^2)}\right]\boldsymbol{E}_{\perp}(\omega) \tag{5.127}$$

and the related current density $\boldsymbol{j}^{(P)}_{\perp}(\omega)$

$$\boldsymbol{j}^{(P)}_{\perp}(\omega) = i\omega ne^2\left[\frac{1}{m_e\,(\omega_{ce}^2 - \omega^2)} + \frac{1}{M_i\,(\omega_{ci}^2 - \omega^2)}\right]\boldsymbol{E}_{\perp}(\omega). \tag{5.128}$$

For a slowly varying electric field $\boldsymbol{E}_{\perp}(\omega)$, $\omega \ll \omega_c$, equations (5.127), (5.128) reduce to

$$\boldsymbol{P}_{\perp}(\omega) = \frac{\rho}{B_0^2}\,\boldsymbol{E}_{\perp}(\omega) \tag{5.129}$$

$$\boldsymbol{j}^{(P)}_{\perp}(\omega) = i\omega\,\frac{\rho}{B_0^2}\,\boldsymbol{E}_{\perp}(\omega), \tag{5.130}$$

or, in t-representation,

$$\boldsymbol{P}_{\perp}(t) = \frac{\rho}{B_0^2}\,\boldsymbol{E}_{\perp}(t) \tag{5.131}$$

$$j_\perp^{(P)}(t) = \frac{\rho}{B_0^2}\frac{d}{dt}E_\perp(t), \qquad (5.132)$$

where ρ is the mass density. Equations (5.131), (5.132) imply that while the current associated with the polarization drift vanishes for the static field, the polarization remains. It can be shown that the induced static polarization gives rise to the lowering of the total energy of the particles by

$$\Delta\mathcal{E} = -\int P_\perp dE_\perp = -\frac{\rho}{2B_0^2}E_\perp^2. \qquad (5.133)$$

The results accomplished by (5.131), (5.133) are identical to those obtained from the exact solution of the Schroedinger equation for the charged particle in the perpendicular static electric and magnetic fields. According to this solution, the presence of an electric field E_1 perpendicular to B gives rise to the shift in the equilibrium position of the Landau oscillator, equal to

$$\frac{m}{eB^2}E_\perp$$

and to the energy gain of

$$\left(-\frac{m}{2B^2}E_\perp^2\right).$$

Being multiplied by particle density n and summed over two types of carriers, these quantities result in (5.131), (5.133).

Let us summarize the results contained in (5.117)–(5.128). The neutral system of charged particles, performing the cyclotron rotation in the plane perpendicular to the constant magnetic field B, responds to the applied electric field $E = E_0\exp i\omega t$ by creating the induced macroscopic dipole moment $P(\omega)\exp i\omega t$ (polarization) and the related current $j(\omega)\exp i\omega t$, with $j(\omega) = i\omega P(\omega)$. The response is linear – the coefficients $P(\omega)$, $j(\omega)$ are proportional to E_0 in the sense that an increase or a decrease in the magnitude of E_0 results in the proportional increase of decrease in the magnitude of $P(\omega)$ and $j(\omega)$. The response is not, however, isotropic; the linear response has the form

$$j_i(\omega) = \sum_j \sigma_{ij}(\omega)E_j(\omega) \qquad (5.134)$$

$$P_i(\omega) = \sum_j \alpha_{ij}(\omega)E_j(\omega), \qquad (5.135)$$

where

$i, j = x, y, z;$
$\sigma_{xx} = \sigma_{yy} \neq \sigma_{zz};\ \alpha_{xx} = \alpha_{yy} \neq \alpha_{zz}$
$\sigma_{xy} = -\sigma_{yx} \neq 0;\ \alpha_{xy} = -\alpha_{yx} \neq 0$
$\sigma_{xz} = \sigma_{zx} = \sigma_{yz} = \sigma_{zy} = 0;\ \alpha_{xz} = \alpha_{zx} = \alpha_{yz} = \alpha_{zy} = 0.$

This anisotropy of response is due entirely to the presence of the constant magnetic field: if $B \to 0$, $\sigma_{xx} = \sigma_{yy} \to \sigma_{zz}$; $\sigma_{xy} = -\sigma_{yx} \to 0$. The coefficients in (5.134), (5.135) form second-rank tensors of conductivity ($\sigma_{ij}(\omega)$) and polarizability ($\alpha_{ij}(\omega)$),

$$\sigma_{ij}(\omega) = i\omega \alpha_{ij}(\omega).$$

The components of these tensors depend on the frequency; this dependence is sometimes called frequency dispersion. The frequency dispersion of the components of both tensors clearly reflects the effect of the magnetic field:

1. The zz components do not depend on B and are smooth functions of ω; their frequency dispersion is completely determined by the motion of free charges, accelerated by the harmonically oscillating-in-time electric field. The response becomes weaker in proportion as the frequency increases, since the charges, owing to inertia, cease to follow the fast oscillations of the field.

2. The frequency dispersion of the xx (yy) and xy (yx) components is much more complicated. All the components depend on the magnetic field; the diagonal components are even, while the off-diagonal are odd function of B. This dependence becomes weak at high frequencies ($\omega \to \omega_{ce}$), when the off diagonal components tend to zero, and the diagonal approach the value of the zz-component. In the domain of low frequency ($0 < \omega < \omega_{ce}$) the response for the motion in the xy-plane has a resonance character – all the transverse components of the tensors diverge at the frequencies of the applied field equal to the cyclotron frequencies of the charges, ω_{ci} and ω_{ce} as $(\omega - \omega_c)^{-1}$. This behavior of response functions is typical of the bounded systems, performing, to use the language of classical mechanics, a finite periodic motion with an eigenfrequency ω_0 (harmonic oscillator, electrons in the discrete energy states in atoms and molecules[3] etc.). If the frequency ω of the applied field is, or is close to being, in resonance with the eigenfrequency of the systems, two major effects ensue:

1) the energy of the field is very rapidly transferred to the system, giving rise to the infinite growth of the amplitude of oscillations in the classical picture (if damping is not taken into account) or to the quantum transition between the states in resonance in quantum description.

2) a mixing of the applied field (electromagnetic wave) with the eigenmode, creating new collective excitations of the hybrid mechanical-electromagnetic nature. The form of the transverse components of the tensors $\sigma_{ij}(\omega)$ and $\alpha_{ij}(\omega)$ implies that the cyclotron rotations of electrons and ions serve

---

[3] In the last case the eigenfrequency ω_{mn} corresponds to the energy difference between the states:

as eigenmodes of the mechanical system. Thus, we can expect that the electromagnetic field, interacting with these modes, shall give rise to transitions between Landau levels (cyclotron resonance), and to the propagation of waves quite different from the electromagnetic waves in vacuum or in a weakly dispersive medium (magnetoplasma waves).

Two final remarks are in order. First, we have seen in the previous sections that the spatial non-uniformity of the electromagnetic field in the case of periodic non-homogeneity results in the spatial dispersion of the response functions – that is, the dependence of these functions on the wavelength of perturbation. Thus, the spatially non-uniform and time-dependent electromagnetic field – for instance, an electromagnetic wave – induces a response, characterized by response functions like σ and α, with frequency and spatial dispersion. Secondly, the linear response is, strictly speaking, only an approximation – in most cases a very good one, but an approximation nevertheless. Consider, for example, the time-dependent electric field which should produce an additional time- and space-dependent magnetic field according to the Maxwell equation (1.19). The magnetic field, in turn, would change the cyclotron frequency ω_c and the response tensors. Thus, these tensors themselves become field-dependent and the Maxwell equations become non-linear. The non-linear optical effects, like the second harmonic generation, soliton formation, optical bistability, etc., as well as non-linear plasma oscillations and instabilities originate from this field-dependence of the response coefficients.

6 Current Instabilities

6.1 General Approach

All kinds of electromagnetic wave generation and amplification in any frequency band are based on a phenomenon of instability. Assume that there exist some undisturbed, or basic state of a physical system. Consider the evolution in time of small deviations from this state. The system can be identified as stable, if these deviations do not increase with time, and in such a case the system cannot function as an oscillator or amplifier. In the opposite case, when the small deviations increase with time, the system is unstable, and it can be used as an oscillator or amplifier. In the further analysis we mainly follow [11, 12].

Instabilities can be divided into two classes:

1. absolute instabilities;
2. convective instabilities.

The absolutely unstable system can serve as an oscillator, while the system with convective instabilities can be used as an amplifier. It is known that the stability problem for a system with the finite number of degrees of freedom can be described adequately by solving a system of linear differential equations, while the evaluation of a resulting steady state in general case requires the solution of a non-linear differential equation system. Stability of a system with distributed parameters is analyzed using a system of linear differential equations in partial derivatives describing the small deviations from some undisturbed (basic) state. It is meaningful to formulate a problem concerning a propagation of deviations caused by an external source through a stable or a convective unstable system. The spatial and temporal evolution of these small deviations in general case is called a wave process, since instabilities at an initial stage have a wave-like character even in unstable systems.

Two kinds of a problem exist for wave processes in a system with distributed parameters (continuous media).

1. The problem with the initial conditions, or the stability problem analogous to the eigen-mode problem for a system with a finite number of degrees of freedom.

2. The problem with the boundary conditions, or the amplification problem which is analogous to the stimulated oscillation problem for a system with a finite number of degrees of freedom.

Consider the first class of problems. In this case, the state of the system in any moment of time is determined by its equations of motion provided that a state of the system at some given moment of time is known. If the initial perturbation at any fixed point of space is decaying with time, the undisturbed state of the system is stable. If, on the contrary, the initial perturbation is increasing with time and evolves in space in such a way that it vanishes at any fixed point of space after a sufficiently large time interval, the system is called convective unstable. Finally, if the initial perturbation increases infinitely in time at any fixed point of space, the system is called absolutely unstable.

For the second class of problems one finds the state of the system driven by an external signal at any point of space using the equations of motion and the boundary conditions. It may happen that the external signal increases as it moves away from the boundary into the system. In such a case a spatial amplification occurs. The opposite case, when the signal attenuates with the depth of the system, corresponds to the "non-transparency" regime. It should be noted that the decay of the signal is not necessarily caused by dissipation.

The standard approach to the analysis of the instability and amplification problems is characterized by the following procedure. The fundamental equations describing the system are linearized near some undisturbed solution. The plane wave elementary solutions $\exp i(\omega t - \bm{k r})$ are substituted into the linearized system. Here $\bm{k}$ and ω are a wave vector and a frequency of a perturbation, respectively. The next step is to find the values of k and ω which correspond to a non-trivial solution of the system under the suitable boundary conditions. In general case, the so-called dispersion equation is obtained

$$D(\bm{k}, \omega) = 0, \tag{6.1}$$

which yields the implicit dependence between $\vec{k}$ and ω. If solutions $\omega(\bm{k})$ of (6.1) exist such that

$$\mathrm{Im}\,\omega(\bm{k}) < 0$$

for some real values of k, the undisturbed state of the system is unstable. If, on the contrary,

$$\mathrm{Im}\,\omega(\bm{k}) > 0$$

for all k, the undisturbed state is characterized as stable.

In the case when the elementary solutions obtained constitute the complete set, the temporal evolution of any perturbation which arise at the moment $t = 0$ can be described using the expansion in the complete system of elementary wave solutions. The solutions with $\mathrm{Im}\omega(\boldsymbol{k}) < 0$ result in the increasing of the perturbation, and the instability is actually realized.

Seemingly, the spatial amplification problem can be analyzed similarly to the temporal instability with the only difference that the dispersion equation solutions $k(\omega)$ should be obtained. The roots $k(\omega)$ with

$$\mathrm{Im} k(\omega) > 0$$

are needed for the existence of a spatial amplification. However, it turned out that this relationship is a necessary, but not sufficient condition of the spatial amplification.

The procedure mentioned above possesses the following essential disadvantages. It cannot yield the sufficient condition of the spatial amplification and determine the type of instability: whether it is a convective instability, or an absolute one. It is hardly possible to choose immediately the right kind of dependence $k(\omega)$ or $\omega(\boldsymbol{k})$, and either quantity is supposed to be real: k or ω. Typically, in the case of an absolute instability the process is increasing in time propagating through over the system. A convective instability increases in space moving out of the system. In real physical systems in some cases a convective instability can be transformed into an absolute one due to the change of some parameters of the system. On the other hand, in systems with a comparatively small active length and in the absence of reflection all perturbations can leave the system without reaching a considerable magnitude, so that such a system operates as an amplifier only.

The criteria of the spatial amplification, convective and absolute instability have been formulated for some practically important cases under the following additional conditions.

1. $D(\boldsymbol{k}, \omega)$ is a polynomial in ω and $\boldsymbol{k}$ with real coefficients.
2. The limiting phase velocities

$$v_f = \lim_{\omega \to \infty} \frac{\omega}{k(\omega)}$$

are different and finite for any solution of dispersion (6.1).

6.2 Criteria of Instability Classification

6.2.1 Criteria of Absolute and Convective Instability

We investigate homogeneous two-dimensional (t, z) systems where t is time and z is a coordinate. Suppose that small perturbations of some initial undisturbed state of a physical system are described by the following system of differential equations

6 Current Instabilities

$$\frac{\partial f_i}{\partial t} + \sum_{j=1}^{n} \left[\alpha_{ij} \frac{\partial f_j}{\partial z} + \beta_{ij} f_j \right] = 0, \qquad (6.2)$$

where $f_i(z,t)$ are some functions characterizing the wave process in the system. It may be, for example, components of electromagnetic field vectors $\boldsymbol{E}$ and $\boldsymbol{B}$, electron velocity, charge density, displacement vector, etc. For the sake of simplicity the matrices α_{ij} and β_{ij} are assumed to be constant, but in general case they may contain operators acting on the transverse coordinates x, y. At this stage of the analysis we ignore the effect of the boundaries assuming the system to be infinite in the z direction.

We solve the initial condition problem where the values of the functions f_i at the moment $t = 0$ as well as their Fourier transforms are supposed to be known:

$$f_i(k,0) = \int_{-\infty}^{\infty} f_i(z,0) \exp ikz \, dz, \qquad (6.3)$$

where the real quantity k has the form

$$k = \frac{2\pi}{\lambda}$$

and λ is the wavelength.

We are looking for the solution of the system of (6.2) with the initial conditions (6.3) in the interval $0 < t < \infty$, $-\infty < z < \infty$. Therefore we use the Laplace transformation on time and Fourier transformation on the z coordinate which yield the system of linear algebraic equations

$$\sum_{j=1}^{n} \left[i \left(\omega \delta_{ij} - k \alpha_{ij} \right) + \beta_{ij} \right] f_j(k,\omega) = f_i(k,0) \qquad (6.4)$$

$$i = 1, 2, ..., n.$$

System of (6.4) without the right-hand side has non-trivial a solution when its determinant equals zero. It will be shown below that the roots of determinant coincide with the poles of the Green function corresponding to inhomogeneous system of (6.4).

Solving system of (6.4) and carrying out the inverse Fourier transformation we obtain the solution of system of differential equations (6.2) obeying the initial conditions (6.3). We have

$$f_i(z,t) = \frac{1}{(2\pi)^2} \sum_{j=1}^{n} \int_{\gamma_\omega} d\omega \int_{-\infty}^{\infty} dk D_{ij}(k,\omega) f_j(k,0) \exp i(\omega t - kz). \qquad (6.5)$$

The elements of the matrix $D_{ij}(k,\omega)$ are determined by the following expression

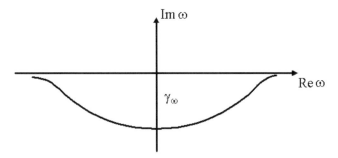

Fig. 6.1. The integration contour γ_ω is situated in the lower half-plane of the complex variable ω, below all singularities of the integrand

$$D_{ij}(k,\omega) = \frac{\Delta_{ij}(k,\omega)}{D(\omega,k)}. \tag{6.6}$$

They represent the rational fractions with respect to the variables k, ω, where the degree of a numerator $\Delta_{ij}(k,\omega)$ is less than the degree n of the denominator $D(\omega,k)$ by at least unity. Here $\Delta_{ij}(k,\omega)$ are the cofactors of the corresponding elements of the main determinant $D(\omega,k)$. The integration contour γ_ω is situated in the lower half-plane of the complex variable ω, below, than all singularities of the integrand (Fig. 6.1).

The solution (6.5) can be expressed in terms of the Green function $D_{ij}(z,t)$ as follows

$$f_i(z,t) = \sum_{j=1}^{n} \int_{-\infty}^{\infty} dz' D_{ij}(z-z',t) f_i(z',0), \tag{6.7}$$

where

$$D_{ij}(z,t) = \frac{1}{(2\pi)^2} \int_{\gamma_\omega} d\omega \int_{-\infty}^{\infty} dk \frac{\Delta_{ij}(k,\omega)}{D(\omega,k)} \exp i(\omega t - kz). \tag{6.8}$$

It is easy to see that the Green function (6.8) depends entirely on the features of the physical system considered, but it does not depend on the initial conditions. The instability type of the system is determined by the asymptotic behavior of the Green function (6.8) at $t \to \infty$ since the functions $f_i(z',0)$ are finite: $|f_i(z',0)| \leq M$. Using the theorem of residues we carry out the integration on ω in (6.8) and find

$$D_{ij}(z,t) = \frac{i}{2\pi} \int_{-\infty}^{\infty} dk \sum_{S} \frac{\Delta_{ij}(k,\omega_S(k))}{D'_\omega(\omega_S(k),k)} \exp i(\omega_S(k)t - kz), \tag{6.9}$$

where the index S stands for the number of the S-th root of the dispersion relation

$$D(\omega, k) = 0 \qquad (6.10)$$

and

$$D'_\omega = \frac{\partial D}{\partial \omega}.$$

If $\mathrm{Im}\,\omega_S(k) > 0$ for any real k and for all S, the system is stable. If for at least one S $\mathrm{Im}\,\omega_S(k) < 0$ for some real k, the instability occurs since the corresponding term in the sum (6.9) would increase with time $\exp|\mathrm{Im}\,\omega_S(k)|t$ [11]. It should be noted that for all S and for any real k the condition

$$\mathrm{Im}\,\omega_S(k) \leq M$$

is assumed to be met, where M is a finite positive number. The quantity

$$-\mathrm{Im}\,\omega_S(k) = |\mathrm{Im}\,\omega_S(k)|$$

is called a temporal increment. Consider the type of a possible instability. The exponential

$$\exp i\omega_S t = \exp\left(i\mathrm{Re}\,\omega_S t + |\mathrm{Im}\,\omega_S|\, t\right)$$

at $t \to \infty$ is a product of a rapidly oscillating function $\exp i\,(\mathrm{Re}\,\omega_S t)$ and smooth function $\exp(|\mathrm{Im}\,\omega_S|\,t)$. Consequently, a possibility exists that at $t \to \infty$ this exponential would finally vanish at any fixed z, and in such a case the instability would be a convective one.

The integration in relationship (6.8) can be carried out using the theorem of residues. The result is determined by the poles of the integrand, i.e. by the solutions of the dispersion equation (6.10) in the plane of the complex variable k. To evaluate the integral in (6.8) we should close the contour of integration either in the upper half-plane, or in the lower one. The choice of the integration path depends on the position of poles $k(\omega)$ which in general case can be situated in any part of the complex plane. For the sake of definiteness, suppose that the contour γ_ω is closed in the lower half-plane ω. For the large $|\omega|$ and finite v_S the following asymptotic expansion is valid

$$k_S(\omega) = \frac{\omega}{v_S} + \alpha_S + \frac{\beta_S}{\omega} + \dots \qquad (6.11)$$

Clearly, for $v_S < 0$ the pole $k_S(\omega)$ situated in the upper half-plane, and for $v_S > 0$ the opposite case takes place. For $z < 0$ the integration contour in (6.8) can be closed in the upper half-plane k, while for $z > 0$ it would be closed in the lower half-plane. Then the Green function (6.8) takes the form

$$D_{ij}(z, t) = \frac{i}{2\pi} \int_{\gamma_\omega} d\omega \exp i\,(\omega t - k_S(\omega) z) \qquad (6.12)$$

$$\times \left[\eta_+(z) \sum_{v_S > 0} \frac{\Delta_{ij}(k_S(\omega), \omega)}{D'_k(\omega, k_S(\omega))} - \eta_-(z) \sum_{v_S < 0} \frac{\Delta_{ij}(k_S(\omega), \omega)}{D'_k(\omega, k_S(\omega))} \right],$$

6.2 Criteria of Instability Classification

where

$$\eta_+(z) = \left\{\begin{array}{ll} 1, & z > 0, \\ 0, & z < 0 \end{array}\right\}$$

and

$$\eta_-(z) = \left\{\begin{array}{ll} 0, & z > 0, \\ 1, & z < 0 \end{array}\right\}$$

$$D'_k = \frac{\partial D(\omega, k)}{\partial k}.$$

The first sum in expression (6.12) contains the terms which correspond to the waves propagating in the positive direction of the z-axis. We call such waves co-propagating ones. The second sum includes the waves which propagate in the negative direction, i.e. the counter-propagating ones. The only possible singularities of the Green function $D_{ij}(z,t)$ Laplace transform in the integrands (6.12) are the branching points of the functions $k_S(\omega)$ where two such functions belonging to the different classes coincide. These branching points are determined by the following equations

$$D(\omega, k) = 0, \ D'_k(\omega, k) = 0, \ D'_\omega(\omega, k) \neq 0.$$

The detailed analysis based on the theory of function of complex variables shows that in the absence of branching points between the real axis and the contour γ_ω the Green function $D_{ij}(z,t) \to 0$ for $t \to \infty$ at any fixed z. In the opposite case, where the branching points of $k_S(\omega)$ exist in the lower half-plane between the real axis and the contour γ_ω the Green function behavior for $t \to \infty$ is determined by the following equation

$$D_{ij}(z,t) \sim \left[\sum_n c_n \frac{t^{-\lambda_n - 1}}{\Gamma(-\lambda_n)}\right] \exp i\omega t, \qquad (6.13)$$

where c_n are some constants, $\Gamma(x)$ is the gamma-function; $1/\Gamma(-\lambda_n) = 0$ if λ_n is integer. It is seen from (6.13) that $D_{ij}(z,t) \to \infty$ as $t \to \infty$, since $\text{Im}\,\omega < 0$. This situation corresponds to the case of the absolute instability. The criterium of such an instability can be formulated as follows.

The instability is absolute, if two functions $k_S(\omega)$ possess a branching point in the lower half-plane of the complex variable ω, coincide at this point, and the limiting phase velocities

$$\lim_{\omega \to \infty} \frac{\omega}{k_S(\omega)} = v_S$$

of these functions have different signs. In other words, the absolute instability occurs when two roots of the dispersion equation $k(\omega)$ coinciding at the branching point in the lower half-plane correspond to counter-propagating waves [11].

Previously we have analyzed the behavior of the Green function $G_{ij}(z,t)$ at $t \to \infty$ and fixed z. Consider now the behavior of the perturbation travelling along the system with the velocity v, i.e. $z = vt$. In this case the Green function $G_{ij}(z - vt, t)$ increases with time in such a way that a convective instability is realized. The integration contour γ_ω in (6.12) would coincide with the real axis. We choose from the sum in (6.12) the term corresponding to the root $k_S(\omega)$ with the positive imaginary part $\text{Im} k_S(\omega) > 0$ for $z > 0$:

$$\varphi_S^{ij} = \frac{1}{2\pi i} \int_{-\infty}^{\infty} \frac{\Delta_{ij}(k_S(\omega), \omega)}{D'_k(\omega, k_S(\omega))} \exp i (\omega t - k_S(\omega) vt) \, d\omega \qquad (6.14)$$

$$= \int_{-\infty}^{\infty} \varphi_S^{ij}(\omega) \exp it \left[\omega - v k_S(\omega)\right] d\omega.$$

Since $\text{Im} k_S(\omega) > 0$ for some frequency interval it has a maximum inside this interval. The frequency ω_0 corresponding to this maximum is determined by the condition

$$\frac{d \text{Im} k_S(\omega)}{d\omega} = 0.$$

Let

$$v^{-1} = \frac{d \text{Re} k_S(\omega)}{d\omega} \bigg|_{\omega=\omega_0}.$$

Then the function $\lambda_S(\omega) = \omega - k_S(\omega) v$ would have a saddle point ω_0 on the real axis, since the following relationship takes place

$$\frac{d\lambda_S}{d\omega}\bigg|_{\omega=\omega_0} = 1 - v \times \left[\frac{d \text{Re} k_S(\omega)}{d\omega} + i \frac{d \text{Im} k_S(\omega)}{d\omega}\right]\bigg|_{\omega=\omega_0} = 0. \qquad (6.15)$$

Evaluating the integral (6.14) by the steepest descent method we get the asymptotic solution

$$\varphi_S^{ij}(vt, t)_{t \to \infty} \sim \frac{1}{\sqrt{t}} \exp i (\omega_0 - v k_S(\omega_0)) t \qquad (6.16)$$

$\text{Im} k_S(\omega) > 0$ and $v > 0$, since the group velocity of a co-propagating wave is positive. Therefore, $\varphi_S^{ij} \to \infty$ as $t \to \infty$. As a result, the perturbations would increase with time in the coordinate system travelling with a velocity

$$v = \left(\frac{d \text{Re} k_S(\omega)}{d\omega}\right)^{-1}\bigg|_{\omega=\omega_0}.$$

Such a situation corresponds to a convective instability [11].

We have shown that the procedure determination of an instability type determination is in principle based on the analysis of a complex variable function $k_S(\omega)$ that in general case can have some branches. The solution of such a problem in general case is hardly possible being very complicated analytically. However, in some practically important cases the problem can be solved explicitly.

6.2.2 Criterion of a Spatial Amplification

We start with the system of (6.2) describing the wave propagation in some medium. We look for a solution in an interval $z > 0$, $t > 0$. It is known that the boundary problem is the most appropriate one for the analysis of an amplification process. Suppose that a signal $f_i(0, t)$ enters the medium at $z = 0$ and at the moment $t = 0$. We are interested in the evolution of this signal with the z-coordinate. Assuming that the functions $f_i(0, t)$ are limited

$$f_i(0, t) \leqslant M \exp(-\alpha t); \quad \alpha > 0$$

we expand them into the Fourier–Laplace integral which has the form

$$f_i(0, t) = \frac{1}{2\pi} \int_{-\infty}^{\infty} f_i(0, \omega) \exp i\omega t \, d\omega.$$

The reverse Fourier–Laplace transform

$$f_i(0, \omega) = \frac{1}{2\pi} \int_{-\infty}^{\infty} f_i(0, t) \exp(-i\omega t) \, dt$$

shows that the functions $f_i(0, \omega)$ are regular in the lower half-plane of the complex variable ω. Suppose additionally that $f_i(0, z) = 0$ which simply means we exclude the possibility of an absolute or convective instability which are considered above. The time interval $0 < t < \infty$ is chosen because of the two reasons [11].

1. Such a choice makes it possible to investigate transition processes which accompany the signal propagation in the medium and to find a steady state limited solution of system of (6.2) at a given point of the medium as $t \to \infty$. In the case of an absolute instability such a solution does not exist, as it was mentioned above.
2. The determination of the initial moment of time $t = 0$ permits the application of the causality principle which states that for any sufficiently small t sufficiently large z always exist where the solution of system of (6.2) identically equals zero since the signal has not yet reached those regions:

$$f_i(z,t) \underset{z\to\infty}{\to} 0, \quad t = \text{const}.$$

Using this principle one can obtain the spatial amplification criterion and formulate the conditions which the functions $f_i(0,\omega)$ have to obey. Applying to the system (6.2) the Laplace transformation on space and time we find

$$\sum_{j=1}^{n}\left[i\left(\omega\delta_{ij}-k\alpha_{ij}\right)+\beta_{ij}\right]f_j(k,\omega)=\varphi_i(0,\omega), \qquad (6.17)$$

where we take into account that $f_i(z,0)=0$ and

$$\varphi_i(0,\omega)\equiv\sum_{j=1}^{n}\alpha_{ij}f_j(0,t).$$

The system of (6.17) coincides with the system (6.4), and its solution has the form

$$f_j(k,\omega)=\sum_{j=1}^{n}D_{ij}(k,\omega)\varphi_j(0,\omega), \qquad (6.18)$$

where $D_{ij}(k,\omega)$ is determined by expression (6.6). The reverse transformation of the solution (6.18) yields

$$f_i(z,t)=\frac{1}{(2\pi)^2}\int_{\gamma_\omega}d\omega\int_{\gamma_k}dk\, f_i(k,\omega)\exp i(\omega t-kz). \qquad (6.19)$$

Here the integration contour γ_k is situated higher than all poles of the function $f_i(k,\omega)$; the integration contour γ_ω is situated lower than all singularities of the integrand including the branching points of the function $k_S(\omega)$. The detailed analysis using the theory of function of complex variable shows that solution (6.19) satisfies the system of (6.2) and the boundary conditions. However, in general, it does not satisfy the causality principle. Indeed, the integration over k results in the following expression

$$f_i(z,t)=\frac{1}{2\pi i}\sum_{j=1}^{n}\sum_{S=1}^{n}\int_{\gamma_\omega}d\omega\,\frac{\Delta_{ij}(k_S(\omega),\omega)}{D'_k(\omega,k_S(\omega))}\varphi_j(0,\omega)\exp i(\omega t-k_S(\omega)z), \qquad (6.20)$$

where S stands for the number of the root of the dispersion relation $D(\omega,k)=0$. It has been shown that the terms with the negative phase velocity v_S such that

$$\lim_{\omega\to\infty}\frac{k_S(\omega)}{\omega}=\frac{1}{v_S}<0 \qquad (6.21)$$

can appear in the sum (6.20), which means that a finite signal would exist far from the input to the medium even at very small t. In order to satisfy simultaneously the causality principle and the boundary conditions such terms must be excluded from the sum (6.20):

$$\sum_{j=1}^{n} \Delta_{ij}\left(k_{S}\left(\omega\right),\omega\right)\varphi_{j}\left(0,\omega\right)=0, \quad v_{S}<0. \tag{6.22}$$

It is seen from (6.22) that some boundary conditions, i.e. some functions $\varphi_i(0,\omega)$ are not independent which results in the existence of reflected waves. It can be shown rigorously that for each $i=1,...,n$ the number of the independent conditions (6.22) does not surpass the number of counter-propagating waves. Finally, we have

$$f_i(z,t) = \frac{1}{2\pi i}\sum_{j=1}^{n}\sum_{S}{}'\int_{\gamma_\omega} d\omega \frac{\Delta_{ij}\left(k_S\left(\omega\right),\omega\right)}{D'_k\left(\omega,k_S\left(\omega\right)\right)}\varphi_j(0,\omega)\exp i\left(\omega t - k_S\left(\omega\right)z\right), \tag{6.23}$$

where prime at the sum over S means that the terms with the negative v_S corresponding to the reflected waves are omitted. Clearly, at the input $z=0$ of the medium filling the half-space $z>0$ only the co-propagating, or diverging, waves with $v_S>0$ can be excited. In the opposite case, when the medium is filling the half-space $z<0$ the solution (6.23) would contain only terms with negative phase velocity $v_S<0$. The medium is assumed to be infinite and homogeneous where reflected waves cannot exist.

Finally, we should carry out integration over ω in expression (6.23). For this purpose, we should displace the integration contour γ_ω along the real axis that is possible only in the case where the branching points are absent in the lower half-plane of the complex variable ω, or where at least, the branches of $k_S(\omega)$ at such a point coincide which corresponds to the co-propagating waves. Then $k_S(\omega)$ can be considered as a function of a real variable ω only, and we have

$$f_i(z,t) = \frac{1}{2\pi i}\sum_{v_S>0}\int_{-\infty}^{\infty} F_{iS}(\omega)\exp i\left(\omega t - k_S\left(\omega\right)z\right)d\omega. \tag{6.24}$$

If $\mathrm{Im} k_S(\omega) > 0$ for some S, then the corresponding terms in the series (6.24) would increase in space with an increase of the z-coordinate, and the spatial amplification occurs. In the opposite case, where $v_S > 0$ but $\mathrm{Im} k_S(\omega) < 0$ the corresponding wave decays with a distance from the boundary and vanishes at infinity. Such a regime is called a non-transparency. Now we formulate the criterion of the spatial amplification.

6 Current Instabilities

If among the roots $k_S(\omega)$ of the dispersion equation

$$D(\omega, k) = 0$$

there exist the ones obeying the following conditions

$$\lim_{\omega \to \infty} \frac{k_S(\omega)}{\omega} = \frac{1}{v_S} > 0, \quad Im k_S(\omega) > 0$$

then the waves corresponding to these $k_S(\omega)$ would be amplified in the range of real frequencies ω where $Im k_S(\omega) > 0$ [11].

Obviously, in the half-space $z < 0$ the waves with $v_S < 0$ and $Im k_S(\omega) < 0$ would be amplified. In any case, the signs of v_S and $Im k_S(\omega)$ should coincide. In the opposite case, the non-transparency regime takes place. The quantity $\alpha = Im k_S(\omega)$ is called a coefficient of amplification, or a spatial increment.

Now we can make a conclusion that in order to solve the problem of stability of a system and to determine the type of instability, it is necessary to construct an adequate dispersion equation, to solve it, and to classify the waves into the co-propagating and counter-propagating ones. In some practically important cases the explicit solution of a problem is hardly possible. In such a situation it is worth to investigate the asymptotic behavior of the dispersion equation for sufficiently large ω and k.

6.3 Analysis of Dispersion Equation for Two Coupled Waves

We illustrate the theory presented above with the instructive example of a coupled waves behavior [11, 12]. As it was already mentioned, in many cases the explicit solution of a dispersion equation is hard to obtain. In real physical systems, however, a small parameter λ can exist such that the dispersion equation can be expanded in powers of this parameter keeping the first order term only as follows

$$\prod_{i=1}^{n} [k - k_i(\omega)] = \lambda f(k, \omega), \qquad (6.25)$$

where $k_i(\omega)$ is a dispersion relation for the i-th wave for $\lambda = 0$. The small correction for the first wave linear in λ can be found immediately by separating the first wave dispersion relation and substituting in the right-hand side the approximate value of $k \approx k_1(\omega)$

$$k \approx k_1(\omega) + \lambda \frac{f(k_1(\omega), \omega)}{\prod_{i=2}^{n} [k_1(\omega) - k_i(\omega)]}. \qquad (6.26)$$

6.3 Analysis of Dispersion Equation for Two Coupled Waves

The corrections of higher orders in λ can be obtained by the recursion process. Unfortunately, this approach is invalid at the so-called matching point where some dispersion curves can coincide:

$$k_1(\omega) = k_i(\omega), \quad i = 1, 2, \ldots. \tag{6.27}$$

In such a case, the interaction between the waves would be strong despite the small value of the parameter λ. For the sake of simplicity we limit our analysis with two waves. Then we have

$$[k - k_1(\omega)][k - k_2(\omega)] = \lambda \frac{f(k(\omega),\omega)}{\prod_{i=3}^{n}[k(\omega) - k_i(\omega)]}. \tag{6.28}$$

We call the waves obeying condition (6.27) the coupled ones. Firstly, we substitute the values of k and ω at the matching point into the right-hand side of (6.28). Such a case is called the weakly coupled waves approximation. Then, we expand $k_{1,2}(\omega)$ in powers near the matching point (k_0, ω_0) as follows:

$$k_{1,2}(\omega) = k_0 + \frac{\partial k_{1,2}}{\partial \omega}\bigg|_{\omega_0}(\omega - \omega_0) + \ldots.$$

Choosing the origin at the matching point we finally obtain:

$$\left(k - \frac{\omega}{v_1}\right)\left(k - \frac{\omega}{v_2}\right) = A, \tag{6.29}$$

where $v_{1,2}$ and A are the group velocities and a right-hand side of (6.28) calculated in the matching point.

$$v_{1,2} = \left(\frac{\partial k_{1,2}}{\partial \omega}\right)^{-1}_{\omega_0}.$$

The four different combinations of the sign of A and the wave propagation direction are possible.

1. Both waves propagate in the same direction, $v_1 v_2 > 0$, and $A > 0$;
2. The waves are co-propagating, $v_1 v_2 > 0$, and $A < 0$;
3. The waves are counter-propagating, $v_1 v_2 < 0$, and $A > 0$;
4. The waves are counter-propagating, $v_1 v_2 < 0$, and $A < 0$.

Consider the behavior of the dispersion equation solution in all these cases. The solution of (6.28) has the form

$$\omega = \frac{1}{2}k(v_1 + v_2) \pm \sqrt{\frac{1}{4}k^2(v_1 - v_2)^2 + Av_1 v_2}, \tag{6.30}$$

or

$$k = \frac{\omega}{2}\left(\frac{1}{v_1} + \frac{1}{v_2}\right) \pm \sqrt{\frac{\omega^2}{4}\left(\frac{1}{v_1} - \frac{1}{v_2}\right)^2 + A}. \tag{6.31}$$

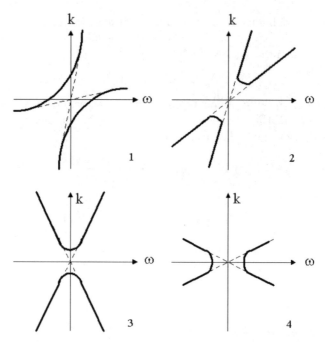

Fig. 6.2. The dispersion curves for two coupled waves. (**1**) $v_1 v_2 > 0; A > 0$. The system is stable. (**2**) $v_1 v_2 > 0; A < 0$. Convective instability, spatial amplification. (**3**) $v_1 v_2 < 0; A > 0$. Absolute instability. (**4**) $v_1 v_2 < 0; A < 0$. Non-transparency regime

1. The analysis of (6.30) and (6.31) shows that in the case 1 real values of ω correspond to real values of k, and vice versa. Consequently, the system is stable and cannot amplify signals. The wave coupling results in the eliminating of degeneracy at the matching point. The dispersion curve is presented in Fig. 6.2(1).
2. In the case 2 where $v_1 v_2 > 0$, and $A < 0$, in the frequency interval $|\omega| \leq 2\sqrt{|A| v_1 v_2}/|v_1 - v_2|$ the complex conjugate values of k correspond to real values of ω, while in the interval $|k| \leq 2\sqrt{|A| v_1 v_2}/(v_1 - v_2)$ the complex conjugate values of ω occur. In this case the spatial amplification takes place, since both waves propagate in the same direction, and one of them possesses $\mathrm{Im} k > 0$. On the other hand, it is seen from (6.30) that the branching points of ω belong to the real axis which should result in a convective instability. The dispersion curve is presented in Fig. 6.2(2).
3. In the case 3 where $v_1 v_2 < 0$, and $A > 0$, the real values of k correspond to the real values of ω according to (6.31). In contrast to that, in the interval $k < 2\sqrt{|A| v_1 v_2}/|v_1 - v_2|$ the complex values of ω correspond to the real values of k. The function $k(\omega)$ has the branching point

$$\omega = -i\frac{2\sqrt{|A|}\,|v_1 v_2|}{|v_1 - v_2|}$$

and the waves are counter-propagating. Then, the absolute instability is realized. The dispersion curve is presented in Fig. 6.2(3).

4. In the case 4 where $v_1 v_2 < 0$, and $A < 0$ the real values of ω correspond to the real values of k. The system is stable. However, in the interval $|\omega| \leq 2\sqrt{|A|}\,|v_1 v_2|/|v_1 - v_2|$ the complex values of k correspond to the real values of ω, and for the co-propagating wave $\mathrm{Im}\,k < 0$, for the counter-propagating wave $\mathrm{Im}\,k > 0$, that gives rise to the non-transparency regime. Assume for the sake of definiteness that $v_1 > 0$. Then (6.30) yields for the co-propagating wave

$$k_1(\omega) = \frac{\omega}{2}\left(\frac{1}{v_1} + \frac{1}{v_2}\right) + \sqrt{\frac{\omega^2}{4}\left(\frac{1}{v_1} - \frac{1}{v_2}\right)^2 - |A|}$$

$$\lim_{\omega \to \infty} \frac{k_1(\omega)}{\omega} = \frac{1}{v_1} > 0.$$

Indeed, circumventing the branching point $\omega = 2\sqrt{|A|}\,|v_1 v_2|/|v_1 - v_2|$ in the lower half-plane we find that

$$\mathrm{Im}\,k_1(\omega) = \exp\left(-i\frac{\pi}{2}\right)\sqrt{|A| - \frac{\omega^2}{4}\left(\frac{1}{v_1} - \frac{1}{v_2}\right)^2} < 0.$$

Consequently, the co-propagating wave decays. For the counter-propagating wave $v_2 < 0$, $\mathrm{Im}\,k_2(\omega) > 0$ which means that the counter-propagating wave also decays. The dispersion curve is presented in Fig. 6.2(4).

6.4 Transformation of Instabilities

The type of an instability in an infinite system is determined by the distribution of a function $k(\omega)$ branching points in the plane of a complex variable ω. Clearly, their position is defined by parameters of the system which makes it possible to change the coordinates of the branching points by variation of the control parameters. Thus, a convective instability may be transformed into an absolute one and vice versa [11]. Consider, for example, the role of dissipation. In our previous analysis the dissipation was neglected. Now we take it into account for the counter-propagating wave, while the dissipation of the co-propagating wave is still ignored. Let the dissipation coefficient of the counter-propagating wave is $\nu > 0$. Then (6.29) takes the form

$$\left(k - \frac{\omega}{v_1}\right)\left(k - \frac{\omega}{v_2} - i\nu\right) = A. \qquad (6.32)$$

Solving equation (6.32) we obtain for the branching points

$$\omega = i \left| \frac{v_1 v_2}{v_1 - v_2} \right| \left(\nu \pm 2\sqrt{A} \right).$$

If $\nu > 2\sqrt{A}$, then both branching points are situated in the upper half-plane of the complex variable ω which corresponds to the case of a convective instability. Consequently, the absolute instability can be stabilized by introducing a sufficiently large dissipation. In such a case the system can operate as an amplifier.

Consider now the influence of boundaries on the transformation of one type of instability into another [11]. The presence of the boundaries can affect essentially the instability type by means of reflected waves creating a feedback between the input and the output of the system. As a result, a convective unstable system can be transformed into an absolute unstable one. The same transition can be also stimulated by an inhomogeneity of a physical system. Assume that system has finite dimension in the $z$direction: $0 < z < L$. The initial conditions are supposed to be known. The boundary conditions should be formulated in such a way that their number at $z = 0$ is equal to the number of co-propagating waves, while their number at $z = L$ is equal to the number of counter-propagating waves. They have the form

$$\sum_{j=1}^{n} r_{ij} f_j(z_m, t) = \varphi_i(z_m, t), \quad z_m = 0, L. \tag{6.33}$$

Here r_{ij} are some constants and $\varphi_i(z_m, t)$ are either assumed to be known, or found by solving the differential equations in the intervals $z < 0$ and $z > L$. Using the standard procedure described in previous sections we find

$$\sum_{j=1}^{n} \left(i\omega \delta_{ij} + \beta_{ij} + \alpha_{ij} \frac{\partial}{\partial z} \right) f_j(z_m, \omega) = f_i(z, 0) \tag{6.34}$$

$$\sum_{j=1}^{n} r_{ij} f_j(z_m, \omega) = \varphi_i(z_m, \omega) \text{ for } z_m = 0, L. \tag{6.35}$$

The solution of the inhomogeneous system of (6.34) consists of the particular solution $f_i^0(z, \omega)$ of this system and the general solution of the corresponding homogeneous system having the form:

$$f_i^H(z, \omega) = \sum_{S=1}^{n} a_{iS} C_S(\omega) \exp\left[-i k_S(\omega) z\right],$$

where $C_S(\omega)$ are some arbitrary constants to be determined from the boundary conditions (6.35); $k_S(\omega)$ are the roots of the dispersion equation (6.10);

a_{iS} are the co-factors of the matrix $(i\omega\delta_{ij} + \beta_{ij} - ik_S(\omega)\alpha_{ij})$. The general solution of the system of (6.34) has the form

$$f_i(z,\omega) = \sum_{S=1}^{n} a_{iS} C_S(\omega) \exp[-ik_S(\omega)z] + f_i^0(z,\omega). \qquad (6.36)$$

Substituting expression (6.36) into (6.35) we obtain the system of inhomogeneous linear algebraic equations for $C_S(\omega)$:

$$\sum_{S=1}^{n}\sum_{l=1}^{n} r_{il} a_{lS} C_S(\omega) \exp[-ik_S(\omega)z_m] \equiv \sum_{S} d_{iS} C_S(\omega) \qquad (6.37)$$

$$= \varphi_i(z_m,\omega) - \sum_{l=1}^{n} r_{il} f_l^0(z_m,\omega).$$

The solution of (6.37) can be written in the form

$$C_S(\omega) = \frac{\Delta_S(\omega)}{D(\omega)}, \qquad (6.38)$$

where

$$D(\omega) = \det\|d_{iS}\|$$

and

$$d_{iS} = \exp[-ik_S(\omega)z_m] \sum_{l=1}^{n} r_{il} a_{lS}. \qquad (6.39)$$

$\Delta_S(\omega)$ is obtained from $D(\omega)$ by changing of the S-th column to the right-hand side of (6.37). Substituting the results (6.38) into (6.36) and carrying out the inverse Laplace transformation we obtain

$$f_i(z,t) = \frac{1}{2\pi} \int_{\gamma_\omega} \frac{\psi_i(z,\omega)}{D(\omega)} \exp i\omega t \, d\omega, \qquad (6.40)$$

where

$$\psi_i(z,\omega) = \sum_{S=1}^{n} \frac{a_{iS}\Delta_S(\omega)}{D(\omega)} \exp(-ik_S z) + f_i^0(z,\omega) D(\omega).$$

As a rule, the functions $\psi_i(z,\omega)$ depending on the initial and boundary conditions do not possess singularities for finite ω, and all singularities of the integrand are determined by the roots of $D(\omega)$. Choosing the integration contour γ_ω along the real axis we can present the last integral as follows

$$f_i(z,t) = \frac{1}{2\pi} \int_{-\infty}^{\infty} \frac{\psi_i(z,\omega)}{D(\omega)} \exp i\omega t\, d\omega + \sum_S i a_S \exp i\omega_S t, \qquad (6.41)$$

where ω_S are the zeros of the function $D(\omega)$ in the lower half-plane, and a_S are the residues of the function $\psi_i(z,\omega)/D(\omega)$ at the points ω_S. According to the asymptotic behavior of Fourier integrals the first term in the right-hand side of (6.41) vanishes as $t \to \infty$. If the function $D(\omega)$ does not possess zeros in the lower half-plane, then the second term in the right-hand side of (6.41) also equals zero, and $f_i(z,t) \to 0$ as $t \to \infty$. Consequently, the system is stable. If, on the contrary, the zeros ω_S of the function $D(\omega)$ exist in the lower half-plane, then an absolute instability arises since

$$f_i(z,t) \underset{t \to \infty}{\to} \sum_S i a_S(z) \exp(-\mathrm{Im}\,\omega_S t) \exp(i\mathrm{Re}\,\omega_S t) \to \infty.$$

Generally, it is not necessary to calculate explicitly the solution of the system obeying the boundary conditions. In most cases one need only to calculate the quantities a_{iS}, then to construct the matrix d_{iS} (6.39) and, finally, to equate its determinant $D(\omega) = \det \|d_{iS}\|$ to zero. The quantities $k_S(\omega)$ are calculated by solving the dispersion equation (6.10) of the basic system of (6.2).

6.5 Instabilities in Solid State Plasma

In this section we consider some practically important and at the same time instructive examples of absolute and convective instabilities in solid state plasma using the formalism developed above.

6.5.1 Parameters of Solid State Plasma

A conducting medium can be considered as plasma, or an ensemble of moving electric charges positive as well as negative. In general, any macroscopic volume filled with plasma can be electrically charged or neutral. We will consider the neutral, or compensated plasma existing in solids and gases. Clearly, the electrodynamic behavior of plasmas in each of these states differs essentially from one another.

Consider first the characteristics of plasma in solids including metals and semiconductors [13]. The solid state plasma can be defined as moving electrons and/or holes which always exist in any crystal. In the undisturbed, or basic state the solid state sample as a whole is assumed to be electrically neutral. In metals, plasma consists of free electrons which are electrically compensated by positive ions of atomic radicals. In semiconductors with disequilibrium number of electrons and holes the total electric neutrality is provided by the electron–hole compensation and by ionized donor and acceptor

6.5 Instabilities in Solid State Plasma

impurities. A magnetic or electric field applied to such a plasma can create the spatial gradients of a space charge giving rise to a multitude of specific effects some of which will be considered in details below. The sources of plasma in solids can be either natural, i.e. metals, semi-metals and semiconductors themselves, or artificial, i.e. the injection of electrons and holes from p-n junction type contacts, avalanche breakdown, impact ionization of impurity centers, ionizing radiation, etc. We now introduce some physical parameters describing the state of plasma.

The density of plasma is determined by the charge carrier concentration n_e for electrons and n_h for holes. It may change considerably for different materials reaching $\sim 10^{28}$ m^{-3} for metals, while being of an order of magnitude of 10^{14} m^{-3} for very pure semiconductors like Ge and Si at very low temperatures. The solid state plasma density can be easily changed by external fields which represents an important advantage as compared to plasmas in gases.

An effective mass m^* of a charge carrier is usually less than a mass of a free electron m_0 due to the periodic potential of a crystal lattice. For electrons in semiconductors (Bi, $InSb$) the ratio of the effective mass and the mass of a free electron $m_e^*/m_0 \sim 10^{-2}$, while in metals (Cu, Na) the same ratio is close to 1. The effective mass of holes is usually less than the one of electrons.

The basic types of vibrations in plasma are the plasma ones with a frequency ω_p and the cyclotron ones with a frequency ω_c.

$$\omega_p = \sqrt{\frac{e^2 n_0}{\varepsilon_0 \varepsilon^r m^*}}; \quad \omega_c = \frac{eB}{m^*},$$

where B is the external magnetic field induction and n_0 is the charge carrier concentration. The cyclotron oscillations have been analyzed in details in the previous sections. The estimations show that $\omega_c \sim 10^{10}$ s^{-1} for metals and $\sim 10^{12}$ s^{-1} for semiconductors with a small effective mass of a carrier m_e^* for the magnetic field $B \sim 0.1$ tesla. The plasma frequency ω_p changes from $\sim 10^{12}$ s^{-1} for typical semiconductors up to $\sim 10^{15}$ s^{-1} for metals. In this section we will concentrate on the plasma oscillations.

The scattering frequency ν in gas and solid state plasma strongly differs. In gases ν is negligibly small. In solid state, on the contrary, the scattering plays a considerable role in the energy and momentum exchange due to thermal lattice vibrations. Besides that, there exists the scattering on the ionized and neutral impurities. In the case of a dense electron-hole plasma in semiconductors with $n \gtrsim 10^{22}$ m^{-3} it is necessary to take into account the electron-hole scattering. In solid state ν belongs to the frequency range $(10^{10} \div 10^{13})$ s^{-1} as the temperature increases from the liquid helium one $(4.2°K)$ up to $300°K$. The ratio of the scattering frequency and the plasma or cyclotron frequency is an important characteristic of different types of wave interaction.

The crystal lattice interacting with plasma contributes into the total polarization of the system, The lattice polarizability is characterized by the relative dielectric constant ε^r. For the typical semiconductors it is equal to $(10 \div 20)$. For some semi-metals the dielectric constant may reach 100.

The important plasma parameter is the dielectric relaxation frequency ω_R which has the form:

$$\omega_R = \frac{\omega_p^2}{\varepsilon^r \nu} = \frac{\sigma}{\varepsilon_0 \varepsilon^r},$$

where σ is the plasma conductivity. The dielectric relaxation time which is also called the charge screening time has the form: $\tau_R = \omega_R^{-1}$ s. It has the order of magnitude of 10^{-18} s for metals and 10^{-12} s for some semiconductors.

The plasma in metals is described by the Fermi-Dirac statistics with the Fermi temperature $\sim 10^4$ °K, i.e. it is totally degenerated. In contrast to that, in most semiconductors the plasma is non-degenerated, and its equilibrium distribution function is the Maxwell distribution. In an external field the drift of charge carriers and the increase of their effective temperature results in the shifted Maxwell distribution.

Plasma frequency ω_p sets a characteristic length of screening of an external charge in a neutral plasma. Indeed, the uncompensated charge in a plasma yields oscillatory motion of charges with the plasma frequency ω_p. The screening of this deviation from the quasineutrality is defined by the distance, on which an electron with a velocity v_T or v_T in classical or degenerated case, respectively, will have travelled during a period of plasma oscillations It is called the Debye screening length and it has the form: $L_D = v/\omega_p$ where v is an average thermal velocity of charge carriers with the corresponding distribution function. For metals $L_D \sim 10^{-9}$ m, i.e. close to the interelectronic spacing, while for semiconductors $L_D \sim 10^{-6}$ m. The Debye length in semiconductors is much longer than in metals because of the lower electron velocities and much lower than in metals electron densities.

At the finite temperature and in the presence of scattering in a solid state the spatially inhomogeneous distribution of carriers causes the diffusion which is characterized by a diffusion coefficient $D = k_B T/m\nu = v^2/2\nu$. For most solids under the practically important conditions $v \approx 10^6$ m/s, $\nu \approx 10^{13}$ s^{-1} and, consequently, $D \sim 10^{-1}$ m^2/s. The diffusion determines the upper limit of the instability frequency. In such a case the diffusion frequency $\omega_D = v_f^2/D$ where v_f is the phase frequency of the wave excited. Then the wave frequency ω is limited by the following condition

$$\omega^2 < \omega_R \omega_D.$$

In other words, the wavelength λ should be greater than the Debye length

$$\lambda > \lambda_D,$$

while the diffusion time $\tau_D = \omega_D^{-1}$ should be greater than the dielectric relaxation time

$$\tau_D > \tau_R.$$

An important role plays the notion of mobility μ. When an external electric field is applied to a conductor, the system behaves as if, in addition to the purely random motion, the charged particle is being accelerated in the direction of the field $\vec{E}$ during the time τ. Elastic (inelastic) scattering events randomize the momenta (the energies), acquired from the electric field between the collisions.

Averaging this sawtooth variation, one can introduce the average drift velocity $\boldsymbol{v}_{\mathrm{dr}}$

$$v_{i\mathrm{dr}} = \mu_{ik} E_k, \tag{6.42}$$

where the mobility μ_{ik} is in general case a second rank tensor. In the simplest case of an isotropic medium it reduces to a scalar defined as

$$\mu = \frac{e\tau}{m}. \tag{6.43}$$

Equation (6.42) may be generalized by replacing the electric force by other force. A whole class of transport phenomena in plasmas may be investigated by this method, diffusion being most important of them.

Diffusion coefficient D is connected with the mobility μ via the Einstein relation:

$$D = \frac{k_B T}{e} \mu, \tag{6.44}$$

where k_B is the Boltzmann constant.

6.5.2 Quasineutral Oscillations in Semiconductors

We start with the simplest case of a semiconductor with two sorts of carriers: electrons and holes. While in metals the deviation from the quasineutrality results in very high frequency plasma oscillation, in gaseous plasma and in semiconductors, where the screening time τ_{screen} and the screening length L_D are much longer, electrostatic oscillations may be very different from the usual plasma waves. These are the quasineutral oscillations connected with the ambipolar drift in plasmas with two sorts of carriers.

Consider now the dynamics of a charged fluctuation n_1 Fig. 6.3. In a metal, where the electrons are mobile and the ions are fixed, the charge will be screened out in a time interval $\tau_{\mathrm{screen}} \approx \varepsilon_0 \varepsilon^r \sigma^{-1}$. In a semiconductor, however, with two charge species: the electrons with concentration n and the holes with concentration p, the charge neutrality will be restored on a much longer scale. Moreover, if an external electric field $\boldsymbol{E}$ is applied, then in the case where the concentrations of negative and positive charge carriers n and p are not equal, the thermally created charge fluctuation n_1 will propagate

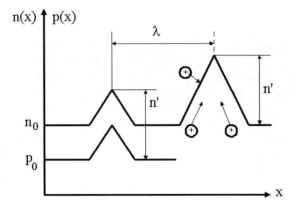

Fig. 6.3. Building up of a quasineutral fluctuation

along the sample as a wave with very slow attenuation, or under certain conditions it may even grow (plasma instabilities). This can be formally shown by solving a simple system of continuity equation for electrons and for holes. The quasineutrality is valid if the oscillation frequency satisfies the condition: $\omega \varepsilon^r \varepsilon_0 \ll \sigma$. Then the displacement current can be neglected, and we use the quasi-static approxiamtion. The system of equations, describing the problem, is:

$$\frac{dn}{dt} + \operatorname{div}\left[-n\mu_n \boldsymbol{E} - D_n \nabla n\right] = 0 \qquad (6.45)$$

$$\frac{dp}{dt} + \operatorname{div}\left[p\mu_p \boldsymbol{E} - D_p \nabla p\right] = 0, \qquad (6.46)$$

where μ_n, D_n and μ_p, D_p are the mobility and diffusion coefficient of the electrons and holes respectively. To find the dispersion relation, we linearize (6.45) and (6.46) near the equilibrium values of the electric field $\boldsymbol{E}_0$ and the electron and hole concentration n_0 and p_0 respectively.

$$\boldsymbol{E} = \boldsymbol{E}_0 - \nabla V, \ |\nabla V| \ll E_0 \qquad (6.47)$$

$$p = p_0 + p_1; \ n = n_0 + n_1; \ p_1 \ll p_0, \ n_1 \ll n_0, \qquad (6.48)$$

where V is the electrostatic potential, and n_1 and p_1 are the fluctuations of electrons and holes concentration. Neglecting the capture of carriers we get that the fluctuations n_1 and p_1 are equal

$$n_1 = p_1. \qquad (6.49)$$

Combining (6.45)–(6.49) we obtain

6.5 Instabilities in Solid State Plasma

$$\frac{dn_1}{dt} + n_0 \mu_n \nabla^2 V - \mu_n (\boldsymbol{E}_0 \cdot \nabla) n_1 - D_n \nabla^2 n_1 = 0 \qquad (6.50)$$

and

$$\frac{dn_1}{dt} - p_0 \mu_p \nabla^2 V + \mu_p (\boldsymbol{E}_0 \cdot \nabla) n_1 - D_p \nabla^2 n_1 = 0. \qquad (6.51)$$

Eliminating the term with the electrostatic potential V we get

$$\frac{dn_1}{dt} (n_0 \mu_n + p_0 \mu_p) + (n_0 - p_0) \mu_n \mu_p (\boldsymbol{E}_0 \cdot \nabla) n_1 -$$

$$- (p_0 \mu_p D_n + n_0 \mu_n D_p) \nabla^2 n_1 = 0. \qquad (6.52)$$

We find a plane wave elementary solution

$$n_1 \sim \exp i (\omega t - \boldsymbol{k r}). \qquad (6.53)$$

Substituting (6.53) into (6.52) we get the dispersion relation

$$\omega = (\boldsymbol{E}_0 \cdot \boldsymbol{k}) \frac{(n_0 - p_0) \mu_n \mu_p}{(n_0 \mu_n + p_0 \mu_p)} + ik^2 \frac{(p_0 \mu_p D_n + n_0 \mu_n D_p)}{(n_0 \mu_n + p_0 \mu_p)}. \qquad (6.54)$$

Obviously, the system is stable, since the imaginary part of the frequency in positive. Now we can introduce the ambipolar mobility μ_a and the ambipolar diffusion coefficient D_a.

$$\mu_a = \frac{(n_0 - p_0) \mu_n \mu_p}{(n_0 \mu_n + p_0 \mu_p)} \qquad (6.55)$$

and

$$D_a = \frac{(p_0 + n_0) \mu_n \mu_p}{(n_0 \mu_n + p_0 \mu_p)} \frac{k_B T}{e}, \qquad (6.56)$$

where we used the Einstein relation

$$D_n = \frac{k_B T}{e} \mu_n; \quad D_p = \frac{k_B T}{e} \mu_p. \qquad (6.57)$$

Dispersion relation (6.54) can be rewritten as follows

$$\omega = (\boldsymbol{k} \cdot \boldsymbol{v}_{\text{dra}}) + ik^2 D_a, \qquad (6.58)$$

where $\boldsymbol{v}_{\text{da}}$ is the ambipolar drift velocity given by

$$\boldsymbol{v}_{\text{dra}} = \mu_a \boldsymbol{E}_0. \qquad (6.59)$$

We can evaluate the ambipolar screening time τ_{screen}^a which has the form

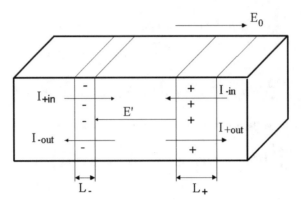

Fig. 6.4. Currents entering and leaving a quasineutral fluctuation

$$\tau^a_{\text{screen}} \simeq \frac{L_D^2}{D_a} = \frac{\varepsilon_0 \varepsilon^r}{\sigma} \frac{D_n}{D_a} = \tau_{\text{screen}} \frac{D_n}{D_a}, \quad (6.60)$$

where we used the definition of the Debye length L_D and the Einstein relation (6.57). Usually the mobility μ_n of electrons is much more than the one of holes μ_p while the equilibrium concentrations of electrons and holes n_0 and p_0 in our case are of the same order of magnitude. Then comparison of (6.56) and (6.57) shows that

$$D_n \gg D_a \quad (6.61)$$

and, indeed, the lifetime τ^a_{screen} of the quasineutral fluctuation defined by the ambipolar diffusion is large compared to the ordinary screening time:

$$\frac{\tau^a_{\text{screen}}}{\tau_{\text{screen}}} \simeq \frac{D_n}{D_a} \gg 1. \quad (6.62)$$

Let us give a transparent qualitative interpretation of this result. If the fluctuation is immobile it will be screened in a diffusive manner. If the fluctuation is moving with a proper velocity however the in- and outcoming currents will balance each other and the fluctuation will propagate as a weakly damped wave, since the currents which are of opposite direction in the rest frame can be equalized in the appropriately chosen moving frame.

To show this consider the currents entering and leaving the fluctuation as it is shown in Fig. 6.4. Assume, for definiteness that $n_0 > p_0$, i.e. $n_0 = p_0 + N_d$ where N_d is the donor concentration. In the frame moving with a velocity v the stationary currents: $I^o_{\text{in}} \simeq e\mu_p p_0 E_0$ and $I^o_{\text{out}} \simeq e\mu_n n_0 E_0$ will change by an equal value

$$\Delta I \simeq en_0 v,$$

while the fluctuation currents $I^1_{\pm\text{in}}$ and $I^1_{\pm\text{out}}$ change in a different way. Since the electric field E^1 is defined in the moving frame, the currents

$I_{-\text{in}}^1 \simeq en_0\mu_n E_n^1$ and $I_{+\text{in}}^1 = e\mu_p p_0 E_p^1$ do not change at all. The changes of the currents $I_{+\text{out}}^1$ and $I_{-\text{out}}^1$ have opposite signs. The fluctuation would not dissipate if the entering and outgoing currents are balanced

$$I_{\pm\text{out}}^1 = I_{\pm\text{in}}^1.$$

Thus in the moving frame for the holes

$$I_{+\text{out}}^1 = e\mu_p \left(E_0 + \frac{v}{\mu_p}\right) n_1; \quad I_{+\text{in}}^1 = e\mu_p p_0 E_p^1$$

and we obtain

$$E_p^1 = \frac{n_1}{p_0}\left(E_0 + \frac{v}{\mu_p}\right).$$

For the electrons

$$I_{-\text{out}}^1 = e\mu_n \left(E_0 - \frac{v}{\mu_n}\right) n_1; \quad I_{-\text{in}}^1 = e\mu_n n_0 E_n^1$$

and

$$E_n^1 = \frac{n_1}{n_0}\left(E_0 - \frac{v}{\mu_n}\right).$$

The obtained expressions show that the fields E_n^1 and E_p^1 are equal if the fluctuation moves with the velocity v equal to the ambipolar drift velocity v_{dra} (6.59):

$$v = v_{\text{dra}} = \mu_a E_0 = \frac{(n_0 - p_0)\mu_n \mu_p}{(n_0\mu_n + p_0\mu_p)} E_0. \tag{6.63}$$

If the mobilities of electrons and holes are different, it follows from (6.63) that the fluctuation moves with the mobility close to that of the minor carriers.

6.5.3 Gradient Instability with Two Types of Carriers

Consider now the instability of the electron–hole plasma in an inhomogeneous semiconductor with two species of charge carriers, i.e. electrons and holes, subjected to an external electric field. We continue to study quasineutral oscillations. We start the analysis with the linearized model. This time we assume that the distribution of electron concentration n_0 and hole concentration p_0 is slightly inhomogeneous, but the inhomogeneity length is much greater than the wavelength λ_{osc} of vibrations excited

$$\frac{\nabla n_0}{n_0}, \frac{\nabla p_0}{p_0} \ll k = \frac{2\pi}{\lambda_{\text{osc}}} \tag{6.64}$$

and the recombination rate is small, such that the capture of carriers can be neglected. In such a case the system of (6.45)–(6.49) can still be applied. However, due to the inhomogeneity of the concentrations n_0 and p_0, new terms occur in the linearized equations (6.50) and (6.51) for the deviation n_1. Then (6.50) and (6.51) take the form

$$\frac{dn_1}{dt} + n_0\mu_n \nabla^2 V - \mu_n (\boldsymbol{E}_0 \cdot \nabla) n_1 - D_n \nabla^2 n_1 + \mu_n (\nabla V \cdot \nabla n_0) = 0 \quad (6.65)$$

and

$$\frac{dn_1}{dt} - p_0\mu_p \nabla^2 V + \mu_p (\boldsymbol{E}_0 \cdot \nabla) n_1 - D_p \nabla^2 n_1 - \mu_p (\nabla V \cdot \nabla p_0) = 0. \quad (6.66)$$

Substituting into (6.65) and (6.66) plane wave elementary solution (6.53) for n_1 and V we obtain the system of homogeneous equations

$$\left[i\omega + i\mu_n (\boldsymbol{k}\boldsymbol{E}_0) + D_n k^2\right] n_1 - \left[n_0\mu_n k^2 + i\mu_n (\boldsymbol{k}\nabla n_0)\right] V = 0 \quad (6.67)$$

and

$$\left[i\omega - i\mu_p (\boldsymbol{k}\boldsymbol{E}_0) + D_p k^2\right] n_1 + \left[p_0\mu_p k^2 + i\mu_p (\boldsymbol{k}\nabla p_0)\right] V = 0. \quad (6.68)$$

The compatibility condition for this system yields the dispersion relation

$$\omega = (\boldsymbol{k}\boldsymbol{E}_0) \frac{\mu_n \mu_p (n_0 - p_0)}{(p_0 \mu_p + n_0 \mu_n)} + i\delta \quad (6.69)$$

$$\delta = \frac{k^4 (D_n p_0 \mu_p + D_p n_0 \mu_n) - (\boldsymbol{k}\boldsymbol{E}_0) \mu_n \mu_p [(\boldsymbol{k}\nabla p_0) - (\boldsymbol{k}\nabla n_0)]}{k^2 (p_0 \mu_p + n_0 \mu_n)}. \quad (6.70)$$

Using the ambipolar mobility (6.55) and diffusion coefficient (6.56) we rewrite (6.69) and (6.70):

$$\omega = \mu_a (\boldsymbol{k}\boldsymbol{E}_0) + i\delta = \omega_0 + i\delta \quad (6.71)$$

$$\delta = k^2 D_a - \frac{(\boldsymbol{k}\boldsymbol{E}_0) \mu_n \mu_p [(\boldsymbol{k}\nabla p_0) - (\boldsymbol{k}\nabla n_0)]}{k^2 (p_0 \mu_p + n_0 \mu_n)} \quad (6.72)$$

and

$$\omega_0 = \mu_a (\boldsymbol{k}\boldsymbol{E}_0). \quad (6.73)$$

As it was mentioned above, the system is unstable when the imaginary part δ (6.72) of the frequency ω (6.71) is negative. Indeed, in such a case the exponent of the elementary solution (6.53) acquires a positive term

$$i(-i)|\delta| > 0 \quad (6.74)$$

6.5 Instabilities in Solid State Plasma

and the deviation of the electron concentration can increase with time as

$$n_1 \sim \exp|\delta|\, t. \tag{6.75}$$

The imaginary part δ is negative for electric fields E greater than a threshold value E_{0cr} which is given by

$$(\mathbf{k}\mathbf{E}_{0cr}) = k^2 D_a \frac{k^2 (p_0 \mu_p + n_0 \mu_n)}{\mu_n \mu_p [(\mathbf{k} \nabla n_0) - (\mathbf{k} \nabla p_0)]}, \tag{6.76}$$

or in the simplest case when $\mathbf{E}_0 \parallel \mathbf{k} \parallel \nabla n_0$

$$E_{0cr} = k^2 D_a \frac{(n_0 \mu_n + p_0 \mu_p)}{\mu_n \mu_p (|\nabla n_0| - |\nabla p_0|)}. \tag{6.77}$$

Relation (6.77) shows that the threshold electric field E_{0cr} is finite only in the case of an inhomogeneous distribution of carrier concentration in a semiconductor sample. Obviously, the value of E_{0cr} is fixed only for linear spatial dependence of n_0 and p_0. In a real situation it may correspond to a slow varying exponential distribution of impurities with a coordinate in a sample where only first terms of the expansion power series could be kept. Comparing relations (6.72) and (6.77) we get

$$|\delta| = k^2 D_a \left[\frac{(\mathbf{k}\mathbf{E}_0)}{(\mathbf{k}\mathbf{E}_{0cr})} - 1 \right] \text{ for } (\mathbf{k}\mathbf{E}_0) > (\mathbf{k}\mathbf{E}_{0cr}). \tag{6.78}$$

Investigate now the type of the instability using the formalism developed in previous sections. Without the loss of generality we use the one-dimensional case (6.77). Then dispersion relation (6.71) for the electric field higher than the threshold of excitation reduces to

$$\omega = \mu_a k E_0 - ik^2 D_a \left(\frac{E}{E_{0cr}} - 1 \right). \tag{6.79}$$

The roots $k_{1,2}(\omega)$ of (6.79) are given by

$$k_{1,2}(\omega) = \frac{1}{2iD_a \left(\frac{E_0}{E_{0cr}} - 1 \right)} \left[\mu_a E_0 \pm \sqrt{(\mu_a E_0)^2 - 4\omega i D_a \left(\frac{E_0}{E_{0cr}} - 1 \right)} \right]. \tag{6.80}$$

They possess the branching point ω_b in the lower half-plane of the complex variable ω:

$$\omega_b = -i \frac{(\mu_a E_0)^2}{4 D_a \left(\frac{E_0}{E_{0cr}} - 1 \right)}. \tag{6.81}$$

6 Current Instabilities

The limiting phase velocities $v_{1,2}$ of the travelling oscillation are calculated directly using (6.79), (6.80) and taking into account that for a small supercriticality

$$\left(\frac{E_0}{E_{0cr}} - 1\right) \ll 1 \tag{6.82}$$

$$|\delta| \ll \omega_0. \tag{6.83}$$

Finally we get

$$v_{1,2} = \lim_{\omega \to \infty} \frac{\omega}{k_{1,2}(\omega)} = \mp \mu_a E_0. \tag{6.84}$$

Conditions (6.81) and (6.84) correspond to the case of an absolute instability [11, 12]. Generally, that means that fluctuations (6.75) can increase infinitely in each point of the semiconductor sample. However, it should be noted that the analysis developed is valid only in the framework of a linearized theory, i.e. for infinitesimal deviations from an equilibrium state and small supercriticality (6.82). Usually, in real situations nonlinear effects result in a saturation of excited oscillations for sufficiently large time intervals.

In order to investigate the system evolution accurately we need to take into account nonlinear terms in the expansion (6.48) and in (6.45) and (6.46). The nonlinear terms $-\text{div}(n_1 \nabla V)$ and $-\text{div}(n_2 \nabla V)$ emerge from $\text{div}(nE)$.

$$\text{div}(nE) = \text{div}[(n_0 + n_1 + n_2)(\boldsymbol{E}_0 - \nabla V)]$$

$$= \text{div}(n_1 \boldsymbol{E}_0) + \text{div}(n_2 \boldsymbol{E}_0) - \text{div}(n_0 \nabla V) - \text{div}(n_1 \nabla V) - \text{div}(n_2 \nabla V). \tag{6.85}$$

We will analyze the essentially nonlinear model of the instability which makes it possible to describe the temporal evolution of the deviation amplitude [14]. For this reason, we should take into account the second approximation deviation $n_2, p_2 \ll n_1$. The electrostatic potential V can be eliminated by virtue of the linear approximation equation (6.67)

$$V = Cn_1; \quad C = \frac{[i\omega + i\mu_n(\boldsymbol{k}\boldsymbol{E}_0) + D_n k^2]}{[n_0 \mu_n k^2 + i\mu_n(\boldsymbol{k}\nabla n_0)]}. \tag{6.86}$$

Then we equate the quantities with the same frequency and wave vector, and (6.50) takes the form

$$\frac{dn_1}{dt} + n_0 \mu_n C \nabla^2 n_1 - \mu_n (\boldsymbol{E}_0 \cdot \nabla) n_1 - D_n \nabla^2 n_1 + \mu_n C (\nabla n_1 \cdot \nabla n_0)$$

$$= -\mu_n \text{div}[n_2 \nabla (Cn_1)], \tag{6.87}$$

6.5 Instabilities in Solid State Plasma

where n_2 is determined by the following equation

$$\frac{dn_2}{dt} - \mu_n \left(\boldsymbol{E}_0 \cdot \nabla\right) n_2 - D_n \nabla^2 n_2 = -\mu_n \mathrm{div}\left[n_1 \nabla \left(Cn_1\right)\right]. \tag{6.88}$$

The deviation n_1 is sought to be

$$n_1 = N\left(t\right) \exp i \left(\omega_0 t - \boldsymbol{kr}\right) + \mathrm{c.c.}, \tag{6.89}$$

where c.c. stands for a complex conjugate operation, the essentially real frequency ω_0 is defined by (6.73), and the amplitude $N(t)$ is assumed to be varying slowly

$$\frac{dN\left(t\right)}{dt} \ll \omega_0 N\left(t\right). \tag{6.90}$$

Substituting (6.90) into (6.88) we obtain the second deviation n_2

$$n_2 = \frac{\mu_n C k^2}{G\left(\boldsymbol{k}, \omega\right)} N^2\left(t\right) \exp 2i \left(\omega_0 t - \boldsymbol{kr}\right) + \mathrm{c.c.}, \tag{6.91}$$

where

$$G\left(\boldsymbol{k}, \omega\right) = i\omega_0 + i\mu_n \left(\boldsymbol{k}\boldsymbol{E}_0\right) + 2D_n k^2. \tag{6.92}$$

Finally, we combine relations (6.87)–(6.91) and (6.86) and separate the terms with the basic frequency ω_0. Then, taking into account that for electric fields larger than the critical one

$$\omega = \omega_0 - i\left|\delta\right| \tag{6.93}$$

we obtain the following equation for the slowly varying amplitude $N(t)$

$$\frac{dN\left(t\right)}{dt} - \left|\delta\right| N\left(t\right) = -\frac{\mu_n^2 k^4 \left|C\right|^2}{G\left(\boldsymbol{k}, \omega\right)} \left|N\left(t\right)\right|^2 N\left(t\right). \tag{6.94}$$

Obviously, all other terms in the left-hand side of (6.94) vanish due to dispersion relation (6.69). The amplitude $N(t)$ is in general case a complex quantity and can be written as

$$N\left(t\right) = \left|N\left(t\right)\right| \exp i\theta. \tag{6.95}$$

Then (6.94) can be separated into two equations, for the magnitude $\left|N(t)\right|$ and phase θ respectively, by virtue of equating real and imaginary parts in both sides. We have

$$\frac{d\left|N\left(t\right)\right|}{dt} - \left|\delta\right| \left|N\left(t\right)\right| = -\left|N\left(t\right)\right|^3 \frac{\mu_n^2 k^4 \left|C\right|^2}{\left|G\left(\boldsymbol{k}, \omega\right)\right|^2} \mathrm{Re} G\left(\boldsymbol{k}, \omega\right) \tag{6.96}$$

222   6 Current Instabilities

and

$$\frac{d\theta}{dt} = |N(t)|^2 \frac{\mu_n^2 k^4 |C|^2}{|G(\mathbf{k},\omega)|^2} \operatorname{Im} G(\mathbf{k},\omega), \qquad (6.97)$$

where

$$\operatorname{Re} G(\mathbf{k},\omega) = 2D_n k^2; \quad \operatorname{Im} G(\mathbf{k},\omega) = \omega_0 + \mu_n(\mathbf{k}\mathbf{E}_0). \qquad (6.98)$$

The factor in the right-hand side of (6.96) and (6.97) can be evaluated as follows

$$\frac{\mu_n^2 k^4 |C|^2}{|G(\mathbf{k},\omega)|^2} \simeq \frac{[\omega_0 + \mu_n(\mathbf{k}\mathbf{E}_0)]^2 + (D_n k^2)^2}{n_0^2 \left\{[\omega_0 + \mu_n(\mathbf{k}\mathbf{E}_0)]^2 + 4(D_n k^2)^2\right\}} \sim \frac{a}{n_0^2}, \qquad (6.99)$$

where a is a dimensionless numerical constant, and the small quantities $\mu_n(\mathbf{k}\nabla n_0)$ and $|\delta|$ are neglected. By definition, $(\mathbf{k}\nabla n_0) \ll k^2 n_0$, and $|\delta| \ll \omega_0$ for small supercriticality $(E_{0\mathrm{cr}} - E_0)$. It has been mentioned above that ambipolar mobility μ_a and diffusion coefficient are small compared to μ_n and D_a respectively. Therefore, $\omega_0 < \mu_n(\mathbf{k}\mathbf{E}_0) < D_n k^2$. As a result, it is seen from (6.99) that

$$\frac{1}{4} < a < \frac{2}{5}.$$

Equations (6.96) and (6.97) take the form

$$\frac{d|N(t)|}{dt} = |\delta| |N(t)| \left[1 - \frac{2D_n k^2 a}{n_0^2 |\delta|} |N(t)|^2\right] \qquad (6.100)$$

and

$$\frac{d\theta}{dt} = |N(t)|^2 \frac{[\omega_0 + \mu_n(\mathbf{k}\mathbf{E}_0)] a}{n_0^2}. \qquad (6.101)$$

Equation (6.100) shows that there exist two stationary solutions corresponding to $N = 0$ and

$$N = N_0 = n_0 \sqrt{\frac{|\delta|}{2D_n k^2 a}} = \frac{n_0}{\sqrt{2a}} \sqrt{\left[\frac{(\mathbf{k}\mathbf{E}_0)}{(\mathbf{k}\mathbf{E}_{0\mathrm{cr}})} - 1\right] \frac{\mu_p(n_0 + p_0)}{(n_0 \mu_n + p_0 \mu_p)}}. \qquad (6.102)$$

In the first stationary state oscillations are absent, while in the second stationary state determined by relation (6.102) oscillations have a finite amplitude proportional to the square root of the supercriticality $(E_{0\mathrm{cr}} - E_0)$ and to the total concentration of electrons. Equation (6.100) has a closed solution which is given by

$$|N(t)| = N_0 \frac{\exp|\delta|t}{\sqrt{\left(N_0^2/|N(0)|^2\right) - 1 + \exp 2|\delta|t}}, \qquad (6.103)$$

where $|N(0)|$ is an initial level of oscillations at $t = 0$. The analysis of solution (6.103) shows that the oscillations reach the stationary level N_0 at large $t > |\delta|^{-1}$

$$t \to \infty, \quad |N(t)| \to N_0. \qquad (6.104)$$

This final state does not depend on the initial conditions. However, the transition path is defined by the ratio $N_0^2/|N(0)|^2$. The curve (6.103) possesses an inflection point $t = t_0$ given by

$$t_0 = \frac{1}{|\delta|} \ln \left[\frac{N_0}{|N(0)|} \sqrt{1 - |N(0)|^2/N_0^2} \right] \qquad (6.105)$$

The sharp increase of the oscillations starts at $t > t_0$ which is logarithmically large for small initial level $|N(0)|$.

Consider now the behavior of the phase θ. In the stationary state $N = N_0$ (6.102) at $t \to \infty$ we have

$$t \to \infty, \quad \frac{d\theta}{dt} = N_0^2 \frac{[\omega_0 + \mu_n(\boldsymbol{kE}_0)]a}{n_0^2} = \frac{|\delta|[\omega_0 + \mu_n(\boldsymbol{kE}_0)]a}{2D_n k^2 a} = \Delta\omega. \qquad (6.106)$$

That means that the constant correction $\Delta\omega$ (6.106) to the basic frequency ω_0 emerges due to the nonlinearity of the process. Substituting (6.103) into (6.101) we immediately get

$$\theta = \frac{N_0^2 a[\omega_0 + \mu_n(\boldsymbol{kE}_0)]}{2n_0^2|\delta|} \ln\left[\left(N_0^2/|N(0)|^2\right) - 1 + \exp 2|\delta|t\right]. \qquad (6.107)$$

The integration constant in (6.107) can be omitted due to the arbitrary choice of the initial phase of oscillations. Comparison of expressions (6.106) and (6.107) shows that at $t \gg t_0$ the phase of the oscillation complex amplitude increases linearly with time, i.e. in the stationary regime the complex amplitude has a constant modulus and oscillates with respect to its level with the low frequency $\Delta\omega$ (6.106).

Typically, $N_0^2/|N(0)|^2 \gg 1$, and the solutions (6.103) and (6.107) can be rewritten as follows

$$|N(t)| \simeq N_0 \frac{\exp|\delta|t}{\sqrt{\left(N_0^2/|N(0)|^2\right) + \exp 2|\delta|t}} \qquad (6.108)$$

and

$$\theta \simeq \frac{N_0^2 a[\omega_0 + \mu_n(\boldsymbol{kE}_0)]}{2n_0^2|\delta|} \ln\left[\left(N_0^2/|N(0)|^2\right) + \exp 2|\delta|t\right]. \qquad (6.109)$$

We studied comprehensively the nonlinear model of the quasineutral oscillations using the methods discussed in the first part of the chapter and we can make the following general conclusions. In the linearized model at the critical value (6.76) of the electric field applied the absolute instability occurs due to the inhomogeneity of the carrier concentration. The increment is proportional to the square root of the so-called supercriticality $(E_{0cr} - E_0)$. The account of the nonlinear corrections to the model results in the saturation effects and the finite level of vibrations as time tends to infinity. In the stationary state that occurs at large t the constant small correction $\Delta\omega$ to the basic frequency emerges due to the nolinearity of the process.

6.5.4 Instability in a Semiconductor with Three Types of Carriers

Consider a semiconductor containing electrons and two species of holes, light and heavy ones. Assume that the electric field E_0 is applied along the z-axis of a sample. The equations describing the problem have the form:

$$\frac{\partial n_-^{tot}}{\partial t} + \text{div}\left(-\mu_- n_-^{tot} E - D_- \nabla n_-^{tot}\right) = 0 \qquad (6.110)$$

$$\frac{\partial n_{1,2}^{tot}}{\partial t} + \text{div}\left(\mu_{1,2} n_{1,2}^{tot} E - D_{1,2} \nabla n_{1,2}^{tot}\right) = 0. \qquad (6.111)$$

Here $n_-^{tot}, n_{1,2}^{tot}, \mu_-, \mu_{1,2}$ and $D_-, D_{1,2}$ are the carrier densities, the mobilities and the diffusion coefficients of electrons and two types of holes, respectively. The collision processes are ignored. We linearize the system of (6.110), (6.111) on small deviations $n_{-,1,2}, \nabla V$ from the equilibrium state

$$n_{-,1,2} = n_{-,1,2}^{tot} - n_{-,1,2}^0 \ll n_{-,1,2}^0; \quad \nabla V = E_0 - E \ll E_0, \qquad (6.112)$$

where φ is the electrostatic potential. The quasineutrality condition for the perturbations yields

$$n_- = n_1 + n_2. \qquad (6.113)$$

Linearization of the system of (6.110), (6.111) results in the following equations

$$\frac{\partial n_-}{\partial t} + \text{div}\left\{-\mu_-\left(n_- E_0 - n_-^0 \nabla V\right) - D_- \nabla n_-\right\} = 0 \qquad (6.114)$$

$$\frac{\partial n_{1,2}}{\partial t} + \text{div}\left\{\mu_{1,2}\left(n_{1,2} E_0 - n_{1,2}^0 \nabla V\right) - D_{1,2} \nabla n_{1,2}\right\} = 0. \qquad (6.115)$$

All perturbations are assumed to be plane waves.

$$n_{-,1,2}, V = c_m \exp i\left(kz - \omega t\right), \ m = 1, ..., 4. \qquad (6.116)$$

6.5 Instabilities in Solid State Plasma

The constants c_m are determined from the system of algebraic equations obtained by substituting quantities (6.116) into (6.114) and (6.115). The condition of non-trivial solution existence for the system of linear algebraic equations yields the following dispersion equation

$$\omega^2 + \omega \left(ik^2 D_{\text{eff}} - kE_0\mu_{\text{eff}}\right) - (kE_0)^2 \overline{\mu^2} - k^4\overline{D^2} - ik^3\overline{\mu D}E_0 = 0, \quad (6.117)$$

where

$$\mu_{\text{eff}} = \frac{\mu_- n_-^0 \left(\mu_1 + \mu_2\right) + \mu_1 n_1^0 \left(\mu_2 - \mu_-\right) + \mu_2 n_2^0 \left(\mu_1 - \mu_-\right)}{\left(\mu_- n_-^0 + \mu_1 n_1^0 + \mu_2 n_2^0\right)} \quad (6.118)$$

$$D_{\text{eff}} = \frac{\mu_- n_-^0 \left(D_1 + D_2\right) + \mu_1 n_1^0 \left(D_2 + D_-\right) + \mu_2 n_2^0 \left(D_1 + D_-\right)}{\left(\mu_- n_-^0 + \mu_1 n_1^0 + \mu_2 n_2^0\right)} \quad (6.119)$$

$$\overline{\mu^2} = \mu_1 \mu_2 \mu_- \frac{n_1^0 + n_2^0 - n_-^0}{\left(\mu_- n_-^0 + \mu_1 n_1^0 + \mu_2 n_2^0\right)}. \quad (6.120)$$

The diffusion type terms $ik^2 D_{\text{eff}}$, $ik^3 \overline{\mu D} E_0$ describing the perturbation decay, and the higher order term $k^4 \overline{D^2}$ can be neglected as small compared to the electric field terms. Indeed, using the Einstein relation between a carrier mobility μ and a diffusion coefficient D

$$\frac{D}{\mu} = \frac{k_B T}{e},$$

where k_B is the Bolzmann constant and e is the electron charge, we have

$$\frac{k^2 D_{\text{eff}}}{kE_0\mu_{\text{eff}}} \sim \frac{k^3 \overline{\mu D} E_0}{(kE_0\mu_{\text{eff}})^2} \sim \frac{k_B T}{\mathcal{E}_{\text{el}}} \ll 1 \quad (6.121)$$

$$\frac{k^4 \overline{D^2}}{(kE_0\mu_{\text{eff}})^2} \sim \frac{k^4 D_{\text{eff}}^2}{(kE_0\mu_{\text{eff}})^2} \sim \left(\frac{k_B T}{\mathcal{E}_{\text{el}}}\right)^2 \ll 1, \quad (6.122)$$

where $\mathcal{E}_{\text{el}} \sim eE_0 k^{-1}$ is the electrostatic energy. Finally, we have two solutions corresponding to two branches of oscillation

$$\omega_{1,2} = \frac{1}{2}kE_0\mu_{\text{eff}} \left\{1 \pm \sqrt{1 + 4\frac{\overline{\mu^2}}{\mu_{\text{eff}}^2}}\right\}. \quad (6.123)$$

It easy to see that the system is stable since the solution (6.123) is real. The term $\overline{\mu^2}/\mu_{\text{eff}}^2$ can be negative when

$$n_-^0 > n_1^0 + n_2^0. \quad (6.124)$$

However, in such a case it can be easily shown that $\left|\overline{\mu^2}/\mu_{\text{eff}}^2\right| < 1/4$, and the sum under the sign of radical in (6.123) remains positive. Combining relationships (6.118), (6.120) and (6.124) we find that

$$\left|\frac{\overline{\mu^2}}{\mu_{\text{eff}}^2}\right| < \mu_1\mu_2\mu_- \frac{n_-^0 \mu_- n_-^0}{\left[\mu_- n_-^0 (\mu_1 + \mu_2)\right]^2} = \frac{\mu_1\mu_2}{(\mu_1 + \mu_2)^2} < \frac{1}{4}, \quad (6.125)$$

since

$$a + \frac{1}{a} \geq 2.$$

Clearly, the oscillation decays exponentially with time decrement $\Gamma = k^2 D_{\text{eff}}$ due to the diffusion term dropped earlier for the sake of simplicity, but this relaxation process appeared to be slow.

6.5.5 Parametric Excitation of the Instability

Consider now the parametric excitation of a quasi-neutral oscillation when the external electric field applied to the sample contains a comparatively small ac component with a frequency which is very close to the doubled eigen frequency of the system $2\omega_0$. Both dc and ac fields are parallel to the z-axis, and they have the form:

$$E = E_0 + E_{\text{ac}} \cos\left(2\omega_0 t + \xi t\right); \quad \frac{E_{\text{ac}}}{E_0} \ll 1, \quad \frac{\xi}{\omega_0} \ll 1. \quad (6.126)$$

We express the carrier densities $n_{-,1,2}$ in terms of the electrostatic potential $V(z,t)$ keeping the operators $\partial/\partial t$ and $\partial/\partial z$ instead of $-i\omega$ and ik, since this time we do not have any preliminary assumptions on the temporal and spatial dependence of $V(z,t)$. We start with the explicit form of (6.114) and (6.115).

$$\left(\frac{\partial}{\partial t} - \mu_- E \frac{\partial}{\partial z}\right) n_- = -\mu_- n_-^0 \frac{\partial^2 V}{\partial z^2} \quad (6.127)$$

$$\left(\frac{\partial}{\partial t} + \mu_1 E \frac{\partial}{\partial z}\right) n_1 = \mu_1 n_1^0 \frac{\partial^2 V}{\partial z^2}. \quad (6.128)$$

Applying the operators $(\partial/\partial t + \mu_1 E \partial/\partial x)$ and $(\partial/\partial t - \mu_- E \partial/\partial x)$ to (6.127) and (6.128), respectively, and excluding $n_2 = n_- - n_1$ we obtain

$$\left(\frac{\partial}{\partial t} + \mu_1 E \frac{\partial}{\partial z}\right)\left(\frac{\partial}{\partial t} - \mu_- E \frac{\partial}{\partial z}\right) \mu_2 n_2^0 \frac{\partial^2 V}{\partial z^2}$$

$$= \left(\frac{\partial}{\partial t} + \mu_2 E \frac{\partial}{\partial z}\right) \left\{-\mu_- n_-^0 \left(\frac{\partial}{\partial t} + \mu_1 E \frac{\partial}{\partial z}\right) - \mu_1 n_1^0 \left(\frac{\partial}{\partial t} - \mu_- E \frac{\partial}{\partial z}\right)\right\} \frac{\partial^2 V}{\partial z^2}. \tag{6.129}$$

Multiplying the operators, changing the sequence of differentiation and taking into account relationships (6.118)–(6.120) we obtain the following equation for the electrostatic potential $V(z,t)$

$$\left(\frac{\partial^2}{\partial t^2} + \mu_{\mathrm{eff}} E \frac{\partial^2}{\partial t \partial z} - \overline{\mu^2} E^2 \frac{\partial^2}{\partial z^2}\right) V = 0. \tag{6.130}$$

We find the electrostatic potential $V(z,t)$ in the form

$$V(z,t) = V_0(t) \exp i(kz - \omega t). \tag{6.131}$$

Substituting expression (6.131) into (6.130) we define the frequency ω excluding the first temporal derivative from (6.130) which gives

$$\omega = \frac{k \mu_{\mathrm{eff}} E_0}{2}. \tag{6.132}$$

Then (6.130) takes the form

$$\frac{\partial^2 V_0(t)}{\partial t^2} + \frac{(k \mu_{\mathrm{eff}} E_0)^2}{4} \left[1 + 4 \frac{\overline{\mu^2}}{\mu_{\mathrm{eff}}^2}\right] (1 + h \cos(2\omega_0 t + \xi t)) V_0(t) = 0. \tag{6.133}$$

Equation (6.133) is the Mathieu equation which possesses either the stable periodical solutions, or unstable ones, depending on the interval of parameters

$$h = 2 \frac{E_{\mathrm{ac}}}{E_0}, \quad \omega_0 = \frac{k \mu_{\mathrm{eff}} E_0}{2} \sqrt{1 + 4 \frac{\overline{\mu^2}}{\mu_{\mathrm{eff}}^2}}. \tag{6.134}$$

In our case the periodic function

$$\omega^2(t) = \omega_0^2 [1 + h \cos(2\omega_0 t + \xi t)]$$

only slightly differs from the eigen frequency ω_0 of (6.133) due to the smallness of h. Therefore we shall determine the conditions of an instability excitation using the expansion in powers of a small parameter ξ [15]. The solution is sought in the form

$$V_0(t) = a(t) \cos\left(\omega_0 t + \frac{\xi}{2} t\right) + b(t) \sin\left(\omega_0 t + \frac{\xi}{2} t\right). \tag{6.135}$$

The amplitudes are supposed to be slowly varying:

$$\frac{\partial a}{\partial t}, \frac{\partial b}{\partial t} \sim \xi a, \xi b \ll \omega_0 a, \omega_0 b.$$

This condition means that the energy exchange between the modes occurs during the time intervals that are much larger than the basic oscillation period $2\pi/\omega_0$. Substituting the solution (6.135) into (6.133) we retain only the terms linear in ξ. The rapidly oscillating terms of the type $\cos 3\left(\omega_0 t + \frac{\xi}{2}t\right)$, $\sin 3\left(\omega_0 t + \frac{\xi}{2}t\right)$ are also omitted. Equating the coefficients at the sine and cosine functions we obtain the system of equations for the slowly varying amplitudes $a(t)$ and $b(t)$. They have the form:

$$2\frac{\partial a}{\partial t} + \left(\frac{h\omega_0}{2} + \xi\right)b = 0 \tag{6.136}$$

$$2\frac{\partial b}{\partial t} + \left(\frac{h\omega_0}{2} - \xi\right)a = 0. \tag{6.137}$$

We find the solution in the form $a, b \sim \exp st$. Substituting these solutions into (6.136) and (6.137) we get

$$2s + \left(\frac{h\omega_0}{2} + \xi\right)b = 0 \tag{6.138}$$

$$\left(\frac{h\omega_0}{2} - \xi\right)a + 2s = 0. \tag{6.139}$$

This system has a non-trivial solution when its determinant is equal to zero which yields the values of s.

$$s^2 = \frac{1}{4}\left[\left(\frac{h\omega_0}{2}\right)^2 - \xi^2\right] = \frac{1}{4}\left[\frac{(k\mu_{\text{eff}}E_{\text{ac}})^2}{4}\left(1 + 4\frac{\mu^2}{\mu_{\text{eff}}^2}\right) - \xi^2\right]. \tag{6.140}$$

Obviously, the parametric instability can occur when $s^2 > 0$, or

$$-\frac{k\mu_{\text{eff}}E_{\text{ac}}}{2}\sqrt{1 + 4\frac{\mu^2}{\mu_{\text{eff}}^2}} < \xi < \frac{k\mu_{\text{eff}}E_{\text{ac}}}{2}\sqrt{1 + 4\frac{\mu^2}{\mu_{\text{eff}}^2}}. \tag{6.141}$$

The inequality (6.141) establishes the relation between the parameters of the problem: the ac field amplitude E_{ac}, the dc field E_0, implicitly, by means of (6.134), the eigen frequency of the system ω_0 and the frequency deviation ξ. In order to provide the increase of the instability the increment s should compensate all kinds of losses in the system, i.e. $s > \Gamma = k^2 D_{\text{eff}}$. The instability excitation process described above is called a parametric resonance. That means that any deviation from the equilibrium state of the system can under some conditions on the amplitude and frequency of an external field lead to a rapidly increasing perturbation. The identification of a type of the instability in general case requires some additional data, such as the influence of boundaries, the value of a dissipation constant, initial conditions etc. These questions have been considered in details in previous sections. For this reason, here we limit the analysis with the determination of the instability excitation conditions.

6.5.6 Helicoidal Instability in Semiconductors

The helicoidal, or screw instability in semiconductors can be observed experimentally as current oscillations in an external circuit containing a semiconductor sample with an electron-hole plasma created by means of a light radiation or carriers injection when parallel sufficiently strong electric and magnetic fields are applied. Two other necessary conditions for the helicoidal instability to occur are the limited transverse dimension of the sample and the presence of positively and negatively charged carriers. The helicoidal instability in plasma has been firstly observed by the authors [16]. Theoretically, this phenomenon was explained by Kadomtsev and Nedospasov [17], and then generalized to the case of solid state plasma [18, 19].

We will analyze below the specific features of the helicoidal instability, determine the interval of convective and absolute instability and estimate the order of magnitude of characteristic frequencies of generated or amplified waves.

Consider the semiconductor cylindrical sample of a radius a which is subjected to the parallel electric and magnetic fields $\boldsymbol{E}_0$ and $\boldsymbol{B}_0$:

$$\boldsymbol{E}_0 = (0, 0, E_0), \quad \boldsymbol{B}_0 = (0, 0, B_0). \tag{6.142}$$

The geometry of the problem is shown in the Fig. 6.5. The cylinder is assumed to be sufficiently long to make end effects negligible so that we may simplify the z variation of the carrier number densities and velocities [18]. In the steady state the plasma will occupy the available volume with a distribution

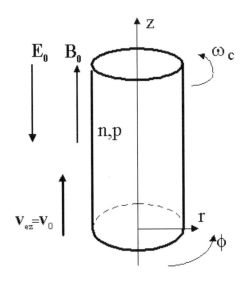

Fig. 6.5. The semiconductor cylindrical sample in the parallel magnetic and electric fields $\boldsymbol{B}_0, \boldsymbol{E}_0$

230   6 Current Instabilities

in density depending on the sources and sinks for the plasma. The sink for the plasma will be assumed to be primarily at the surface, with some volume recombination also allowable, so that the plasma density will have a maximum at the center and decreases with distance in the direction towards the surface [18]. We suppose that the plasma vanishes at the surface of the cylinder $r = a$. The bulk recombination and trapping of carriers are ignored. A plasma of this distribution, carrying an axial current, will be stable in form at low currents where the self-magnetic field of the current applies a force much smaller than the diffusion force which maintains the steady state [18]. Any perturbations will be opposed by diffusion, and they will be damped. If an external magnetic field $\boldsymbol{B}_0$ is applied in the axial direction, instabilities may occur. A helicoidal perturbation in carrier number density causes a helicoidal distortion in the current with azimuthal component. This current will cause a driving Lorentz force of magnitude jB_0 in the direction either towards or away from the surface, depending on the sense of the helicoidal perturbation with respect to the direction of $\boldsymbol{B}_0$. The strength of this driving force will depend on j and B_0, and one may expect that at sufficiently large values of the product jB_0 this force will be larger than the diffusion force and the perturbations will grow [18]. We neglect the internal magnetic field caused by the current in the semiconductor as small as compared to the external field.

In the hydrodynamic approximation when the effective frequencies of hole and electron collisions $\nu_\pm \gg \omega$ the basic system of equations has the form

$$\frac{e}{m_-}\boldsymbol{E} + \frac{e}{m_-}[\boldsymbol{v}_- \times \boldsymbol{B}_0] + \nu_-\boldsymbol{v}_- + \frac{k_B T}{n_0 m_-}\nabla n = 0 \tag{6.143}$$

$$\frac{e}{m_+}\boldsymbol{E} + \frac{e}{m_+}[\boldsymbol{v}_+ \times \boldsymbol{B}_0] - \nu_+\boldsymbol{v}_+ - \frac{k_B T}{p_0 m_+}\nabla p = 0 \tag{6.144}$$

$$\operatorname{div}\boldsymbol{E} = -\frac{1}{\varepsilon_0 \varepsilon^r}e(n-p) \tag{6.145}$$

$$\operatorname{rot}\boldsymbol{E} = 0 \tag{6.146}$$

$$-e\frac{\partial n}{\partial t} + \operatorname{div}\boldsymbol{j}_- = 0 \tag{6.147}$$

$$e\frac{\partial p}{\partial t} + \operatorname{div}\boldsymbol{j}_+ = 0, \tag{6.148}$$

where, as usual, n_0, p_0 are the equilibrium number densities and n, p are their deviations for electrons and holes, respectively; $m_\pm, \boldsymbol{j}_\pm, \boldsymbol{v}_\pm$ are the mass, current density and velocity of holes and electrons, respectively. Introduce the dimensionless variables

6.5 Instabilities in Solid State Plasma

$$\bm{h} = \frac{\bm{B}_0}{B_0}, \quad y = \mu B_0, \quad b = \frac{m_+ \nu_+}{m_- \nu_-}. \tag{6.149}$$

Then, substituting (6.149) into (6.143) and (6.144) we obtain

$$\bm{v}_- + y\left[\bm{v}_- \times \bm{h}\right] = -\mu \bm{E} - \frac{D}{n_0}\nabla n \tag{6.150}$$

$$\bm{v}_+ - \frac{y}{b}\left[\bm{v}_+ \times \bm{h}\right] = \frac{\mu \bm{E}}{b} - \frac{D}{bp_0}\nabla p, \tag{6.151}$$

where is the electron mobility μ and the diffusion coefficient D have the form

$$\mu = \frac{e}{m_- \nu_-}; \quad D = \frac{k_B T}{\mu e}. \tag{6.152}$$

Solving equations (6.150) and (6.151) with respect to $\bm{v}_\pm$ we find

$$\bm{v}_- = \frac{\left[-\mu \bm{E} - D\frac{\nabla n}{n_0} + y\left(\mu \bm{E} + D\frac{\nabla n}{n_0}\right) \times \bm{h} - y^2 \bm{h}\left(\mu E_z + \frac{D}{p_0}\right)\frac{\partial n}{\partial z}\right]}{1 + y^2} \tag{6.153}$$

$$\bm{v}_+ = \frac{\left[b^2\left(\mu \bm{E} - D\frac{\nabla p}{p_0}\right) + by\left(\mu \bm{E} - D\frac{\nabla p}{p_0}\right) \times \bm{h} + y^2 \bm{h}\left(\mu E_z - \frac{D}{p_0}\right)\frac{\partial p}{\partial z}\right]}{b(b^2 + y^2)}. \tag{6.154}$$

In the stationary regime, when the fields obey (6.142) and the carrier number densities have their equilibrium values (6.153) and (6.154) yield

$$\bm{v}_- = -h\mu E_0, \quad \bm{v}_+ = \frac{h\mu E_0}{b}, \tag{6.155}$$

which means that the constant current $\bm{j}_0$ is flowing through the sample:

$$\bm{j}_0 = e\mu E_0\left(n_0 + \frac{p_0}{b}\right)\bm{h}. \tag{6.156}$$

Assuming the deviations from the equilibrium state to be small we linearize the continuity equations for electrons and holes (6.147) and (6.148) which results in the following equations

$$\frac{\partial n}{\partial t} - \mu E_0 \frac{\partial n}{\partial z} + n_0 \mathrm{div}\bm{v}_- = 0 \tag{6.157}$$

$$\frac{\partial p}{\partial t} + \frac{\mu E_0}{b}\frac{\partial p}{\partial z} + p_0 \mathrm{div}\bm{v}_+ = 0. \tag{6.158}$$

6 Current Instabilities

We solve the system of (6.153)–(6.158) in the cylindrical coordinate system with the boundary conditions describing the situation when the electron and hole currents vanish at the surface of the cylinder:

$$j_r^- = j_r^+ = 0, \quad r = a, \tag{6.159}$$

or

$$v_r^- = v_r^+ = 0, \quad r = a, \tag{6.160}$$

since

$$\boldsymbol{j} = -en\boldsymbol{v}.$$

Combining the conditions (6.160) and the expressions (6.153), (6.154) in the cylindrical coordinates we obtain

$$\left(\frac{\partial}{\partial r} - \frac{y}{r}\frac{\partial}{\partial \varphi}\right)\left(\mu V - D\frac{n}{n_0}\right) = 0 \tag{6.161}$$

$$\left(b\frac{\partial}{\partial r} + \frac{y}{r}\frac{\partial}{\partial \varphi}\right)\left(\mu V + D\frac{p}{p_0}\right) = 0, \quad r = a, \tag{6.162}$$

where V is the electrostatic potential, such that

$$\boldsymbol{E} = -\nabla V. \tag{6.163}$$

Excluding the electric field $\boldsymbol{E}$ by means of (6.163) we finally obtain the system of equations for the variables n, p and V.

$$\frac{\partial n}{\partial t} - \mu E_0 \frac{\partial n}{\partial z} + \frac{n_0}{1+y^2}\left(\Delta + y^2 \frac{\partial^2}{\partial z^2}\right)\left(\mu V - D\frac{n}{n_0}\right) = 0 \tag{6.164}$$

$$\frac{\partial p}{\partial t} + \frac{\mu E_0}{b}\frac{\partial p}{\partial z} - \frac{p_0}{b(b^2+y^2)}\left(b^2\Delta + y^2 \frac{\partial^2}{\partial z^2}\right)\left(\mu V + D\frac{p}{p_0}\right) = 0 \tag{6.165}$$

$$\Delta V = \frac{1}{\varepsilon_0 \varepsilon^r} e(n-p). \tag{6.166}$$

The radial part of the solution of the system of (6.164)–(6.166) is expressed in terms of the m-th order Bessel functions of an imaginary argument $I_m(x)$ specific for the cylindrical coordinate. The solution is sought to be

$$n = n_0 C_1 I_m(k\beta r) \exp i(\omega t - m\varphi - kz) \tag{6.167}$$

6.5 Instabilities in Solid State Plasma

$$p = p_0 C_2 I_m (k\beta r) \exp i (\omega t - m\varphi - kz) \tag{6.168}$$

$$V = \frac{D}{\mu} C_3 I_m (k\beta r) \exp i (\omega t - m\varphi - kz), \tag{6.169}$$

where m is integer: $m = 1, 2, \ldots$. The substitution of the solution (6.167) into the system of (6.164)–(6.166) yields the homogeneous system of algebraic equations for the constants $C_{1,2,3}$:

$$C_1 \left[i(\Omega + \Lambda) + 1 - \frac{\beta^2}{1 + y^2} \right] - C_3 \left(1 - \frac{\beta^2}{1 + y^2} \right) = 0 \tag{6.170}$$

$$\Omega_c (C_1 - \gamma C_2) - (\beta^2 - 1) C_3 = 0 \tag{6.171}$$

$$C_2 \left[i \left(\Omega - \frac{\Lambda}{b} \right) - \beta^2 \frac{b}{b^2 + y^2} + \frac{1}{b} \right] - \left(\beta^2 \frac{b}{b^2 + y^2} - \frac{1}{b} \right) C_3 = 0, \tag{6.172}$$

where

$$\gamma = \frac{p_0}{n_0}, \quad \Omega = \frac{\omega}{k^2 D}, \quad \Lambda = \frac{k \mu E_0}{k^2 D}, \quad \Omega_c = \frac{\omega_c}{k^2 D}, \quad \omega_c = \frac{e n_0 \mu}{\varepsilon_0 \varepsilon^r}. \tag{6.173}$$

For the samples with high conductivity $\Omega_c \gg \Omega, \Lambda$, and the terms of the order of magnitude $\Omega/\Omega_c, \Lambda/\Omega_c$ can be neglected. The existence of a non-trivial solution of the system of (6.170)–(6.172) yields the double-square dispersion equation which is too involved to be presented here. The dispersion equation possesses two complex roots $\beta_{1,2}$ with the different real parts. The corresponding solution has the form

$$n = p = n_0 \left[S_1 C_3^{(1)} I_m (k\beta_1 r) + S_2 C_3^{(2)} I_m (k\beta_2 r) \right] \exp i (\omega t - m\varphi - kz) \tag{6.174}$$

$$V = \frac{D}{\mu} \left[C_3^{(1)} I_m (k\beta_1 r) + C_3^{(2)} I_m (k\beta_2 r) \right] \exp i (\omega t - m\varphi - kz), \tag{6.175}$$

where

$$S_{1,2} \equiv S(\beta_{1,2}), \quad S(\beta) = \frac{1 + y^2 - \beta^2}{i(\Omega + \Lambda)(1 + y^2) + 1 + y^2 - \beta^2}.$$

The solution (6.174), (6.175) should obey the boundary conditions (6.161), (6.162) which gives the second dispersion relation, this time the one which connects directly k and ω. We are limiting the further analysis with the case when $|k\beta_{1,2} a| \ll 1$ and the Bessel functions can be expanded in powers keeping only two first terms. Then the dispersion equation can be written as follows

$$(k - k_0)^2 - \alpha (\omega - \omega_0) = 0, \tag{6.176}$$

where

$$k_0 = \frac{i\mu E_0 \varphi_2}{2(1+\gamma) D\varphi_1}, \quad \alpha = -\frac{i\varphi_3}{(1+\gamma) D\varphi_1} \tag{6.177}$$

$$\omega_0 = i \left[\frac{4|m|(1+|m|)(1+\gamma) D}{a^2 \varphi_3} + \frac{(\mu E_0)^2 \varphi_2^2}{4(1+\gamma) D\varphi_1 \varphi_3} \right], \tag{6.178}$$

where $\varphi_{1,2,3}$ are some sufficiently complicated complex functions of $|m|$, b, y, γ.

In order to determine the condition of an instability we investigate in detail (6.176). This is the equation of the parabolic type. Taking into account expressions (6.177) and (6.178) we get

$$\mathrm{Im}\omega = \mathrm{Im}\omega_0 + \frac{\left[-(k - \mathrm{Re}k_0)^2 \mathrm{Im}\alpha - 2\mathrm{Re}\alpha \mathrm{Im}k_0 (k - \mathrm{Re}k_0) + \mathrm{Im}\alpha (\mathrm{Im}k_0)^2 \right]}{|\alpha|^2}. \tag{6.179}$$

Then analysis of (6.179) shows that $\mathrm{Im}\omega$ has the minimum when

$$k = \mathrm{Re}k_0 - \frac{\mathrm{Re}\alpha \mathrm{Im}k_0}{\mathrm{Im}\alpha}. \tag{6.180}$$

The minimal value of $\mathrm{Im}\omega$ has the from:

$$(\mathrm{Im}\omega)_{\min} = \mathrm{Im}\omega_0 + \frac{(\mathrm{Im}k_0)^2}{\mathrm{Im}\alpha}. \tag{6.181}$$

The system is unstable when $(\mathrm{Im}\omega)_{\min} < 0$ which takes place when

$$\mathrm{Im}\omega_0 < \frac{(\mathrm{Im}k_0)^2}{|\mathrm{Im}\alpha|}, \quad \mathrm{Im}\alpha < 0. \tag{6.182}$$

The branching point $\omega = \omega_0$ of the function

$$k(\omega) = k_0 \pm \sqrt{\alpha(\omega - \omega_0)} \tag{6.183}$$

is situated in the lower half-plane of the complex variable ω when $\mathrm{Im}\omega_0 < 0$. It can be also shown that the branching point corresponds to the waves propagating in the opposite direction. Hence, in such a case an absolute instability occurs. When

6.5 Instabilities in Solid State Plasma

$$0 < \mathrm{Im}\,\omega_0 < \frac{(\mathrm{Im}\,k_0)^2}{|\mathrm{Im}\,\alpha|} \tag{6.184}$$

the convective instability takes place, and the system is stable when

$$\mathrm{Im}\,\omega_0 > \frac{(\mathrm{Im}\,k_0)^2}{|\mathrm{Im}\,\alpha|}. \tag{6.185}$$

Using the explicit expressions (6.177) and (6.178) for ω_0, k_0 and α we obtain the threshold values of the electric field for the convective instability

$$\left(\frac{\mu E_0 a}{D}\right)^2 > \left(\frac{\mu E_0 a}{D}\right)^2_{\text{threshold conv}} = \Psi_1(y, \alpha, b, m) \tag{6.186}$$

and for the absolute instability

$$\left(\frac{\mu E_0 a}{D}\right)^2 > \left(\frac{\mu E_0 a}{D}\right)^2_{\text{threshold abs}} = |\Psi_2(y, \alpha, b, m)|. \tag{6.187}$$

It can be shown that the function $\Psi_2(y, \alpha, b, m)$ should be negative. The explicit expressions of the complicated functions $\Psi_{1,2}(y, \alpha, b, m)$ are too involved, and we do not present them here. The numerical estimations show that the oscillations occur at low frequencies. For the typical values $D = 10^{-2}\,\mathrm{m^2 s^{-1}}$, $a = 10^{-3}\,\mathrm{m}$, $\mathrm{Re}\,\omega_0 \sim (10^5 \div 10^6)\,\mathrm{s^{-1}}$.

7 Waves in Plasma

7.1 Hydrodynamic Model

In the previous chapter we considered the quasi-static instabilities in the solid state plasma described by the continuity equations, equations of motion and expressions for the current density for each type of charged particles. The longitudinal quasi-static electric field caused by charge carriers oscillations has been determined by the Poisson equation. However, this approach is inadequate for the study of electromagnetic wave propagation in plasma since it does not include a variable magnetic field. A description of wave processes in plasma requires the self-consistent solution of the complete system of the Maxwell equations and the equations describing the behavior of charged particles. An electric field $\boldsymbol{E}$ and a magnetic field $\boldsymbol{B}$ in plasma are unknown and depend on the distribution and motion of charged particles. At the same time, the motion of the charged particles themselves is determined by the fields. In general case, the problem is extremely complicate taking into account a number of particles in a unit volume, an arbitrary spatial and temporal dependence of the variables, non-locality, various types of interaction, possible non-linear effects, etc. However, the well established realistic approximations exist which permit the successful explanation of wave phenomena in plasma.

We start with the so-called hydrodynamic approximation where the plasma is considered as two or more mutually penetrating liquids [13, 21]. Each liquid consists of charged particles of the one type. In the simplest case there exist a liquid of positively charged ions and another one of electrons. The ion and electron components of plasma interact by virtue of collisions and through generated by ions and electrons electric and magnetic fields. Hence interactions in plasma exist even in the absence of collision. In the hydrodynamic model a motion of infinitesimal but macroscopic elements of the liquid volume is considered. The velocity distribution for each type of particle is assumed to be Maxwellian.

The Maxwell equations for the plasma should be modified in such a way that the electric induction $\boldsymbol{D}$ and the magnetic field intensity $\boldsymbol{H}$ be excluded since the Lorentz force playing the key role in the plasma problems depends on the electric field intensity $\boldsymbol{E}$ and the magnetic induction $\boldsymbol{B}$. The medium itself is assumed to be non-magnetic and isotropic with the relative permeability $\mu^r = 1$ and the relative permittivity due to the crystal lattice ε_L^r. Then

$$\boldsymbol{B} = \mu_0 \boldsymbol{H}, \quad \boldsymbol{D} = \varepsilon_0 \varepsilon_L^r \boldsymbol{E} \tag{7.1}$$

and equations (1.16)–(1.19) take the form:

$$\mathrm{div}\boldsymbol{B} = 0 \tag{7.2}$$

$$\mathrm{curl}\boldsymbol{E} = -\frac{\partial \boldsymbol{B}}{\partial t} \tag{7.3}$$

$$\mathrm{div}\boldsymbol{E} = \frac{\rho}{\varepsilon_0 \varepsilon_L^r} \tag{7.4}$$

$$\mathrm{curl}\boldsymbol{B} = \mu_0 \left(\boldsymbol{j} + \varepsilon_0 \varepsilon_L^r \frac{\partial \boldsymbol{E}}{\partial t} \right). \tag{7.5}$$

The contribution of the charged particles is determined by the charge density ρ and the current density $\boldsymbol{j}$ in the right-hand side of (7.4) and (7.5). Unlike the material equations (7.1), the relation between the current density $\boldsymbol{j}$ and the electric field $\boldsymbol{E}$ cannot be written directly since the plasma conductivity σ should be calculated in a self-consistent way for each specific problem by virtue of the equation of motion of a charged particle in the framework of the hydrodynamic model, or by virtue of the so-called Boltzmann–Vlasov equation for the distribution function in the framework of the kinetic theory which will be considered in a separate section.

Plasma may contain some types of positive and negative charge carriers. For instance, the plasma in semiconductors consists of electrons and holes with different effective mass, as we have seen in the previous chapter. The gaseous plasma may contain electrons and different species of positive ions. In metals, the electron component of plasma is essential while the ions composing the crystal lattice are fixed. Combining (7.4) and (7.5) we get the continuity equation for the charge density and the current density for each type of charged particles:

$$\mathrm{div}\boldsymbol{j}_j + \frac{\partial \rho_j}{\partial t} = 0, \tag{7.6}$$

where the subscript j stands for the carrier type: electrons, holes, ions etc.

The equation of motion of a single charged particle with a charge e and an effective mass m^* in the electromagnetic field (5.5) in the case of plasma should be completed with the terms specific for the collective hydrodynamic motion. The particles in any hydrodynamic flow participate in a momentum exchange due to collisions and in a chaotic thermal motion. The term responsible for the particle collisions connected with the momentum exchange is defined by the scattering frequency ν or by the corresponding relaxation time $\tau = \nu^{-1}$. The thermal processes in a hydrodynamic flow result in the pressure force $\boldsymbol{F}$ which is determined by the hydrodynamic pressure P in a liquid:

$$\boldsymbol{F} = -\mathrm{grad}P. \tag{7.7}$$

The pressure for an isothermal compression process is defined according to thermodynamics as follows:

$$P = nk_B T, \tag{7.8}$$

where n is a number density of particles, k_B is the Boltzmann constant and T is the temperature of the plasma. It should be noted that in the presence of a constant uniform magnetic field $\boldsymbol{B}_0$ a plasma can be characterized by two temperatures $T_\parallel$ and $T_\perp$, parallel and perpendicular to the magnetic field respectively. As a result, two values of the pressure should be introduced

$$P_\parallel = nk_B T_\parallel, \quad P_\perp = nk_B T_\perp. \tag{7.9}$$

Then the usual scalar pressure becomes a stress tensor P_{ik}:

$$P_{xx} = P_{yy} = P_\perp, \quad P_{zz} = P_\parallel, \quad P_{ik} = 0 \text{ for } i \neq k, \tag{7.10}$$

where the magnetic field $\boldsymbol{B}_0$ is assumed to be directed along the z-axis. In such a case the force expression (7.7) is replaced by

$$F_i = -\frac{\partial P_{ik}}{\partial x_k}. \tag{7.11}$$

Finally, in hydrodynamic model the ordinary temporal derivative must be replaced by the so-called convective derivative since in general case the velocity depends on spatial coordinates [21]:

$$\frac{d\boldsymbol{v}}{dt} = \frac{\partial \boldsymbol{v}}{\partial t} + \frac{\partial x}{\partial t}\frac{\partial \boldsymbol{v}}{\partial x} + \frac{\partial y}{\partial t}\frac{\partial \boldsymbol{v}}{\partial y} + \frac{\partial z}{\partial t}\frac{\partial \boldsymbol{v}}{\partial z} = \frac{\partial \boldsymbol{v}}{\partial t} + (\boldsymbol{v} \cdot \nabla)\boldsymbol{v}. \tag{7.12}$$

Taking into account all these considerations we write the equation of motion for j-th type of charged particles:

$$m_j^* \left[\frac{\partial \boldsymbol{v}_j}{\partial t} + (\boldsymbol{v}_j \cdot \nabla)\boldsymbol{v}_j\right] = e_j \{\boldsymbol{E} + [\boldsymbol{v}_j \times \boldsymbol{B}]\} - \frac{1}{n_j}\mathrm{grad}P - m_j^*\frac{\boldsymbol{v}_j}{\tau}. \tag{7.13}$$

Here the pressure is assumed to be isotropic. The current density $\boldsymbol{j}_j$ and the charge density ρ_j of each type of carriers are related to their velocity $\boldsymbol{v}_j$ and the number density n_j

$$\boldsymbol{j}_j = e_j n_j \boldsymbol{v}_j; \quad \rho_j = e_j n_j. \tag{7.14}$$

The charge e_j is defined as follows $e_j = -|e|$ for electrons and $|e|$ for holes or positive ions. Here $|e| = 1.602 \times 10^{-19}$ coulomb is the absolute value of the electron charge. Equations (7.2)–(7.6), (7.13) and (7.14) compose the complete system describing the hydrodynamic model of plasma. This model makes it possible, in principle, to evaluate a plasma response to any given perturbation. Simultaneous solution of (7.2)–(7.6), (7.13) and (7.14) yields a self-consistent set of electromagnetic fields and charged particles motion in the hydrodynamic approximation [13, 21].

7.2 Waves in One Component Electron Plasma

7.2.1 General Dispersion Relations

Consider the conditions of the propagation of an electromagnetic wave

$$\boldsymbol{E} = \boldsymbol{E}_0 \exp i \left(\boldsymbol{k} \cdot \boldsymbol{r} - \omega t \right) \tag{7.15}$$

in plasma subject to the constant uniform magnetic field $\boldsymbol{B}_0$ which is assumed to be directed along the z-axis. The wave vector $\boldsymbol{k}$ belongs to the xz plane having an angle Δ with the magnetic field. The geometry of the problem is shown in Fig. 7.1.

In order to proceed with the analysis of (7.2)–(7.6) we should establish the relation between the current density $\boldsymbol{j}$ and the electric field (7.15). In the linear approximation in a sufficiently weak field we have the general relation:

$$j_i \left(\boldsymbol{r}, t \right) = \int \sigma_{ik} \left(\boldsymbol{r}, \boldsymbol{r}', t - t' \right) E_k \left(\boldsymbol{r}', t' \right) d^3 r' dt', \tag{7.16}$$

where the conductivity σ_{ik} is a second rank tensor in the presence of a magnetic field even in an isotropic medium. In an unlimited translation invariant medium all quantities depend on the difference of the spatial coordinates:

$$\sigma_{ik} \left(\boldsymbol{r}, \boldsymbol{r}', t - t' \right) = \sigma_{ik} \left(\boldsymbol{r} - \boldsymbol{r}', t - t' \right). \tag{7.17}$$

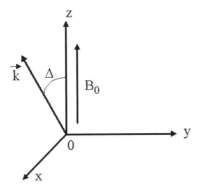

Fig. 7.1. Direction $\boldsymbol{k}$ of the wave propagation with respect to a magnetic field $\boldsymbol{B}_0$

Assuming that

$$j(r,t) = j(k,\omega)\exp i(k\cdot r - \omega t) \qquad (7.18)$$

and substituting (7.15) and (7.17) into (7.16) we obtain:

$$j_i(k,\omega) = \sigma_{ik}(k,\omega)E_k(k,\omega), \qquad (7.19)$$

where $\sigma_{ik}(k,\omega)$ is the Fourier transform of the conductivity (7.17):

$$\sigma_{ik}(k,\omega) = \int \sigma_{ik}(r,t)\exp[-i(k\cdot r - \omega t)]d^3rdt. \qquad (7.20)$$

Applying the curl operation to (7.3), using (7.5) and substituting (7.15), (7.18) and (7.19) we obtain

$$[k\times[k\times E_0]] + k_0^2\varepsilon^r\cdot E_0 = 0, \qquad (7.21)$$

where

$$k_0^2 = \frac{\omega^2}{c^2} \qquad (7.22)$$

and ε^r is the tensor of the complex relative permittivity which has the form:

$$\varepsilon_{ik}^r = \varepsilon_L^r \delta_{ik} - \frac{\sigma_{ik}(k,\omega)}{i\omega\varepsilon_0}. \qquad (7.23)$$

Relation (7.23) shows that in order to solve vector equation (7.21) with respect to the normal modes propagated in plasma it is necessary to determine the explicit form of the conductivity tensor $\sigma_{ik}(k,\omega)$. Some general features of $\sigma_{ik}(k,\omega)$ can be formulated by using the symmetry considerations. According to the Onsager principle, the conductivity tensor in the presence of the magnetic field B_0 transforms with the change of the axial vector B_0 sign as follows [22]:

$$\sigma_{ik}(B_0,k,\omega) = \sigma_{ki}(-B_0,k,\omega) = \sigma_{ki}(-B_0,-k,\omega). \qquad (7.24)$$

In the geometry chosen the axes z,y,x are parallel to the vectors

$$B_0,\ [B_0\times k],\ [[B_0\times k]\times k],$$

respectively. Consequently, the change of the sign of the vectors B_0 and k results in the change of the direction of the x- and z-axes while the direction of y-axis remains constant. In such a case in an isotropic medium the sign of σ_{xz} and of the diagonal elements σ_{ii} do not change while the elements σ_{xy} and σ_{yz} change their sign to the opposite:

$$\sigma_{xy} = -\sigma_{yx};\ \sigma_{yz} = -\sigma_{zy};\ \sigma_{xz} = \sigma_{zx}. \qquad (7.25)$$

7.2 Waves in One Component Electron Plasma

Thus the number of independent components of the tensor σ_{ik} reduces to six. The system of homogeneous equations (7.21) with the permittivity tensor (7.23) and the conductivity (7.25) has non-trivial solutions $E_{x,y,z}$ if its determinant vanishes. The latter condition yields the dispersion of normal modes propagated in plasma. In general case it is a very complicated equation of the third order. The problem can be essentially simplified for the case of the wave propagation in the directions parallel and perpendicular to the magnetic field $\boldsymbol{B}_0$: $\boldsymbol{k} \parallel \boldsymbol{B}_0$ and $\boldsymbol{k} \perp \boldsymbol{B}_0$. It is known that the oblique propagation does not give qualitatively new effects [22]. Therefore we limit our study with two these cases. Consider first the case of the parallel propagation when $\boldsymbol{k} \parallel \boldsymbol{B}_0$. Then

$$k = k_z \tag{7.26}$$

and for the medium with a rotational symmetry with respect to the z-axis we have

$$\sigma_{xx} = \sigma_{yy}, \quad \sigma_{xz} = \sigma_{yz} = 0. \tag{7.27}$$

The first condition (7.27) means that the x- and y-axes are equivalent in the chosen geometry; the second condition manifests the independence of electron motion along the magnetic field and in the plane xy. Combining the results (7.23) and (7.25)–(7.27) and substituting them into the system (7.21) we obtain the condition of the existence of a non-trivial solution [22]:

$$\begin{vmatrix} \varepsilon^r_{xx} - k^2/k_0^2 & \varepsilon^r_{xy} & 0 \\ -\varepsilon^r_{xy} & \varepsilon^r_{xx} - k^2/k_0^2 & 0 \\ 0 & 0 & \varepsilon^r_{zz} \end{vmatrix} = 0. \tag{7.28}$$

The calculation of the determinant yields the general dispersion relation for the wave propagated in plasma along the magnetic field. It reads

$$\left(\frac{k^2}{k_0^2} - \varepsilon_+\right)\left(\frac{k^2}{k_0^2} - \varepsilon_-\right)\varepsilon^r_{zz} = 0, \tag{7.29}$$

where

$$\varepsilon_\pm = \varepsilon^r_{xx} \pm i\varepsilon^r_{xy}. \tag{7.30}$$

Dispersion relation (7.29) permits three possible modes determined by the following conditions [22]

$$\varepsilon^r_{zz} = 0 \tag{7.31}$$

and

$$\frac{k^2}{k_0^2} = \varepsilon_\pm. \tag{7.32}$$

The detailed analysis of the solutions (7.31) and (7.32) requires the explicit form of the conductivity tensor $\sigma_{ik}(\boldsymbol{k}, \omega)$. However, the direct substitution of (7.31) and (7.32) into basic system (7.21) permits some conclusions about the character of the modes.

Dispersion relation (7.31) corresponds to the longitudinal mode with the electric field E_z parallel to the propagation direction. This mode is called plasmon [21, 22].

Dispersion relation (7.32) results in the two purely transverse modes with the right and left circular polarization since the ratio of the electric field components $E_{x,y}$ has the form:

$$\frac{E_x}{E_y} = \pm i. \tag{7.33}$$

These modes are called helicons [22]. Helicon is a low frequency wave excitation which can propagate almost without decay in metals subject to a strong magnetic field. The electric field of the helicon with the right circular polarization rotates in the direction of the electron motion in the magnetic field, while the electric field of the left circularly polarized helicon moves opposite to it.

Consider now the case of the wave propagated in the direction perpendicular to the magnetic field along the x-axis: $\boldsymbol{k} \perp \boldsymbol{B}_0$, $k = k_x$. Then the system of field equations (7.21) possesses a non-trivial solution under the following condition:

$$\begin{vmatrix} \varepsilon^r_{xx} & \varepsilon^r_{xy} & 0 \\ -\varepsilon^r_{xy} & \varepsilon^r_{xx} - k^2/k_0^2 & 0 \\ 0 & 0 & \varepsilon^r_{zz} - k^2/k_0^2 \end{vmatrix} = 0. \tag{7.34}$$

The evaluation of the determinant (7.34) yields

$$\left[\varepsilon^r_{xx} \left(\varepsilon^r_{xx} - \frac{k^2}{k_0^2} \right) + \left(\varepsilon^r_{xy} \right)^2 \right] \left(\varepsilon^r_{zz} - \frac{k^2}{k_0^2} \right) = 0. \tag{7.35}$$

Equation (7.35) gives two possible modes with the following dispersion relations:

$$\frac{k^2}{k_0^2} = \varepsilon^r_{zz} \tag{7.36}$$

and

$$\frac{k^2}{k_0^2} = \varepsilon^r_{xx} + \frac{\left(\varepsilon^r_{xy}\right)^2}{\varepsilon^r_{xx}}. \tag{7.37}$$

The analysis of (7.21) with $k = k_x$ shows that dispersion relation (7.36) corresponds to a purely transverse mode polarized along the magnetic field:

$E = E_z$. This mode is called an ordinary wave, similarly to the optics of anisotropic media [21, 22]. Dispersion relation (7.37) describes an extraordinary wave polarized in the xy plane perpendicular to the magnetic field. The extraordinary wave contains both the transverse component E_y and the longitudinal one E_x [21, 22]. We will show below that the extraordinary wave is elliptically polarized.

The dispersion relations for plasmons, helicons, ordinary and extraordinary waves (7.31), (7.32), (7.36) and (7.37) have been obtained under some general assumptions on the form of the conductivity tensor $\sigma_{ik}(\mathbf{k}, \omega)$. The more detailed analysis of the wave behavior is impossible without the calculation of $\sigma_{ik}(\mathbf{k}, \omega)$ in explicit form which is based on the dynamics of electrons in plasma. We pass now to this procedure. Once the conductivity tensor is calculated under some appropriate approximation we will return to the separate analysis of each mode in plasma mentioned above.

7.2.2 Conductivity Tensor in a Local Approximation

In order to calculate the conductivity tensor $\sigma_{ik}(\mathbf{k}, \omega)$ in the framework of the hydrodynamic model we should solve the equation of motion (7.13) with the constant uniform magnetic field $\mathbf{B}_0$ and variable electric field $\mathbf{E}(\mathbf{r}, t)$ of an electromagnetic wave studied in the previous section. In general case this problem is very complicated. However, for the sake of simplicity we may use the local approximation neglecting the conductivity tensor dependence on the wave vector $\mathbf{k}$, or in other words the spatial dispersion. By doing so we ignore the non-locality effects and assume the time-dependent electric field to be spatially uniform [22]. The validity of such an approach and its limitations should be discussed separately for each type of waves. Then relation (7.19) reduces to

$$j_i(\omega) = \sigma_{ik}(\omega) E_k(\omega). \tag{7.38}$$

The spatial derivatives of the velocity and pressure in (7.13) can be dropped. It should be noted that in the absence of the average drift velocity v_0 of plasma as a whole the term $(\mathbf{v} \cdot \nabla)\mathbf{v}$ is a small quantity of the second order with respect to deviations from an equilibrium state and can be neglected even in a spatially inhomogeneous case. We consider the electron plasma neglecting the ion contribution. Then (7.13) takes the form for the velocity components:

$$\frac{\partial v_x}{\partial t} + \omega_c v_y + \frac{v_x}{\tau} = -\frac{|e|}{m^*} E_{0x} \exp(-i\omega t) \tag{7.39}$$

$$\frac{\partial v_y}{\partial t} - \omega_c v_x + \frac{v_y}{\tau} = -\frac{|e|}{m^*} E_{0y} \exp(-i\omega t) \tag{7.40}$$

$$\frac{\partial v_z}{\partial t} + \frac{v_z}{\tau} = -\frac{|e|}{m^*} E_{0z} \exp(-i\omega t), \quad (7.41)$$

where ω_c is the cyclotron frequency

$$\omega_c = \frac{|e| B_0}{m^*}. \quad (7.42)$$

Equations of the type (7.39)–(7.41) have been investigated above. Using the previous results and omitting the exponential factors we write the velocity amplitudes [21]:

$$v_x(\omega) = -\frac{|e|\tau}{m^*\left[(\omega_c\tau)^2 + (1 - i\omega\tau)^2\right]} [E_{0x}(1 - i\omega\tau) - E_{0y}\omega_c\tau] \quad (7.43)$$

$$v_y(\omega) = -\frac{|e|\tau}{m^*\left[(\omega_c\tau)^2 + (1 - i\omega\tau)^2\right]} [E_{0x}\omega_c\tau + E_{0y}(1 - i\omega\tau)] \quad (7.44)$$

$$v_z(\omega) = -\frac{|e|\tau}{m^*(1 - i\omega\tau)} E_{0z}. \quad (7.45)$$

Substituting expressions (7.43)–(7.45) into the current density j (7.14) and comparing the result with (7.38) we obtain for the non-zero conductivity tensor components:

$$\sigma_{xx}(\omega) = \sigma_{yy}(\omega) = \frac{i\varepsilon_0 \omega_p^2}{(\omega + i/\tau)} \left[1 - \frac{\omega_c^2}{(\omega + i/\tau)^2}\right]^{-1} \quad (7.46)$$

$$\sigma_{xy}(\omega) = -\sigma_{yx}(\omega) = \frac{\varepsilon_0 \omega_p^2 \omega_c}{(\omega + i/\tau)^2} \left[1 - \frac{\omega_c^2}{(\omega + i/\tau)^2}\right]^{-1} \quad (7.47)$$

$$\sigma_{zz} = \frac{i\varepsilon_0 \omega_p^2}{(\omega + i/\tau)}, \quad (7.48)$$

where ω_p is the plasma frequency

$$\omega_p = \sqrt{\frac{n_0 e^2}{\varepsilon_0 m^*}} \quad (7.49)$$

and n_0 is the equilibrium concentration of the conduction electrons. Finally, we insert the conductivity $\sigma_{ik}(\omega)$ (7.46)–(7.48) into the permittivity tensor (7.23) and get [22]

$$\varepsilon_{xx}^r(\omega) = \varepsilon_{yy}^r(\omega) = \varepsilon_L^r - \frac{\omega_p^2}{\omega(\omega+i/\tau)}\left[1 - \frac{\omega_c^2}{(\omega+i/\tau)^2}\right]^{-1} \qquad (7.50)$$

$$\varepsilon_{xy}^r(\omega) = -\varepsilon_{xy}^r(\omega) = i\frac{\omega_p^2 \omega_c}{\omega(\omega+i/\tau)^2}\left[1 - \frac{\omega_c^2}{(\omega+i/\tau)^2}\right]^{-1} \qquad (7.51)$$

$$\varepsilon_{zz}^r(\omega) = \varepsilon_L^r - \frac{\omega_p^2}{\omega(\omega+i/\tau)}. \qquad (7.52)$$

Further analysis of the waves in plasma in hydrodynamic approximation is based on the expressions for the relative permittivity tensor $\varepsilon_{ik}^r(\omega)$ (7.50)–(7.52).

7.2.3 Plasmons

We begin our study with plasma oscillations as the simplest and at the same time fundamental type of excitation in plasma. For the sake of definiteness we consider now electron plasma in metals where the lattice part of the permittivity is unity:

$$\varepsilon_L^r = 1. \qquad (7.53)$$

Then, substituting (7.52) and (7.53) into dispersion relation (7.31) and neglecting the collisions which corresponds to an infinite relaxation time $\tau \to \infty$ we obtain the natural result:

$$\omega = \omega_p = \sqrt{\frac{n_0 e^2}{\varepsilon_0 m^*}}. \qquad (7.54)$$

This mode corresponds to the longitudinal electrostatic oscillations in a plasma with one kind of mobile charges, or *plasmons*. Plasmons are the most fundamental low lying excitations in a plasma in the absence of an external magnetic field. The plasma oscillations are caused by deviations of electrons from their equilibrium state with respect to the uniformly distributed ions that locally break the electric neutrality of the system. The ions can be assumed to be fixed since the frequency of electron perturbations is very high: it can reach almost light range frequencies $\omega_p \sim 10^{15}\,\text{s}^{-1}$ in metals where the electron density is very high ($n_e \approx 10^{22}\,\text{cm}$) [13]. As a result, a spatially periodic electric field emerges tending to restore the neutrality state. The plasma oscillations can be viewed as a plane capacitor (Fig. 7.2), with an electric field $E = n_0 e z/\varepsilon_0$, where z is the thickness of layer of displaced electron gas where the abundance of electrons exists. This field accelerates the charges according to the equation of motion:

$$m^* \frac{d^2 z}{dt^2} = -\frac{n_0 e^2}{\varepsilon_0} z, \qquad (7.55)$$

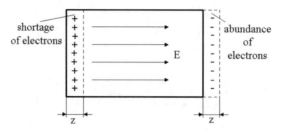

Fig. 7.2. Plasma oscillations

which yields the electric field oscillating with the plasma frequency ω_p (7.54)

$$E = E_{0z} \exp(-i\omega_p t). \tag{7.56}$$

The longitudinal plasma oscillations with the electric field $E = E_z$ are electrostatic since the variable magnetic field does not occur due to the condition

$$\mathrm{curl}\, \boldsymbol{E} = 0. \tag{7.57}$$

The plasma frequency depends only on the electron concentration being independent of $\boldsymbol{k}$. The group velocity $\boldsymbol{v}_{\mathrm{gr}} = d\omega_p/d\boldsymbol{k} = 0$, and the perturbation does not propagate. The reason is that we neglected the spatial dispersion of the plasma permittivity, i.e. the dependence of ε_{ik}^r on the wave vector $\boldsymbol{k}$. Obviously, the conductivity tensor $\sigma_{ik}(\omega)$ (7.46)–(7.48) evaluated in the uniform field approximation does not include dynamical effects caused by a spatial inhomogeneity of plasma.

Consider the conditions under which the plasma oscillations may become propagated modes, or plasma waves. The following phenomena can be responsible for an essential spatial inhomogeneity of plasma [21].

1. The thermal diffusion of electrons creates the spatial inhomogeneity of the pressure P which gives rise to the plasma oscillations propagation, or plasma waves. We still assume that the average drift velocity v_0 of electrons is absent, and the spatially inhomogeneous nonlinear term $(\boldsymbol{v} \cdot \nabla)\boldsymbol{v}$ is neglected as small.
2. The plasma may drift as a whole with the average drift velocity v_0 along the magnetic field direction. Then the spatially inhomogeneous term $v_0 \partial v/\partial z$ of the first order in small deviations emerges in the equation of motion (7.13). The drift velocity v_0 parallel to the uniform magnetic field $\boldsymbol{B}_0$ does not contribute to the Lorentz force, and plasmons remain independent of the magnetic field.

We start with the case of the thermal motion. Linearizing equations (7.4), (7.6), (7.13), and (7.14) in small deviations of the electron number density

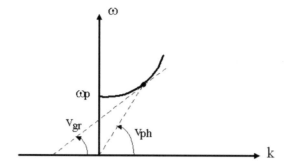

Fig. 7.3. The dispersion curve of the Langmuir waves

$n_e \ll n_0$ and v near their equilibrium values, taking into account that for the adiabatic compression [21]

$$\nabla P = 3k_B T \nabla n_e, \qquad (7.58)$$

neglecting the collision term, and bearing in mind that in our case the electric field $\boldsymbol{E}$ and the electron velocity $\boldsymbol{v}$ are parallel to the z-axis and all quantities depend only on the z coordinate we obtain the following system of equations:

$$\frac{\partial E_z}{\partial z} = -\frac{|e|\, n_e}{\varepsilon_0} \qquad (7.59)$$

$$\frac{\partial n_e}{\partial t} + n_0 \frac{\partial v_z}{\partial z} = 0 \qquad (7.60)$$

$$m^* \frac{\partial v_z}{\partial t} = -|e|\, E_z - \frac{1}{n_0} 3k_B T \frac{\partial n_e}{\partial z}. \qquad (7.61)$$

We set E_z, n_e and v_z proportional to $\exp i(kz - \omega t)$ and obtain from the condition of a non-trivial solution existence for (7.59)–(7.61) the plasma wave dispersion relation [21]:

$$\omega^2 = \omega_p^2 + \frac{3}{2} k^2 v_{\text{therm}}^2, \qquad (7.62)$$

where the thermal velocity v_{therm} of electrons has the form:

$$v_{\text{therm}}^2 = \frac{2k_B T}{m^*}. \qquad (7.63)$$

The dispersion relation (7.62) is shown in Fig. 7.3.

Unlike the plasma oscillations (7.54) dispersion relation (7.62) describe the plasma waves, or the Langmuir waves with the finite group velocity v_{gr} and phase velocity $v_{\text{ph}} = \omega/k$:

$$v_{gr} = \frac{d\omega}{dk} = \frac{3}{2}v_{therm}^2 \frac{k}{\omega} = \frac{3}{2}\frac{v_{therm}^2}{v_{ph}}. \tag{7.64}$$

It is easy to see that the group velocity v_{gr} (7.64) of the plasma waves is always less than the vacuum light velocity c. Indeed, the explicit form of v_{ph} is

$$v_{ph} = \sqrt{\frac{3}{2}} \frac{\omega v_{therm}}{\sqrt{\omega^2 - \omega_p^2}}. \tag{7.65}$$

Substituting (7.65) into (7.64) we obtain

$$v_{gr} = \sqrt{\frac{3}{2}} v_{therm} \left(1 - \frac{\omega_p^2}{\omega^2}\right)^{1/2} < \sqrt{\frac{3}{2}} v_{therm}. \tag{7.66}$$

Obviously, the thermal velocity v_{therm} for the non-relativistic case is always much less than the vacuum light velocity c.

For large k the plasma wave propagates with the thermal velocity (7.63) of electrons. For small k the group velocity is less than the thermal velocity because for large wavelengths the density gradient in the wave is small, and only small momentum is transported with the thermal motion.

Consider now the case of the plasma drift the average drift velocity v_0 along the z direction [13]. This time we neglect the thermal diffusion of electrons. The geometry of the problem remains the same as well as (7.59). In (7.60) and (7.61) the new terms emerge due to the drift velocity: $v_0 \partial n_e/\partial z$ and $v_0 \partial v_z/\partial z$ respectively. We get

$$\frac{\partial n_e}{\partial t} + n_0 \frac{\partial v_z}{\partial z} + v_0 \frac{\partial n_e}{\partial z} = 0 \tag{7.67}$$

and

$$m^* \frac{\partial v_z}{\partial t} + m^* v_0 \frac{\partial v_z}{\partial z} = -|e|E_z. \tag{7.68}$$

The standard procedure applied to (7.59), (7.67) and (7.68) yields the following dispersion relation:

$$k_{1,2} = \frac{\omega \pm \omega_p}{v_0}. \tag{7.69}$$

The roots $k_{1,2}$ correspond to the slow and fast waves of the space charge in the "cold" collisionless electron plasma. The group velocity of the both waves is, obviously, the drift velocity v_0. The phase velocities $v_{1,2ph}$ of the slow and fast waves, respectively, are

$$v_{1,2ph} = \frac{\omega}{k_{1,2}} = \frac{v_0}{1 \pm \omega_p/\omega}. \tag{7.70}$$

7.2 Waves in One Component Electron Plasma

In order to make our model more realistic and evaluate the decay characteristics of the plasma waves we must take into account collisions, assuming the relaxation time τ to be finite. For this reason we keep the term $m^*\boldsymbol{v}/\tau$ in equation of motion (7.13). Then (7.68) takes the form:

$$m^*\frac{\partial v_z}{\partial t} + m^* v_0 \frac{\partial v_z}{\partial z} + \frac{m^* v}{\tau} = -|e|E_z \qquad (7.71)$$

and the dispersion relation (7.69) becomes [13]:

$$k_{1,2} = \frac{1}{v_0}\left\{\omega + \frac{i}{2\tau} \pm \omega_p\sqrt{1 - \frac{1}{4(\omega_p\tau)^2}}\right\}. \qquad (7.72)$$

It is seen from (7.72) that collisions result in the decay of the plasma waves with a distance. For $\omega_p\tau > 1/2$ both modes have the damping coefficient $(2v_0\tau)^{-1}$. The phase velocities are given by

$$v_{1,2\mathrm{ph}} = v_0\left[1 \pm \frac{\omega_p}{\omega}\sqrt{1 - \frac{1}{4(\omega_p\tau)^2}}\right]^{-1}. \qquad (7.73)$$

In the opposite case when $\omega_p\tau < 1/2$ dispersion relation (7.72) takes the form:

$$k_{1,2} = \frac{\omega}{v_0} + \frac{i}{2\tau v_0}\left[1 \pm \sqrt{1 - 4(\omega_p\tau)^2}\right]. \qquad (7.74)$$

Then both waves possess the same phase velocity equal to the drift velocity v_0 and different damping coefficients

$$\mathrm{Im}\,k_{1,2} = \frac{1}{2\tau v_0}\left[1 \pm \sqrt{1 - 4(\omega_p\tau)^2}\right]. \qquad (7.75)$$

In the case of a small relaxation time $2\omega_p\tau \ll 1$ we obtain from (7.75):

$$\mathrm{Im}\,k_1 \approx (v_0\tau)^{-1} \gg \frac{\omega}{v_0} \qquad (7.76)$$

and

$$\mathrm{Im}\,k_2 \approx \frac{\omega_p^2\tau}{v_0} \ll \frac{\omega}{v_0}. \qquad (7.77)$$

Inequality (7.76) shows that the mode corresponding to k_1 becomes overdamped and actually ceases to exist since its damping coefficient surpasses the real part of the wave number. The second mode with the wave number k_2 still has a sufficiently large lifetime $(\omega_p^2\tau)^{-1} \geq \omega^{-1}$ and therefore can propagate.

It can be shown that the simultaneous account of both collisions (a finite relaxation time τ) and the thermal diffusion (a spatial inhomogeneity of the pressure P) result in the increasing of damping for both modes and decreasing of the phase velocity for the mode with a stronger decay.

7.2.4 Ordinary and Extraordinary Electromagnetic Waves

Consider the transverse waves defined by the dispersion relations (7.36) and (7.37). These waves are transverse and possess the variable magnetic field $B \exp i(\mathbf{k} \cdot \mathbf{r} - \omega t)$ determined by the Maxwell equation (7.3) where this time the left-hand side is not zero. The geometry of the problem is shown in Fig. 7.4.

We start with the ordinary wave. Substituting (7.52) and (7.53) into (7.36) we and neglecting electron collisions we get

$$\frac{k^2}{k_0^2} = 1 - \frac{\omega_p^2}{\omega^2}, \tag{7.78}$$

which yields [21]

$$\omega^2 = \omega_p^2 + k^2 c^2. \tag{7.79}$$

Obviously the ordinary wave can propagate only for $\omega > \omega_p$. The plasma frequency serves as a so-called cut-off frequency for the ordinary wave. The dispersion curve of the ordinary wave is shown in Fig. 7.5.

The phase velocity v_{ph} of the ordinary wave is greater than the vacuum light velocity:

$$v_{\mathrm{ph}}^2 = c^2 + \frac{\omega_p^2}{k^2} > c^2. \tag{7.80}$$

However, the group velocity v_{gr} cannot surpass c. Indeed, it has the form:

$$v_{\mathrm{gr}} = \frac{d\omega}{dk} = \frac{c^2}{v_{\mathrm{ph}}} = c^2 \left[c^2 + \frac{\omega_p^2}{k^2} \right]^{-1/2} < c. \tag{7.81}$$

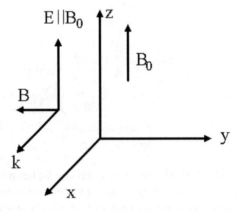

Fig. 7.4. Ordinary electromagnetic wave in plasma

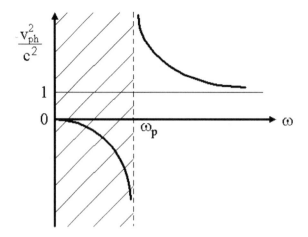

Fig. 7.5. The dispersion curve of the ordinary wave in plasma

In the geometry considered with $k = k_x$ and $E = E_z$ the magnetic field of the ordinary wave is parallel to the y-axis. Using (7.3) we obtain

$$B = B_y = -\frac{k}{\omega}E_z = -\frac{\sqrt{\omega^2 - \omega_p^2}}{\omega c}E_z. \tag{7.82}$$

The ordinary wave is linearly polarized. It easy to see from the expressions obtained that the characteristics of the ordinary wave do not depend on the uniform magnetic field $\boldsymbol{B}_0$. It turned out that the dispersion relation of this wave coincides with the one of the so-called fast electromagnetic wave propagated in plasma without the constant magnetic field. Indeed, the Maxwell equations (7.3), (7.5) for $E = E_{0z}\exp i\,(kx - \omega t)$ and the equation of motion (7.13) for $B_0 = 0$ yield:

$$kE = -\omega B;\ ikB = \mu_0 j_z - \frac{i\omega}{c^2}E;\ -i\omega m^* v_z = -|e|\,E. \tag{7.83}$$

Excluding the current density $j_z = -|e|\,n_0 v_z$ from the second equation (7.83) we immediately obtain from the consistency condition of (7.83) the dispersion relation (7.79).

Consider now the extraordinary wave. Combining (7.37), (7.50), (7.51) and (7.53) and assuming the relaxation time to be infinite $\tau \to \infty$ we obtain:

$$\frac{k^2}{k_0^2} = 1 - \frac{\omega_p^2\left(\omega^2 - \omega_p^2\right)}{\omega^2\left(\omega^2 - \omega_p^2 - \omega_c^2\right)}. \tag{7.84}$$

The relation between the components $E_{x,y}$ of the extraordinary wave can be derived from the wave equation (7.21) by substituting the permittivity tensor components (7.50) and (7.51).

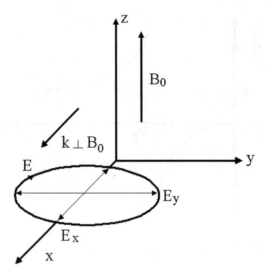

Fig. 7.6. The elliptically polarized extraordinary wave in plasma

$$\frac{E_y}{E_x} = -i\frac{\omega\left(\omega^2 - \omega_p^2 - \omega_c^2\right)}{\omega_p^2 \omega_c}. \tag{7.85}$$

Equations (7.84) and (7.85) clearly show that, unlike the ordinary wave, both the dispersion and the polarization of the extraordinary wave essentially depend on the constant magnetic field $\boldsymbol{B}_0$ through the cyclotron frequency ω_c. It is seen from (7.85) that the extraordinary wave is elliptically polarized. The geometry of its polarization and propagation is shown in Fig. 7.6.

The magnetic field of the extraordinary wave is parallel to the z-axis. Equations (7.3) and (7.84) result in the following relation:

$$B = B_z = \frac{k}{\omega}E_y = \frac{1}{c}\left[1 - \frac{\omega_p^2\left(\omega^2 - \omega_p^2\right)}{\omega^2\left(\omega^2 - \omega_p^2 - \omega_c^2\right)}\right]^{1/2} E_y. \tag{7.86}$$

The E_x component does not contribute to the magnetic field since it is parallel to the wave vector $\boldsymbol{k}$. The phase velocity $v_{\rm ph}$ of the extraordinary wave is easily found from the dispersion relation (7.84):

$$v_{\rm ph}^2 = \frac{\omega^2}{k^2} = c^2\left[1 - \frac{\omega_p^2\left(\omega^2 - \omega_p^2\right)}{\omega^2\left(\omega^2 - \omega_p^2 - \omega_c^2\right)}\right]^{-1}. \tag{7.87}$$

The group velocity $v_{\rm gr}$ of the extraordinary wave is given by

$$v_{\rm gr} = \frac{d\omega}{dk} = \frac{c^2}{v_{\rm ph}}\frac{\left(\omega^2 - \omega_p^2 - \omega_c^2\right)^2}{\left[\left(\omega^2 - \omega_p^2 - \omega_c^2\right)^2 + \left(\omega_p\omega_c\right)^2\right]}. \tag{7.88}$$

Dispersion relation (7.84) possesses the resonance frequency and the cut-off frequency [21]. The resonance frequency corresponds to the case where the wave number k becomes infinite which takes place under the following condition:

$$k \to \infty, \quad \omega^2 = \omega_h^2 = \omega_p^2 + \omega_c^2. \tag{7.89}$$

Here ω_h is a so-called upper hybrid frequency [21]. At the upper hybrid frequency both the group and the phase velocities vanish, as it is seen from (7.87) and (7.88) which means that the extraordinary wave cannot propagate. Moreover, (7.85) shows that

$$E_y = 0 \text{ at } \omega = \omega_h, \tag{7.90}$$

which results in vanishing of the magnetic field of the wave, too. Thus, only the electric field component E_x remains, and the extraordinary wave at the resonance frequency ω_h reduces to the longitudinal electrostatic wave propagating perpendicularly to the uniform magnetic field.

Consider now the cut-off frequency that corresponds to $k = 0$ and an infinite wavelength. Under this condition dispersion relation (7.84) gives:

$$\omega^2 \left(\omega^2 - \omega_p^2 - \omega_c^2 \right) = \omega_p^2 \left(\omega^2 - \omega_p^2 \right) \tag{7.91}$$

and after some algebra we obtain two real solutions $\omega_{R,L}$ corresponding to the cut-off of the right and left elliptically polarized modes respectively:

$$\omega_{R,L} = \frac{1}{2} \left[\left(\omega_c^2 + 4\omega_p^2 \right)^{1/2} \pm \omega_c \right]. \tag{7.92}$$

Analysis shows that for the frequencies $\omega > \omega_R$ and $\omega_L < \omega < \omega_h$ the extraordinary wave is propagated, since $k^2 > 0$. In this regime for $\omega > \omega_R$ and $\omega_L < \omega < \omega_p$ the phase velocity is greater than the vacuum light velocity: $v_{\text{ph}} > c$. For intermediate frequencies $\omega_p < \omega < \omega_h$ the phase velocity is less than c. For $\omega_h < \omega < \omega_R$ and $\omega < \omega_L$ we have the non-transparency regime since $k^2, v_{\text{ph}}^2 < 0$. The dispersion curve of the extraordinary wave is presented in Fig. 7.7.

7.2.5 Helicons

Consider the propagation of helicon waves in electron plasma in metals, in the local regime. The helicon dispersion relation determined by (7.30), (7.32), (7.50) and (7.51) takes the form:

$$\frac{k^2}{k_0^2} = \varepsilon_\pm = 1 - \frac{\omega_p^2}{\omega \left(\omega \mp \omega_c + i/\tau \right)}. \tag{7.93}$$

The propagation direction and polarizations of the waves (7.93) are shown in Fig. 7.8.

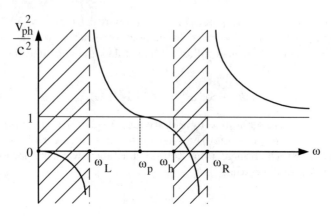

Fig. 7.7. The dispersion curve of the extraordinary wave in plasma

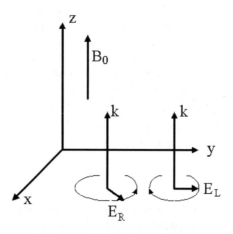

Fig. 7.8. The right and left circularly polarized waves propagated parallel the constant magnetic field $\boldsymbol{B_0}$

The left circularly polarized mode corresponding to

$$\frac{k_-^2}{k_0^2} = \varepsilon_- = 1 - \frac{\omega_p^2}{\omega\left(\omega + \omega_c + i/\tau\right)} \tag{7.94}$$

does not possess a resonance since it rotates in the opposite direction with respect to electrons rotating with the cyclotron frequency around the constant magnetic field $\boldsymbol{B_0}$. Its cut-off frequency coincides with the cut-off frequency ω_L (7.92) of the left elliptically polarized extraordinary wave propagated perpendicular to the magnetic field. Therefore, the mode (7.94) can be propagated only for high frequencies

$$\omega > \frac{1}{2}\left[\left(\omega_c^2 + 4\omega_p^2\right)^{1/2} - \omega_c\right]. \tag{7.95}$$

We are interested in low frequency excitations and neglect the left circularly polarized wave (7.94).

The right circularly polarized wave, or the helicon wave is determined by

$$\frac{k_+^2}{k_0^2} = \varepsilon_+ = 1 - \frac{\omega_p^2}{\omega(\omega - \omega_c + i/\tau)} \qquad (7.96)$$

and possesses both the resonance at the cyclotron frequency $\omega = \omega_c$ and the cut-off frequency ω_R (7.92) which is greater than ω_L. Generally, the propagation regime of the mode (7.96) may take place at $\omega > \omega_R$ and $\omega < \omega_c$. However, it can be shown in the framework of the kinetic theory that the helicons can exist only at low frequencies $\omega \ll \omega_c$. This more adequate approach which accounts for the spatial dispersion effects will be discussed below. At $\omega \ll \omega_c$ helicons are described by virtue of the "local" permittivity tensor (7.50) and (7.51) with a sufficient accuracy. Qualitatively, the helicon propagation can be described as follows [23]. When the mean free path of electrons is sufficiently large and the frequency ω of the wave is low enough, the electrons affected by the Lorentz force $-|e|[\boldsymbol{v} \times \boldsymbol{B}_0]$ would drift in the direction perpendicular to the plane formed by the uniform magnetic field $\boldsymbol{B}_0$ and the electric field $\boldsymbol{E}$ of the wave. The current $\boldsymbol{j}$ created by such a drift is in fact a so-called Hall current. It is perpendicular to the electric field $\boldsymbol{E}$ of the wave and causes no dissipation. The other processes leading to the wave propagation in a metal, i.e. the induction of the variable magnetic field by virtue of the Ampère law and the induction of the variable electric field $\boldsymbol{E}$ of the wave by virtue of the Faraday law, are also non-dissipative. Consequently, the electromagnetic energy is conserved and in the absence of collisions the wave does not attenuate [23]. The mechanism of a helicon wave propagation is illustrated in Fig. 7.9.

Near the resonance $\omega \to \omega_c$ the condition of spatial homogeneity $k \to 0$ that is essential for the local approximation ceased to exist since $k \to \infty$. Obviously, in the frequency range of interest $\omega \ll \omega_p$. The plasma frequency reaches a value of $\omega_p \sim (10^{15} \div 10^{16})$ s^{-1} for typical metals due to the large concentration of conduction electrons. Thus the lattice part of the relative permittivity can be neglected. Then, in the collisionless limiting case $\tau \to \infty$ the helicon dispersion relation (7.96) takes the form [22]:

$$\frac{k^2}{k_0^2} = \frac{\omega_p^2}{\omega \omega_c}, \qquad (7.97)$$

or

$$\omega = \frac{\omega_c c^2}{\omega_p^2} k^2 = \frac{c^2 B_0 \varepsilon_0}{e n_{0e}} k^2. \qquad (7.98)$$

The phase velocity v_{ph}^H of the helicon is given by

$$v_{ph}^H = c \frac{\sqrt{\omega \omega_c}}{\omega_p} = c \sqrt{\frac{\omega B_0 \varepsilon_0}{e n_{0e}}}. \qquad (7.99)$$

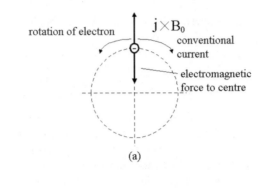

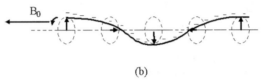

Fig. 7.9. (a) The force diagram for the helicon wave. The Lorentz force $[\mathbf{j} \times \mathbf{B}_0]$ and the elastic of the magnetic line should balance each other. (b) The helicon wave represented as an electrically charged string performing helicoidal oscillations

It is seen from expression (7.99) that the phase velocity of the helicon does not depend on the electron effective mass, and it is proportional to the square root of the magnetic field and frequency. It was mentioned above that the helicon frequencies small compared to the cyclotron and plasma frequencies ω_c and ω_p. Thus, $v_{\text{ph}}^H \ll c$, and helicons in metals are very slow excitations due to the inverse dependence on the large plasma frequency which is the largest characteristic frequency of the problem. For a typical metal $\omega_p \sim 10^{16}$ s^{-1}, and for $\omega \sim 10^7$ s^{-1}, $\omega_c \sim 10^{11}$s^{-1} we get that $v_{\text{ph}}^H \sim 30$ m/s. The only velocity comparable with the helicon velocity in a solid is the sound velocity.

Consider more comprehensively the polarization of helicon waves. We introduce the right and left polarized vectors $E_{R,L}$ as follows [21]

$$E_{R,L} = E_x \pm iE_y. \tag{7.100}$$

Then, obviously,

$$E_x = \frac{1}{2}(E_R + E_L),\ E_y = \frac{1}{2i}(E_R - E_L). \tag{7.101}$$

Substituting the electric field components (7.101) and the permittivity (7.50), (7.51) and (7.53) into the wave equation (7.21) we derive the equations for the circular components $E_{R,L}$

$$G(\omega) E_R = 0,\ F(\omega) E_L = 0, \tag{7.102}$$

where

$$G(\omega) = \omega^2 - c^2 k^2 - \frac{\omega \omega_p^2}{\omega + \omega_c + i/\tau} \tag{7.103}$$

and

$$F(\omega) = \omega^2 - c^2 k^2 - \frac{\omega \omega_p^2}{\omega - \omega_c + i/\tau}. \tag{7.104}$$

It is seen that (7.102) are consistent when

$$G(\omega) = 0, \; E_L = 0, \tag{7.105}$$

or

$$F(\omega) = 0, \; E_R = 0. \tag{7.106}$$

Comparison of these conditions with the dispersion relations (7.94) and (7.96) that the pair of equations (7.105) defines the left circularly polarized mode, while the second one indeed determines the dispersion relation and the polarization of the right circularly polarized helicon mode. The electric field components $E_{x,y}$ of the right circularly polarized helicon are related by [21]

$$\frac{E_x}{E_y} = -i. \tag{7.107}$$

The magnetic field $\boldsymbol{B}_H$ of the helicon wave evaluated from the Maxwell equation (7.3) is perpendicular to the propagation direction and it is given by

$$B_{Hx} = -\frac{k}{\omega} E_y, \; B_{Hy} = \frac{k}{\omega} E_x. \tag{7.108}$$

The remarkable fact about the helicon is that its magnetic field is much larger than the magnetic field of an ordinary electromagnetic wave with the same electric field. Indeed, usual electromagnetic waves possess a magnetic field $B \sim E/c$, while $B_H \sim E/v_{\text{ph}}^H \gg E/c$. Hence, the field of the helicon wave is mainly magnetic.

To summarize, the helicon is a low-lying circularly polarized magnetic excitation in a solid-state plasma under a strong magnetic field. It arises from the suppression of the currents along the electric field of the wave due to a strong external magnetic field and maintenance of the transverse currents that generate a time dependent magnetic field, which is sufficient to maintain self-sustaining oscillations. It is essentially a radio frequency (RF) Hall effect [23].

Experimentally helicons have been observed both in the standing wave and in the travelling wave regimes. For example, the helicon standing wave with the frequency

$$\omega \approx 2 \times 10^2 \text{ s}^{-1}$$

has been excited by a pulse high frequency external electric field in a cylindrical sample subject to the uniform magnetic field $B_0 = 3$ tesla which gives the value of the wave number $k \approx 10$ cm^{-1} according to (7.97).

In the typical experiments with traveling waves a source of the electromagnetic energy excites eddy currents on the surface of a metal plate of the given thickness placed in the uniform magnetic field. The currents give rise to the oscillating helicon modes which transfer the energy to the other side of the plate where the detector is placed. Comparing the phases of the transferred wave and the reference signal one can investigate the dispersion relation of the helicon wave. Since helicons are slow waves, of almost purely magnetic nature, penetrating into a metal much beyond the skin layer, they present an excellent tool to study the behavior of conduction electrons in time and spatially dependent magnetic fields [22].

7.3 Waves in Two Component Plasma

7.3.1 Ion Acoustic Waves

Until now we considered only the one component electron plasma assuming the ions to be fixed. However being charged particles the ions of plasma also can interact through the electric field. The mass of ions is large, and therefore they can participate only in low frequency processes. In such a case the so-called plasma approximation can be applied [21]. Essentially, this approximation means that although plasma obeys the neutrality condition for the concentration n_e of electrons and n_i of ions

$$n_e = n_i = n \tag{7.109}$$

at the same time the right-hand side of the Poisson equation (7.4) is not zero:

$$\operatorname{div} \boldsymbol{E} \neq 0. \tag{7.110}$$

Unlike other electrodynamics problems where the electric field is calculated by using the Poisson equation with the given charge distribution ρ, in plasma problems the electric field $\boldsymbol{E}$ is usually determined by virtue of the equation of motion (7.13). Then the charge density deviations can be evaluated from (7.4). Such an approach is based on the plasma tendency to remain neutral in any case. Indeed, the motion of ions results in the electron distribution perturbation, and then the electric field would react in a self-consistent manner in order to support the plasma neutrality. Thus, applying the plasma approximation one can exclude from the complete system of equations describing the hydrodynamic model the Poisson equation (7.4) and one of the variables by using (7.109).

The plasma approximation is similar to the quasineutrality considered in the previous chapter, but has a more general meaning. The former is valid for sufficiently slow wave processes while the latter concerns only the static regime. If a wave perturbation is so slow that both ions and electrons can respond to it, then with a good accuracy the Poisson equation can be replaced with (7.109). Evidently, the plasma approximation is not valid in the case of high frequency waves discussed above where only electrons can participate in fast processes. Then, as we have already seen, the electric field $\boldsymbol{E}$ is evaluated by virtue of the Maxwell equations (7.3) and (7.5).

We start with the application of the plasma approximation to the analysis of the ion oscillations in the absence of the external uniform magnetic field $\boldsymbol{B}_0$. In such a case the equation of motion (7.13) reduces to [21]:

$$Mn\left[\frac{\partial \boldsymbol{v}_i}{\partial t} + (\boldsymbol{v}_i \cdot \nabla)\boldsymbol{v}_i\right] = |e|\,n\boldsymbol{E} - \nabla P. \tag{7.111}$$

Here M is the mass of the ion and the subscript i stands for ions. Taking into the electrostatic character of the process considered we express the electric field $\boldsymbol{E}$ in terms of the electrostatic potential V:

$$\boldsymbol{E} = -\nabla V. \tag{7.112}$$

The pressure P still has the form (7.58) since the ion oscillations can be considered as adiabatic. Then substituting (7.58) and (7.112) into (7.111), linearizing (7.111) with respect to small quantities $n_1 \ll n_0$, V, v_i, and expressing all variables in terms of plane waves

$$n_1, V, v_i \sim \exp i\,(kz - \omega t) \tag{7.113}$$

we obtain

$$-i\omega M n_0 v_i = -ik(|e|\,n_0 V + 3k_B T_i n_1). \tag{7.114}$$

Consider now the behavior of electrons. Their mass is negligibly small as compared to ions. Then according to the equation of motion (7.13) they should be accelerated up to very high energies under the influence of the electrostatic force $(-|e|\,n\boldsymbol{E})$. However, electrons cannot leave behind a positively charged region of uncompensated ions. For this reason, the electrostatic must be balanced by the compression force due to the gradient of pressure in electron gas while the electron mass in (7.13) can be neglected: $m_e^* \to 0$. Then the equation of motion (7.13) for electrons yields:

$$|e|\frac{\partial V}{\partial z} = \frac{k_B T_e}{n_e}\frac{\partial n_e}{\partial z}, \tag{7.115}$$

where T_e is the temperature of the electron gas, the charge of electron is $-e$ and electron gas is supposed to be isothermal such that $P = n_e k_B T_e$. The integration of (7.115) with the equilibrium condition in the absence of the field

$$V = 0, \quad n_e = n_0 \tag{7.116}$$

we get the Boltzmann distribution for electrons:

$$n_e = n_0 \exp\left(\frac{|e|V}{k_B T_e}\right). \tag{7.117}$$

For small deviations of the potential the distribution (7.117) can be expanded in powers of $(|e|V/k_B T_e)$ neglecting the terms of the higher orders. Then we have

$$n_e = n_0 \left(1 + \frac{|e|V}{k_B T_e}\right). \tag{7.118}$$

Comparing (7.116) and (7.118) it is easy to see that the small deviation n_1 is proportional to the potential V

$$n_1 = n_0 \frac{|e|V}{k_B T_e}. \tag{7.119}$$

Finally, the linearized continuity equation (7.6) gives for ions

$$i\omega n_1 = ikn_0 v_i. \tag{7.120}$$

The system of homogeneous equations (7.114), (7.119) and (7.120) has a solution when its determinant vanishes which yields the dispersion relation for the so-called ion acoustic waves:

$$\omega = k\sqrt{\frac{k_B T_e + 3k_B T_i}{M}}. \tag{7.121}$$

The phase and group velocities of these waves coincide being the sound velocity v_s in plasma [21]:

$$v_s = v_{\text{ph}} = v_{\text{gr}} = \frac{\omega}{k} = \sqrt{\frac{k_B T_e + 3k_B T_i}{M}}. \tag{7.122}$$

Unlike the plasma Langmuir waves where the frequency is constant and the deviations due to a thermal motion are small, the ion waves possess the constant velocity, their frequency increases linearly with the wave number and they cannot exist without a thermal motion.

We should emphasize the basic difference between electron plasma waves and ion acoustic ones. In the former case ions do not participate in oscillations due to their large mass, and the plasma is essentially one component. In the latter type of waves electrons cannot be considered as fixed. On the contrary, they permanently follow the ions tending to compensate the electric field created by ion clouds. The electron screening is not total due to their thermal motion which gives rise to the small potential $k_B T_e/|e|$. For this reason, the term with the electron gas temperature appears in (7.122). This term is responsible for the existence of the ion acoustic waves even at very low temperatures T_i of the ion gas. It is well known that in a neutral gas sound waves vanish as $T \to 0$ since the sound velocity $v_s \sim \sqrt{T}$. In plasma the sound velocity at $T_i = 0$ takes the form:

$$v_s = \sqrt{\frac{k_B T_e}{M}}. \tag{7.123}$$

In the limiting case (7.123) the sound velocity is determined by the electron temperature since the electric field of the wave is proportional to it, and by the ion mass as a measure of the ion "liquid" inertia.

Finally, evaluate the accuracy of the plasma approximation. Suppose that the neutrality condition (7.109) is not satisfied, and the deviation of electron concentration n_{1e} differs from the one of the ion concentration n_{1i}. Then the electric field is determined by the Poisson equation (7.4) which yields:

$$\varepsilon_0 \mathrm{div} \boldsymbol{E} = \varepsilon_0 k^2 V = |e|(n_{1i} - n_{1e}). \tag{7.124}$$

Combining (7.124) with (7.111), (7.119) and (7.120) we obtain the following expression instead of dispersion relation (7.121):

$$\omega = \frac{k}{\sqrt{M}} \left[k_B T_e \left(1 + k^2 L_D^2\right)^{-1} + 3 k_B T_i \right]^{1/2}, \tag{7.125}$$

where L_D is the Debye length

$$L_D = \sqrt{\frac{\varepsilon_0 k_B T_e}{n_0 e^2}}. \tag{7.126}$$

The dispersion relation (7.125) differs from (7.121) only by factor $\left(1 + k^2 L_D^2\right)^{-1}$ in the first term in brackets. This factor, in turn, differs from unity by a quantity $k^2 L_D^2 \ll 1$ since usually the Debye length is very small. Consequently, the error is $\sim \left(2\pi L_D^2/\lambda^2\right) \ll 1$, and the plasma approximation is valid for all wavelengths except very short ones.

In the limiting case of short waves when $k^2 L_D^2 \gg 1$ the dispersion (7.125) becomes

$$\omega^2 = \frac{n_0 e^2}{\varepsilon_0 M} = \Omega_p^2, \tag{7.127}$$

where Ω_p is the ion plasma frequency. Expression (7.127) shows that for very large k the ion acoustic wave frequency is constant, and the wave itself reduces to low frequency plasma oscillations.

7.3.2 Electrostatic Ion Waves

Consider the ion acoustic wave behavior in the presence of the constant uniform magnetic field $\boldsymbol{B}_0$ [21]. We choose as usual the z-axis along the magnetic

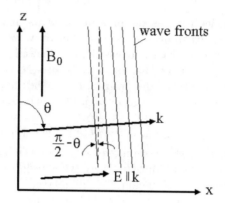

Fig. 7.10. Electrostatic ion-cyclotron waves propagating almost perpendicularly to the magnetic field $\boldsymbol{B}_0$.

field while the wave propagation direction is almost perpendicular to the magnetic field. The geometry of the problem is presented in Fig. 7.10.

For the sake of simplicity we set $T_i = 0$. It has been shown above that the ion wave can exist even at a zero temperature of the ion "liquid". The waves are assumed to be electrostatic, i.e. their electric field is parallel to the propagation direction:

$$\boldsymbol{E} \parallel \boldsymbol{k}, \quad \mathrm{curl}\,\boldsymbol{E} = 0, \quad \boldsymbol{E} = -\nabla V. \tag{7.128}$$

The deviation ϑ of the propagation direction from the right angle is assumed to be small

$$\vartheta = \frac{\pi}{2} - \theta \ll 1, \tag{7.129}$$

in such a way that the small z-components of the electric field $\boldsymbol{E}$ and the wave vector $\boldsymbol{k}$ can be neglected in the equations describing the ion motion. Ions having the large mass cannot be displaced in the z direction on a considerable distance during the wave period. Electrons due to their small mass can drift along the uniform magnetic field and contribute to the Debye screening even for small angles ϑ greater then the critical angle ϑ_0:

$$\vartheta_0 \sim \sqrt{\frac{m_e^*}{M}}. \tag{7.130}$$

Equation of motion (7.13) for the ions in the magnetic field $\boldsymbol{B}_0$ takes the form:

$$M \frac{\partial \boldsymbol{v}_i}{\partial t} = -|e| \nabla V + |e| [\boldsymbol{v}_i \times \boldsymbol{B}_0]. \tag{7.131}$$

In the plane wave representation the solution of (7.131) is given by

7.3 Waves in Two Component Plasma   263

$$v_{ix} = \frac{|e|\,k}{M\omega} V \left(1 - \frac{\Omega_c^2}{\omega^2}\right)^{-1}, \qquad (7.132)$$

where Ω_c is the ion cyclotron frequency:

$$\Omega_c = \frac{|e|\,B_0}{M}. \qquad (7.133)$$

Solving simultaneously (7.109), (7.119), (7.120) and (7.132) we get the dispersion relation for the ion cyclotron waves which has the form:

$$\omega^2 = \Omega_c^2 + k^2 \frac{k_B T_e}{M} = \Omega_c^2 + k^2 v_s^2. \qquad (7.134)$$

In the ion cyclotron waves ions oscillate similarly to the ion acoustic waves, but this time they are influenced additionally by the Lorentz force which results in the term Ω_c^2 in the dispersion relation (7.134). Electrons are screening the electric field that occur at the oscillations. It is the electron Debye screening effect, that provides the dispersion relation of ion acoustic waves (7.121). In the presence of the uniform magnetic field $\mathbf{B}_0$ electrons should cover a large distance along its direction in order to provide a screening for the ion cyclotron waves.

In the case when the wave vector $\mathbf{k}$ is strictly perpendicular to the uniform magnetic field $\mathbf{B}_0$, i.e. $\vartheta = 0$ electrons cannot evolve along the magnetic field direction and provide the neutrality of the plasma. In such a case the Boltzmann distribution function (7.117) cannot be applied, and the electron concentration n_e must be evaluated directly from the equation of motion (7.13). Neglecting the thermal motion and thus dropping a spatially inhomogeneous term ∇P and substituting the electric field (7.128) we obtain for the electron velocity v_{ex}:

$$v_{ex} = -\frac{|e|\,k}{m_e^*\omega} V \left(1 - \frac{\omega_c^2}{\omega^2}\right)^{-1}. \qquad (7.135)$$

The continuity equation (7.6) gives for the electrons

$$n_{e1} = \frac{k}{\omega} n_0 v_{ex}. \qquad (7.136)$$

Then comparing the plasma approximation condition (7.109), (7.120) and (7.136) we conclude that the velocities (7.132) of the ions and (7.135) of the electrons are equal:

$$v_{ex} = v_{ix}, \qquad (7.137)$$

which yields

$$-m_e^* \left(1 - \frac{\omega_c^2}{\omega^2}\right) = M \left(1 - \frac{\Omega_c^2}{\omega^2}\right) \qquad (7.138)$$

and

$$\omega = \sqrt{\omega_c \Omega_c} \equiv \omega_l. \tag{7.139}$$

The frequency ω_l (7.139) of the electrostatic ion oscillations is called the lower hybrid frequency. The frequency (7.139) does not depend on the wave vector, and the oscillations perpendicular to the magnetic field cannot propagate.

7.3.3 Alfven Waves

Consider now the ion wave propagating along the uniform magnetic field $\boldsymbol{B}_0$. Assume that the wave is linearly polarized along the x-axis:

$$k = k_z, \quad E = E_x = E_{0x} \exp i(kz - \omega t). \tag{7.140}$$

The geometry of the problem is shown in Fig. 7.11.

We return to the Maxwell equations (7.3) and (7.5) where we include the displacement current into the total current density $\boldsymbol{j}$ containing the contributions of both positive and negative charge carriers. Then we obtain the wave equation which takes the form for a plane wave (7.140)

$$\varepsilon_0 \left(\omega^2 - c^2 k^2 \right) E = -i\omega n_0 e \left(v_{ix} - v_{ex} \right). \tag{7.141}$$

The electron velocity in general has been calculated above and it is defined by (7.43)–(7.45). For the electric field (7.140), $\omega_c \tau \gg 1$ and low frequencies $\omega \ll \omega_c$ these equations yield:

$$v_{ex} = \frac{i|e|}{m_e^* \omega} \frac{\omega^2}{\omega_c^2} E_x, \quad v_{ey} = -\frac{|e|}{m_e^* \omega_c} E_x = -\frac{E_x}{B_0}, \quad v_{ez} = 0. \tag{7.142}$$

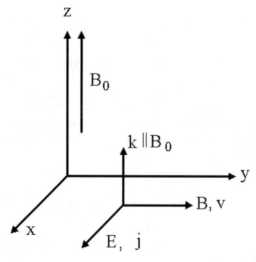

Fig. 7.11. Alfven wave propagating along the uniform magnetic field $\boldsymbol{B}_0$

The component v_{ex} can be neglected as small compared to v_{ey}.

$$\left|\frac{v_{ex}}{v_{ey}}\right| = \frac{\omega}{\omega_c} \ll 1. \tag{7.143}$$

Electrons drift essentially along the y-axis. The ion velocity in the electric field (7.140) and magnetic field $\boldsymbol{B}_0$ can be easily obtained from the equation of motion (7.13) neglecting collisions and spatial inhomogeneity. We then write

$$v_{ix} = \frac{i|e|}{M\omega} E_x \left(1 - \frac{\Omega_c^2}{\omega^2}\right)^{-1}, \quad v_{iy} = \frac{|e|}{M\omega} E_x \frac{\Omega_c}{\omega} \left(1 - \frac{\Omega_c^2}{\omega^2}\right)^{-1}. \tag{7.144}$$

Inserting v_{ix} (7.144) into (7.141) and recalling expression (7.127) for the ion plasma frequency Ω_p we obtain the dispersion relation of the so-called Alfven wave, or magnetohydrodynamic wave [21]

$$\omega^2 - c^2 k^2 = \Omega_p^2 \left(1 - \frac{\Omega_c^2}{\omega^2}\right)^{-1}. \tag{7.145}$$

In the limiting case $\omega \ll \Omega_c$ which is typical for the Alfven waves equation (7.145) reduces to

$$\frac{\omega^2}{k^2} = c^2 \left(1 + \frac{n_0 M}{\varepsilon_0 B_0^2}\right)^{-1}. \tag{7.146}$$

Equation (7.146) determines the phase velocity v_{ph} of the electromagnetic wave propagated in a dielectric with the effective relative permittivity $\varepsilon_{\text{eff}}^r$ which is given by

$$\varepsilon_{\text{eff}}^r = 1 + \frac{n_0 M}{\varepsilon_0 B_0^2}. \tag{7.147}$$

For typical values of the carrier concentration n_0 and the magnetic field B_0 the second term in the right-hand side of (7.147) is large compared to unity or, in the case of a solid state plasma, to the lattice part of permittivity ε_L^r

$$\frac{n_0 M}{\varepsilon_0 B_0^2} \gg 1. \tag{7.148}$$

Then the phase velocity v_A of the Alfven wave, or the Alfven velocity takes the form:

$$\frac{\omega}{k} = v_A = \frac{B_0}{\sqrt{\mu_0 n_0 M}}. \tag{7.149}$$

This is a velocity of the Alfven wave propagation in the direction parallel to the uniform magnetic field $\boldsymbol{B}_0$. The magnetic field of the Alfven wave is given by

$$B_y = \frac{k}{\omega} E_x = \frac{\sqrt{\mu_0 n_0 M}}{B_0} E_x, \qquad (7.150)$$

which results in small deviations of the magnetic field force lines along the y-axis. While a deviation moves along the z-axis with the phase velocity v_A, the force line itself evolves along the y-axis with the velocity $v_A |B_y/B_0|$. On the other hand, plasma as a whole moves along the y-axis with the hydrodynamic velocity $|E_x/B_0|$ since under the condition $\omega \ll \Omega_c$ the ion velocity v_{iy} (7.144) coincide with the electron velocity v_{ey} (7.142). Taking into account relations (7.149) and (7.150) one can see that the velocity of the plasma equals to the velocity of the magnetic field force line. Generally, the Alfven wave can be characterized as the collective oscillations of plasma and magnetic field force lines in such a manner that the force lines are "frozen" into the plasma. Experimental results confirmed the linear dependence of the Alfven wave velocity on the magnetic field and show that in the hydrogen plasma for $B_0 \sim 10$ tesla and $n_0 \sim 6 \times 10^{21}$ m^{-3} the phase velocity of the Alfven wave $v_A \sim 3 \times 10^5$ m/s [21].

The Alfven wave with the velocity v_A (7.149) studied above is called the fast, or ordinary Alfven wave. In the case of the oblique wave propagation with respect to the uniform magnetic field B_0 the second, or slow extraordinary Alfven wave emerges with the dispersion relation [13]

$$\frac{\omega_{\text{slow}}}{k_{\text{slow}}} = v_{\text{slow}} = \frac{B_0 \cos\theta}{\sqrt{\mu_0 n_0 M}}, \qquad (7.151)$$

where θ is the angle between the wave vector $\boldsymbol{k}$ and $\boldsymbol{B}_0$ in the xz plane. It can be shown that the slow wave is linearly polarized with the electric field lying in the xz plane and perpendicular to $\boldsymbol{B}_0$ since the component E_z is negligibly small: $E_z \sim E_x (\omega/\Omega_c)^2$.

In the case of the solid state plasma the Alfven waves be observed in semimetals with a sufficiently large concentration of conduction electrons and holes to provide the low phase velocity. The holes play a role of positive charge carriers instead of ions mentioned above. In this case the ion mass M must be replaced by the sum of electron and hole effective masses $(m_h^* + m_e^*)$. The phase velocity measured in Bi at $B_0 = 1$ tesla appeared to be 10^7 m/s because m_h^* is obviously much less than the ion mass M [13].

7.3.4 Magnetosonic Waves

Consider the low frequency electromagnetic waves propagated in the perpendicular direction with respect to the uniform magnetic field $\boldsymbol{B}_0$. This time the wave vector $\boldsymbol{k}$ is parallel to the y-axis

$$k = k_y. \qquad (7.152)$$

The directions of the wave electric field $\boldsymbol{E}$ and the uniform magnetic field $\boldsymbol{B}_0$ remain the same as in the previous section.

7.3 Waves in Two Component Plasma

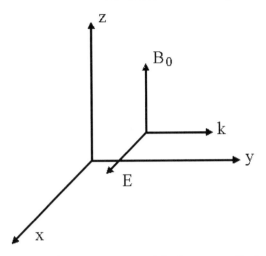

Fig. 7.12. A magnetosonic wave propagated in the perpendicular direction with respect to the uniform magnetic field $\boldsymbol{B}_0$

$$E = E_x = E_{0x} \exp i\,(ky - \omega t). \tag{7.153}$$

The geometry of the problem is presented in Fig. 7.12.

In this geometry the plasma velocity component $\boldsymbol{v} \sim [\boldsymbol{E} \times \boldsymbol{B}_0]$ parallel to the propagation direction would exist. Therefore we must keep the spatially inhomogeneous term ∇P responsible for the longitudinal compression of the plasma. The x and y projections of the equation of motion (7.13) for ions give for the velocity components in the geometry shown in Fig. 7.12

$$v_{ix} = \frac{i\,|e|}{M\omega}\,(E_x + v_{iy}B_0) \tag{7.154}$$

$$v_{iy} = -\frac{i\,|e|}{M\omega}\,v_{ix}B_0 + \frac{k\gamma_i k_B T_i}{M\omega n_0}\,n_1, \tag{7.155}$$

where the constant γ_i is known from thermodynamics:

$$\gamma_i = \frac{C_p}{C_V}. \tag{7.156}$$

It relates a pressure and a density in the equation of a state

$$P = \text{const} \cdot n^\gamma, \quad \nabla P = \frac{\gamma P}{n}\,\nabla n \tag{7.157}$$

and C_P, C_V are the specific heats at the constant pressure and volume respectively. The continuity equation (7.6) yields for ions:

$$n_1 = n_0 \frac{k}{\omega} v_{iy}. \tag{7.158}$$

In order to solve wave equation (7.141) we need to know the component v_{ix}. Eliminating v_{iy} and n_1 we get from (7.154), (7.155) and (7.158):

$$v_{ix}\left[1 - \frac{\Omega_c^2}{\omega^2}\left(1 - \frac{k^2\gamma_i k_B T_i}{\omega^2 M}\right)^{-1}\right] = \frac{i\,|e|}{M\omega} E_x. \tag{7.159}$$

A similar procedure permits the calculation of the electron velocity v_{ex}, that in the limiting case $\omega^2 \ll \omega_c^2$, $\omega^2 \ll k^2 v_{\text{therm},e}^2$ takes the form

$$v_{ex} \simeq -\frac{ik^2}{\omega B_0^2} \frac{\gamma_e k_B T_e}{|e|} E_x. \tag{7.160}$$

Combining (7.141), (7.159) and (7.160), and assuming that $\omega^2 \ll \Omega_c^2$ we obtain after some algebra

$$\omega^2\left(1 + \frac{c^2}{v_A^2}\right) = c^2 k^2 \left(1 + \frac{\gamma_e k_B T_e + \gamma_i k_B T_i}{M v_A^2}\right). \tag{7.161}$$

The numerator of the second term in the right-hand side of (7.161) can be easily expressed in terms of the ion acoustic wave velocity v_s (7.122) where we used $\gamma_e = 1$ and $\gamma_i = 3$. Therefore we can write

$$\gamma_e k_B T_e + \gamma_i k_B T_i = M v_s^2 \tag{7.162}$$

and

$$\frac{\omega^2}{k^2} = c^2 \frac{v_s^2 + v_A^2}{c^2 + v_A^2}. \tag{7.163}$$

Expression (7.163) is the dispersion relation of the magnetosonic wave propagated perpendicular to the uniform magnetic field $\boldsymbol{B}_0$ [21]. In the limiting case $B_0 \to 0$, $v_A \to 0$ the magnetosonic wave reduces to an ion acoustic wave. If $k_B T \to 0$, $v_s \to 0$, then the forces determined by the pressure become negligibly small, and the wave transforms into a modified Alfven wave.

7.4 Excitation of Transverse Waves in Indium Antimonide

In this section we consider the Cerenkov excitation of helicon waves in indium antimonide located in a strong magnetic field [24]. The radiation of such waves in a frequency range $(10^9 \div 10^{11})$ Hz from samples of indium antimonide placed in a strong electric field $\boldsymbol{E}$ and, generally speaking, a magnetic field $\boldsymbol{B}$ had been observed experimentally as early as in late sixties (see, for instance, [25]). This phenomenon represents an instructive and practically important example of the application of the theory of instabilities to the description of excitation of electromagnetic waves in a two component plasma

7.4 Excitation of Transverse Waves in Indium Antimonide

in a semiconductor. This radiation occurs for a magnetic field $\boldsymbol{B} = 0$ or for $\boldsymbol{B} \parallel \boldsymbol{E}$ or for $\boldsymbol{B} \perp \boldsymbol{E}$. However, if the magnetic field is so strong that $\mu_- B > 1$ where $\mu_\pm$ are the mobilities of holes and electrons, respectively, then, first, a lower constant current density $\boldsymbol{j}_0$ is necessary for excitation of transverse waves than at $\boldsymbol{B} = 0$, and, second, the amplitude of the oscillations is much greater. We limit ourselves to the case $\boldsymbol{B} \parallel \boldsymbol{j}_0$ and to values of magnetic field

$$\frac{1}{\mu_+} > B > \frac{1}{\sqrt{\mu_+ \mu_-}}.$$

We shall consider one of the possible reasons for excitation of transverse waves. In the presence of a magnetic field, of the "magnetizing" electron and the "nonmagnetizing" hole, weakly attenuated helicon waves exist in the crystal with frequency

$$\omega = c^2 k^2 \frac{\varepsilon_0 B}{en} = \frac{k^2 B}{\mu_0 en},$$

where e is the charge of the electron, n is the concentration of electrons that under circumstances is assumed to be equal to the one of holes, $\boldsymbol{k}$ is the wave vector, we take into account that $c^2 = (\varepsilon_0 \mu_0)^{-1}$. If the drift velocity of the carriers v_{dr} becomes greater than the velocity of helicon waves $kB/\mu_0 en$, then "Cerenkov" radiation of helicon waves takes place. This radiation is analogous to the well-known emission of sound when the drift velocity of carriers is greater than the sound velocity. The condition

$$v_{\text{dr}} > k \frac{B}{\mu_0 en},$$

or

$$j_0 > k \frac{B}{\mu_0}$$

and the expression for the frequency differs from the condition obtained below, first by the fact that in place of a wave vector $\boldsymbol{k}$ with all three components there is only the single component of $\boldsymbol{k}$ along the cylindrical axis in the obtained expression and, second, by a numerical factor less than unity. These differences are connected with the boundary conditions on the lateral surfaces of the cylinder and with the relations between the kinetic coefficients. Here the magnetic field of the current (the "self-field") B_s is less than the external field, by a factor greater than R/L where R is the radius of the cylindrical sample and L is its length. Therefore, its effect can be neglected.

The set of equations describing the problem inside the sample has the form

$$\frac{\partial \boldsymbol{B}}{\partial t} = -\text{curl}\boldsymbol{E} \qquad (7.164)$$

$$\mathrm{curl}\boldsymbol{H} = \boldsymbol{j} \tag{7.165}$$

$$\boldsymbol{E} = \eta\boldsymbol{j} + \eta_1\left[\boldsymbol{j} \times \boldsymbol{H}\right] + \eta_2\boldsymbol{H}\left(\boldsymbol{j}\cdot\boldsymbol{H}\right). \tag{7.166}$$

The kinetic coefficients η, η_1, η_2 are given for example in [26]. It can be shown that they are determined only by the hole mobility. The holes do not heat up at not too high a concentration $n < 3\cdot 10^{15}\mathrm{cm}^{-3}$. Therefore, we can disregard the dependence of the kinetic coefficients on the field. Also, we shall not take into account the inhomogeneity of the concentration along the radius of the cylinder, due to the pinch effect, since the calculation shows that this inhomogeneity is negligible. Finally, neglect of the displacement current and of the temporal dispersion is valid up to frequencies of 10^{12}Hz which is much higher than the frequencies obtained below. Outside the sample, (7.165) changes to

$$\mathrm{curl}\boldsymbol{H}' = \frac{\partial \boldsymbol{D}'}{\partial t} \tag{7.167}$$

and on the surface of the cylinder $r = R$ it is necessary that all three components of the magnetic field be continuous. The fields should vanish as the distance from the axis of the cylinder $r \to \infty$.

We linearize (7.164)–(7.166) and introduce a cylindrical set of coordinates (r, φ, z) and set

$$\boldsymbol{E}', \boldsymbol{H}', \boldsymbol{j}' \sim f(r)\exp i\left(kz + m\varphi - \omega t\right). \tag{7.168}$$

We shall see below that $B_s = \mu_0 j_0 r/2 \ll B_0$. Therefore, we shall neglect B_s in (7.166). Furthermore, we shall neglect the constant current $j_{0\varphi}$ originating in the sample, since $j_{0\varphi}/j_{0z} \sim B_s/B_0 \ll 1$. We note that for the linearization, one must take into account the dependence of η on H^2: $\eta \approx -\eta_2 H^2$.

The detailed calculations can be found in [24]. We omit the complicated explicit expressions. The linearized system of (7.164)–(7.168) has the solution expressed in terms of the first kind Bessel functions $J_m(x)$ which is finite as $r \to 0$,

$$H'_z = C_1 J_m\left(r\sqrt{\kappa_1^2 - k^2}\right) + C_2 J_m\left(r\sqrt{\kappa_2^2 - k^2}\right), \tag{7.169}$$

where $C_{1,2}$ are the constants and $\kappa_{1,2}$ are the roots of the dispersion relation obtained from the compatibility condition for (7.164)–(7.168). The components H'_r and H'_φ can be determined by using H'_z taking into account that $\mathrm{div}\boldsymbol{H}' = 0$. We do not present here their complicated explicit expressions.

Outside the sample, the solution of the system leads to

$$H'_z \sim C_3 N_m\left(r\sqrt{\frac{\omega^2}{c^2} - k^2}\right) + C_4 J_m\left(r\sqrt{\frac{\omega^2}{c^2} - k^2}\right),$$

7.4 Excitation of Transverse Waves in Indium Antimonide

where $N_m(x)$ is the Neumann function. As we shall see below, $\omega \ll c/k$, and therefore

$$J_m\left(r\sqrt{\frac{\omega^2}{c^2} - k^2}\right) \to \infty, \ r \to \infty,$$

which leads to $C_4 = 0$. Using $\text{div}\,\boldsymbol{H}' = 0$, we find, finally

$$H'_r = C_3 \frac{\partial}{\partial r} N_m\left(r\sqrt{\frac{\omega^2}{c^2} - k^2}\right), \ H'_\varphi = iC_3 \frac{m}{r} N_m\left(r\sqrt{\frac{\omega^2}{c^2} - k^2}\right) \quad (7.170)$$

$$H'_z = -C_3 \frac{i}{k}\left(\frac{\omega^2}{c^2} - k^2\right) N_m\left(r\sqrt{\frac{\omega^2}{c^2} - k^2}\right). \quad (7.171)$$

The boundary conditions for $r = R$ lead to a homogeneous system for the determination of $C_{1,2,3}$ with the coefficients determined by (7.169)–(7.171). The condition of the non-zero solution of this system gives the dispersion equation. It has the form of a sum of products of three Bessel and Neumann functions of the arguments

$$\left(R\sqrt{\kappa_1^2 - k^2}\right), \ \left(R\sqrt{\kappa_2^2 - k^2}\right), \ \left(R\sqrt{\frac{\omega^2}{c^2} - k^2}\right)$$

and we shall not write it down, because of its complexity. The exact solution is possible only numerically; we shall solve it approximately. We write, as usual, $k = \pi p/L$, where $p = 1, 2, \dots$ Then we find $kR = \pi Rp/L < 1$ for not too large p. In practice, $L/R \approx 10 \div 30$, so that $p < 3 \div 10$. Further, we take it into account that one of the roots $\kappa_{1,2}$ of dispersion equation is much larger than the other. Let $\kappa_1^2 \ll \kappa_2^2$, then $\kappa_2 R \gg 1$. Using the approximate expressions of the Bessel and Neumann functions for small and large arguments, we obtain, for $m \neq 0$,

$$J_m\left(R\sqrt{\kappa_1^2 - k^2}\right) = \frac{1}{2}\mathcal{B}(kR), \quad (7.172)$$

here $\mathcal{B}$ is a number much less than unity. We shall be interested in the smallest root of (7.172), since it is easy to see that even for $R\sqrt{\kappa_1^2 - k^2} \approx 3.8$ i.e., close to the first non-zero root of J_1, the critical current is such that $\mu_0 j_0 R/2\mathcal{B} \gg 1$ and our consideration is invalid. If $R\sqrt{\kappa_1^2 - k^2} \ll 1$, then, expanding the Bessel function in a series and substituting κ_1^2, we find the condition at which the frequency is real

$$j_0 = j_{0\text{tr}} = k\frac{\mathcal{B}}{\mu_0}\left(1 + \frac{3 + 2\mathcal{B}}{\mathcal{B}}\frac{\eta_0}{\eta}\right)^{-1}\sqrt{1 + \mathcal{B}} \quad (7.173)$$

and the frequency itself is:

$$\omega = k^2 \frac{H}{en} \frac{\sqrt{1+B}\,(3+2B)}{(3+2B+\eta/\eta_0)}. \tag{7.174}$$

As it is seen from (7.173), $\mu_0 j_0 R/2B \ll 1$, i.e., our neglect of B_s and $j_{0\varphi}$ is justified. It is also seen that the expression for the frequency is close to the corresponding one for helicon waves in the unbounded specimen. It is also seen from (7.173) that carrier velocity at which $\mathrm{Im}\,\omega > 0$, should be greater than the velocity of the helicon wave, in agreement with the qualitative picture mentioned in the beginning of the section.

We estimate the critical current and frequency in indium antimonide for the concentration $n \approx 10^{20}\mathrm{m}^{-3}$, the electron mobility $\mu_- \approx 30\mathrm{m}^2/\mathrm{V\,s}$, $\mu_-/\mu_+ \approx 36$, $L = 0.01\mathrm{m}$ and magnetic field $B_0 \approx 0.3\mathrm{tesla}$. Here $j_{0cr} \approx 3 \cdot 10^7 \mathrm{a/m}^2$, $\omega \approx 2 \cdot 10^9 \mathrm{Hz}$, which is close to the experimental data. Here $\omega/ck < 1$.

The calculation set forth makes it possible to draw a number of conclusions. If

$$\frac{1}{\sqrt{\mu_-\mu_+}} > B > \frac{1}{\mu_-},$$

then the kinetic coefficients depend not only on the hole mobility but also on the electron mobility. In this case, it is necessary to take into account the heating of the electrons. However, if

$$\left(1+\partial \ln \eta/\partial \ln E^2\right), \left(1+\partial \ln \eta_2/\partial \ln E^2\right)$$

are close to unity, which is the case in indium antimonide, the result of the calculation is almost unchanged; therefore, the expressions (7.173) and (7.174) are true in the interval of values of the magnetic field

$$\frac{1}{\mu_+} > B > \frac{1}{\mu_-}.$$

If the magnetic field is so strong that

$$B > \frac{1}{\mu_+},$$

i.e., the holes are magnetized, then at equal carrier concentrations there are no helicon waves and Alfven waves are generated, the velocity of which is much higher. Therefore, the current necessary for their radiation is much greater. Similarly, if the carrier concentrations are strongly different, then, as calculation shows, the critical current increases in the ratio $(\eta/\eta_2 H^2)$, which is greater than unity. Therefore excitation should be expected in certain materials with high ratios of mobilities. Thus, in addition to indium antimonide, there is bismuth, where for $L = 0.01\mathrm{m}$, $B = 0.01\mathrm{tesla}$, and $T \approx 4°\mathrm{K}$, $j_{0cr} \approx 10^6 \mathrm{a/m}^2$, $\omega \approx 3 \cdot 10^4 \mathrm{Hz}$.

7.5 Kinetic Theory

7.5.1 The Boltzmann–Vlasov Equation

There exist phenomena in plasma that cannot be described adequately by virtue of the hydrodynamic model. In such a case we must use a distribution function $f(\mathbf{r}, \mathbf{v}, t)$ of time t, generalized coordinates $\mathbf{r}$ and velocities $\mathbf{v}$ for each type of a particle. The distribution function $f(\mathbf{r}, \mathbf{v}, t)$ measures the number of particles dn in a unit volume at the point $\mathbf{r}$ and at the moment of time t that possess the velocity components in the range [21]

$$(v_x, \ v_x + dv_x), \ (v_y, \ v_y + dv_y), \ (v_z, \ v_z + dv_z)$$

$$dn = f(\mathbf{r}, \mathbf{v}, t) \, dv_x dv_y dv_z. \tag{7.175}$$

Then the total number of particles of a given type is

$$n(\mathbf{r}, t) = \int_{-\infty}^{\infty} f(\mathbf{r}, \mathbf{v}, t) \, d^3 v. \tag{7.176}$$

The dimensionality of the distribution function $f(\mathbf{r}, \mathbf{v}, t)$ normalized according to (7.176) is $\mathrm{m}^{-6}\,\mathrm{s}^3$. The distribution function obeys the fundamental equation which is called the Boltzmann transport equation. It describes the evolution of the distribution function in time t and in the space of coordinates $\mathbf{r}$ and momenta $\mathbf{p} = m\mathbf{v}$, where m is a particle mass. This is a kinetic approach. The Boltzmann equation claims that the total rate of change of the distribution function df/dt in the phase space equals to its rate of change by virtue of collisions $C(f)$. The Boltzmann equation in general case has the form

$$\frac{df}{dt} = C(f). \tag{7.177}$$

The total derivative df/dt is expressed in terms of partial derivatives with respect to time t and the variables of the 6-dimensional phase space of the generalized coordinates x_i and generalized momenta p_i. The collision term $C(f)$, in particular, may be written as $(df/dt)_{\text{col}}$. Then the Boltzmann equation becomes [12]

$$\frac{\partial f}{\partial t} + \mathbf{v} \cdot \nabla f + \frac{\partial \mathbf{p}}{\partial t} \cdot \nabla_{\mathbf{p}} f = \left(\frac{\partial f}{\partial t}\right)_{\text{col}}, \tag{7.178}$$

where $\nabla_{\mathbf{p}}$ means the gradient operation in the space of momenta, and the right-hand part of (7.178) represents a change of the distribution function due to collisions with particles of any kind. The temporal derivative of the momentum $\mathbf{p}$ is the force $\mathbf{F}$ acting on a particle as we have seen above. In a plasma placed in a magnetic field, or magnetoactive plasma the force $\boldsymbol{F}$ is the Lorentz force.

$$\frac{\partial \boldsymbol{p}}{\partial t} = \boldsymbol{F} = e\left(\boldsymbol{E} + [\boldsymbol{v} \times \boldsymbol{B}]\right). \tag{7.179}$$

Substituting (7.179) into (7.178) we obtain the Boltzmann–Vlasov equation [12, 13, 21]

$$\frac{\partial f}{\partial t} + \boldsymbol{v} \cdot \nabla f + e\left(\boldsymbol{E} + [\boldsymbol{v} \times \boldsymbol{B}]\right) \cdot \nabla_{\boldsymbol{p}} f = \left(\frac{\partial f}{\partial t}\right)_{\text{col}}. \tag{7.180}$$

The equations of the hydrodynamic model of plasma studied above can be easily derived from (7.180) [13, 21]. In the hydrodynamic model we operate with particle velocities. For this reason, in the further derivations we replace a momenta space with a velocity space. First, we integrate both sides of (7.180) over the velocity space taking into account that

$$\nabla_{\boldsymbol{p}} f = \frac{1}{m} \nabla_{\boldsymbol{v}} f. \tag{7.181}$$

We write

$$\int \frac{\partial f}{\partial t} d^3 v + \int \boldsymbol{v} \cdot \nabla f d^3 v + \frac{e}{m} \int \left(\boldsymbol{E} + [\boldsymbol{v} \times \boldsymbol{B}]\right) \cdot \nabla_{\boldsymbol{v}} f d^3 v$$

$$= \int \left(\frac{\partial f}{\partial t}\right)_{\text{col}} d^3 v. \tag{7.182}$$

The first term in the left-hand side of (7.182) is simply the temporal derivative of the number density, or concentration $n(\boldsymbol{r}, t)$ of particles according to (7.176)

$$\int \frac{\partial f}{\partial t} d^3 v = \frac{\partial n}{\partial t}. \tag{7.183}$$

The second term contains the spatial derivatives which can be interchanged with the integration over the velocity space:

$$\int \boldsymbol{v} \cdot \nabla f d^3 v = \nabla \cdot \int f \boldsymbol{v} d^3 v. \tag{7.184}$$

The hydrodynamic velocity of a liquid $\boldsymbol{v}_H$ is by definition the average velocity $\overline{\boldsymbol{v}}$ of a particle which is given by

$$\boldsymbol{v}_H = \overline{\boldsymbol{v}} = \frac{1}{n} \int f \boldsymbol{v} d^3 v \tag{7.185}$$

and finally

$$\int \boldsymbol{v} \cdot \nabla f d^3v = \nabla \cdot (n\boldsymbol{v}_H). \tag{7.186}$$

The term with the electric field $\boldsymbol{E}$ can be transformed into a surface integral and thus vanishes as the integration surface tends to infinity.

$$\int \boldsymbol{E} \cdot \nabla_v f d^3v = \int \frac{\partial}{\partial \boldsymbol{v}} (\boldsymbol{E}f) d^3v = \int_{S \to \infty} \boldsymbol{E}f \cdot d\boldsymbol{s} = 0. \tag{7.187}$$

The term with the magnetic field $\boldsymbol{B}$ can be rewritten as follows

$$\int [\boldsymbol{v} \times \boldsymbol{B}] \cdot \nabla_v f d^3v = \int \frac{\partial}{\partial \boldsymbol{v}} \cdot [f\boldsymbol{v} \times \boldsymbol{B}] d^3v - \int f \frac{\partial}{\partial \boldsymbol{v}} \cdot [\boldsymbol{v} \times \boldsymbol{B}] d^3v = 0. \tag{7.188}$$

The first integral in the right-hand side of (7.188) transforms into the surface one similarly to (7.187) and vanishes since the Maxwellian distribution function decreases faster than v^a, $a > 0$ as $v \to \infty$. The second integral is zero because the vector $[\boldsymbol{v} \times \boldsymbol{B}]$ is perpendicular to $\partial/\partial\boldsymbol{v}$. The collision term in the right-hand side of (7.182) equals zero due the conservation of a total number of particles. Combining the results (7.183) and (7.186)–(7.188) we obtain the continuity equation (7.6) for any type of a particle

$$\frac{\partial n}{\partial t} + \text{div}\,(n\boldsymbol{v}_H) = 0. \tag{7.189}$$

This equation expresses the conservation of total number of particles.

In order to derive from the Boltzmann-Vlasov equation the equation of motion (7.13) we multiply (7.180) by $m\boldsymbol{v}$ and then integrate both sides over the velocity space. Equation (7.180) takes the form

$$m \int \boldsymbol{v}\frac{\partial f}{\partial t} d^3v + \int m\boldsymbol{v}\,(\boldsymbol{v}\cdot\nabla)f d^3v + e\int \boldsymbol{v}\,(\boldsymbol{E} + [\boldsymbol{v}\times\boldsymbol{B}])\cdot\nabla_v f d^3v$$

$$= m \int \boldsymbol{v}\left(\frac{\partial f}{\partial t}\right)_{\text{col}} d^3v. \tag{7.190}$$

Comparing the first term in the left-hand side of (7.190) with (7.186) we write immediately

$$m\int \boldsymbol{v}\frac{\partial f}{\partial t} d^3v = \frac{\partial}{\partial t} m\,(n\boldsymbol{v}_H) = m\left(\boldsymbol{v}_H \frac{\partial n}{\partial t} + n\frac{\partial \boldsymbol{v}_H}{\partial t}\right). \tag{7.191}$$

Consider now the second integral in the left-hand side of (7.190). Taking into account that $\boldsymbol{v}$ is an independent variable we can write

$$\int \boldsymbol{v}\,(\boldsymbol{v}\cdot\nabla)f d^3v = \int \nabla\cdot(f\boldsymbol{v}\boldsymbol{v}) d^3v = \nabla\cdot\int f\boldsymbol{v}\boldsymbol{v} d^3v. \tag{7.192}$$

By definition of an average value we have

$$\nabla \cdot \int f \boldsymbol{vv} d^3 v = \nabla \cdot n\overline{\boldsymbol{vv}}. \tag{7.193}$$

The particle velocity $\boldsymbol{v}$ represents a sum of an average hydrodynamic velocity $\boldsymbol{v}_H$ and a thermal velocity $\boldsymbol{v}_T$

$$\boldsymbol{v} = \boldsymbol{v}_H + \boldsymbol{v}_T. \tag{7.194}$$

Substitute (7.194) into (7.193). Obviously

$$\overline{\boldsymbol{v}}_H = \boldsymbol{v}_H, \quad \overline{\boldsymbol{v}}_T = 0 \tag{7.195}$$

and

$$\nabla \cdot n\overline{\boldsymbol{vv}} = \nabla \cdot (n\boldsymbol{v}_H \boldsymbol{v}_H) + \nabla \cdot (n\overline{\boldsymbol{v}_T \boldsymbol{v}_T}) =$$

$$= \boldsymbol{v}_H \nabla \cdot (n\boldsymbol{v}_H) + n(\boldsymbol{v}_H \cdot \nabla) \boldsymbol{v}_H + \nabla \cdot (n\overline{\boldsymbol{v}_T \boldsymbol{v}_T}). \tag{7.196}$$

It is known from the hydrodynamics that the quantity $mn\overline{\boldsymbol{v}_T \boldsymbol{v}_T}$ is a stress tensor P_{ik} (7.10)

$$P_{ik} = mn\overline{v_{Ti} v_{Tk}}, \tag{7.197}$$

that in the simplest isotropic case reduces to a scalar pressure P.

Consider the integral containing the Lorentz force. It can be rewritten as follows

$$\int \boldsymbol{v} \left(\boldsymbol{E} + [\boldsymbol{v} \times \boldsymbol{B}] \right) \cdot \nabla_v f d^3 v$$

$$= \int \frac{\partial}{\partial \boldsymbol{v}} \cdot \{ f \boldsymbol{v} \left(\boldsymbol{E} + [\boldsymbol{v} \times \boldsymbol{B}] \right) \} d^3 v - \int f \boldsymbol{v} \frac{\partial}{\partial \boldsymbol{v}} \left(\boldsymbol{E} + [\boldsymbol{v} \times \boldsymbol{B}] \right) d^3 v$$

$$- \int f \left(\boldsymbol{E} + [\boldsymbol{v} \times \boldsymbol{B}] \right) \cdot \frac{\partial}{\partial \boldsymbol{v}} \boldsymbol{v} d^3 v. \tag{7.198}$$

The first integral in the right-hand side of (7.198) reduces to the surface integral and vanishes for the same reasons as integral (7.188). The integrand of the second integral is zero because the electric field $\boldsymbol{E}$ is independent of $\boldsymbol{v}$ and the vector $[\boldsymbol{v} \times \boldsymbol{B}]$ is perpendicular to $\partial/\partial \boldsymbol{v}$. Finally, the quantity $\partial \boldsymbol{v}/\partial \boldsymbol{v}$ in the integrand of the third integral is a unit tensor. As a result, we get

$$\int \boldsymbol{v} \left(\boldsymbol{E} + [\boldsymbol{v} \times \boldsymbol{B}] \right) \cdot \nabla_v f d^3 v = - \int f \left(\boldsymbol{E} + [\boldsymbol{v} \times \boldsymbol{B}] \right) d^3 v. \tag{7.199}$$

Once more, one can see that the remaining integral in the right-hand side of (7.199) is by definition a product of a particle number density (concentration) n and an average Lorentz force:

$$\int f\left(\boldsymbol{E}+[\boldsymbol{v}\times\boldsymbol{B}]\right)d^{3}v = n\left(\boldsymbol{E}+[\boldsymbol{v}_{H}\times\boldsymbol{B}]\right). \quad (7.200)$$

The right-hand side of (7.190) represents the momentum change due to collisions of different type and can be approximately expressed in term of the average hydrodynamic velocity $\boldsymbol{v}_H$ and a phenomenological relaxation time τ

$$m\int \boldsymbol{v}\left(\frac{\partial f}{\partial t}\right)_{\text{col}}d^{3}v \cong -mn\frac{\boldsymbol{v}_{H}}{\tau}. \quad (7.201)$$

Combining the results (7.191), (7.196), (7.197), (7.200) and (7.201) we obtain

$$m\boldsymbol{v}_{H}\left[\frac{\partial n}{\partial t}+\nabla\cdot(n\boldsymbol{v}_{H})\right]+mn\frac{\partial \boldsymbol{v}_{H}}{\partial t}+mn\left(\boldsymbol{v}_{H}\cdot\nabla\right)\boldsymbol{v}_{H}+\nabla\cdot P_{ik}$$

$$-en\left(\boldsymbol{E}+[\boldsymbol{v}_{H}\times\boldsymbol{B}]\right) = -mn\frac{\boldsymbol{v}_{H}}{\tau}. \quad (7.202)$$

The sum in square brackets in the left-hand side of (7.202) reduces to zero according to the continuity equation (7.189). Then (7.202) becomes

$$m\left[\frac{\partial \boldsymbol{v}_{H}}{\partial t}+\left(\boldsymbol{v}_{H}\cdot\nabla\right)\boldsymbol{v}_{H}\right] = e\left(\boldsymbol{E}+[\boldsymbol{v}_{H}\times\boldsymbol{B}]\right)-\frac{1}{n}\nabla\cdot P_{ik}-m\frac{\boldsymbol{v}_{H}}{\tau}. \quad (7.203)$$

Equation (7.203) describes a momentum flow. It coincides with the equation of motion of a charged particle (7.13) derived in the framework of the hydrodynamic model.

7.5.2 Landau Damping

To illustrate the application of the kinetic approach we consider the so-called Landau damping effect. L.D. Landau theoretically predicted (1946) that the energy dissipation occurs even in a collisionless plasma [12, 21]. The Landau damping differs fundamentally from dissipation processes in ordinary medium since it is independent of collisions. It does not involve an increase of the entropy and, consequently, represents a thermodynamically reversible process. The mechanism of the Landau damping is connected with a spatial dispersion, i.e. with the dependence of the plasma permittivity ε_{ik}^{r} on the wave vector $\boldsymbol{k}$ of a perturbation. The electrons possessing a velocity v in the direction of the electric wave propagation which is equal to the phase velocity $v_{\text{ph}} = \omega/k$ of the wave are moving in phase with the wave. The electric field of the wave is stationary with respect to these electrons, and therefore they can exchange energy with the wave. The time average value of the energy for the electrons with $v = \omega/k$ is not zero, in contrast to other electrons with respect to which the wave field is frequently oscillates resulting in the zero contribution into the time average energy. Electrons with a velocity slightly below v_{ph} can get the energy from the wave, while electrons having a velocity slightly higher than v_{ph} can return the energy to the wave. In plasma some electrons are moving faster than other. In the case of the Maxwellian distribution the number of slow electrons is greater than the number of fast electrons.. Hence, the number of electrons taking the energy from the wave is larger than the number of electrons contributing it to the wave, the wave loses energy and attenuates.

Let us study the Landau damping quantitatively. Suppose that in an equilibrium state plasma is spatially homogeneous and possesses the distribution function $f_0(\boldsymbol{v})$. External magnetic and electric fields are assumed to be absent. Collisions are neglected in such a way that

$$\left(\frac{\partial f}{\partial t}\right)_{\text{col}} = 0. \tag{7.204}$$

The distribution function $f(\boldsymbol{r}, \boldsymbol{v}, t)$ of the disturbed plasma is sought to be

$$f(\boldsymbol{r}, \boldsymbol{v}, t) = f_0(\boldsymbol{v}) + \delta f(\boldsymbol{r}, \boldsymbol{v}, t). \tag{7.205}$$

We limit our analysis with the electron motion assuming that the heavy ions are fixed. Then the Boltzmann–Vlasov equation (7.180) for the electron distribution function (7.205) takes the form

$$\frac{\partial \delta f}{\partial t} + \boldsymbol{v} \cdot \nabla \delta f - \frac{|e|}{m_e^*} \boldsymbol{E} \cdot \frac{\partial f_0}{\partial \boldsymbol{v}} = 0. \tag{7.206}$$

In the plane wave representation for a one-dimensional case

$$\delta f, E_x \sim \exp i(kx - \omega t). \tag{7.207}$$

Substituting (7.207) into (7.206) we obtain

$$\delta f = i \frac{|e|}{m_e^*} E_x \frac{1}{\omega - kv_x} \frac{\partial f_0}{\partial v_x}. \tag{7.208}$$

We must use additionally the Poisson equation (7.4) relating the electric field in plasma to the charge density perturbation ρ which cannot be described anymore with (7.14). Instead it must be expressed in terms of the distribution function for the given type of a particle:

$$\rho = e \int f d^3 v. \tag{7.209}$$

7.5 Kinetic Theory

In an unperturbed plasma, the electron charge density is balanced at every point by ion charges (or, for instance, by holes in semiconductors). Consequently

$$\int f_0 d^3v = 0 \qquad (7.210)$$

and in the disturbed plasma the charge density is determined by the deviation δf of the distribution function f from its equilibrium value f_0. For electrons we have

$$\rho = -|e| \int \delta f d^3 v. \qquad (7.211)$$

It is convenient to introduce the new distribution function f_{0N} normalized to unity using relation (7.176)

$$f(\mathbf{r}, \mathbf{v}, t) = n(\mathbf{r}, t) f_{0N}(\mathbf{r}, \mathbf{v}, t); \quad \int f_{0N}(\mathbf{r}, \mathbf{v}, t) d^3 v = 1. \qquad (7.212)$$

Then substituting (7.53), (7.208), (7.211), and (7.212) into (7.4) we obtain

$$i\varepsilon_0 k E_x = -i \frac{e^2 n_0}{m_e^*} E_x \int \frac{1}{\omega - k v_x} \frac{\partial f_{0N}}{\partial v_x} d^3 v. \qquad (7.213)$$

In the integrand only f_{0N} depends on the three velocity components. Consequently, we can introduce the one-dimensional distribution function with respect to v_x only

$$f_{0N}(v_x) = \int f_{0N}(\mathbf{v}) dv_y dv_z \qquad (7.214)$$

and finally we get the dispersion equation for plasma waves in the implicit form:

$$1 = \frac{\omega_p^2}{k^2} \int_{-\infty}^{\infty} dv_x \frac{1}{v_x - \omega/k} \frac{\partial f_{0N}(v_x)}{\partial v_x}. \qquad (7.215)$$

We should emphasize that relationship (7.215) is valid for any equilibrium distribution function including the Maxwellian one. The integrand in (7.215) has a singularity, a pole at the point

$$v_x = \frac{\omega}{k} = v_{\text{ph}} \qquad (7.216)$$

It is this singularity, that results in the essential modification of plasma wave dispersion relation. The integral in (7.215) is evaluated by virtue of the theory of complex variable functions. The path of integration should be chosen below the real axis in the plane of complex variable v_x since the imaginary part of

280   7 Waves in Plasma

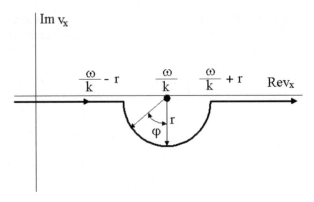

Fig. 7.13. Integration contour in a plane of complex variable v_x for the evaluation of the Landau damping

the frequency ω should be negative: $\mathrm{Im}\,\omega < 0$. It should be noted that the permittivity in general may possess singularities only in the lower half-plane of complex variable as mentioned above.

The approximate solution of the dispersion equation (7.215) in a closed form can be obtained for large phase velocities and weak damping when an absolute value of $\mathrm{Im}\,\omega$ is assumed to be small. The integration contour for such a case is shown in Fig. 7.13.

The integral in the right-hand side of (7.215) can be decomposed into two parts: the term I resulting from the integration along the real axis and the term R resulting from the integration around the pole $v_x = \omega/k$.

$$\int_{-\infty}^{\infty} dv_x \frac{1}{v_x - \omega/k} \frac{\partial f_{0N}(v_x)}{\partial v_x} = I + R, \tag{7.217}$$

where

$$I = P \int_{-\infty}^{\infty} dv_x \frac{1}{v_x - \omega/k} \frac{\partial f_{0N}(v_x)}{\partial v_x} \tag{7.218}$$

and

$$R = \left[\frac{\partial f_{0N}(v_x)}{\partial v_x}\right]_{v_x = \omega/k} \int_0^{\pi} \frac{ir \exp i\varphi}{r \exp i\varphi} d\varphi = i\pi \left[\frac{\partial f_{0N}(v_x)}{\partial v_x}\right]_{v_x = \omega/k}. \tag{7.219}$$

Here P stands for the principal value according to Cauchy.

Evaluation of I (7.218) shows that for large phase velocities the contribution due to $v_x > \omega/k$ can be neglected since both the equilibrium distribution function and its derivative are negligibly small at this part of the integration contour. Then taking the integral by parts we find

7.5 Kinetic Theory

$$I = \left[\frac{f_{0N}(v_x)}{v_x - \omega/k}\right]_{-\infty}^{\infty} - \int_{-\infty}^{\infty}\frac{-f_{0N}(v_x)}{(v_x - \omega/k)^2} = \int_{-\infty}^{\infty}\frac{f_{0N}(v_x)}{(v_x - \omega/k)^2}. \quad (7.220)$$

Expression (7.220) is the average value of $(v_x - \omega/k)^{-2}$. Expanding the latter quantity in powers of kv_x/ω and taking into account that the terms with odd powers vanish as a result of the averaging we get

$$I \simeq \frac{k^2}{\omega^2}\left(1 + 3\frac{k^2}{\omega^2}\overline{v_x^2}\right). \quad (7.221)$$

Substituting (7.219) and (7.221) into dispersion relation (7.215) we obtain

$$\omega^2 = \omega_p^2\left(1 + 3\frac{k^2}{\omega^2}\overline{v_x^2}\right) + \frac{\omega_p^2\omega^2}{k^2}i\pi\left[\frac{\partial f_{0N}(v_x)}{\partial v_x}\right]_{v_x=\omega/k}. \quad (7.222)$$

For the small thermal contribution we may replace ω^2 with ω_p^2 in the right-hand side of (7.222). For the Maxwellian distribution function

$$f_{0N} = \left(\frac{m}{2\pi k_B T}\right)^{3/2}\exp\left(-\frac{v^2}{v_{therm}^2}\right) \quad (7.223)$$

we find that

$$\overline{v_x^2} = \frac{k_B T_e}{m_e^*}. \quad (7.224)$$

Now it is easy to see that the real part of dispersion relation (7.222) coincides with the dispersion relation of the plasma wave (7.62) obtained in the framework of the hydrodynamic model.

The imaginary part of the right-hand side (7.222) is small compared to ω_p^2. Then we can write omitting the small real term describing the thermal contribution

$$\omega \simeq \omega_p\left\{1 + i\frac{\pi}{2}\frac{\omega_p^2}{k^2}\left[\frac{\partial f_{0N}(v_x)}{\partial v_x}\right]_{v_x=\omega/k}\right\}. \quad (7.225)$$

It is seen from (7.225) that the imaginary part of the frequency Imω may be positive or negative depending on the sign of $\partial f_{0N}(v_x)/\partial v_x$ at $v_x = \omega/k$. In general, the distribution functions with the positive derivative in some interval of velocities can exist. Such a function must have a local minimum and two maxima. Then Im$\omega > 0$ and an instability occurs. The particles are "trapped" in the potential well of the wave. However, the trapping process is essentially non-linear. It corresponds to the so-called non-linear Landau damping, and its complete analysis is out of the limits of this work. An example of a distribution function with $\partial f_{0N}(v_x)/\partial v_x > 0$ is presented in Fig. 7.14.

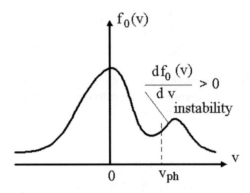

Fig. 7.14. The distribution function $f_0(v)$ with a minimum and a region where $\partial f_0(v)/\partial v > 0$. In this region an instability can occur

For the Maxwellian distribution function (7.223) the damping constant equal to the imaginary part of (7.225) is given by:

$$\mathrm{Im}\,\omega \simeq -\omega_p\sqrt{\pi}\exp\left(-\frac{3}{2}\right)\left(\frac{\omega_p}{kv_{\text{therm}}}\right)^3 \exp\left(-\frac{\omega_p^2}{k^2 v_{\text{therm}}^2}\right). \tag{7.226}$$

The damping constant (7.226) is negative which corresponds to the attenuation of plasma wave. Hence, the remarkable fact is established that the damping exists in a collisionless plasma.

It should be noted that in the limiting case $\omega/kv_{\text{therm}} \gg 1$, or in other words, recalling that the Debye length $L_D \sim v_{\text{therm}}/\omega_p$, for small kL_D the damping constant is exponentially small since in the case of the Maxwellian distribution only an exponentially small fraction of the electrons have velocities $v_x \gg v_{\text{therm}}$. However, for $kL_D \sim 1$ it can become considerable due to the perturbation of the distribution function δf caused by the wave.

7.5.3 Permittivity Tensor in a Non-local Case

In the kinetic theory the plasma behavior is described by the self-consistent system of the Maxwell equations (7.2)–(7.5) and the Boltzmann–Vlasov equation (7.180). The current density similarly to the charge density (7.209) is determined by the distribution function [12]

$$\boldsymbol{j} = e\int \boldsymbol{v} f d^3 v. \tag{7.227}$$

In an unperturbed plasma, the current density (7.227) is zero due to the isotropy of plasma. Hence, in the plasma perturbed by electric and magnetic fields the current density is expressed in terms of the deviation δf of the distribution function f from its equilibrium value f_0

$$\delta f = f - f_0. \tag{7.228}$$

Then (7.227) takes the form for electrons [22]

$$\boldsymbol{j} = -|e| \int \boldsymbol{v} \delta f d^3 v. \tag{7.229}$$

The collision term can be approximated by virtue of a relaxation time τ

$$\left(\frac{\partial f}{\partial t}\right)_{col} = \frac{f_0 - f}{\tau} = -\frac{\delta f}{\tau}. \tag{7.230}$$

In order to evaluate the conductivity tensor of the magnetoactive plasma we must therefore solve the Boltzmann–Vlasov equation with respect to δf and then evaluate the integral (7.229). Substituting (7.228) in (7.180) we get

$$\frac{\partial \delta f}{\partial t} + \boldsymbol{v} \cdot \nabla \delta f + \frac{\delta f}{\tau} - |e|\, [\boldsymbol{v} \times \boldsymbol{B}] \cdot \nabla_p \delta f = |e|\, \boldsymbol{E} \cdot \nabla_p f_0 + |e|\, [\boldsymbol{v} \times \boldsymbol{B}] \cdot \nabla_p f_0. \tag{7.231}$$

The second term in the right-hand side of (7.231) vanishes in the case where f_0 is the equilibrium distribution function depending only on a particle energy $\mathcal{E}_{0p}$. Therefore its gradient in the momenta space is proportional to the velocity

$$\nabla_p f_0 = \frac{\partial f_0}{\partial \mathcal{E}_{0p}} \frac{\partial \mathcal{E}_{0p}}{\partial \boldsymbol{p}} \equiv \boldsymbol{v}\, \frac{\partial f_0}{\partial \mathcal{E}_{0p}}. \tag{7.232}$$

Otherwise, for instance, in the case of a plasma drift this term should be kept. For the equilibrium distribution function f_0 and the plane wave perturbations

$$\delta f, \boldsymbol{E} \sim \exp i\,(\boldsymbol{k} \cdot \boldsymbol{r} - \omega t) \tag{7.233}$$

(7.231) reduces to

$$\left(-i\omega + i\boldsymbol{k} \cdot \boldsymbol{v} + \frac{1}{\tau}\right)\delta f - |e|\, [\boldsymbol{v} \times \boldsymbol{B}] \cdot \nabla_p \delta f = |e|\, \boldsymbol{E} \cdot \nabla_p f_0. \tag{7.234}$$

We assume that wave vector $\boldsymbol{k}$ of the plane wave perturbation belongs to the xz plane and has the angle Δ with the magnetic field $\boldsymbol{B}_0$ which is parallel to the z-axis. It is convenient to define the particle velocity $\boldsymbol{v}$ in the spherical coordinates (v, θ, φ). The geometry of the problem is shown in Fig. 7.15.

Then the direct calculation shows that the magnetic part of the Lorentz force in the left-hand side of (7.234) has the form for an electron with a charge $-|e|$

$$-|e|\, [\boldsymbol{v} \times \boldsymbol{B}] \cdot \nabla_p \delta f = \omega_c \frac{\partial \delta f}{\partial \varphi} \tag{7.235}$$

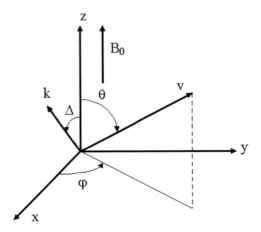

Fig. 7.15. The coordinate system for a perturbation in a magnetoactive plasma. The perturbation is propagated in the $\boldsymbol{k}$ direction. The particle velocity is defined by the spherical coordinates (v, θ, φ)

and (7.234) becomes

$$\omega_c \frac{\partial \delta f}{\partial \varphi} - i\left(\omega + \frac{i}{\tau} - kv\cos\Delta\cos\theta - kv\sin\Delta\sin\theta\cos\varphi\right)\delta f$$

$$= |e|\, \boldsymbol{E}\cdot\boldsymbol{v}\frac{\partial f_0}{\partial \mathcal{E}_{0p}}. \tag{7.236}$$

In order to proceed with the solution of (7.236) we need to define the equilibrium distribution function f_0. For the sake of definiteness we consider electron plasma in metal at the temperature near to zero which represents a degenerate fermion system. In such a case we can write [22]

$$\frac{\partial f_0}{\partial \mathcal{E}_{0p}} = -\delta\left(\mathcal{E}_{0p} - \mathcal{E}_F\right), \tag{7.237}$$

where $\mathcal{E}_F$ is the Fermi energy and the electron kinetic energy is given by

$$\mathcal{E}_{0p} = \frac{p^2}{2m_e^*}. \tag{7.238}$$

Here m_e^* is the electron effective mass. It should be noted that in a classic plasma or in a semiconductor plasma the equilibrium distribution f_0 can be Maxwellian. Then the velocity $\boldsymbol{v}$ is not fixed at the Fermi surface. However equation (7.236) and the relation (7.19) remain to be valid for the Maxwellian distribution function.

The presence of the δ-function in the right-hand side of (7.237) results in the localization of the particle velocity $\boldsymbol{v}$ on the Fermi surface [22]

$$v_F = \frac{p_F}{m_e^*} = \sqrt{\frac{2\mathcal{E}_F}{m_e^*}}. \qquad (7.239)$$

Hence the Boltzmann–Vlasov equation (7.236) with expression (7.237) in the right-hand side describes the motion of the velocity vector v on the Fermi surface. Indeed, this is a first order differential equation with respect to the azimuthal angle φ. In order to solve (7.236) we introduce the auxiliary function Ψ as follows

$$\delta f = -\Psi \frac{\partial f_0}{\partial \mathcal{E}_{0p}}. \qquad (7.240)$$

Inserting (7.240) into (7.236) we obtain the solution [22]

$$\Psi = \frac{|e|}{\omega_c} \int_{-\infty}^{\varphi} \boldsymbol{E} \cdot \boldsymbol{v}' d\varphi' \times \exp i \left[a \left(\varphi - \varphi' \right) - b \left(\sin \varphi - \sin \varphi' \right) \right], \qquad (7.241)$$

where

$$a = \frac{\omega - kv_F \cos \Delta \cos \theta + i/\tau}{\omega_c} \qquad (7.242)$$

and

$$b = \frac{kv_F \sin \Delta \sin \theta}{\omega_c}. \qquad (7.243)$$

We substitute (7.237)–(7.241) into the current density expression (7.229) and can identify the conductivity tensor components $\sigma_{ik}(\boldsymbol{k}, \omega)$ omitting the exponential factor (7.233). The calculations are very involved and we do not present them. We should only note that the central point of the procedure is the well known expansion of the factor $\exp(ib \sin \varphi)$ in the series of the Bessel functions $J_m(z)$

$$\exp(ib \sin \varphi) = \sum_{m=-\infty}^{m=\infty} J_m(b) \exp im\varphi. \qquad (7.244)$$

As a result of the substitution of (7.244) into (7.241) and (7.229), it appears to be necessary to evaluate the following integrals

$$I_1 = \int_{-\infty}^{\varphi} \exp i \left(b \sin \varphi' - a\varphi' \right) d\varphi' \qquad (7.245)$$

$$I_2 = \int_{-\infty}^{\varphi} \exp i \left(b \sin \varphi' - a\varphi' \right) \times \sin \varphi' d\varphi' \qquad (7.246)$$

$$I_3 = \int_{-\infty}^{\varphi} \exp i \left(b \sin \varphi' - a\varphi'\right) \times \cos \varphi' d\varphi'. \tag{7.247}$$

The integral I_1 (7.245) can be expressed directly in terms of the Bessel functions $J_m(z)$ by virtue of the expansion (7.244)

$$I_1 = \int_{-\infty}^{\varphi} \exp i \left(b \sin \varphi' - a\varphi'\right) d\varphi' = \sum_{m=-\infty}^{m=\infty} J_m(b) \frac{i}{a-m} \exp i \left(m-a\right)\varphi. \tag{7.248}$$

The integral I_2 (7.246) is expressed in terms of the derivatives

$$J'_m(b) = \partial J_m(b)/\partial b$$

of the Bessel functions

$$I_2 = \frac{\partial I_1}{\partial b} = \sum_{m=-\infty}^{m=\infty} \frac{\partial J_m(b)}{\partial b} \frac{1}{a-m} \exp i \left(m-a\right)\varphi. \tag{7.249}$$

Finally, the integral I_3 (7.247) can be reduced to I_1 by calculating by parts:

$$I_3 = \int_{-\infty}^{\varphi} \frac{1}{ib} \exp\left(-ia\varphi'\right) d\left[\exp ib \sin \varphi'\right]$$

$$= \frac{1}{ib} \exp\left(ib \sin \varphi - ia\varphi\right) + \frac{a}{b} I_1. \tag{7.250}$$

Using (7.248)–(7.250) one can carry out the integration and get the final result. The components $\sigma_{ik}(\boldsymbol{k},\omega)$ have the form [22]

$$\sigma_{xx} = iN \sum_{m=0}^{\infty} m^2 \int_0^{\pi} \frac{a J_m^2(b) \sin^3 \theta}{b^2 (a^2 - m^2)} d\theta \tag{7.251}$$

$$\sigma_{yy} = iN \sum_{m=0}^{\infty} \int_0^{\pi} \frac{a \left[J'_m(b)\right]^2 \sin^3 \theta}{(\delta_{m0} + 1)(a^2 - m^2)} d\theta \tag{7.252}$$

$$\sigma_{zz} = iN \sum_{m=0}^{\infty} \int_0^{\pi} \frac{a J_m^2(b) \cos^2 \theta \sin \theta}{(\delta_{m0} + 1)(a^2 - m^2)} d\theta \tag{7.253}$$

$$\sigma_{xy} = -\sigma_{yx} = N \sum_{m=0}^{\infty} \int_0^{\pi} \frac{a^2 J_m(b) J'_m(b) \sin^3 \theta}{b(\delta_{m0}+1)(a^2-m^2)} d\theta \qquad (7.254)$$

$$\sigma_{yz} = -\sigma_{zy} = -N \sum_{m=0}^{\infty} \int_0^{\pi} \frac{a J_m(b) J'_m(b) \sin^2 \theta \cos \theta}{(\delta_{m0}+1)(a^2-m^2)} d\theta \qquad (7.255)$$

$$\sigma_{zx} = \sigma_{xz} = iN \sum_{m=0}^{\infty} m^2 \int_0^{\pi} \frac{J_m^2(b) \sin^2 \theta \cos \theta}{b(a^2-m^2)} d\theta, \qquad (7.256)$$

where

$$N = \frac{3n_0 e^2}{m_e^* \omega_c} \qquad (7.257)$$

and n_0 is an electron concentration.

Consider some important characteristics of the conductivity tensor (7.251)–(7.256) [22].

1. At the finite values of k and $\Delta = \pi/2$

$$a = \frac{\omega}{\omega_c} + i\frac{1}{\omega_c \tau} \qquad (7.258)$$

and each component $\sigma_{ik}(\mathbf{k}, \omega)$ manifests an infinite number of resonances at frequencies $\omega = m\omega_c$, $m = 1, 2, 3, ...$ corresponding to the condition

$$(\text{Re}\, a)^2 = m^2. \qquad (7.259)$$

2. In the long wavelength limit $kv_F \ll 1$ the quantity $b \to 0$ and the quantity b^l suppresses the given singularity $\omega = m'\omega_c$. Obviously, the exponent l increases with the rise of m'.

3. There exist an infinite number of areas containing the singularities of all components $\sigma_{ik}(\mathbf{k}, \omega)$. The boundaries of these areas are determined by the condition

$$\omega - m\omega_c = kv_F \sin \Delta. \qquad (7.260)$$

The boundaries correspond to the emergence of a cyclotron attenuation shifted due to the Doppler effect. In the case of the propagation parallel to the magnetic field $\mathbf{B}_0$ when $\mathbf{k} \parallel \mathbf{B}_0$ and $\Delta = 0$ the areas reduce to the lines, and the resonance absorption occurs only at the points $\omega = m\omega_c$. The attenuation areas are shown in Fig. 7.16.

4. For $\mathbf{k} \parallel \mathbf{B}_0$ the constant $b = 0$ and the components $\sigma_{yz}, \sigma_{zx} = 0$.

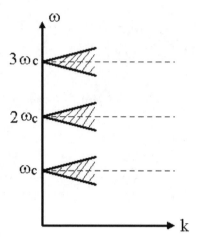

Fig. 7.16. The areas of the cyclotron attenuation. The attenuation occurs for the frequencies inside the dashed areas

The conductivity tensor $\sigma_{ik}(\boldsymbol{k},\omega)$ has been also calculated for the plasma described by the classic statistics. In such a case all integrals have been obtained in a closed form for $\boldsymbol{k} \perp \boldsymbol{B}_0$. for example, the component σ_{xx} has the form

$$\sigma_{xx} = -i\frac{n_0 e^2 (\omega + i/\tau)}{m_e^* \omega_c^2 \lambda_{\rm cl}}$$

$$\times \left\{ [\exp(-\lambda_{\rm cl}) I_0(\lambda_{\rm cl}) - 1] + 2a^2 \sum_{m=1}^{\infty} \frac{\exp(-\lambda_{\rm cl}) I_m(\lambda_{\rm cl})}{a^2 - m^2} \right\} \quad (7.261)$$

where

$$\lambda_{\rm cl} = \frac{k^2 k_B T}{m_e^* \omega_c^2}$$

and $I_m(\lambda_{\rm cl})$ is the Bessel function of the imaginary argument.

7.5.4 Helicon Dispersion in a Non-local Case

We discussed above the helicon waves in the local approximation. Now we will study the peculiarities in the helicon dispersion relation due to the non-locality effects by using the conductivity tensor $\sigma_{ik}(\boldsymbol{k},\omega)$ (7.251)–(7.256) derived in the framework of the kinetic theory [22].

We start with the case of the parallel propagation where $\boldsymbol{k} \parallel \boldsymbol{B}_0$ and $\Delta = 0$. Then, as it was mentioned above $b = 0$, and the components σ_{yz}, σ_{zx} of the conductivity tensor vanish. The series of the Bessel function reduces to the one term independent of θ. Integration over θ can be carried out in a closed form. Then combining relations (7.23) and (7.30) and substituting the values of σ_{xx} and σ_{xy} for $b = 0$ we obtain

$$\varepsilon_+ = \varepsilon_{xx} + i\varepsilon_{xy} \simeq \frac{3}{4} \frac{\omega_p^2}{\omega k v_F} \left[(1 - Z^2) \ln\left(\frac{Z+1}{Z-1}\right) + 2Z \right], \qquad (7.262)$$

where

$$Z = -\frac{\omega - \omega_c + i/\tau}{k v_F}. \qquad (7.263)$$

In the long wavelength limit where $k \to 0$ and therefore $|Z| \to \infty$ the logarithm in (7.262) can be expanded in powers of Z^{-1}. Then the helicon dispersion relation (7.96) becomes

$$\frac{k_+^2}{k_0^2} = -\frac{\omega_p^2}{\omega(\omega - \omega_c + i/\tau)} \left[1 + \frac{1}{5Z^2} + \ldots \right]. \qquad (7.264)$$

The first term of this expansion coincides with the helicon dispersion (7.96) in the local approximation. The corrections occur as $|Z|$ decreases. It is seen from (7.262) that for $\omega_c \tau \to \infty$ and $Z > 1$ the permittivity ε_+ is real. When Z becomes less than unity the value of the logarithm changes by $i\pi$ which corresponds to the cyclotron damping. That means that for

$$\left| \frac{k v_F}{\omega_c - \omega} \right| > 1 \qquad (7.265)$$

a fraction of electrons situated at the Fermi surface moves along the z-axis with such a velocity that the electric field of the helicon wave transforms with respect to them into a stationary one. The edge of the cyclotron damping is given by

$$\frac{k v_F}{\omega_c} = 1 \qquad (7.266)$$

with the wave vector k defined by the general dispersion relation

$$\frac{k^2}{k_0^2} = \varepsilon_+ (\mathbf{k}, \omega). \qquad (7.267)$$

It is known that these results are in good accord with experimental data.

Consider now the propagation of helicon waves under the angle $0 < \Delta < \pi/2$ with respect to the uniform magnetic field $\mathbf{B}_0$. In general case of the oblique propagation wave equation (7.21) results in the dispersion relation

$$Ak^4 - Bk^2 + C = 0, \qquad (7.268)$$

where

$$A = \varepsilon_{zz} \cos^2 \Delta, \quad B = \frac{\omega^2}{c^2}[\varepsilon_{zz}(\varepsilon_{xx} + \varepsilon_{yy} \cos^2 \Delta)$$

$$+ (\varepsilon_{xy} \sin \Delta - \varepsilon_{yz} \cos \Delta)^2], \quad C = \frac{\omega^4}{c^4}\varepsilon_{zz}\varepsilon_{xy}^2. \qquad (7.269)$$

In the case where $\omega_c \tau \gg 1$, and $k v_F/\omega_c = kR \ll 1$ the Bessel function in expressions (7.251)–(7.256) can be expanded and the integration over θ can be carried out. Then, dropping the terms of order $(kR)^2$ and smaller, and substituting the results in (7.23) and (7.269) we find the dispersion relation of the helicon wave with the account of the spatial dispersion [22]

$$k^2 \simeq k_H^2 (1 + i\Gamma), \qquad (7.270)$$

where

$$k_H^2 = \frac{\omega \omega_p^2}{c^2 \omega_c \cos \Delta} \qquad (7.271)$$

and

$$\Gamma = \frac{1}{\omega_c \tau \cos \Delta}\left[1 - \frac{3}{2}\left(1 - \frac{\pi^2}{16}\right)\sin^2 \Delta\right] + \frac{3\pi}{16}kR\sin^2 \Delta. \qquad (7.272)$$

The latter expression has been obtained under the additional condition

$$kR\omega_c \tau \gg 1. \qquad (7.273)$$

The real part k_H^2 of (7.270) is the dispersion relation of a helicon wave in a local limit. For the parallel propagation $\Delta = 0$ it coincides with (7.97). The damping factor Γ consists of two parts. The first term in (7.272) with the coefficient $(\omega_c \tau \cos \Delta)^{-1}$ describes the collision damping. The second term would exist even in a collisionless plasma.

Consider the mechanism of the collisionless helicon damping. In the case of the oblique propagation the helicon wave is not purely transverse, which results in two specific mechanisms of damping. The Landau damping discussed above is due to the small longitudinal component of the wave electric field. However, it has been shown that for helicons this type of damping is negligly small since the helicon phase velocity is much less than the Fermi velocity of carriers. The essential damping of obliquely propagating helicons connected with a considerable magnetic field of the wave. The variable magnetic field $\boldsymbol{B}_H$ being added to the static magnetic field $\boldsymbol{B}_0$ forms a periodic system of so-called magnetic mirrors considered above. This effect is called a magnetic Landau damping. We present a simplified model of the magnetic Landau damping [27]. We begin with the equation of motion of a carrier with a charge e and a mass m along the z-axis

$$m\frac{dv_z}{dt} = eE_z + e\left[\bm{v} \times \bm{B}\right]_z. \tag{7.274}$$

The magnetic part of the Lorentz force for the field slowly varying in space can be written as follows

$$e\left[\bm{v} \times \bm{B}\right]_z \simeq \mu_m \frac{\partial B_z}{\partial z}, \tag{7.275}$$

where μ_m is a magnetic moment of the carrier:

$$\mu_m = \frac{mv_\perp^2}{2B_0}. \tag{7.276}$$

The velocity $v_\perp$ is perpendicular to $\bm{B}_0$. The force (7.275) is experienced by a charged particle with the magnetic moment (7.276) when it moves adiabatically in a magnetic field whose strength varies slowly with position. The variable magnetic field B_{Hz} of the helicon alternately adds to and subtracts from the static magnetic field $\bm{B}_0$. As a result, the gyrating particle is placed in a moving periodic magnetic mirror. Its interaction with that mirror is especially strong when the velocity of the particle is equal to the velocity of the moving magnetic mirror along the z direction

$$v_z = \frac{\omega}{k_z}. \tag{7.277}$$

The particles with the velocity just less than ω/k_z will be accelerated by the mirror and by then will extract the energy from the wave. The particles with v_z slightly greater than ω/k_z will be slowed down. In thermal equilibrium there are more slowly moving particles than the fast ones and $\partial f_0/\partial v < 0$. Consequently, the interaction will result in the damping of the wave. For helicon waves in a metal, the magnetic Landau damping is the predominant damping mechanism since a helicon in metal manifests mainly a magnetic field, and the electric effects are small compared to the magnetic effects. Indeed, the ratio of the electric and magnetic forces is given by [27]

$$\frac{eE_z}{\mu_m\left(\partial B_{Hz}/\partial z\right)} = \frac{eE_z}{\mu_m k_z B_{Hz}} = \frac{e\omega}{\mu_m k_z k_x}\frac{E_z}{E_y} \sim \frac{\omega}{kv_F}. \tag{7.278}$$

In the limiting case of a large mean free path l ($kl = kR\omega_c\tau \gg 1$) the rate at which the particles gain energy $\mathcal{E}$ from the field due to a one-dimensional force F in z direction is given by [27]

$$\frac{d}{dt}\langle\mathcal{E}\rangle_{z_0,v_0} = -\frac{\pi}{2mk_z^2}\omega nF^2\left[\frac{\partial f_0}{\partial v}\right]_{v=\omega/k_z}. \tag{7.279}$$

Let the equilibrium distribution function f_0 for a simple metal at low temperatures be the one-dimensional Fermi distribution

$$f_0 = \frac{3}{4v_F}\left(1 - \frac{v_z^2}{v_F^2}\right), \quad v_z \leq v_F, \text{ and } f_0 = 0, \quad v_z \geq v_F. \tag{7.280}$$

Then combining (7.275), (7.276) and the local helicon dispersion relation for the oblique propagation we rewrite (7.279)

$$\frac{d}{dt}\langle \mathcal{E}\rangle_{z_0,v_0} = \frac{B_0^2}{2\mu_0}\left[\frac{3\pi}{8}kR\sin^2\Delta\right]. \tag{7.281}$$

The rate of gain of average particle kinetic energy is equal to the rate of decrease of electromagnetic energy W of the helicon wave. The electric part of energy is negligibly small as compared to the magnetic part. Then we get

$$\frac{dW}{dt} = -2\mathrm{Im}\omega W = -\frac{B_0^2}{2\mu_0}\left[\frac{3\pi}{8}kR\sin^2\Delta\right] \tag{7.282}$$

and the imaginary part $\mathrm{Im}\omega$ of the helicon frequency is [27]

$$\frac{\mathrm{Im}\omega}{\omega} = \frac{3\pi}{16}kR\sin^2\Delta. \tag{7.283}$$

The attenuation factor (7.283) coincides with the collisionless term in (7.272). Thus we come to a conclusion that the magnetic Landau damping is essential in the helicon decay [27]. Experimental results showed that dispersion relation (7.272) describes the helicon oblique propagation with a high accuracy [22]. Indeed, the contribution of the collision term into the damping is negligibly small in comparison with the magnetic Landau damping term.

8 Electrodynamics of Liquid Crystals

8.1 Classification and Fundamental Properties of Liquid Crystals

It is well known that in crystals the centres of gravity of the atoms or molecules are located on a three-dimensional periodic lattice. In the liquid, the centres of gravity are not ordered in this sense. As a result, the states of matter differ by their mechanical properties; for example, a liquid flows easily. Certain organic materials do not show a single transition from solid to liquid, but rather a cascade of transitions involving new phases. The mechanical properties and the symmetry properties of these new phases are intermediate between those of a liquid and those of a crystal. They have been called liquid crystals, or mesomorphic phases. From the point of view of electrodynamics, liquid crystals are strongly anisotropic, dispersive and, in general case, inhomogeneous media. Therefore, the study of liquid crystals behavior in external magnetic and electric fields and of electromagnetic wave propagation in liquid crystals is very instructive and presents some interesting illustrations for the formalism developed in previous chapters. We consider first very briefly some essential features of liquid crystals which are necessary for the further analysis.

There exist four types of liquid crystals [28].

1. Thermotropic mesophases consist of rod-like or disk-like small organic molecules. In such materials the phase transitions occur due to the temperature change.
2. Lyotropic mesophases consist of rod-like molecules in a liquid substrate. The induce phase transitions are induced by the change of the rods concentration.
3. The main-chain or side-chain polymers are thermotropic mesogens.
4. Amphiphilic compounds may give rise to associations and to mesomorphic behavior, either in the presence of a selective solvent or as a pure phase. They may be lyotropic or thermotropic.

In the present review we have restricted our attention to the major types of thermotropic mesophases. There exist two main types of thermotropic liquid crystals: nematic liquid crystals (NLC) and smectic liquid crystals

Fig. 8.1. The arrangement of molecules in the nematic liquid crystal. The vector n is the director

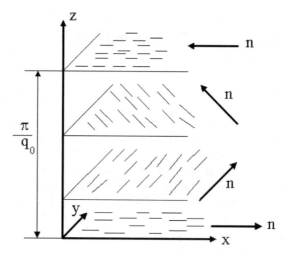

Fig. 8.2. The arrangement of molecules in the cholesteric liquid crystal (a helical structure)

(SLC). These two types are found only when the constituent molecules, or group of molecules (the building blocks) are strongly elongated. Depending upon the nature of the building blocks, and upon the external parameters (temperature, solvents, external magnetic and electric field, light radiation, etc.) one can observe a wide variety of phenomena and transitions amongst liquid crystals.

In NLC molecules are positionally disordered, but orientationally ordered in such a way that their longer axes are aligned preferentially along one direction. Their arrangement is presented in Fig. 8.1.

If we dissolve in a nematic liquid a molecule which is chiral (i.e. different from its mirror image), we find that the structure undergoes a helical distortion. The same distortion is also found with pure cholesteric esters which are also chiral. For this reason, the helical phase is called a cholesteric liquid crystal (CLC). CLC is a specific type of NLC. The structure of CLC is shown in Fig. 8.2.

In SLC the positional order exists in one or two dimensions. In the main practical case, we have positional order in one direction only; the system can

8.1 Classification and Fundamental Properties of Liquid Crystals

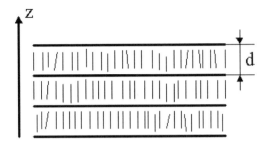

Fig. 8.3. The arrangement of molecules in a smectic A liquid crystal; the thickness of a layer equals d

be viewed as a set of two-dimensional liquid layers stacked on each other with a well-defined spacing. The corresponding phase is called smectic A liquid crystal (Fig. 8.3).

The main features of a nematic phase are as follows.

1. The centres of gravity of the molecules have no long-range order. The correlations in position between the centres of gravity of neighboring molecules are similar to those existing in a conventional liquid. Nematics do flow like liquids.
2. There is some order in the direction of the molecules. They tend to be parallel to some common axis, labelled by a unit vector n called director. This is reflected in all macroscopic tensor properties: for example, optically, nematic is a uniaxial medium with the optical axis along n. In all known cases, there appears to be complete rotational symmetry around the axis n.
3. The direction of n is arbitrary in space; in practice it is imposed by minor forces such as the guiding effect of the walls of the container.
4. The states of director n and $-n$ are indistinguishable.
5. Nematic phases occur only with materials which do not distinguish between right and left.

From a crystallographic point of view, the properties (2), (4) and (5) may be summarized by the symbol $D_{\infty h}$ in the Schonflies notation.

Locally, a cholesteric is very similar to a nematic material. Again, the centres of gravity have no long-range order, and the molecular orientation shows a preferred axis labelled by a director n. However n is not constant in space. The preferred conformation is helical. If we call z-axis the helical axis, we have the following structure for n:

$$n_x = \cos(q_0 z + \phi) \tag{8.1}$$

$$n_y = \sin(q_0 z + \phi) \tag{8.2}$$

$$n_z = 0. \tag{8.3}$$

Both the helical axis z and the value of ϕ are arbitrary; we see here another type of broken symmetry. The structure is periodic along z, and (since the states n and $-n$ are again equivalent) the spatial period P_0 is equal to one half of the pitch:

$$P_0 = \frac{\pi}{|q_0|}. \tag{8.4}$$

Typical value of P_0 are in the 3000 Å range, i.e. much larger than the molecular dimensions. Since P_0 is comparable to an optical wavelength, the periodicity results in Bragg scattering of light beams. Both the magnitude and the sign of q_0 are meaningful. The sign distinguishes between right- and left-handed helices; a given sample at a given temperature always produces helices of the same sign.

Smectic (from the Greek $\sigma\mu\eta\gamma\mu\alpha$ = soap) is the name coined by G. Friedel for certain mesophases with mechanical properties reminiscent of soaps. From a structural point of view, all smectics are layered structures, with a well-defined spacing. Smectics are thus more ordered than nematics. For a given material, the smectic phases always occur at temperatures below the nematic domain. The three main types of smectic phases are defined by letters A, B, and C. The major characteristics of smectic A phase are as follows.

1. A layer structure (with layer thicknesses close to the full length of the constituent molecules).
2. Inside each layer, the centers of gravity show no long range order; each layer is a two-dimensional liquid. Properties (1) and (2) together define a remarkable type of one-dimensional ordering, which is in fact quite singular.
3. The system is optically uniaxial, the optical axis being the normal Oz to the plane of the layers. There is a complete rotational symmetry around Oz.
4. The directions z and $-z$ are equivalent.

Properties (3) and (4) lead to a symmetry (D_∞) in the Schonflies notation. Note the difference between nematic ($D_{\infty h}$) and smectic A (D_∞). It can be shown that the requirement of constant interlayer thickness imposes the condition

$$\text{curl } \boldsymbol{n} = 0 \tag{8.5}$$

for all macroscopic deformations of smectics. The helical arrangement with

$$\text{curl } \boldsymbol{n} = -q_0 \boldsymbol{n} \neq 0 \tag{8.6}$$

is thus forbidden.

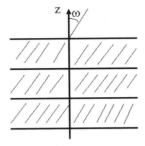

Fig. 8.4. The tilted arrangement of molecules in a smectic C liquid crystal; tilt angle is ω

The structure of a smectic C is defined as follows:

1. Each layer is still a two-dimensional liquid.
2. The material is optically biaxial. The most natural interpretation of these features amounts to assuming that, in a smectic C, the long molecular axis is tilted with respect to the normal z of the layers (Fig. 8.4).

This interpretation is substantiated by a number of X-ray experiments, which give a layer thickness $d = l\cos\omega$, where l is the length of the molecule, and ω is the tilt angle. If molecules are tilted in the xy plane, the principal axes of the dielectric tensor are two orthogonal directions in the xz-plane plus the y direction. Assuming that no ferroelectricity is present, we see that the symmetry elements of a smectic C are a two-fold axis (y) and a plane normal to it (xz), corresponding to the point group C_{2h}. In both the A and the C type of smectics, each layer behaves as a two-dimensional liquid.

In the smectic B liquid crystal the layers appear to have the periodicity and rigidity of a two-dimensional solid.

8.2 Liquid Crystals in a Static Magnetic Field

8.2.1 Continuum Theory of Liquid Crystals

Quantitatively the liquid crystalline state of matter is successfully described by the so-called continuum elastic theory which is based on the phenomenological expression of the elastic part of liquid crystal free energy F_d in terms of a director gradients [28, 29]. The variations of the director are slow on the molecule scale a:

$$a\nabla n \ll 1. \tag{8.7}$$

The thorough analysis of all possible terms in $\mathcal{F}_d$ of order $(\nabla \boldsymbol{n})^2$ shows that some of the spatial derivatives of the director $\partial n_\alpha/\partial x_\beta$ vanish due to the symmetry considerations and the unitary length of the director, and part of them contribute only to the surface terms, and therefore can be neglected. As a result, the distortion energy density $\mathcal{F}_d$ can be written in the following form [28, 29]:

$$\mathcal{F}_d = \frac{1}{2}K_1 \left(\operatorname{div} \boldsymbol{n}\right)^2 + \frac{1}{2}K_2 \left(\boldsymbol{n} \cdot \operatorname{rot} \boldsymbol{n}\right)^2 + \frac{1}{2}K_3 \left(\boldsymbol{n} \times \operatorname{rot} \boldsymbol{n}\right)^2. \qquad (8.8)$$

Equation (8.8) is the fundamental formula of the continuum theory for nematics. The constants K_i ($i = 1, 2, 3$) introduced into (8.8) are associated with three basic types of deformation, respectively:

1. K_1: conformations with $\operatorname{div}\boldsymbol{n} \neq 0$, or splay;
2. K_2: conformations with $\boldsymbol{n} \cdot \operatorname{rot}\boldsymbol{n} \neq 0$, or twist;
3. K_3: conformations with $\boldsymbol{n} \times \operatorname{rot}\boldsymbol{n} \neq 0$, or bend.

It is possible to generate deformations which are pure splay, pure twist, or pure bend. Thus each constant K_i must be positive; if not, the undistorted nematic conformation would not correspond to a minimum of the free energy $\mathcal{F}_d$. The elastic constants K_i have the dimension of energy/cm, or dynes, since $\mathcal{F}_d$ is an energy per cm^3, $\boldsymbol{n}$ is dimensionless. By a purely dimensional argument, we expect the K's to be of order U/a where U is a typical interaction energy between molecules, while a is a molecular dimension. Taking $U \sim 2$ kcal/mole and $a \simeq 14$ Å we expect $K_i \sim 1.4 \cdot 10^{-13}$ erg/$1.4 \cdot 10^{-7}$cm= 10^{-6} dynes= 10^{-11}newton. This is indeed the correct order of magnitude; for PAA at $120°C$ the measured elastic constants are $K_3 = 1.7 \cdot 10^{-11}$newton, $K_1 = 0.7 \cdot 10^{-11}$newton, $K_2 = 0.43 \cdot 10^{-11}$newton. It is useful to estimate the magnitude of the distortion energy, per molecule, for a typical distortion taking place in a distance l: this will be roughly

$$\mathcal{F}_d a^3 \sim \frac{K_i}{l^2} a^3 \sim U \left(\frac{a}{l}\right)^2.$$

It is seen that in the continuum limit $a \ll l$ it represents only a small fraction of the total energy. In many cases the full form of (8.8) is still too complex to be of practical use; either because the relative values of the three elastic constants K_i are unknown, or because the equilibrium equations derived from (8.8) are too difficult to solve. In such cases, a further approximation is often useful; this amounts to assuming all three elastic constants equal

$$K_1 = K_2 = K_3 = K. \qquad (8.9)$$

This is known as the one-constant approximation.

In contrast to nematic liquid crystals where the terms linear in the gradients of $\boldsymbol{n}$ are incompatible with a equilibrium conformation, in a cholesteric the equilibrium state is twisted, and such terms in the free energy are possible. There are two terms which are linear in the spatial derivatives of $\boldsymbol{n}$ and rotationally invariant: $\operatorname{div}\boldsymbol{n}$ and $(\boldsymbol{n} \cdot \operatorname{rot}\boldsymbol{n})$. Terms proportional to $\operatorname{div}\boldsymbol{n}$

cannot occur in $\mathcal{F}_d$ because the states n and $-n$ are indistinguishable. On the other hand, the pseudoscalar quantity $(n \cdot \mathrm{rot} n)$ may appear in $\mathcal{F}_d$, provided that the molecules are different from their mirror images. Adding to this usual nematic terms from (8.8) we obtain a distortion energy of the form

$$\mathcal{F}_d = \frac{1}{2} K_1 (\mathrm{div} n)^2 + \frac{1}{2} K_2 (n \cdot \mathrm{rot} n + q_0)^2 + \frac{1}{2} K_3 (n \times \mathrm{rot} n)^2. \quad (8.10)$$

In equation (8.10) we find a term linear in the gradients, namely

$$K_2 q_0 n \cdot \mathrm{rot} n \quad (8.11)$$

and a constant term $K_2 q_0^2/2$. The meaning of (8.10) becomes more transparent if we consider the situation of pure twist

$$n_x = \cos\theta(z), \quad n_y = \sin\theta(z), \quad n_z = 0. \quad (8.12)$$

Then (8.10) reduces to

$$\mathcal{F}_d = \frac{1}{2} K_2 \left(\frac{\partial \theta}{\partial z} - q_0 \right)^2. \quad (8.13)$$

It is seen from (8.13) that the equilibrium distortion corresponds to a helix of wave vector $\partial\theta/\partial z = q_0$. Equation (8.10) is the correct form for the distortion free energy when both ∇n and q_0 are small on the molecular scale a. In all practical situations $q_0 a \sim 10^{-3}$, and the corrections are not important.

Consider, finally, the continuum description of a smectic A phase. Let us take an ideal single-domain sample of a smectic A, with parallel and equidistant layers (interval a, direction of the normal to the layers z). Let the n-th layer is displaced by a certain amount $u_n(x, y)$; u_n is the fundamental variable. For the cases of slow spatial variations in which we are interested we can substitute for the discrete index n the continuous variable $z = na$

$$u_n(x, y) \to u(x, y, z). \quad (8.14)$$

In the unperturbed state the molecules were normal to the layers. In the perturbed state they do not remain exactly normal to the new plane of the layers. If we associate a unit director n to the optical axis, it will have the components

$$n_x = -\frac{\partial u}{\partial x} \ll 1, \quad n_y = -\frac{\partial u}{\partial y} \ll 1. \quad (8.15)$$

Equations (8.15) show that at each point n is normal to the layers. Consequently,

$$n \cdot \mathrm{rot} n = 0. \quad (8.16)$$

It is seen that the twist deformation which was allowed in nematics becomes forbidden in smectics A. In the distorted material the mass density ϱ_m will usually differ from its unperturbed value ϱ_{m0}

$$\varrho_m = \varrho_{m0}\left[1 - \vartheta\left(\mathbf{r}\right)\right]. \tag{8.17}$$

However, for the static distortions of interest here $\vartheta\left(\mathbf{r}\right)$ will adjust itself to minimize the free energy in a given $u\left(\mathbf{r}\right)$: thus in the absence of external pressures which would change ρ it need not be treated as an independent variable. It may be shown that

$$\vartheta = m\frac{\partial u}{\partial z}, \tag{8.18}$$

where m is a dimensionless constant, characteristic of the material which may in principle be positive or negative. Let us now write the free energy $\mathcal{F}$ (per unit volume of the unperturbed system) as a function of the derivatives of $u\left(\mathbf{r}\right)$. It is assumed that $\mathrm{grad}\,u$ is small; physically, this means that we consider layers which are not very tilted from (x,y) plane. As regards symmetry, for the sake of simplicity we postulate that in undisturbed smectic the directions z and $-z$ are equivalent (no ferroelectricity): this appears to be correct in all clear-cut cases. Finally, we obtain the following form for F [28, 29]:

$$\mathcal{F} = \mathcal{F}_0 + \frac{1}{2}B\left(\frac{\partial u}{\partial z}\right)^2 + \frac{1}{2}K_1\left(\frac{\partial^2 u}{\partial x^2} + \frac{\partial^2 u}{\partial y^2}\right)^2, \tag{8.19}$$

where $\mathcal{F}_0$ is the unperturbed free energy. The second term represents an elastic energy for compression of the layers. The estimations and the experimental results show that the elastic constant B is of order of $10^8 \mathrm{erg/cm}^3$. It is important to understand that in the absence of any orienting field there cannot be any term in (8.19) proportional to $(\partial u/\partial x)^2$ or $(\partial u/\partial y)^2$. It is thus necessary to include certain higher order terms, and in particular

$$\frac{1}{2}K_1\left(\frac{\partial^2 u}{\partial x^2} + \frac{\partial^2 u}{\partial y^2}\right)^2 = \frac{1}{2}K_1\left(\mathrm{div}\,\mathbf{n}\right)^2. \tag{8.20}$$

where the (8.15) are used. Equation (8.20) shows that this term is a splay energy, identical in form to this in a nematic. When the variations of u take place only in the layer plane, i.e. $\partial u/\partial z = 0$ the B term drops out and the elastic effects depend on K_1 alone. The static elastic properties of a smectic A, unlike nematics and cholesterics, are described by two constants: B (dimension energy length^{-3}) and K_1 (dimension energy length^{-1}).

8.2.2 Nematic Liquid Crystal in a Static Magnetic Field

Consider the effect of a magnetic field $\mathbf{H}$ on a nematic liquid crystal. We can write down the magnetization $\mathbf{M}$ induced by $\mathbf{H}$ for an arbitrary angle between $\mathbf{H}$ and $\mathbf{n}$ [28, 29]:

8.2 Liquid Crystals in a Static Magnetic Field

$$M = \chi_\perp H + (\chi_\parallel - \chi_\perp)(H \cdot n) n. \tag{8.21}$$

Here both magnetic susceptibilities (parallel and perpendicular to the undisturbed director) $\chi_\parallel$ and $\chi_\perp$ are negative (liquid crystals are diamagnetic) and small ($\sim 10^{-7}$ to 10^{-6} in CGS units). It should be noted that the magnetic susceptibility has different values in the MKS system of units and in the CGS system of units. In the MKS system the magnetic susceptibility has the form [3]

$$(\chi_{ik})_{\text{MKS}} = \mu^r_{ik} - 1.$$

In the CGS system the magnetic susceptibility is [4]

$$(\chi_{ik})_{\text{CGS}} = \frac{\mu^r_{ik} - 1}{4\pi}.$$

The relative magnetic permeability μ^r_{ik} is an observable tabulated quantity characteristic for a medium, and it remains the same in any system of units. Therefore

$$(\chi_{ik})_{\text{MKS}} = 4\pi (\chi_{ik})_{\text{CGS}}.$$

In usual nematics the difference χ_a is positive:

$$\chi_a = \chi_\parallel - \chi_\perp > 0. \tag{8.22}$$

Using (8.8), (8.21) and (8.22) we obtain the free energy per cm^3:

$$\mathcal{F} = \mathcal{F}_d - \int_0^H M d H = \mathcal{F}_d - \frac{1}{2}\chi_\perp H^2 - \frac{1}{2}\chi_a (H \cdot n)^2. \tag{8.23}$$

Note that in the MKS system of units the magnetic part of the free energy density has the form

$$-\frac{1}{2}\mu_0 \left[\chi_\perp H^2 + \chi_a (H \cdot n)^2 \right].$$

Traditionally, the electrodynamics of liquid crystals in presented in the CGS system of units [28,29], and we mainly follow these books. However, we at the same time write the key expressions in the MKS system of units. The term $\frac{1}{2}\chi_\perp H^2$ is independent of the molecular orientation and may be omitted in all the cases to be discussed below. The last term is the interesting one. Note that for $\chi_a > 0$ this term is minimized when n is collinear with H.

We start the analysis of a nematic liquid crystal behavior in a magnetic field with the simplest case of the competing effects of a wall and a magnetic field on the alignment of a nematic cell. We consider first the case of a sufficiently thick cell which can be assumed to be semi-infinite. Let us take the plane of the wall as the (xz) plane, nematic lying in the region $y > 0$. We assume that at the wall there is one easy direction for the molecules ($\pm 0 z$) and

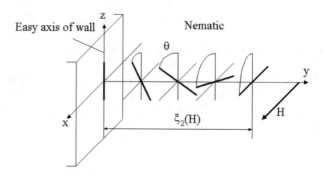

Fig. 8.5. Competition between wall-alignment and field-alignment. A case of a pure twist

that strong anchoring prevails. The magnetic field $\boldsymbol{H}$ is along the x direction, i.e. normal to the wall's easy axis, but in the plane of the wall. If we go far enough from the wall (y large and positive) we will find the nematic aligned along $\boldsymbol{H}$ according to (8.23). Closer to the wall, there will be a transition layer, where the nematic molecules stay parallel to the (xz) plane, but make a variable angle $\theta(y)$ with the z direction. This is a situation of pure twist which is shown in Fig. 8.5.

We take the director $\boldsymbol{n}$ in the form

$$n_x = \sin\theta(y),\ n_y = 0,\ n_z = \cos\theta(y). \tag{8.24}$$

Substituting (8.24) into (8.8) we immediately obtain

$$\mathcal{F}_d = \frac{1}{2} K_2 \left(\frac{\partial\theta}{\partial y}\right)^2 \tag{8.25}$$

and

$$\mathcal{F} = \frac{1}{2} K_2 \left(\frac{\partial\theta}{\partial y}\right)^2 - \frac{1}{2}\chi_a H^2 \sin^2\theta(y). \tag{8.26}$$

We minimize $\int \mathcal{F} d\boldsymbol{r}$, where $\mathcal{F}$ is given by (8.26). The condition of equilibrium is

$$-\frac{\partial \mathcal{F}}{\partial \theta} + \frac{\partial}{\partial y}\left(\frac{\partial \mathcal{F}}{\partial \frac{\partial \theta}{\partial y}}\right) = 0. \tag{8.27}$$

It yields

$$K_2 \frac{d^2\theta}{dy^2} + \chi_a H^2 \sin\theta\cos\theta = 0. \tag{8.28}$$

Let us define a so-called magnetic coherence length $\xi_2(H)$ through

8.2 Liquid Crystals in a Static Magnetic Field

$$\xi_2(H) = \frac{1}{H}\left(\frac{K_2}{\chi_a}\right)^{\frac{1}{2}}, \qquad (8.29)$$

or in the MKS system of units

$$\xi_2(H) = \frac{1}{H}\left(\frac{K_2}{\mu_0 \chi_a}\right)^{\frac{1}{2}}.$$

In terms of $\xi_2(H)$ (8.28) takes the form

$$\xi_2^2 \frac{d^2\theta}{dy^2} + \sin\theta\cos\theta = 0. \qquad (8.30)$$

It can be solved explicitly. Multiplying by $d\theta/dy$ we get

$$\xi_2^2 \frac{d}{dy}\left[\frac{1}{2}\left(\frac{d\theta}{dy}\right)^2\right] + \frac{d}{dy}\left(-\frac{1}{2}\cos^2\theta\right) = 0. \qquad (8.31)$$

The integration of (8.31) yields

$$\xi_2^2 \left(\frac{d\theta}{dy}\right)^2 = \cos^2\theta + \text{const.} \qquad (8.32)$$

Far from the wall ($y \to \infty$) we expect $\theta = \pi/2$ and $d\theta/dy = 0$. Thus the integration constant in (8.32) must vanish. Then, the equation becomes a kind of the static sine-Gordon (sG) equation which has exact soliton solutions, i.e. the solutions in the form of localized waves [30]. We find

$$\xi_2 \frac{d\theta}{dy} = \pm\cos\theta. \qquad (8.33)$$

Both choices of sign are permissible, corresponding to a right-handed or a left-handed transition region. These conformations are related to each other by a mirror reflection in the $y0z$ plane. Choosing, for instance, the $+$ sign in (8.33) we have

$$\frac{dy}{\xi_2} = \frac{d\theta}{\cos\theta}. \qquad (8.34)$$

After some algebraic transformations we obtain from (8.34) the following solution

$$\tan\left(\frac{\pi}{4} - \frac{\theta}{2}\right) = \exp\left(-\frac{y}{\xi_2}\right). \qquad (8.35)$$

The integration constant in (8.32) has been chosen to ensure that for $y = 0$ at the wall $\theta = 0$ as required. The solution (8.35) can be considered as the magnetic field induced static kink soliton [30]. Equation (8.35) shows that the

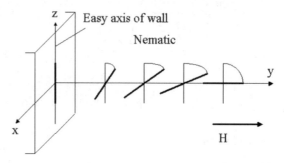

Fig. 8.6. Competition between wall-alignment and field-alignment. The case shown involves a mixture of bending and splay

thickness of the transition layer is essentially $\xi_2(H)$. Taking $K_2 = 10^{-6}$ dynes, $\chi_a = 10^{-7}$ in CGS units, and $H = 10^4$ oersteds, we get from relationship (8.29) $\xi_2(H) \sim 3\mu m$. Here a liquid crystal manifests one of its fascinating properties: with a rather weak external perturbation we can induce distortions on a scale comparable to an optical wavelength. The competition between the effects of a wall and the effects of a field may occur in a number of different geometries. For example, the field $\boldsymbol{H}$ can be normal to the wall, along the y direction. In this case the distortion is a combination of bend and splay. The algebra is more involved, but the same general features are found: the effects of the wall decrease exponentially at large distances, and the thickness of the transition layer is a certain weighted mean of the lengths

$$\xi_1(H) = \frac{1}{H}\left(\frac{K_1}{\chi_a}\right)^{\frac{1}{2}}, \quad \xi_3(H) = \frac{1}{H}\left(\frac{K_3}{\chi_a}\right)^{\frac{1}{2}} \tag{8.36}$$

related to the elastic constants K_1 and K_3 for splay and bend. The orientation of the molecules in such a situation is presented in Fig. 8.6.

The three lengths ξ_i ($i = 1, 2, 3$) are usually of comparable magnitude. In the one constant approximation ($K_i = K$) they become equal:

$$\xi(H) = \frac{1}{H}\left(\frac{K}{\chi_a}\right)^{\frac{1}{2}}, \tag{8.37}$$

or in the MKS system of units

$$\xi(H) = \frac{1}{H}\left(\frac{K}{\mu_0 \chi_a}\right)^{\frac{1}{2}}.$$

It can be shown that in general case the nematic alignment is perturbed only in a region of linear dimensions ξ.

We apply the method developed above to the analysis of the behavior of a finite thickness nematic sample under magnetic field. Consider a nematic

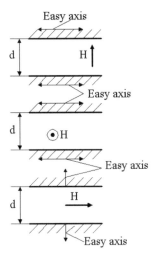

Fig. 8.7. The Frederiks transition for a nematic slab under a magnetic field H. At low H the molecules are parallel to the easy axis of the wall. For $H > H_c$ the molecules near the center of the slab rotate towards H

single crystal of thickness $d \sim 20$ µm, oriented between two solid plates. At the surface of the plates we assume strong anchoring. The easy direction imposed by both surfaces may be in plane of the surfaces or normal to it. A magnetic field H is applied normal to the easy axis. There will be a transition, at a certain critical value H_c of the field, between the unperturbed conformation and the distorted conformation. A transition of this type was detected optically and by Fredericks in 1927, and then it was called after him: the Fredericks transition. The geometry of the problem is shown in Fig. 8.7.

Fredericks showed that the critical field H_c was inversely proportional to the sample thickness d.

$$Hd = \text{const.} \tag{8.38}$$

Typical values were in the range of $H \sim 10^4$ œrsteds=1 tesla for $d = 10$ µm. Taking into account this result the critical field can be derived by a following argument. Starting from the unperturbed state $n = n_0$ we consider a slight deflection

$$n = n_0 + \delta n(r), \tag{8.39}$$

where $\delta n(r)$ is perpendicular to n_0 since $n^2 = 1$, and is parallel to H (since this is the direction in which the molecules are solicited). It is natural to assume that the distortion depends only on z where z is normal to the slab. The distortion energy (8.8) reduces to

$$\mathcal{F}_d = \frac{1}{2} K_i \left(\frac{\partial \delta n}{\partial z} \right)^2 \tag{8.40}$$

for the case i, where $K_i = K_{1,2,3}$, for the splay, twist, and bend conformations, respectively. The magnetic energy gives a contribution

$$\mathcal{F}_M = -\frac{1}{2}\chi_a H^2 \delta n^2, \tag{8.41}$$

or in the MKS system of units

$$\mathcal{F}_M = -\frac{1}{2}\mu_0 \chi_a H^2 \delta n^2.$$

Since we have assumed strong anchoring, at both boundaries $z = 0$ and $z = d$ δn must vanish. It is then convenient to analyze δn in a Fourier series

$$\delta n = \sum_q \delta n \sin qz \tag{8.42}$$

$$q = \nu \frac{\pi}{d}, \quad \nu = \text{positive integer}.$$

Substituting expression (8.42) into (8.40) and (8.41), combining the results and integrating over the thickness one obtains (per cm^2 of slab)

$$\mathcal{F}_{\text{tot}} = d\left(\mathcal{F}_{d\text{tot}} + \mathcal{F}_{M\text{tot}}\right) = \frac{d}{4}\sum_q \delta n_q^2 \left(K_i q^2 - \chi_a H^2\right). \tag{8.43}$$

If we want the unperturbed state to be stable, the increase in free energy $\mathcal{F}_{\text{tot}}$ must be positive for all values of the parameters δn_q.

$$\chi_a H^2 < K_i q^2. \tag{8.44}$$

The smallest value of q is $q = \pi/d$ corresponding to a distortion of half-wavelength d. Thus the threshold field $H_{c,i}$ corresponds to

$$\chi_a H_{c,i}^2 = K_i \left(\frac{\pi}{d}\right)^2 \tag{8.45}$$

$$H_{c,i} = \left(\frac{\pi}{d}\right)\left(\frac{K_i}{\chi_a}\right)^{\frac{1}{2}},$$

or in the MKS system of units

$$H_{c,i} = \left(\frac{\pi}{d}\right)\left(\frac{K_i}{\mu_0 \chi_a}\right)^{\frac{1}{2}}.$$

Equation (8.45) may also be stated in terms of the magnetic coherence length. Namely, at the critical field the coherence length $\xi_i(H_{c,i})$ is equal to π/d. The result (8.45) shows that the $1/d$ dependence found experimentally by Fredericks is correct. It also gives the principle of a simple determination of the three elastic constants. We should underline that the analysis developed above is valid only if the following conditions are met:

1. strong anchoring prevails;
2. the easy direction is normal or parallel to the slab.

8.2 Liquid Crystals in a Static Magnetic Field

We consider now in details the general case of deformation of a nematic crystal which is sandwiched between two glass plates and subjected to a magnetic field $\boldsymbol{H}$ that is above a critical strength and parallel to the z-axis [29]. This time we assume that the molecule orientation is not restricted to the plane parallel or perpendicular to the walls. In such a situation the three-dimensional director $\boldsymbol{n}$ has the form:

$$n_x = \cos\theta(z)\cos\phi(z), \quad n_y = \cos\theta(z)\sin\phi(z), \quad n_z = \sin\theta(z). \tag{8.46}$$

The free energy density derived from (8.8) and (8.23) then takes the form (in the CGS system of units)

$$\mathcal{F} = \frac{1}{2}\left(K_1\cos^2\theta + K_3\sin^2\theta\right)\left(\frac{d\theta}{dz}\right)^2$$

$$+\frac{1}{2}\cos^2\theta\left(K_2\cos^2\theta + K_3\sin^2\theta\right)\left(\frac{d\phi}{dz}\right)^2 - \frac{1}{2}\chi_a H^2\sin^2\theta \tag{8.47}$$

Minimizing the free energy (8.47) with respect to the angles $\theta(z)$ and $\phi(z)$ we obtain two exact equations describing the equilibrium conformation of a twisted nematic cell of a thickness d.

$$f(\theta)\frac{d^2\theta}{dz^2} + \frac{1}{2}\frac{df(\theta)}{d\theta}\left(\frac{d\theta}{dz}\right)^2 - \frac{1}{2}\frac{dg(\theta)}{d\theta}\left(\frac{d\phi}{dz}\right)^2 + \chi_a H^2\sin\theta\cos\theta = 0 \tag{8.48}$$

and

$$g(\theta)\frac{d^2\phi}{dz^2} + \frac{dg(\theta)}{d\theta}\frac{d\theta}{dz}\frac{d\phi}{dz} = 0, \tag{8.49}$$

where

$$f(\theta) = K_1\cos^2\theta + K_3\sin^2\theta, \quad g(\theta) = \left(K_2\cos^2\theta + K_3\sin^2\theta\right)\cos^2\theta$$

and ϕ obeys the following boundary conditions:

$$\phi(0) = -\phi_0, \quad \phi(d) = \phi_0.$$

The integration of (8.48) and (8.49) yields

$$f(\theta)\left(\frac{d\theta}{dz}\right)^2 + g(\theta)\left(\frac{d\phi}{dz}\right)^2 + \chi_a H^2\sin^2\theta = C_1 \tag{8.50}$$

$$g(\theta)\frac{d\phi}{dz} = C_2, \tag{8.51}$$

where C_1 and C_2 are the integration constants. Combining results (8.50) and (8.51) we find

$$f(\theta)\left(\frac{d\theta}{dz}\right)^2 + \frac{B^2}{g(\theta)} + \chi_a H^2 \sin^2\theta = C_1. \tag{8.52}$$

Because of strong anchoring at the walls, $\theta = 0$ at $z = 0$ and d, while θ_m, the maximum value of θ, occurs at the midplane $z = d/2$. Since $d\theta/dz = 0$ at $z = d/2$

$$C_1 = \frac{C_2^2}{g(\theta_m)} + \chi_a H^2 \sin^2\theta_m. \tag{8.53}$$

Substituting (8.53) into (8.52) we obtain

$$f(\theta)\left(\frac{d\theta}{dz}\right)^2 + C_2^2\left(\frac{1}{g(\theta)} - \frac{1}{g(\theta_m)}\right) + \chi_a H^2\left(\sin^2\theta - \sin^2\theta_m\right) = 0, \tag{8.54}$$

or, finally,

$$z = \int_0^\theta \{N(\psi)\}^{\frac{1}{2}}\, d\psi, \tag{8.55}$$

where

$$N(\psi) = \frac{f(\psi)}{\chi_a H^2\left(\sin^2\psi - \sin^2\theta_m\right) + C_2^2\left[\frac{1}{g(\psi)} - \frac{1}{g(\theta_m)}\right]}. \tag{8.56}$$

Similarly, using the result (8.56) and the boundary conditions, we derive from (8.51)

$$\phi = -\phi_0 + \int_0^\theta \{N(\psi)\}^{\frac{1}{2}} \frac{C_2}{g(\psi)}\, d\psi. \tag{8.57}$$

Therefore

$$d = 2\int_0^{\theta_m} \{N(\psi)\}^{\frac{1}{2}}\, d\psi \tag{8.58}$$

and

$$\phi_0 = \int_0^{\theta_m} \{N(\psi)\}^{\frac{1}{2}} \frac{C_2}{g(\psi)}\, d\psi. \tag{8.59}$$

Taking the limit $\theta_m \to 0$, we have $\theta \to 0$, $f(\theta) \to K_1$, $g(\theta) \to 0$, $d\phi/dz \to 2\phi_0/d$, $C_2 \to 2K_2\phi_0/d$ and

$$d = 2 \int_0^{\frac{\pi}{2}} \left(\frac{K_1}{\chi_a H^2 - (4\phi_0^2/d^2)(K_3 - 2K_2)} \right)^{\frac{1}{2}} d\lambda, \qquad (8.60)$$

where $\lambda = \sin\psi / \sin\theta_m$. The critical field H_c for the deformation to occur has the form

$$H_c = \left(\frac{2}{\chi_a^{\frac{1}{2}} d} \right) \left[K_1 \left(\frac{\pi}{2} \right)^2 + (K_3 - 2K_2) \phi_0^2 \right]^{\frac{1}{2}}, \qquad (8.61)$$

or in the MKS system of units

$$H_c = \left(\frac{2}{(\mu_0)^{\frac{1}{2}} \chi_a^{\frac{1}{2}} d} \right) \left[K_1 \left(\frac{\pi}{2} \right)^2 + (K_3 - 2K_2) \phi_0^2 \right]^{\frac{1}{2}}.$$

The results of the exact solution of the system (8.48)–(8.49) permit the determination of K_2 by measuring H_c provided that K_1 and K_3 are known.

The domain formation in an infinite sample of a nematic liquid crystal subjected to a magnetic field has also been studied theoretically, and it has been shown that under some conditions a helical structure with a field dependent period can exist [31].

8.2.3 Cholesteric Liquid Crystal in a Static Magnetic Field

For the analysis of the cholesteric liquid crystals behavior in a magnetic field the same procedure can be applied as in the previous section. We combine the energy contributions (8.10) and (8.23) and take the director in the form (8.12). It should be noted that the diamagnetic susceptibilities $\chi_\parallel$ and $\chi_\perp$ in cholesterics are considerably smaller than in conventional nematics: $|\chi_a|$ is of order 10^{-9} c.g.s. units. The reason is that there are no benzene rings in cholesterol esters. Also χ_a is usually negative and in a magnetic field the director tends to be aligned normal to $\boldsymbol{H}$. To obtain cholesterics with positive χ_a, the simplest procedure is to dissolve chiral molecules in a conventional nematic like MBBA.

A bulk cholesteric sample with negative χ_a minimizes its energy by putting the helical axis $\boldsymbol{q}_0$ along the field; then the director $\boldsymbol{n}$ is normal to the field at all points. No distortion energy is required and the helical pitch is independent of the field as it is seen from (8.23). This very simple effect has been observed with a.c. electric fields of frequency higher than ~ 1 kHz in mixtures of MBBA and cholesterol esters. A finite frequency is required to eliminate the convective instabilities usually observed in MBBA. This experiment allows to transform a polydomain sample into a well-ordered planar texture.

310   8 Electrodynamics of Liquid Crystals

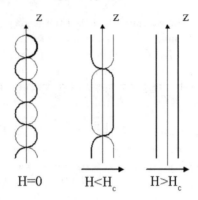

Fig. 8.8. Untwisting of a cholesteric spiral by magnetic field $\boldsymbol{H}$ ($\chi_a > 0$)

Consider a cholesteric with positive χ_a. Let us start with a thick cholesteric sample, so that wall effects may be neglected. In low fields H the helical structure is undistorted. The susceptibility measured along the helical axis is the perpendicular susceptibility $\chi_\perp$; when measured normal to the helical axis it is the average $(\chi_\perp + \chi_\parallel)/2$. If $\chi_\parallel > \chi_\perp$ this average is larger than $\chi_\perp$. The system will tend to adjust in order to display the maximum susceptibility; the helical axis is thus normal to the applied field. This is well confirmed experimentally. Let us now look for internal distortions of the helical structure. The regions of the helix where the molecules are favorably aligned along the field tend to expand if the field becomes strong enough. The regions with an unfavorable orientation of molecules with respect to the field cannot contract very much, since this would require too much twist energy. As a result, the pitch P increases with field. At even higher fields this leads to a succession of 180° walls separating large regions of the favorable molecular orientation. Each wall has a finite thickness, of order $2\xi_2(H)$ where ξ_2 is defined by (8.29). The distance between walls

$$L = \frac{1}{2} P(H) \tag{8.62}$$

is now much larger than ξ_2. Finally, at a certain critical field H_c the walls become infinitely separated, $P \to \infty$ and we obtain a nematic structure. This cholesteric-nematic transition has been observed experimentally under magnetic field as well as under electric one. The detailed studies showed that the critical field is inversely proportional to the unperturbed pitch $P(0)$. In practice, to have reasonable critical field values, one always has to work in mixtures of large pitch. The untwisting of a cholesteric spiral is illustrated by Fig. 8.8. The exact solution for a behavior of a cholesteric helix under a magnetic field has been obtained by P.G. de Gennes (1968) in his pioneered work [32]. The director is taken in the form (8.12), the elastic and magnetic parts of the free energy density have the form (8.10) and (8.23), respectively, and the magnetic field $\boldsymbol{H}$ is parallel to the y-axis. The total free energy takes

8.2 Liquid Crystals in a Static Magnetic Field

the form

$$\mathcal{F} = \frac{1}{2} \int \left[K_2 \left(\frac{d\theta}{dz} - q_0 \right)^2 - \chi_a H^2 \sin^2 \theta \right] d\theta. \tag{8.63}$$

By using the standard procedure of the free energy minimization with respect to the angle $\theta(z)$ the equation (8.30) has been obtained. The first integral has the form

$$\xi^2 \left(\frac{d\theta}{dz} \right)^2 + \sin^2 \theta = \frac{1}{k^2} = \text{const.} \tag{8.64}$$

However, considering an infinite cholesteric crystal with the ideally periodic structure, no boundary conditions of a homogeneous type has been introduced. Instead, we calculate the period P of the structure corresponding to the variation of the angle $\Delta\theta = \pi$, which yields

$$P = \int_0^\pi d\theta \frac{d\theta}{dz} = 2\xi k \int_0^{\frac{\pi}{2}} \frac{d\theta}{\sqrt{1 - k^2 \sin^2 \theta}} = 2\xi k \mathcal{K}(k), \tag{8.65}$$

where $\mathcal{K}(k)$ is the complete elliptic integral of the first kind. The value of the constant k must correspond to the minimum of the free energy. Therefore, we must substitute the expression of the pitch P into the free energy $\mathcal{F}$ and minimize it once more, this time with respect to k. The calculations give the following results.

$$\mathcal{F} = \frac{1}{2} K_2 q_0^2 \left[1 - \frac{2\pi}{q_0 z} + \frac{2h J_0}{q_0 z} - h^2 (1 + B) \right] \int dz, \tag{8.66}$$

where

$$(1 + B) = k^{-2}, \quad h = (\xi q_0)^{-1}, \quad J_0 = 2 \int_0^{\frac{\pi}{2}} (B + \cos^2 \theta)^{\frac{1}{2}} d\theta = \frac{4}{k} \mathsf{E}(k) \tag{8.67}$$

and

$$\mathsf{E}(k) = \int_0^{\frac{\pi}{2}} (1 - k^2 \sin^2 \theta)^{\frac{1}{2}} d\theta \tag{8.68}$$

is the complete elliptic integral of the second kind. The condition

$$\frac{\partial \mathcal{F}}{\partial B} = 0$$

leads to the relations

$$J_0 = 2\frac{\pi}{h}, \quad \frac{\partial J_0}{\partial B} = zq_0 h$$

Taking into account that $\pi/q_0 = P_0/2$ we finally obtain

$$\frac{P}{P_0} = \left(\frac{2}{\pi}\right)^2 \mathsf{K}(k)\,\mathsf{E}(k) \tag{8.69}$$

and

$$h = \frac{\pi}{2}\frac{k}{\mathsf{E}(k)}. \tag{8.70}$$

When $z \to \infty$, $k \to 1$, $\mathsf{E}(k) \to 1$, $\mathsf{K}(k) \to \infty$ and $H \to H_c$ so that $h = \pi/2$ or

$$H_c = \frac{1}{2}\pi q_0 \left(\frac{K_2}{\chi_a}\right)^{\frac{1}{2}}, \tag{8.71}$$

or in the MKS system of units

$$H_c = \frac{1}{2}\pi q_0 \left(\frac{K_2}{\mu_0 \chi_a}\right)^{\frac{1}{2}},$$

which is the critical field at which the structure becomes nematic. The variation of pitch with magnetic field strength predicted by (8.69) has been verified experimentally. It has been also confirmed that H_c is inversely proportional to P_0 (the pitch of the undistorted structure). For example, in a typical measurement using nematic PAA doped with a small amount of cholesteril acetate H_c was 8.3 kgauss corresponding to the induction $B = 0.83$ tesla for $P_0 = 26\mu\text{m}$ [29]. The experimental and theoretical results concerning the pitch dependence on the magnetic field fairly coincide [28].

Consider now the situation when a magnetic field is parallel to the helical axis [32]. In this case the director takes the form

$$n_x = \cos\theta(z)\sin\varphi, \quad n_y = \sin\theta(z)\sin\varphi, \quad n_z = \cos\varphi. \tag{8.72}$$

The analysis shows that the minimum of the free energy occurs in two cases:

1. $\varphi = 0$ which corresponds to the condition

$$\mathcal{F} = \frac{1}{2}K_2 q_0^2 \left[1 - h^2\right] \int dz.$$

2. $\varphi = \pi/2$ which corresponds to the condition $\mathcal{F} = 0$.

The first case is realized since it corresponds to the lowest energy minimum [32] which means that the helical axis rotates by the 90° angle.

The incorporating of the director gradients into the magnetic susceptibility tensor of a cholesteric results in the additional terms in the free energy density proportional to $(\boldsymbol{n}\cdot\boldsymbol{H})(\boldsymbol{H}\cdot\text{rot}\boldsymbol{n})$ and $H^2(\boldsymbol{n}\cdot\text{rot}\boldsymbol{n})^2$ [33]. The

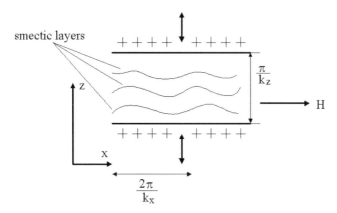

Fig. 8.9. The Helfrich-Hurault transition for smectics A: the transition can be induced in principle by a magnetic field $\boldsymbol{H}$. In practice, it is more convenient to apply a mechanical tension to the limiting plates

analysis showed that due to these terms the re-entrance transition from the nematic structure into the cholesteric one would be possible at sufficiently high magnetic fields $H_r \gg H_c$. According to the estimations, the re-entrance field H_r should be approximately $30H_c$ corresponding to the induction 30 tesla which can be achieved.

8.2.4 Helfrich–Hurault Effect in Smectic A Liquid Crystals

Consider a smectic A liquid crystal placed between two glass plates; the molecules are aligned along z, the unperturbed layers are parallel to the xy plane of the plates. We add on this system a magnetic field $\boldsymbol{H}$ in the x direction, and we assume that χ_a is positive.

To minimize the magnetic energy the system would like to rotate its optical axis. But the layers are strongly clamped at both walls. In the case of a nematic one could expect, above some field threshold, a bend distortion of the Fredericks type [28]. However, the characteristic feature of the smectic A crystal elastic energy (8.19) is the presence of a term describing the layers compression $B(\partial u/\partial z)^2$, associated with a large elastic constant $B \sim (10^7 \div 10^8)$ erg·cm^{-3}. Considering the behavior of smectics A in a magnetic field we must take into account this term, the term (8.20) connected with a director, and a magnetic energy term. In the case of a smectic A crystal, using (8.15) we can express the magnetic energy in terms of tangential layer deformation $\partial u/\partial x$. Finally, we have [28]

$$\mathcal{F} = \frac{1}{2}B\left(\frac{\partial u}{\partial z}\right)^2 + \frac{1}{2}K_1\left[\left(\frac{\partial^2 u}{\partial x^2}\right) + \left(\frac{\partial^2 u}{\partial y^2}\right)\right]^2 - \frac{1}{2}\chi_a H^2 \left(\frac{\partial u}{\partial x}\right)^2. \quad (8.73)$$

We assume that the layers undergo a periodic distortion along the x direction. Then the fundamental smectic variable, a layer displacement $u(x, z)$ takes the

form
$$u(x, z) = u_0(z) \cos kx, \qquad (8.74)$$

where k is a certain wave vector, the optimal value of which will be obtained later. The displacement $u(x, z)$ must vanish on both plates. For small amplitude distortions just above threshold we can take $u_0(z)$ as a sine wave vanishing both for $z = 0$ and $z = d$

$$u_0(z) = u_0 \sin k_z z, \qquad k_z = \frac{\pi}{d}. \qquad (8.75)$$

This distortion of the layers corresponds to an optical axis locally defined by

$$n_x = -\frac{\partial u}{\partial x} = \varsigma \sin k_z z \sin kx, \quad n_y = 0, \quad n_z \simeq 1. \qquad (8.76)$$

Combining three last equations, substituting them into the free energy expression and averaging over the sample thickness we obtain

$$\langle \mathcal{F} \rangle = \frac{\varepsilon^2}{8} B \left[\left(\frac{k_z}{k} \right)^2 + k^2 \lambda^2 \right] - \frac{\varsigma^2}{8} \chi_a H^2, \qquad (8.77)$$

where we have taken into account that

$$\langle \sin^2 \theta \rangle = \langle \cos^2 \theta \rangle = \frac{1}{2}, \quad \lambda = \sqrt{\frac{K_1}{B}},$$

and $\langle \rangle$ means the operation of averaging over the sample thickness. When the overall coefficient of ς^2 in the equation for $\langle \mathcal{F} \rangle$ is positive for all k values, the unperturbed arrangement is stable. Instability will occur first for that value of k at which the first bracket in $\langle \mathcal{F} \rangle$ is minimum which corresponds to

$$k^2 = \frac{k_z}{\lambda} = \frac{\pi}{\lambda d}. \qquad (8.78)$$

The optimal wavelength of the distortion $(2\pi/k)$ is equal to the geometric mean of the sample thickness d and of the microscopic length b. The threshold field H_c is obtained when the elastic and magnetic terms in $\langle \mathcal{F} \rangle$ cancel exactly. We find [28]

$$\chi_a H_c^2 = 2\pi \frac{B\lambda}{d} = 2\pi \frac{K_1}{\lambda d}, \qquad (8.79)$$

or in the MKS system of units

$$\chi_a H_c^2 = 2\pi \frac{K_1}{\mu_0 \lambda d}.$$

It is seen that H_c is proportional to $d^{-\frac{1}{2}}$ which is quite different from the conventional Fredericks transition where $H_c \sim d^{-1}$. Here the field decreases more slowly with sample thickness. Using the estimations $K_1 = 10^{-6}$ dynes, $\lambda = 20$ Å, $d = 1$ mm, $\chi_a = 10^{-7}$ in CGS units, we find $H_c \sim 60$ kgauss which corresponds to the induction 60 tesla which is, in principle, achievable. We can reduce the value of H_c to a more convenient range, say ~ 20 kgauss, using smectic monocrystals of thickness 1 cm. Such samples have been prepared. However, there is one serious difficulty: above the threshold field H_c the distortion amplitude ς remains very small since the undulation of the layers is strongly limited by the requirement of nearly constant interlayer thickness. In such a situation we have a ghost transition which is formally present but hard to observe [28].

8.3 Electrohydrodynamic Instabilities in Nematic Liquid Crystals

8.3.1 Domain Formation in an External Electric Field

In the previous section we considered the orientational effects in liquid crystals subjected to the static magnetic field. It seems that the similar effects would occur when an electric field $\boldsymbol{E}$ is applied. The dielectric contribution to the liquid crystal free energy density has the form [28, 29, 34]:

$$\mathcal{F}_{\text{diel}} = -\frac{1}{8\pi}\varepsilon_a \left(\boldsymbol{n} \cdot \boldsymbol{E}\right)^2, \tag{8.80}$$

or in the MKS system of units

$$\mathcal{F}_{\text{diel}} = -\frac{1}{2}\varepsilon_0 \varepsilon_a \left(\boldsymbol{n} \cdot \boldsymbol{E}\right)^2,$$

where $\varepsilon_a = \varepsilon_\parallel - \varepsilon_\perp$ is the dielectric constant anisotropy, $\varepsilon_\parallel$ and $\varepsilon_\perp$ are the dielectric constants along and perpendicular to the unperturbed director $\boldsymbol{n}$. For $\varepsilon_a > 0$ a reorientation of the director similar to the magnetic case occurs when the electric field is applied along the z-axis which is perpendicular to the walls of the nematic cell and director $\boldsymbol{n}$. The threshold field for the dielectric realignment is analogous to the threshold magnetic field for the Fredericks transition and has the form [34]

$$E_{\text{th}} = \frac{\pi}{d}\sqrt{\frac{4\pi K_1}{\varepsilon_a}}, \tag{8.81}$$

or in the MKS system of units

$$E_{\text{th}} = \frac{\pi}{d}\sqrt{\frac{K_1}{\varepsilon_0 \varepsilon_a}}.$$

However, when $\varepsilon_a < 0$, some interesting complications arise [34]. It has been found that at low frequencies the director tends to align parallel to the applied

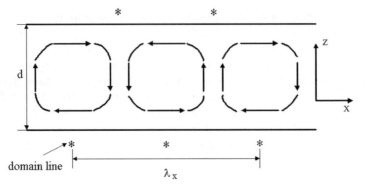

Fig. 8.10. Cross-section of the flow pattern in the WDM. The periodicity $\lambda_z \simeq 2d$. In adjacent flow cells the vorticity is opposite. The asterisks denote the domain lines

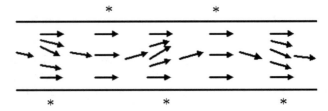

Fig. 8.11. Director pattern in the WDM, corresponding with the flow pattern

electric field, contrary to the expectations. At high frequencies this anomalous alignment also occurred, although at much higher electric fields. This perturbed state above the threshold electric field is characterized by regular parallel striations, perpendicular to the original direction of the director with a period about equal to the thickness of the sample. This domain pattern is known as the Williams domain mode (WDM) [34]. The visible domain pattern is closely connected with the existence of a regular pattern of cellular fluid motion. A schematic drawing of the flow pattern is given in Fig. 8.10.

Unlike ordinary liquid, in liquid crystals specific hydrodynamic modes exist caused by the director motion. The disturbed director pattern corresponding to the flow of liquid crystal as a whole is presented in Fig. 8.11.

When the voltage increased the liquid crystal becomes turbulent and is capable of strong light scattering. It is therefore called dynamic scattering mode. The optical pattern of the perturbed state in the high frequency region is characterized by periodic parallel striations of much shorter period (a few micrometers) than the classical Williams domains. Above threshold these striations move and bend and give rise to the so-called chevron pattern [34]. This regime corresponds to the fast-turnoff mode. At still higher fields turbulence also occurs [34].

The systematic study of these anomalous alignment effects showed that they are due to the anisotropy of the conductivity of nematic liquid crystals. When the liquid crystal possessing the anisotropic conductivity is in an electric field a space charge distribution can exist. These charges may interact with the applied field, giving rise to forces not accounted for by the anisotropy of the dielectric constant. The way in which the anomalous alignment occurs is rather complex, representing an interplay of several mechanisms. The space charges interact with the applied field, which depending on the sign of charges, causes material flow along and opposite to the field direction. The flow is therefore accompanied by shear, which in turn exerts a torque on the director. This space-charge-induced torque is counteracted by the elastic and dielectric torques exerted on the director. The cause of the effect of anomalous alignment is therefore essentially hydrodynamic in nature [34].

8.3.2 General System of Equations

In order to describe the electrohydrodynamic instabilities in liquid crystals we have to consider the full set of equations of motion for the director and the fluid. These equations have been derived for the most general case and analyzed in details in [28, 29], and we will only write the simplified equations of motion for the director n and fluid velocity v in an electric field E, which are necessary for the further analysis. We emphasize the fact that the additional equation of motion for the director as the dynamic variable emerges due to the orientational ordering of liquid crystals. This equation describes the balance of torques acting on the director, and it has the form [34]:

$$-\left[n \times \frac{\delta \mathcal{F}}{\delta n}\right] - \gamma_1 \left[n \times \frac{dn}{dt}\right] + \frac{1}{2}\gamma_1 \left[\mathrm{curl}\, v \times n\right] - \gamma_2 \left[n \times \left((A_{ij}) \cdot n\right)\right] = 0. \tag{8.82}$$

Here $\delta \mathcal{F}/\delta n$ is the functional derivative of the total free energy of the liquid crystal which has been already used in the previous section; $\gamma_{1,2}$ are the shear viscosity coefficients of the order of 1 to 10^{-1} poise; the shear rate tensor A_{ij} has the form

$$A_{ij} = A_{ji} = \frac{1}{2}\left(\frac{\partial v_i}{\partial x_j} + \frac{\partial v_j}{\partial x_i}\right). \tag{8.83}$$

The equation of motion for the moving fluid, i.e. the liquid crystal, states [34]

$$\varrho_m \frac{dv}{dt} = -\mathrm{grad}P + \mathrm{div}\overline{t'} + \rho E, \tag{8.84}$$

where ϱ_m is the mass density of the nematic liquid crystal, P is the hydrostatic pressure, ρ is the field induced space charge density, and $\overline{t'}$ is the viscous stress tensor. In the most general case it has the form [28, 29, 34]:

$$t'_{ij} = \alpha_1 n_k n_p A_{kp} n_i n_j + \alpha_2 n_j N_i + \alpha_3 n_i N_j + \alpha_4 A_{ij} + \alpha_5 n_i n_k A_{kj} + \alpha_6 n_j n_k A_{ki}. \tag{8.85}$$

Here N presents the internal motion of the director with respect to the fluid

$$\boldsymbol{N} = \frac{d\boldsymbol{n}}{dt} - \frac{1}{2}[\mathrm{curl}\boldsymbol{v} \times \boldsymbol{n}] \tag{8.86}$$

and α_i are viscosity coefficients associated with the stress. It should be noted that for an isotropic fluid only one term $\alpha_4 A_{ij}$ remains in (8.85). It can be shown that

$$\gamma_1 = \alpha_3 - \alpha_2, \ \gamma_2 = \alpha_6 - \alpha_5, \ \alpha_{2,3} < 0, \ |\alpha_2| \gg |\alpha_3|. \tag{8.87}$$

The order of magnitude of a typical term of the force due to the viscous stress is $\alpha v/l^2$, where $\alpha \sim 10^{-1}$ poise, $l \sim (1 \div 100)$ μm is the thickness of the sample. Therefore the inertial term in the left-hand side of (8.84) $\varrho_m dv/dt \sim \varrho_m \omega v$ can be neglected up to frequencies of the order of 10^6 Hz. Equation (8.84) then takes the form [34]:

$$-\mathrm{grad}P + \mathrm{div}\overline{t'} + q\boldsymbol{E} = 0. \tag{8.88}$$

The nematic liquid crystal can be considered as incompressible, and finally have

$$\mathrm{div}\boldsymbol{v} = 0. \tag{8.89}$$

Equations (8.82), (8.88) and (8.89) describe the behavior of liquid crystal in an external electric field. They are the basis for the treatment of the electrodynamics instabilities. The system must be completed with the equations defining the distribution of the space charge ρ and the change of the electric field caused by the space charge-director coupling, i.e. the continuity equation

$$\frac{d\rho}{dt} + \mathrm{div}\boldsymbol{j} = 0 \tag{8.90}$$

and the Poisson equation (in the MKS system of units)

$$\mathrm{div}\boldsymbol{D} = \rho, \tag{8.91}$$

where the electric current density $j_i = \sigma_{ik} E_k$ and the electric induction $D_i = \varepsilon_0 \varepsilon_{ik} E_k$. The conductivity tensor σ_{ik} has the form:

$$\sigma_{ik} = \sigma_\perp \delta_{ik} + \sigma_a n_i n_k, \ \sigma_a = \sigma_\parallel - \sigma_\perp \tag{8.92}$$

and $\sigma_\parallel, \sigma_\perp$ are the conductivity along and perpendicular the unperturbed director, respectively.

The unperturbed state is assumed to be a uniform planar structure shown in Fig. 8.12. We will consider small thermal deviations from the uniform

8.3 Electrohydrodynamic Instabilities in Nematic Liquid Crystals 319

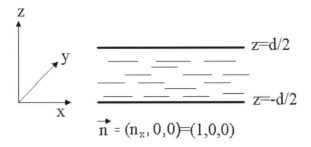

Fig. 8.12. Uniform planar texture

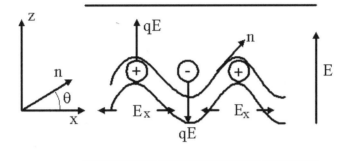

Fig. 8.13. Periodic director pattern with charge separation in a Carr–Helfrich type instability. $n = n(x)$, $\sigma_a > 0$, $\varepsilon_a < 0$

alignment and see whether they grow or damp. In the geometry chosen the director has the components

$$n_x = \cos\theta, \; n_y = 0, \; n_z = \sin\theta. \tag{8.93}$$

The angle of the director orientation θ is supposed to be small, and in the further analysis we keep only terms linear in θ. We also assume that θ is a periodic function of x with a periodicity length $\lambda_x = 2\pi/k_x$ (see Fig. 8.13). As long as the boundary conditions are neglected, all quantities considered depend only on x, not z. If the electric field E_0 is applied along the z-axis, (8.90) and (8.91) take the form, respectively:

$$\frac{d\rho}{dt} + \sigma_\parallel \frac{\partial E_x}{\partial x} + \sigma_a E_0 \frac{\partial \theta}{\partial x} = 0 \tag{8.94}$$

and

$$\frac{\partial E_x}{\partial x} = \frac{\rho}{\varepsilon_0 \varepsilon_\parallel} - \frac{\varepsilon_a}{\varepsilon_\parallel} E_0 \frac{\partial \theta}{\partial x}, \tag{8.95}$$

where E_x is the field along the x-axis, induced by the space charge. Combining (8.94) and (8.95) we obtain the relation between the space charge ρ and the curvature $\partial\theta/\partial x$ determined by the periodic disturbance of the director

$$\frac{d\rho}{dt} + \frac{\sigma_\| \rho}{\varepsilon_0 \varepsilon_\|} + \sigma_\| \left(\frac{\varepsilon_\perp}{\varepsilon_\|} - \frac{\sigma_\perp}{\sigma_\|} \right) E_0 \frac{\partial \theta}{\partial x} = 0. \qquad (8.96)$$

Without an applied field E_0 or with a uniform director pattern one has

$$\rho = \rho_0 \exp\left(-\frac{t}{\tau}\right), \qquad (8.97)$$

where $\tau = \varepsilon_0 \varepsilon_\| / \sigma_\|$ is the dielectric relaxation time. The charge is damped out within a time $\tau \simeq 10^{-3}$s. For nematic liquid crystals the conductivity anisotropy is positive; $\sigma_\| / \sigma_\perp \simeq 3/2$. For a negative dielectric anisotropy $(\varepsilon_\perp / \varepsilon_\| > 1)$ the quantity

$$\sigma_H = \sigma_\| \left(\frac{\varepsilon_\perp}{\varepsilon_\|} - \frac{\sigma_\perp}{\sigma_\|} \right) > 0.$$

We need to find now the equation for the curvature $\partial\theta/\partial x$. Consider the balance of torques acting on the director, described by (8.82). In our case the free energy is the sum of the elastic free energy (8.8), dielectric free energy (8.80) and, in general, the magnetic free energy (8.23), if we also apply a magnetic field $\boldsymbol{H}$ along the x-axis. The functional derivative $\delta F/\delta \boldsymbol{n}$ is equal to the left-hand side of (8.27). Then, using the explicit form of the director (8.93) and quantities A_{ij} (8.83), t'_{ij} (8.85) and $\boldsymbol{N}$ (8.86), and keeping only the first order terms, we obtain after some algebra [34]

$$\eta'' \frac{\partial^2 \theta}{\partial x \partial t} + \left(k_x^2 K_3 + \mu_0 \chi_a H^2 - \varepsilon_0 \varepsilon_a \frac{\varepsilon_\perp}{\varepsilon_\|} E_0^2 \right) \frac{\partial \theta}{\partial x} - \left(\frac{\varepsilon_a}{\varepsilon_\|} + \frac{\alpha_2}{\eta_1} \right) q E_0 = 0, \qquad (8.98)$$

where the viscosity coefficients η'' and η_1 have the form

$$\eta'' = \gamma_1 - \frac{\alpha_2^2}{\eta_1}, \quad \eta_1 = \frac{1}{2}(-\alpha_2 + \alpha_4 + \alpha_5) \qquad (8.99)$$

The conditions for the eventual stability of the system should be derived from the coupled equations for the space charge ρ (8.96) and the curvature $\partial\theta/\partial x$ (8.98). In the case $E_0 = 0$ they reduce to the uncoupled equations. The space charge in such a case has the form (8.97), while the curvature $\partial\theta/\partial x$ evolves according to the expression

$$\frac{\partial \theta}{\partial x} = \left(\frac{\partial \theta}{\partial x} \right)_0 \exp\left(-\frac{t}{\tau_0} \right), \qquad (8.100)$$

where the relaxation time τ_0 is

$$\tau_0 = \frac{\eta''}{(k_x^2 K_3 + \mu_0 \chi_a H^2)}. \qquad (8.101)$$

Any fluctuation of the space charge ρ and the curvature $\partial\theta/\partial x$ decays to zero, as it is seen from (8.97) and (8.100). The system is stable against the fluctuations according to the criteria mentioned above. When the electric field E_0 is applied, ρ and $\partial\theta/\partial x$ are coupled, and stability against fluctuations is no longer obvious. We investigate the excitations both with dc and ac electric fields.

8.3.3 Excitations with a DC Electric Field

We describe the time dependence of the space charge ρ and the curvature $\partial\theta/\partial x$ by the factor $\exp(gt)$ where g is a number which amy be complex. it has been shown above that for $E_0 = 0$ g is real and negative. The fluctuations grow in time when $\mathrm{Re}g > 0$, while the onset of the instability is characterized by $\mathrm{Re}g = 0$. At threshold WDM does not exhibit oscillatory motion. Therefore we assume that $\mathrm{Im}g = 0$. Taking into account these conditions we have at the threshold

$$\frac{d\rho}{dt} = 0, \quad \frac{\partial^2\theta}{\partial x \partial t} = 0. \tag{8.102}$$

Then equations (8.96) and (8.98) take the form [34]:

$$\rho = -\sigma_H \tau E_0 \frac{\partial\theta}{\partial x} \tag{8.103}$$

and

$$\left(k_x^2 K_3 + \mu_0 \chi_a H^2 - \varepsilon_0 \varepsilon_a \frac{\varepsilon_\perp}{\varepsilon_\parallel} E_0^2\right)\frac{\partial\theta}{\partial x} - \left(\frac{\varepsilon_a}{\varepsilon_\parallel} + \frac{\alpha_2}{\eta_1}\right)\rho E_0 = 0. \tag{8.104}$$

The first term in (8.104) is the sum of stabilizing torques of elastic, magnetic and dielectric origin; the second term is a torque that destabilizes the system due to the space charge ρ. It should be noted that both ε_a and α_2 are supposed to be negative, while $\sigma_H \tau$ and η_1 are positive. The destabilizing torque caused by the space charge consists of two parts. The first part $\varepsilon_a \rho E_0/\varepsilon_\parallel$ is a dielectric torque due to the induced transverse field E_x. Since

$$|\varepsilon_a|\frac{\sigma_H \tau}{\varepsilon_\parallel} = \frac{\varepsilon_0|\varepsilon_a|\varepsilon_\perp}{\varepsilon_\parallel}\left(1 - \frac{\varepsilon_\parallel \sigma_\perp}{\varepsilon_\perp \sigma_\parallel}\right) < \frac{\varepsilon_0|\varepsilon_a|\varepsilon_\perp}{\varepsilon_\parallel}, \tag{8.105}$$

the total torque of dielectric origin always stabilizes. The second part, $-\alpha_2 \rho E_0/\eta_1$, due to the induced shear flow is essential for the existence of the instability, which therefore is really hydrodynamic in nature [34]. Substituting expression (8.103) into (8.104) we finally obtain containing only the curvature $\partial\theta/\partial x$

$$\left[k_x^2 K_3 + \mu_0 \chi_a H^2 - \varepsilon_0 \varepsilon_a \frac{\varepsilon_\perp}{\varepsilon_\parallel} E_0^2 + \sigma_H \tau \left(\frac{\varepsilon_a}{\varepsilon_\parallel} + \frac{\alpha_2}{\eta_1}\right) E_0^2\right]\frac{\partial\theta}{\partial x} = 0. \tag{8.106}$$

A nonvanishing curvature $\partial\theta/\partial x$ requires the sum in brackets to be zero. This condition yields the threshold value of the electric field applied for the instability excitation. It is convenient to express the threshold value in terms of the dc voltage $V_{\text{th}} = E_0 d$ applied to the liquid crystal sample, where d is the thickness of the sample. Note that the minimum allowable value for $k_x = \pi/d$. Then we have [28]:

$$V_{\text{th}}^2 = \frac{V_0^2}{(\varsigma^2 - 1)}, \tag{8.107}$$

where

$$\varsigma^2 = \left(1 + \frac{\alpha_2 \varepsilon_\|}{\eta_1 \varepsilon_a}\right)\left(1 - \frac{\varepsilon_\| \sigma_\perp}{\varepsilon_\perp \sigma_\|}\right), \quad V_0^2 = \frac{\pi^2 K_3 \varepsilon_\|}{\varepsilon_0 \varepsilon_\perp |\varepsilon_a|}. \tag{8.108}$$

A positive value of V_{th}^2 or a real value for V_{th} requires $\varsigma^2 > 1$. Equation (8.107) defines a finite threshold voltage V_{th} independent of the sample thickness as long as a magnetic field is weak $\mu_0 \chi_a H^2 \ll k_x^2 K_3$, and the magnetic field term can be neglected [34].

We have considered two cases:

1. Both the dielectric constant anisotropy and the conductivity anisotropy are positive:

$$\varepsilon_a > 0, \; \sigma_a > 0.$$

Then the Fredericks transition occurs.

2. The dielectric constant anisotropy is negative and the conductivity anisotropy is positive:

$$\varepsilon_a < 0, \; \sigma_a > 0.$$

Then, if conditions (8.107) and (8.108) are met, WDM instability would be possible.

Two more cases can exist:

1. when both anisotropies are negative;
2. the dielectric constant anisotropy is positive while the conductivity anisotropy is negative.

In the former situation no instability is expected since both the elastic torque and the dielectric one tend to stabilize the structure. For the latter situation an instability is predicted created mainly by the dielectric torques [28].

8.3.4 Excitation with AC Electric Field

Consider the electrohydrodynamic instabilities due to excitation with an ac electric field

$$E(t) = E_\mathrm{m} \cos \omega t. \tag{8.109}$$

For the sake of simplicity we denote the curvature $\partial \theta / \partial x = \psi$ and rewrite (8.96) and (8.98) [28, 34]:

$$\frac{d\rho}{dt} + \frac{\rho}{\tau} + \sigma_H E(t)\psi = 0 \tag{8.110}$$

and

$$\frac{d\psi}{dt} + \frac{1}{T}\psi + \frac{1}{\eta}\rho E(t) = 0, \tag{8.111}$$

where

$$T = \eta'' \left(k_x^2 K_3 + \mu_0 \chi_a H^2 + \varepsilon_0 |\varepsilon_a| \frac{\varepsilon_\perp}{\varepsilon_\parallel} E_0^2 \right)^{-1}, \quad \eta = \eta'' \left(\frac{|\varepsilon_a|}{\varepsilon_\parallel} - \frac{\alpha_2}{\eta_1} \right)^{-1}. \tag{8.112}$$

Equations (8.110) and (8.111) must be solved in the presence of a given ac field (8.109). The general solutions have the form [28]:

$$\rho(t) = \rho_p(t) \exp st, \quad \psi(t) = \psi_p(t) \exp st, \tag{8.113}$$

where $\rho_p(t)$ and $\psi_p(t)$ are periodic functions of period $2\pi/\omega$, and s is a parameter depending on the field amplitude E_m. The threshold is obtained when the real part of s vanishes. The exact solution of the system of (8.110) and (8.111) is rather complex, especially since the relaxation rate for orientation T^{-1} depends on the instantaneous value of the field $E(t)$ [28]. For the sake of definiteness we two limiting regimes. The detailed analysis is presented in Ref. [34].

Consider first the low frequency range. At low frequencies ω, the fields E near threshold are rather small, and T is rather large: $\omega T > 1$, and the only important Fourier component of $\psi(t)$ is at zero frequency $\overline{\psi}$. Then the charge source, proportional to $-\psi E$ according to (8.110), is a simple sinusoidal wave $-\sigma_H \overline{\psi} E_\mathrm{m} \cos \omega t$. The corresponding charge is given by

$$\rho(t) = q' \cos \omega t + q'' \sin \omega t, \tag{8.114}$$

where

$$q' = -\sigma_H \tau \overline{\psi} E_\mathrm{m} \left(1 + \omega^2 \tau^2\right)^{-1}.$$

Consider now the equation for the average curvature $\overline{\psi}$ obtained by averaging (8.111) over one period. It reduces to

$$\overline{\psi}\left(T\right)^{-1} + \frac{1}{2\eta}q'E_{\mathrm{m}} = 0, \tag{8.115}$$

where $(\overline{T})^{-1}$ is the time-average of the relaxation rate (8.112). This gives a threshold condition [28]:

$$1 = \frac{E_{\mathrm{m}}^2}{2}\frac{\sigma_H \tau \overline{T}}{\eta}\left(1+\omega^2\tau^2\right)^{-1},$$

which can transformed using (8.112) into the following expression

$$\overline{E^2} = \frac{E_{\mathrm{m}}^2}{2} = \frac{\varepsilon_{\parallel}}{\varepsilon_0 \left|\varepsilon_a\right|\varepsilon_{\perp}} \left(K_3 k_x^2 + \mu_0 \chi_a H^2\right) K_3 k_x^2 \frac{1+\omega^2\tau^2}{\varsigma^2 - (1+\omega^2\tau^2)}, \tag{8.116}$$

where the parameter ς^2 is defined in (8.108). The threshold field will correspond to the minimum allowable wave vector k_x. It can be assumed to be π/d [34], as in the previous case. Then we obtain from (8.116):

$$V_{\mathrm{th}}\left(\omega\right) = V_{\mathrm{th}}\left(0\right)\sqrt{\frac{1+\omega^2\tau^2}{\varsigma^2 - (1+\omega^2\tau^2)}}, \tag{8.117}$$

where

$$V_{\mathrm{th}}\left(0\right) = \pi\sqrt{\frac{\varepsilon_{\parallel}\left(K_3 + \mu_0 \chi_a H^2 k_x^{-2}\right)}{\varepsilon_0 \left|\varepsilon_a\right|\varepsilon_{\perp}}}.$$

Equation (8.117) shows that $V_{\mathrm{th}}(\omega)$ increases with ω and finally becomes very large when ω approaches a cut-off frequency

$$\omega_c = \frac{1}{\tau}\sqrt{\varsigma^2 - 1}.$$

A magnetic field applied along the x-axis exerts a stabilizing torque. It is seen from (8.117) that the threshold becomes higher for increasing H. For $\mu_0 \chi_a H^2 > K_3 k_x^2$ the magnetic field takes over and one gets a rather high threshold field [34]. At the threshold, for all frequencies $\omega < \omega_c$, the instability pattern corresponds to a static distortion $\overline{\psi} \neq 0$ and to oscillating charges. For this reason it is referred to as the conducting regime [28].

Now we turn to the high frequency range. The above calculation breaks down when ω reaches ω_c, because in this region the fields become large, ωT reaches values of order of unity, and the curvature ψ becomes time dependent. The situation simplifies in the limiting case $\omega \gg \omega_c$, which implies $\omega\tau > 1$. Then the charge ρ cannot follow the excitation: $\rho \to \overline{\rho}$ [28]. The bulk force $\rho E \to \overline{\rho} E_{\mathrm{m}} \cos \omega t$ is sinusoidal, and the curvature response

$$\Psi(t) = \frac{\partial \theta}{\partial x}\left[\frac{\sigma_H \eta''}{\left(\left|\varepsilon_a\right|/\varepsilon_{\parallel}\right) - (\alpha_2/\eta_1)}\right]^{1/2}$$

8.3 Electrohydrodynamic Instabilities in Nematic Liquid Crystals

is a complicated periodic function of time that can be written as a Fourier series

$$\Psi(t) = \sum_{n=0,1...} (\psi'_n \cos n\omega t + \psi''_n \sin n\omega t). \quad (8.118)$$

Neglecting the higher harmonics of the curvature and using the time averaged space charge we obtain two equations for the amplitudes of the first harmonic ψ'_1 and ψ''_1 [34]:

$$\psi''_1 + \frac{1}{\eta''\omega}\left\{K_3 k_x^2 + \mu_0 \chi_a H^2 + \frac{\varepsilon_0 |\varepsilon_a| \varepsilon_\perp E_m^2}{2\varepsilon_\parallel}\left(\frac{3}{2} - \varsigma^2\right)\right\}\psi'_1 = 0 \quad (8.119)$$

$$-\psi'_1 + \frac{1}{\eta''\omega}\left(K_3 k_x^2 + \mu_0 \chi_a H^2 + \frac{1}{2\varepsilon_\parallel}\varepsilon_0 |\varepsilon_a|\varepsilon_\perp E_m^2\right)\psi''_1 = 0. \quad (8.120)$$

For nonvanishing ψ'_1 and ψ''_1 the determinant of (8.119) and (8.120) should be zero which yields the condition for the threshold field [34]:

$$\frac{1}{(\eta''\omega)^2}\left(K_3 k_x^2 + \mu_0 \chi_a H^2 + \frac{1}{2\varepsilon_\parallel}\varepsilon_0 |\varepsilon_a|\varepsilon_\perp E_m^2\right)$$

$$\times \left\{\frac{\varepsilon_0 |\varepsilon_a|\varepsilon_\perp E_m^2}{2\varepsilon_\parallel}\left(\varsigma^2 - \frac{3}{2}\right) - K_3 k_x^2 - \mu_0 \chi_a H^2\right\} = 1. \quad (8.121)$$

It is seen from (8.121) that a positive value of E_m^2 requires $\varsigma^2 > 3/2$. The threshold field corresponds to the minimum value of E_m^2 that satisfies (8.121). It depends on the elastic and magnetic parts of the free energy which is proportional to k_x^2 for small magnetic field H. For $\varsigma^2 > 2$ the minimal values of E_m^2 and the elastic and magnetic parts of the free energy have the form [34]:

$$\frac{\varepsilon_0 |\varepsilon_a|\varepsilon_\perp E_m^2}{\varepsilon_\parallel \eta''\omega} = \frac{1}{\varsigma^2 - 1}, \quad \frac{K_3 k_x^2 + \mu_0 \chi_a h}{\eta''\omega} = \frac{\varsigma^2 - 2}{\varsigma^2 - 1}. \quad (8.122)$$

The threshold field in this frequency region is proportional to the square root of ω and independent of k_x and the magnetic field H. It is much higher than the corresponding threshold in the low frequency region. The second equation (8.122) shows that the wave vector of the oscillating curvature is independent of the thickness, and it is a function of ω and magnetic field H.

In the high frequency regime $\omega \gg \omega_c$ the molecular pattern oscillates while the charges are static. For this reason it is called the dielectric regime [28]. Originally it has been called the fast turn-off mode because the relaxation time for the curvature when the field is turned off is much smaller than in the conduction regime, and the striation pattern disappears rapidly [28, 34].

The electrohydrodynamic instabilities of the type described have been observed for a range of materials and sample thicknesses, and the experimental results showed a good accord with the theory [28, 34].

8.4 Excitation of Second Sound in a Smectic A Liquid Crystal

The hydrodynamics of smectic A liquid crystals is thoroughly investigated in [28, 29, 35]. Here we discuss very briefly the equations of motion that are necessary for our analysis.

Smectic A liquid crystals being a layered structure possess a translational ordering in the direction perpendicular to the layers, as was mentioned above. The molecules are constrained to lie in parallel planes and the symmetry is broken by displacement of layers in the z direction perpendicular to them. As a result, the dynamics of smectic A liquid crystals is essentially different from the one of nematics considered in the previous section. The role of director reorientation is negligible for the molecules which are strictly bounded within the layers, and equation of motion (8.82) for the director is not considered. The basic dynamical variable is the layer displacement along the z-axis $u(\bm{r},t)$ (8.14). The main contribution to the elastic part of the free energy (8.19) is given by the term $B\left(\partial u/\partial z\right)^2$ determined by the normal deformation of the layer since $B \sim 10^7$ joule/m$^3 \gg K_1 k^2$ even for sufficiently large k, and therefore the orientational part of the free energy can be neglected, if $k_z \neq 0$. At the high frequency regime, which is typical for sound wave propagation the inertial term in the left-hand side of the equation of motion (8.84) must be remained. The restoring force has only z component since the displacement $u(\bm{r},t)$ is parallel to the z-axis. Unlike the nematic liquid crystals, the smectics are characterized by a specific equation which expresses the continuity of the layers in the vertical direction. For sufficiently high frequencies we have [28]:

$$\frac{\partial u}{\partial t} = v_z. \qquad (8.123)$$

The viscous tensor t'_{ij} (8.85) for the smectic A liquid crystal is the same as for the nematic one. Smectic A liquid crystals can be regarded incompressible, and (8.89) is still valid. Taking into account these considerations it has been shown that two types of weakly coupled acoustic waves can propagate in a smectic A liquid crystal [28, 29, 35].

1. One acoustic branch is associated with mass density fluctuations and a velocity essentially independent of the direction of propagation. This is called a "first sound".
2. The other acoustic branch describes changes in the interlayer distance, which take place at a nearly constant mass density. The velocity s for this branch is much smaller than the first sound velocity since the elastic constant B is much smaller than the elastic modulus associated with a bulk compression. The velocity s is strongly dependent on the angle φ between the oscillation wave vector $\bm{k}$ and the z-axis.

8.4 Excitation of Second Sound in a Smectic A Liquid Crystal

$$s = \sqrt{\frac{B}{\varrho_m}} \sin\varphi \cos\varphi. \tag{8.124}$$

This branch is called "second sound". It is seen from (8.124) that the second sound vanishes for the direction of propagation parallel or perpendicular to the layers. The second sound has been observed experimentally [28, 29].

It is reasonable to expect that the excitation of the second sound should occur at lower threshold energy as compared to the ordinary sound. We consider the second sound excitation in the conducting smectic A liquid crystal when the dc electric field is applied, and the excitation in the insulating smectic A liquid crystal subjected to the ac electric field [36].

Consider first the weakly conducting smectic A liquid crystal subjected to the static electric field $\boldsymbol{E}_0 = (E_{0x}, 0, E_{0z})$ where the x- and z-axes are chosen to be parallel and perpendicular to the layers, respectively. Then the free energy density $\mathcal{F}$ has the form:

$$\mathcal{F} = \frac{1}{2} B \left(\frac{\partial u}{\partial z}\right)^2 - \frac{1}{2}\varepsilon_0 \varepsilon_{ik} E_i E_k, \tag{8.125}$$

where the field in the liquid crystal has the form

$$E_i = E_{0i} - \frac{\partial V}{\partial x_i}, \quad \frac{\partial V}{\partial x_i} \ll E_{0i}, \tag{8.126}$$

and V is an electrostatic potential due to the space charge deviation. The dielectric constant tensor is dependent on the layer deformations [28, 37, 38]:

$$\varepsilon_{xx} = \varepsilon_\perp + a_\perp \frac{\partial u}{\partial z}, \quad \varepsilon_{zz} = \varepsilon_\| + a_\| \frac{\partial u}{\partial z}, \quad \varepsilon_{xz} = \varepsilon_{zx} = -\varepsilon_a \frac{\partial u}{\partial x}, \tag{8.127}$$

where $a_\perp$ and $a_\|$ are the constants of the order of magnitude of unity. Combining (8.83)–(8.85), (8.89), (8.123) and (8.125)–(8.127) we obtain the equation of motion for the layer displacement $u(x, z, t)$

$$-\varrho_m \nabla^2 \frac{\partial^2 u}{\partial t^2} + \left[\alpha_1 \frac{\partial^4}{\partial x^2 \partial z^2} + \frac{1}{2}(\alpha_4 + \alpha_{56})\nabla^2 \nabla^2\right] \frac{\partial u}{\partial t} + B \frac{\partial^4 u}{\partial x^2 \partial z^2}$$

$$= -\varepsilon_0 \frac{\partial^2}{\partial x^2} \left\{ \frac{\partial}{\partial z}\left[a_\perp E_{0x}\frac{\partial V}{\partial x} + a_\| E_{0z}\frac{\partial V}{\partial z}\right] - \varepsilon_a \frac{\partial}{\partial x}\left[E_{0x}\frac{\partial V}{\partial z} + E_{0z}\frac{\partial V}{\partial x}\right] \right\}. \tag{8.128}$$

In the high frequency dielectric regime the term qE can be neglected in the first approximation, while the space charge deviation is taken into account by means of the potential V. Equation (8.128) should be supplemented by the Poisson equation and the equation of continuity

$$\mathrm{div}\boldsymbol{D} = e(n - n_0), \quad n - n_0 = n_1 \ll n_0 \qquad (8.129)$$

$$\mathrm{div}\boldsymbol{j} + e\frac{\partial n_1}{\partial t} = 0. \qquad (8.130)$$

Here the current density is $j_i = en\mu_{ik}E_k$, the induction is $D_i = \varepsilon_0\varepsilon_{ik}E_k$, e is the charge, μ_{ik} is the mobility and n_0 is the equilibrium concentration of carriers. We linearize (8.128)–(8.130) in the small quantities

$$n_1, u, V \sim \exp i(k_x x + k_z z - \omega t), \qquad (8.131)$$

and we finally get

$$\left[-\omega^2 - 2i\omega\Gamma + \omega_s^2\right]u$$

$$+\frac{\varepsilon_0 k_x^2}{\rho k^2}\left[a_\perp E_{0x}k_xk_z + a_\| E_{0z}k_z^2 - \varepsilon_a\left(E_{0x}k_xk_z + E_{0z}k_x^2\right)\right]V = 0 \qquad (8.132)$$

$$-\varepsilon_0\left[a_\perp E_{0x}k_xk_z + a_\| E_{0z}k_z^2 - \varepsilon_a\left(E_{0x}k_xk_z + E_{0z}k_x^2\right)\right]u$$

$$+\varepsilon_0\left(\varepsilon_\perp k_x^2 + \varepsilon_\| k_z^2\right)V - en_1 = 0 \qquad (8.133)$$

$$n_0\left(\mu_\perp k_x^2 + \mu_\| k_z^2\right)V + i(\boldsymbol{k}\boldsymbol{v}_d - \omega)n_1 = 0, \qquad (8.134)$$

where ω_s and Γ are the second sound frequency and decay constant, respectively, $\boldsymbol{v}_d$ is the drift velocity of carriers

$$kv_d = k_x\mu_\perp E_{0x} + k_z\mu_\| E_{0z}, \quad \omega_s = \sqrt{\frac{B}{\varrho_m}\frac{k_xk_z}{k}}$$

$$\Gamma = \frac{1}{2\rho}\left[\alpha_1\frac{(k_xk_z)^2}{k^2} + \frac{1}{2}(\alpha_4 + \alpha_{56})k^2\right].$$

For the frequency range $(10^4 \div 10^6)$ Hz the wave vector $k \sim (1 \div 100)$ cm^{-1}, the decay constant $\Gamma \sim (1 \div 10^4)$ s$^{-1} \ll \omega_s$ and can be neglected. The nonvanishing solutions of (8.132)–(8.134) exist when the determinant of the system is zero. This condition yields the dispersion relation:

$$i(\boldsymbol{k}\boldsymbol{v}_d - \omega)\left\{\left(\varepsilon_\perp k_x^2 + \varepsilon_\| k_z^2\right)\left(\omega_s^2 - \omega^2\right) + \frac{\varepsilon_0 k_x^2}{\varrho_m k^2}G^2\right\}$$

8.4 Excitation of Second Sound in a Smectic A Liquid Crystal

$$+ \left(\omega_s^2 - \omega^2\right) \frac{\left(\sigma_\perp^0 k_x^2 + \sigma_\parallel^0 k_z^2\right)}{\varepsilon_0} = 0, \qquad (8.135)$$

where $\sigma_\perp^0 = en_0\mu_\perp, \sigma_\parallel^0 = en_0\mu_\parallel$ and

$$G = a_\perp E_{0x} k_x k_z + a_\parallel E_{0z} k_z^2 - \varepsilon_a \left(E_{0x} k_x k_z + E_{0z} k_x^2\right).$$

We find the solution of dispersion equation (8.135) in the vicinity of the second sound frequency ω_s:

$$\omega = \omega_s + \delta, \ |\delta| \ll \omega_s. \qquad (8.136)$$

Substituting (8.136) into (8.135), neglecting the small quantities of higher order and separating the real and imaginary parts we get the complex value of δ:

$$\delta = \mathrm{Re}\,\delta + i\mathrm{Im}\,\delta.$$

The real part of δ results in the small field induced shift of the sound frequency. The imaginary part that is responsible for a possible instability has the form:

$$\mathrm{Im}\,\delta = (\mathbf{k v}_d - \omega_s) \frac{k_x^2 G^2 \left(\sigma_\perp^0 k_x^2 + \sigma_\parallel^0 k_z^2\right)}{\varrho_m k^2 \omega_s}$$

$$\times \left\{ \frac{\left(\sigma_\perp^0 k_x^2 + \sigma_\parallel^0 k_z^2\right)^2}{\varepsilon_0^2} + \left[\frac{\varepsilon_0 k_x^2 G^2}{\varrho_m k^2 \omega_s} + 2\left(\mathbf{k v}_d - \omega\right)\left(\varepsilon_\perp k_x^2 + \varepsilon_\parallel k_z^2\right)\right]^2 \right\}^{-1}.$$

$$(8.137)$$

The onset of an instability occurs when $\mathrm{Im}\,\delta > 0$, which takes place when

$$\mathbf{k v}_d > \omega_s, \qquad (8.138)$$

which is analogous to the condition for the drift excitation of the ordinary acoustic waves. However, there are two important features typical for the second sound in the smectic A liquid crystal:

1. The threshold is lower than in the ordinary case because the second sound velocity is much slower than that of the ordinary sound.
2. The threshold value of the electric field strongly depends on the sound wave propagation direction and the angle ϑ between the field and the normal to the layers due to the anisotropy of the second sound dispersion relation (8.124).

Inserting $E_{0x} = E_0 \sin\vartheta$, $E_{0z} = E_0 \cos\vartheta$ and (8.124) into the threshold condition (8.138) we obtain

$$E_{0th} = \sqrt{\frac{B}{\varrho_m}} \frac{\sin\varphi\cos\varphi}{\mu_\perp \sin\varphi \sin\vartheta + \mu_\| \cos\varphi \cos\vartheta}. \quad (8.139)$$

Consider now the parametric excitation of the second sound in the insulating smectic A liquid crystal by the ac electric field $\boldsymbol{E}(t) = \boldsymbol{E}_0 \cos\omega_E t$. Accordingly, equation of continuity (8.134) vanishes. From (8.133) we get

$$V = \frac{G}{\left(\varepsilon_\perp k_x^2 + \varepsilon_\| k_z^2\right)} u. \quad (8.140)$$

The solution of (8.128) is sought to be

$$u = U(t) \exp\left[-\Gamma t + i\left(k_x x + k_z z\right)\right]. \quad (8.141)$$

Substituting expressions (8.140) and (8.141) into (8.128) and neglecting small terms $\sim \Gamma^2$ we obtain the following equation for the amplitude $U(t)$:

$$\frac{\partial^2 U}{\partial t^2} + \omega_s^2 \left\{ 1 + \frac{\varepsilon_0 E_0^2 g^2}{B k_z^2 \left(\varepsilon_\perp k_x^2 + \varepsilon_\| k_z^2\right)} \cos^2 \omega_E t \right\} U = 0, \quad (8.142)$$

where

$$g = a_\perp k_x k_z \sin\vartheta + a_\| k_z^2 \cos\vartheta - \varepsilon_a \left(k_x k_z \sin\vartheta + k_x^2 \cos\vartheta\right).$$

Clearly, the parameter $\varepsilon_0 E_0^2 / B \ll 1$ for any reasonable field E_0 since $B \sim 10^7$ joule/m^3. Then we can rewrite (8.142) in the standard form of the Mathieu equation:

$$\frac{\partial^2 U}{\partial t^2} + \omega_s^2 \left\{ 1 + h \cos 2\omega_E t \right\} U = 0, \quad h = \frac{\varepsilon_0 E_0^2 g^2}{2 B k_z^2 \left(\varepsilon_\perp k_x^2 + \varepsilon_\| k_z^2\right)} \ll 1. \quad (8.143)$$

It describes the parametric excitation of the oscillations with the frequency close to ω_s. The unstable solution increasing in time $U(t) \sim \exp pt$, $p > 0$ can exist, if the frequency deviation $\xi = 2(\omega_E - \omega_s)$ belongs to the interval

$$|\xi| < \frac{1}{2} h\omega_s,$$

which yields the threshold value of the field

$$E_{0th}^2 > \left|\frac{\omega_E}{\omega_s} - 1\right| \frac{8 B k_z^2 \left(\varepsilon_\perp k_x^2 + \varepsilon_\| k_z^2\right)}{\varepsilon_0 g^2}. \quad (8.144)$$

For the amplification of the oscillations the increment p should compensate the decay constant Γ. This condition yields

$$p = \frac{1}{2}\sqrt{\left(\frac{1}{2} h\omega_s\right)^2 - \xi^2} > \Gamma.$$

The last inequality defines the level of the ac field that is necessary for the transition to the essentially non-linear regime.

8.5 Electromagnetic Wave Propagation in a Cholesteric Liquid Crystal

The optics of liquid crystals represents a separate field of both theoretical and experimental research. Liquid crystals are characterized by unique optical properties possessing at the same time a strong anisotropy, spatial inhomogeneity, dispersion and high non-linearity. The literature concerning these problems is, indeed, enormous. The state-of-art is reviewed, for example, in [39]. The non-linear optics of smectic A liquid crystals, in particular, is studied in [40]. The comprehensive investigation of these problems is impossible and, generally speaking, is not necessary in the framework of this book. Our objective is more limited: to discuss the basic theory of light propagation in liquid crystals which would be instructive as an exciting example of application of the general methods studied in Chap. 5. From this point of view, the cholesteric liquid crystals are the most attractive species because they are anisotropic, inhomogeneous in space and have a helicoidal structure with a period close to the wavelength of infrared and visible radiation. As for nematic and smectic A liquid crystals, they behave in a linear regime as typical uniaxial crystals. All results obtained for such media in Chap. 5 can be applied to them immediately. We limit our consideration with the propagation of light waves along the helical axis because the theory of propagation inclined to the optical axis is very complicated. The unique optical features of the cholesteric phase clearly manifest in the case of the incidence parallel to the helical axis. We briefly enumerate these features [29].

1. When white light is incident on a planar sample selective reflection takes place, the wavelengths of the reflected maxima varying with angle of incidence according to Bragg's law. In the case of light beam propagation along the helical axis z only one Bragg reflection is observed experimentally while higher order reflections are forbidden. In the case of oblique incidence the reflections are observed which obey the Bragg condition $2P_0 \cos r = m\lambda$, where r is the angle of the refracted beam in the cholesteric sample, λ is a light wavelength, $m = 1, 2, 3...$
2. At normal incidence, the reflected light is strongly circularly polarized. One circular component is almost totally reflected over a spectral range of some 100 Å, while the other passes through practically unchanged. Contrary to usual experience, the reflected wave has the same sense of circular polarization as that of the incident wave. In the case of the oblique incidence, the polarizations are elliptical.
3. Along its optical axis, the medium possesses a very high rotatory power, usually of the order of several thousands of degrees per millimeter. In the neighborhood of the region of reflection, the rotatory dispersion is anomalous and the sign of rotation opposite on opposite sides of the reflected band. The huge optical rotations observed in the cholesteric phase are clearly not due to an intrinsic spectroscopic property of the constituent

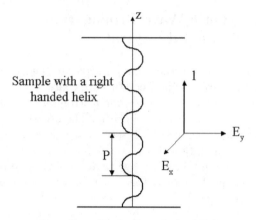

Fig. 8.14. The propagation of an elliptically polarized light wave along the optical axis z of a cholesteric liquid crystal with a right handed helix

molecules since they do not persist in the isotropic phase. Allegedly, they reflect the properties of light waves propagating in a twisted anisotropic medium.

The theory of these phenomena was mainly developed in the works of Mauguin [41], Oseen [42], de Vries [43], Katz [44], Nityananda [45], Belyakov [46]. The results are summarized in [28, 29, 39].

The most general theoretical approach to the optics of cholesteric liquid crystal is based on the exact solution of the wave equation for propagation along the optical axis. The geometry of the problem is shown in Fig. 8.14. We represent the dielectric tensor by a spiralling ellipsoid whose principal axis Oc is always parallel to z; the other two principal axes Oa and Ob (with principal values ε_a and ε_b) spiral around z with a twist angle $q = 2\pi/P$ per unit length [44]. If Oa, Ob are taken to be along x, y at the origin, the tensor ε_{ik} at any point z may be expressed with respect to x, y as

$$\varepsilon_{ik} = \begin{vmatrix} \cos qz & -\sin qz \\ \sin qz & \cos qz \end{vmatrix} \begin{vmatrix} \varepsilon_a & 0 \\ 0 & \varepsilon_b \end{vmatrix} \begin{vmatrix} \cos qz & \sin qz \\ -\sin qz & \cos qz \end{vmatrix} \qquad (8.145)$$

$$= \begin{vmatrix} \varepsilon + \alpha \cos 2qz & \alpha \sin 2qz \\ \alpha \sin 2qz & \varepsilon - \alpha \cos 2qz \end{vmatrix},$$

where

$$\varepsilon_a = n_a^2, \ \varepsilon_b = n_b^2, \ \varepsilon = (\varepsilon_a + \varepsilon_b)/2$$
$$\alpha = (\varepsilon_a - \varepsilon_b)/2 = (n_a + n_b)(n_a - n_b)/2 = n\delta n.$$

The wave equation for propagation along z is

$$\frac{\partial^2 \boldsymbol{E}}{\partial z^2} = -\frac{\omega^2}{c^2} \varepsilon \boldsymbol{E}. \qquad (8.146)$$

8.5 Electromagnetic Wave Propagation in a Cholesteric Liquid Crystal

Equation (8.146) does not have exactly the structure of an eigenvalue problem. However there are some simple features. The operators on both sides are unchanged by a transition of length $P/2$ along z. This implies a Bloch-Floquet theorem: a complete set of solutions can be found, such, that for each of them

$$\left. \begin{vmatrix} E_x \\ E_y \end{vmatrix} \right|_{z+P} = \text{const} \left. \begin{vmatrix} E_x \\ E_y \end{vmatrix} \right|_z .$$

We introduce the new variables

$$E_{\pm} = \frac{1}{\sqrt{2}} \left(E_x \pm i E_y \right), \tag{8.147}$$

where $E_{\pm}$ are right-circular and left-circular for propagation along $+z$ and vice versa for propagation along $-z$, respectively. Substituting expression (8.147) into (8.146) we obtain [44]

$$-\frac{\partial^2 E_+}{\partial z^2} = k_0^2 E_+ + k_1^2 \exp\left(2iqz\right) E_- \tag{8.148}$$

$$-\frac{\partial^2 E_-}{\partial z^2} = k_0^2 E_- + k_1^2 \exp\left(-2iqz\right) E_+,$$

where

$$k_0^2 = \frac{\omega^2}{c^2} \varepsilon, \ k_1^2 = \frac{\omega^2}{c^2} \alpha. \tag{8.149}$$

The solution can be immediately written in the form of the modes

$$E_+ = a \exp\left[i\left(l+q\right)z\right], \ E_- = b \exp\left[i\left(l-q\right)z\right], \tag{8.150}$$

where a, b are two constants, linked by the following relations

$$\left[\left(l+q\right)^2 - k_0^2\right] a - k_1^2 b = 0 \tag{8.151}$$

$$-k_1^2 a + \left[\left(l-q\right)^2 - k_0^2\right] b = 0.$$

These two equations have a non-trivial solution only if the corresponding determinant vanishes

$$\left(-k_0^2 + l^2 + q^2\right)^2 - 4q^2 l^2 - k_1^4 = 0. \tag{8.152}$$

For a given frequency ω, k_0 and k_1 are fixed, and (8.152) gives four possible values of l (real or complex). There are two distant branches $\omega_{\pm}(l)$. Consider first the case when $l = 0$. This gives

$$k_0^2 - q^2 = \pm k_1^2.$$

Using (8.149) we find the following frequencies

$$\omega_+(0) = \frac{cq}{n_a}, \quad \omega_-(0) = \frac{cq}{n_b}, \tag{8.153}$$

where $\omega_+(0)$ corresponds to $a = -b$. From (8.150) this describes a linear wave polarized along the direction $\theta(z) + \pi/2$ (ordinary axis). Similarly $\omega_-(0)$ corresponds to $a = b$, i.e. a linear wave polarized along the local extraordinary axis. The interval $\omega_-(0) < \omega < \omega_+(0)$ can be called the frequency gap. The existence of such a gap shows that cholesteric liquid crystals behave as a one-dimensional photonic crystal [47]. For the frequencies outside of this interval we must retain only the roots which give a positive group velocity $v_g = \partial \omega / \partial l > 0$, if we are interested in waves travelling along the $(+z)$ direction. There are two such roots $l_{1,2}$. Each of them defines an eigenmode of vibration with an eigenvector (a_i, b_i) defined within a multiplicative constant. The analysis of the solutions obtained shows that the electric field associated with the eigenmode is elliptically polarized; at each point the axes of the ellipse coincide with the local optical axes of the cholesteric. A useful parameter to describe the ellipse is the real number

$$\rho = \frac{-a+b}{a+b}.$$

The axial ratio of the ellipse is $|\rho|$, and the sign of ρ gives the sign of rotation of the vibration in the (x,y) plane (at fixed z and increasing t). To calculate ρ in terms of l and ω, we use (8.151). We obtain for the ratio of a and b

$$\frac{a}{b} = \frac{k_1^2}{(l+q)^2 - k_0^2} = \frac{(l-q)^2 - k_0^2}{k_1^2}$$

and

$$\rho = \frac{-2lq}{k_0^2 - l^2 - q^2 - k_1^2}. \tag{8.154}$$

If we define a positive quantity s^2 by

$$s^2 = \sqrt{k_1^4 + 4q^2 l^2}, \tag{8.155}$$

we have

$$\rho = \frac{-2lq}{\pm s^2 - k_1^2}. \tag{8.156}$$

The sign ($\pm$) in (8.156) depends on the branch used for $\omega_\pm(l)$.

8.5 Electromagnetic Wave Propagation in a Cholesteric Liquid Crystal

Consider now the practically important particular cases.

1. **Waveguide regime.** Equations (8.155) and (8.156) show that the mode structure depends critically on the value of the parameter

$$\kappa = \frac{2lq}{k_1^2}. \tag{8.157}$$

Let us consider first the regimes where $\kappa \ll 1$. This can be obtained in two ways: (i) by going to small l values; (ii) by going to very large l values. Then $l \sim k_0$ and

$$\kappa \sim \frac{2q\left(n_a^2 + n_b^2\right)}{k_0\left(n_a^2 - n_b^2\right)} \sim \frac{2q\bar{n}}{k_0\left(n_a - n_b\right)}. \tag{8.158}$$

The criterion $\kappa \ll 1$ is the criterion for the so-called Mauguin limit $\lambda \ll (n_a - n_b) P$. In both cases we find one mode associated with (+) branch which has $\rho \to \infty$, and represents a linear ordinary wave guided by the helix where the polarization remains everywhere parallel to the ordinary axis. The (−) mode has $\rho \to 0$ and represents a linear extraordinary wave, guided in the same way.

2. **Circular regime.**
In many practical cases the two indices n_a and n_b are not very different. Then k_1^2 tends to be small and the parameter κ is large for most frequencies of interest. Then $s^2 \to 2ql$, and it is seen from (8.154) that $\rho \to \pm 1$. In this regime the eigenmodes are nearly circular. In fact, all essential properties can be derived directly from the initial equations (8.150) and (8.151). If we neglect k_1^2 completely (zero order approximation) we see from (8.151) that the l values conncted to the corresponding a_1, b_1 are

$$l_1 = k_0 + q \to a_1 = 0, \ b_1 = 0$$
$$l_2 = k_0 - q \to a_1 = 1, \ b_1 = 0.$$

In the next approximation, we find from (8.152)

$$l_1 = k_0 + q + \frac{k_1^4}{8k_0 q\left(k_0 + q\right)} + O\left(k_1^8\right) \tag{8.159}$$

$$l_2 = k_0 - q + \frac{k_1^4}{8k_0 q\left(k_0 - q\right)} + O\left(k_1^8\right).$$

Since $a_1 \simeq 0$ the first mode is according to (8.150) a circular mode of wave vector $l_1 - q = \omega n_1/c$. Similarly, for the second mode we have a circular wave, of opposite sense, and of wave vector $l_1 + q = \omega n_2/c$, where $n_{1,2}$ are their refractive indices, respectively. The optical rotation per unit length ψ/d is then obtained simply by taking one half of the difference between these two wave vectors

$$\frac{\psi}{d} = \frac{\omega}{2c}(n_1 - n_2) = \frac{k_1^4}{8k_0 \left(k_0^2 - q^2\right)}. \tag{8.160}$$

This formula is often expressed in terms of a reduced wavelength $\lambda' = \lambda/P = q/k_0$. Transforming k_1 by (8.149), one obtains

$$\frac{\psi}{d} = \frac{q}{32} \left(\frac{n_a^2 - n_b^2}{n_a^2 + n_b^2}\right)^2 \frac{1}{\lambda'^2 (1 - \lambda'^2)}. \tag{8.161}$$

Note the following features: (i) very large magnitude of rotation: for $\lambda' = 0.7$ and $\left(n_a^2 - n_b^2\right)/\left(n_a^2 + n_b^2\right) = 0.1$ we get $\psi/d \sim 10^{-3} q$. With a pitch P of $1\mu m$, corresponding to $q = 6 \cdot 10^4$, we expect $\psi/d \sim 3500 \text{ deg} \cdot \text{cm}^{-1}$. (ii) Dispersion anomaly with a change in sign at the Bragg reflection $\lambda' = 1$. Very near the Bragg reflection equation (8.161) breaks down and the waveguide regime takes over. (iii) Optical rotation is proportional to P^3/λ^4 at long wavelengths. (iv) At wavelengths $\lambda \ll P$ (but still too large to be in the Mauguin regime) the rotation becomes proportional to P/λ^2.

3. Bragg reflection.

Let us now choose a frequency inside the gap

$$\frac{cq}{n_b} < \omega < \frac{cq}{n_a}.$$

For such a case, (8.152) has only two real roots $l = \pm l_1$. The other two roots are pure imaginary $l = \pm i |l|$. Consider now a thick slab ($d \to \infty$) attacked from below, at normal incidence, by a light beam of polarization $\boldsymbol{i}(i_x, i_y)$. This will in general induce in the slab two waves: (i) one travelling wave with an amplitude proportional to $\exp(il_1 z)$; according to (8.154) the parameter ρ associated with this wave is close to 1 for a right-hand helix; the wave is circularly polarized. (ii) One evanescent wave with an amplitude $\sim \exp(-|l| z)$. By a suitable choice of the polarization $\boldsymbol{i}, (\boldsymbol{i} = \boldsymbol{i}_R)$ it is possible to extinguish the travelling wave component: this means that a beam of polarization $\boldsymbol{i}_R$ will be totally reflected. We conclude that the gap $[\omega_-(0), \omega_+(0)]$ corresponds to the frequency range for possible Bragg reflections.

The theory of the light propagation in a cholesteric phase can be used as a basis for the analysis of magneto-optical phenomena in cholesterics. The pitch of the cholesteric helix can be governed by a magnetic field as was mentioned above. As a result, the dielectric tensor of the cholesteric depends on the magnetic field strength which permits the influence of the external magnetic field on the light propagation regimes and the measurement of magnetic field induced changes of the pitch by optical methods.

9 Electrodynamics of Superconductors

9.1 Superconducting Current

The phenomenon of superconductivity discovered by Kamerlingh Onnes in 1911 [48] occurs in some metals, compounds and alloys at the temperature close to the absolute zero. The physics and engineering applications of superconductivity have been described in a number of excellent monographs and textbooks such as [4,49–57], where the phenomenological and the microscopic theories are presented as well as the fundamental experimental results. We do not intend to carry out a complete study of the subject. Our purpose is to deliver the results of the pivotal experiments and to explain the superconductor behavior in an external magnetic field using the phenomenological theories of London [58,59], and Ginzburg–Landau [60]. We also present the results of some applied problems thoroughly investigated in [56].

At the so-called critical, or transition temperature T_{cr} the properties of the conduction electrons are changing dramatically on the microscopic level which results in a number of unique macroscopic effects [49–56].

1. Electric resistance to a dc current completely vanishes.
2. A magnetic flux is expelled from a material that is known as the Meissner effect [61].
3. The electronic wave function possesses the long-range order.
4. A persistent current occurs in superconducting rings.
5. A magnetic flux trapped within multiply connected superconductors is quantized.
6. Electron pairs demonstrate tunneling between closely spaced superconductors without any potential difference which is called the Josephson effect [62].

At any temperature T below T_{cr}, the application of a definite magnetic field $H_{\mathrm{cr}}(T)$ destroys the superconductivity and restores the normal resistance corresponding to that field. It turned out that for most of superconductors the following equation is valid with a sufficient accuracy [63]:

$$h_c \simeq 1 - t^2, \qquad (9.1)$$

where the reduced coordinates h_c and t have the form

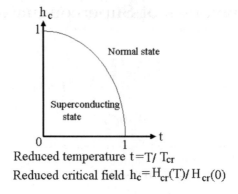

Fig. 9.1. Reduced critical field $h_c = H_{cr}(T)/H_{cr}(0)$ versus reduced temperature $t = T/T_{cr}$

$$h_c = \frac{H_{cr}(T)}{H_{cr}(0)}, \quad t = \frac{T}{T_{cr}}.$$

The curve describing the dependence (9.1) of critical parameters limits the region of a superconducting state existence which can be seen in Fig. 9.1.

These phenomena as well as their far reaching consequences and applications can be explained consistently only in the framework of the quantum theory first developed by Bardeen, Cooper and Schrieffer (BCS theory) [64]. This theory based on the Cooper pair model assumes that in the superconducting state electrons compose the bound states consisting of the pairs of particles with equal and anti-parallel momenta and anti-parallel spins. The ensemble of electron pairs can be characterized as the condensed electron-pair fluid. The properties of the electron pairs responsible for superconduction strongly differ from those of single free electrons. The reason is that the free electrons are fermions, i.e. obey the Fermi statistics, while the paired electrons with a zero spin are more similar to bosons [57]. In particular, it turned out that a gap Δ in the energy spectrum $\mathcal{E}(p)$ of a quasiparticle appears [57]:

$$\mathcal{E}(p) = \sqrt{\Delta^2 + \frac{(p - p_F)^2}{2m}}$$

where p, p_F are the momentum and its value at the so-called Fermi surface, m is the quasiparticle mass. The gap Δ_0 at $T = 0$ is given by

$$\Delta_0 = \widetilde{\mathcal{E}} \exp\left(-\frac{2\pi^2 \hbar^3}{g m p_F}\right)$$

where $\hbar = h/(2\pi)$, h is the Plank constant, $\widetilde{\mathcal{E}} \sim p_F^2/(2m)$, and g is the coupling constant.

The comprehensive description of superconductivity is rather complicated, requires the knowledge of the quantum field theory formalism as a prerequisite, and surpasses the objectives of these book. The fundamentals of this theory can be found, for example, in [57]. However, the above mentioned macroscopic observed effects can be described by virtue of the essentially phenomenological approach. We shall derive the system of electromagnetic equations for a superconductor where the relation between current and electric field plays the essential role. The total electric current j in a superconductor consists of the two components [57]:

$$j = j_n + j_s, \quad (9.2)$$

where a normal current density j_n is due to single electrons existing in a superconducting metal at any finite temperature, and it produces the heat according to the Joule's law; a superconducting, or pair current density j_s caused by the electron-pair fluid is not accompanied by energy dissipation. In our treatment we describe the behavior of the both current components in terms of the so-called two-fluid model. For this purpose we use the equations of the classical electrodynamics incorporating the important result of the quantum theory, namely, the so-called wave function $\Psi(r)$ of the condensed electron pairs which represents the macroscopic ensemble-average many-body wave function with a well-defined amplitude $|\Psi(r)|$ and phase angle $\theta(r)$ [57]:

$$\Psi(r) = |\Psi(r)| \exp i\theta(r). \quad (9.3)$$

It has been shown that the ensemble-average wave function (9.3) provides a sufficiently accurate description of the electron-pair fluid. The appropriate normalization of (9.3) permits the definition of the amplitude $|\Psi(r)|$ in terms of the number density n_s of Cooper pairs at a given point of the superconductor:

$$|\Psi(r)| = \sqrt{n_s}. \quad (9.4)$$

In the further analysis the number density n_s is assumed to be spatially constant which is valid for many practically important cases. The temporal dependence can also be neglected for time variations slower than 10^{12} s^{-1} since the electron relaxation processes are sufficiently rapid

$$|\Psi(r)| \simeq \text{const.} \quad (9.5)$$

The relation between the quantum mechanical wave function (9.3) and the classical density of the superconducting current j_s can be established by using the quantum mechanical expression for the probability density flux j_0 which determines the probability for a given particle to traverse a definite surface in a time unity [57]:

$$j_0 = \frac{i\hbar}{2m^*} [\Psi(r) \nabla \Psi^*(r) - \Psi^*(r) \nabla \Psi(r)], \quad (9.6)$$

where m^* is the effective mass of the electron pair. Substituting (9.3)–(9.5) into (9.6) we obtain [57]

$$\boldsymbol{j}_s = e^* \boldsymbol{j}_0 = \frac{e^* \hbar n_s}{m^*} \nabla \theta, \qquad (9.7)$$

where $e^* = -2e$ is the effective charge of the Cooper pair, and $e = 1.6 \cdot 10^{-19}$ C is the absolute value of the electron charge. Equation (9.7) shows that the superconducting current is caused by the spatial variations of wave function phase. These variations should be sufficiently slow on the characteristic distance, or the so-called coherence length, $\xi_0 \sim \hbar v_F / \Delta_0$ where v_F is a velocity of the electron on the Fermi surface.

In the presence of a magnetic field with a vector potential $\boldsymbol{A}(\boldsymbol{r})$ expression (9.7) for the superconducting current density must be essentially modified in order to account for the field dependence and provide the gauge invariance with respect to the transformation

$$\boldsymbol{A}'(\boldsymbol{r}) = \boldsymbol{A}(\boldsymbol{r}) + \nabla \chi. \qquad (9.8)$$

For this purpose we note that expression (9.6) related to the canonical momentum $\boldsymbol{P}$ of the electron pair as follows [56, 57]:

$$\boldsymbol{P} = \boldsymbol{j}_0 \frac{m^*}{n_s} = \hbar \nabla \theta. \qquad (9.9)$$

On the other hand, in the magnetic field the generalized canonical momentum takes the form in our case:

$$\boldsymbol{P} = m^* \boldsymbol{v}_s + e^* \boldsymbol{A}, \qquad (9.10)$$

where $\boldsymbol{v}_s$ is the velocity of the electron pair. Classically, it can be expressed in terms of the superconducting current density:

$$\boldsymbol{v}_s = \boldsymbol{j}_s \frac{1}{n_s e^*}. \qquad (9.11)$$

Combining (9.9)–(9.11) we obtain [56]:

$$\hbar \nabla \theta = e^* (\Lambda \boldsymbol{j}_s + \boldsymbol{A}), \quad \Lambda = \frac{m^*}{n_s e^{*2}}. \qquad (9.12)$$

Comparison of (9.3) and (9.12) shows that the necessary and sufficient condition of the superconducting current gauge invariance is the following transformation of the electron-pair wave function Ψ:

$$\Psi' = \Psi \exp\left(\frac{ie^*}{\hbar} \chi\right), \text{ or } \theta' = \theta + \frac{e^*}{\hbar} \chi. \qquad (9.13)$$

The direct substitution of (9.8) and (9.13) into (9.12) proves immediately that the superconducting current density remains the same.

Applying the curl operation to the both sides of (9.12) and recalling that curl$A = B$ we get

$$\Lambda \mathrm{curl} j_s + B = 0. \tag{9.14}$$

In the presence of the electric field E the electron-pair fluid is accelerated, and its equation of motion coincides with the equation of motion of a charged particle in an external electric field studied above:

$$m^* \frac{dv_s}{dt} = e^* E. \tag{9.15}$$

Substituting the "classical" current definition (9.11) into equation of motion (9.15) we obtain:

$$\Lambda \frac{dj_s}{dt} = E. \tag{9.16}$$

Equations (9.14) and (9.16) are the so-called London equations obtained by F. and H. London in 1935 [58, 59]. They are adequate for many practically important problems, although their applications have some limitations [56].

1. The London equations are invalid in the case of the strong spatial variation of the magnetic induction at the surface of a superconductor.
2. They can be applied only for weak magnetic fields and do not describe the disappearance of superconductivity in high magnetic fields.

The London equations show in particular that, unlike the relation between current density and electric field for a normal conductor, the magnetic field is responsible for the existence of the current in a superconductor. The electric field affects only the temporal variations of a superconducting current. Equation (9.16) shows that for stationary currents the electric field in a superconductor vanishes.

9.2 The Meissner Effect and Penetration Depth

In 1933 Meissner and Ochsenfeld [61] performed the experiment which appeared to be crucial with respect to the studies of superconductivity. A lead sample was subjected to a comparatively weak magnetic field at the temperature higher than the transition temperature T_{cr}. Then the temperature was reduced lower than T_{cr} in order to make the lead sample superconducting. In the case of the perfect conductor, i.e. the metal sample with the zero resistance, the magnetic flux was supposed to be trapped remaining in the lead after the removal of the field. In contrast to these predictions, the magnetic flux appeared to be expelled from the sample. It was concluded that a weak magnetic field does not penetrate into the superconductor except for the thin layer at the surface of the sample. The induction rapidly decays in a distance of approximately 100 nm from the sample surface.

In this section we clarify the difference in the electric and magnetic properties of a superconductor and a hypothetical perfect conductor [56]. We start with the analysis of a would be perfect conductor behavior in the electric and magnetic field. The model of a perfect conductor presumes that it is a perfectly pure crystal of a normal metal at the absolute zero temperature $T = 0$. Under circumstances the lattice vibrations are absent, and consequently an electric resistance R vanishes due to the elimination of phonon scattering. Such a situation correspond to the infinite electron momentum relaxation time $\tau \to \infty$.

Consider first the electron motion in ac electric field

$$E_x = E_0 \exp i\omega t. \qquad (9.17)$$

The equation of motion for electron in an electric field accounting for the finite relaxation time has the form

$$m\frac{d\langle v_x \rangle}{dt} + m\frac{\langle v_x \rangle}{\tau} = -eE_x, \qquad (9.18)$$

where $\langle v_x \rangle$ is the average electron velocity in the x direction. The solution of (9.18) is

$$\langle v_x \rangle = -\frac{e\tau}{m(1 + i\omega\tau)} E_x, \qquad (9.19)$$

which gives the following value of the ac current density j_x

$$j_x = \frac{ne^2\tau}{m(1 + \omega^2\tau^2)}(1 - i\omega\tau) E_x. \qquad (9.20)$$

In the perfect conductor limiting case $\tau \to \infty$ (9.20) reduces to

$$j_x = -i\frac{ne^2}{m\omega} E_x. \qquad (9.21)$$

Equation (9.21) shows that the real part of the ac conductivity vanishes, and losses do not take place. For the perfect conductor of the length l and cross-sectional area S with the uniform current density distribution (9.21) can be transformed into the relation between current I and voltage V:

$$I = \frac{ne^2 S}{im\omega l} V = \frac{1}{i\omega \mathcal{L}} V, \qquad (9.22)$$

where the self-inductance $\mathcal{L}$ is associated with the inertia of electrons:

$$\mathcal{L} = \frac{ml}{ne^2 S}. \qquad (9.23)$$

9.2 The Meissner Effect and Penetration Depth

We can conclude that an ac electric field can exist in a perfect conductor, while the losses are absent.

Investigate now the magnetic properties of the perfect conductor [56]. Equation of motion (9.18) can be easily generalized to the vector form. Dropping form the very beginning the second term in the left-hand side of (9.18) for the case of the infinite relaxation time and expressing the electron velocity in terms of the current density we obtain

$$\mathbf{E} = \frac{m}{ne^2} \frac{\partial \mathbf{j}}{\partial t}. \tag{9.24}$$

Combining the time derivative of the Maxwell equation where the term with the electric induction can be neglected for low frequencies

$$\text{curl}\mathbf{H} = \mathbf{j}, \tag{9.25}$$

and (9.24) we get

$$\text{curl}\frac{\partial \mathbf{H}}{\partial t} = \frac{ne^2}{m}\mathbf{E}. \tag{9.26}$$

Applying the curl operation to both sides of (9.26) and using the Maxwell equations

$$\text{div}\mathbf{B} = 0 \tag{9.27}$$

$$\text{curl}\mathbf{E} = -\frac{\partial \mathbf{B}}{\partial t} \tag{9.28}$$

and

$$\mathbf{B} = \mu_0 \mathbf{H} \tag{9.29}$$

we finally obtain the equation for the time derivative of the magnetic induction:

$$\nabla^2 \frac{\partial \mathbf{B}}{\partial t} = \frac{\mu_0 n e^2}{m} \frac{\partial \mathbf{B}}{\partial t}. \tag{9.30}$$

We solve (9.30) for one-dimensional perfect conductor filling a half-space $x \geq 0$. Then it takes the form:

$$\left(\frac{\partial^2}{\partial x^2} - \frac{\mu_0 n e^2}{m}\right)\frac{\partial \mathbf{B}}{\partial t} = 0. \tag{9.31}$$

The solution of (9.31) finite at $x \to \infty$ is given by the following expression:

$$\frac{\partial \mathbf{B}}{\partial t} = C \exp\left(-\frac{x}{\kappa}\right), \tag{9.32}$$

where the characteristic length κ is given by

$$\kappa = \sqrt{\frac{m}{\mu_0 n e^2}}. \tag{9.33}$$

Numerical estimation shows that the time derivative (9.32) rapidly decays with distance from the boundary $x = 0$. The characteristic length κ is less than 100 nm which means that the magnetic induction itself is kept constant in the bulk of the perfect conductor [56]. This result contradicts to the Meissner experiment where the magnetic induction $\boldsymbol{B}$ but not its temporal derivative, is expelled from the superconducting sample. We should therefore stress that the properties of a superconductor are very specific and do not simply reduce to the absence of electric resistance near the critical temperature.

Consider the superconductor in an external magnetic field. We use for its description the London equations (9.14) and (9.16) and the Maxwell equations (9.25) and (9.27). According to (9.16) in the case of a dc regime an electric field in the superconductor vanishes, and consequently there are no currents of normal electrons. Then only superconducting current remains in the right-hand side of (9.25)

$$\mathrm{curl}\boldsymbol{H} = \boldsymbol{j}_s. \tag{9.34}$$

Substituting (9.34) into the London equation (9.14) and combining it with (9.29) we find [56, 57]:

$$\frac{\Lambda}{\mu_0}\left[\nabla\left(\nabla\cdot\boldsymbol{B}\right) - \nabla^2\boldsymbol{B}\right] = -\boldsymbol{B}. \tag{9.35}$$

Taking into account the Maxwell equation (9.29), (9.35) is reduced to

$$\nabla^2\boldsymbol{B} - \lambda^{-2}\boldsymbol{B} = 0, \tag{9.36}$$

where

$$\lambda = \sqrt{\frac{\Lambda}{\mu_0}} = \sqrt{\frac{m^*}{\mu_0 n e^{*2}}}. \tag{9.37}$$

Apply the curl operation to the London equation (9.14) and insert $\boldsymbol{j}_s$ from (9.34) and $\boldsymbol{B}$ from (9.29). Then bearing in mind that for a dc current

$$\mathrm{div}\boldsymbol{j}_s = 0 \tag{9.38}$$

we get for the superconducting current density the same equation as (9.36):

$$\nabla^2\boldsymbol{j}_s - \lambda^{-2}\boldsymbol{j}_s = 0. \tag{9.39}$$

For simply connected bodies where

$$\nabla\theta = 0$$

we obtain the similar equation for the vector potential $\boldsymbol{A}$ using (9.34) and the gauge $\mathrm{div}\boldsymbol{A} = 0$:

$$\nabla^2 \boldsymbol{A} - \lambda^{-2}\boldsymbol{A} = 0. \tag{9.40}$$

One can see that the magnetic induction, the superconducting current density and the vector potential inside the superconductor obey the same equation with one characteristic parameter λ having the length dimensionality.

Let the magnetic field $\boldsymbol{B}$ parallel to the z-axis is applied to the one-dimensional superconducting half-space $x \geq 0$. Then (9.36) yields the solution vanishing as $x \to \infty$

$$B_z = B_0 \exp\left(-\frac{x}{\lambda}\right) \tag{9.41}$$

which is in accord with the results of the Meissner experiment that the magnetic field does not penetrate into the superconductor. The quantity λ is called the penetration length. Equations (9.34) and (9.29) give the value of the superconducting current. It is directed along the y-axis and has the form:

$$\boldsymbol{j}_s = (0, j_{sy}, 0), \quad j_{sy} = -\frac{1}{\mu_0 \lambda} B_0 \exp\left(-\frac{x}{\lambda}\right). \tag{9.42}$$

It turned out that both the magnetic field and the current density are concentrated in a very thin surface layer $\lambda \sim 50$ nm of a superconductor in the dc regime, in contrast to the so-called skin-effect in normal conductors when the field does not penetrate into the medium at high frequencies. There is neither current, nor magnetic field in the bulk of a superconductor. It should be emphasized, however, that the Meissner effect permits a penetration of a small part of a magnetic flux into the superconductor. The field and current can penetrate a sample substantially, if the sample dimensions are comparable with the penetration depth. For bodies with the dimensions much larger than the penetration depth the exclusion of the magnetic flux may be considered as complete since the region of penetration is a negligibly small part of the total volume. The solutions (9.41) and (9.42) obtained for a superconducting half-space can be applied to a sample of any possible shape provided that its dimensions are large in comparison with the penetration depth.

In general, the penetration depth λ calculated for $T = 0$ can increase considerably near the transition temperature. Many attempts have been made to get a theoretical dependence $\lambda(T)$ which would agree with the experimental data. The best agreement with the thermal properties of superconductors has been obtained by Gorter and Casimir in 1934 [65]. They assumed that the fraction of the electrons in paired state n_s/n is unity at $T = 0$ and then gradually decreases with temperature up to zero at the transition point T_{cr}. The temperature dependence has been chosen to be [56]

$$\frac{n_s}{n} = 1 - \left(\frac{T}{T_{cr}}\right)^4. \tag{9.43}$$

Substituting (9.43) into (9.37) we obtain for the penetration depth:

$$\lambda(T) = \frac{\lambda(0)}{\sqrt{1-\left(\frac{T}{T_{cr}}\right)^4}} = \sqrt{\frac{m^*}{\mu_0 n e^{*2}\left[1-\left(\frac{T}{T_{cr}}\right)^4\right]}}. \tag{9.44}$$

The London penetration depth is infinite at the transition point T_{cr}, but otherwise it describes the behavior of λ with a sufficient accuracy.

The London electrodynamic equations have serious limitations since they are based on a rather simplified model possessing the following disadvantages.

1. The relations between fields and currents are local and for this reason they are unapplicable for the situations where the spatial variations are sharp.
2. The presence of normal conduction electrons is ignored.
3. The magnitude of the wave function (9.3) is supposed to be independent of the magnetic field which limits the considerations with weak fields.

Consider first the difficulties caused by a possible nonlocality. If the effective size of the electron pairs is greater than the distance over which the vector potential variations are considerable, then the London equation (9.14) requires some modifications. The size of the particles is called coherence length being the spatial interval of the wave function phase coherence. It turned out that the coherence length ξ_0 for many superconductors is greater than the penetration depth λ. The London equation

$$\Lambda \boldsymbol{j}_s = -\boldsymbol{A} \tag{9.45}$$

and the Ohm's law

$$\boldsymbol{j}_n = \sigma \boldsymbol{E} \tag{9.46}$$

for normal metals are formally similar which permits the application of the nonlocal theory in both cases. However, the mechanisms of nonlocality in normal metals and superconductors are different. In a normal metal the momentum exchange is due to the collisions between electrons. The conduction electrons at any point carry with them the momentum got from the electric field in other points and in earlier moments of time, and the current is not determined by the value of the field at the same point. If the spatial variations of the field are sufficiently strong on the electronic mean free path distance l_e, it is necessary to take into account the contributions from the other points introducing an integral relation between the current and the field values. In a superconductor the electron collisions do not occur, and the electronic mean free path l_e cannot serve as a characteristic distance of nonlocality.

9.2 The Meissner Effect and Penetration Depth

Pippard developed a nonlocal theory [66] based on the assumption that in the case of a spatially inhomogeneous field the current response of the pair fluid in a given point would depend on the vector potential contributions from the points situated within a distance of about a coherence length. He also obtained experimental results showing that the intrinsic coherence length ξ_0 was reduced due to the presence of impurities. The empirical relation for the coherence length ξ formulated by Pippard has the form

$$\frac{1}{\xi} = \frac{1}{\xi_0} + \frac{1}{\alpha l_e}, \qquad (9.47)$$

where α is a constant of the order of unity. As for the intrinsic coherence length ξ_0 its value is given by the microscopic theory:

$$\xi_0 = \frac{0.18 \hbar v_F}{k_B T_{\mathrm{cr}}}, \qquad (9.48)$$

where k_B is the Boltzmann constant. The nonlocal relation which replaces the London equation (9.14) has the form [56]:

$$\boldsymbol{j}_s = -\frac{3}{4\pi \xi_0 \Lambda} \int_{V'} \frac{(\boldsymbol{r}-\boldsymbol{r}')\left((\boldsymbol{r}-\boldsymbol{r}')\cdot\boldsymbol{A}(\boldsymbol{r}')\right)\exp\left[-|\boldsymbol{r}-\boldsymbol{r}'|/\xi\right]}{|\boldsymbol{r}-\boldsymbol{r}'|^4}dV'. \qquad (9.49)$$

It is seen from (9.49) that the range of effectiveness of the vector potential $\boldsymbol{A}(\boldsymbol{r}')$ is of the order of magnitude of the coherence length ξ. When ξ decreases due to a small mean free path according to (9.47), expression (9.49) becomes a nearly local relation.

Let us investigate the penetration depth for different situations using (9.49). The superconductors can be classified as follows [56]:

1. Pure superconductors with large coherence lengths ξ_0 such as aluminum.
2. Impure superconductors with ξ which is determined by the mean free path l_e, so that $\xi \sim l_e$.
3. Pure superconductors with small ξ_0.

Pure superconductors with large coherence lengths must be considered in a framework of the nonlocal theory and are called the Pippard superconductors. It can be shown that the penetration depth λ_P in such a case has the form

$$\lambda_P = 0.65 \lambda \left(\frac{\xi_0}{\lambda}\right)^{1/3} \quad \text{for } \xi^3 \gg \xi_0 \lambda^2. \qquad (9.50)$$

The ratio ξ_0/λ can be large and consequently λ_P can be considerably larger than λ. In the nonlocal case the decay of the magnetic induction is not exponential. Nevertheless, the penetration depth λ_P can be used as a measure of the total flux in the superconductor by virtue of the definition

$$\lambda = \frac{1}{B(0)} \int_0^\infty B(x)\,dx, \qquad (9.51)$$

where $B(0)$ is the value of the induction at the surface.

In the impure superconductors and alloys the electronic mean free paths are much smaller than ξ_0. They are mainly determined by the distribution of physical defects, impurities and boundaries of small samples. This condition is called the London limit where $\xi \to l_e$ and $\xi \ll \xi_0$. Then, under the conditions of random scattering of electrons from the superconductor surface and $\xi^3 \ll \xi_0 \lambda^2$ the nonlocal relation (9.49) reduces to the London equation of the type (9.45) containing the ratio of the coherence lengths:

$$\Lambda \boldsymbol{j}_s = -\frac{\xi}{\xi_0} \boldsymbol{A} \text{ for } \xi^3 \ll \xi_0 \lambda^2. \qquad (9.52)$$

The insertion of this ratio into the London equation is equivalent to a reduction of the penetration length for impure superconductors λ_I. Indeed, comparison of (9.52) and (9.36), (9.39) shows that

$$\lambda_I = \lambda \sqrt{\frac{\xi_0}{\xi}} \approx \lambda \sqrt{\frac{\xi_0}{l_e}} \text{ for } \xi^3 \ll \xi_0 \lambda^2 \qquad (9.53)$$

Pure superconductors may have small values of ξ_0 such that $\xi_0 \ll \lambda$. In such a situation the spatial variation of the magnetic field and superconducting current density in the penetration layer is slow compared with ξ_0, the local London equation (9.45) is valid, and the penetration depth is λ. The phenomenological approach used by Pippard has been confirmed by the microscopic theory based on BCS model.

Returning now to the other two limitations of the London approach we note that the role of normal electrons is taken into account in the so-called two-fluid model which will be studied below in the analysis of the ac field phenomena.

The spatial inhomogeneity of the modulus of the wave function (9.3) must be considered in the framework of the Ginzburg–Landau theory [57,60] which will be reviewed separately, or by means of microscopic Gor'kov's theory [67] which is out of the scope of this book. However, there exist many situations where the wave function with the uniform modulus (9.5) is an appropriate approximation.

9.3 Magnetic Flux in a Superconducting Ring

Consider the following remarkable effects which occur in a superconducting ring [4, 56, 57].

9.3 Magnetic Flux in a Superconducting Ring

1. The magnetic flux Φ traversing the nonsuperconducting hole of the ring remains constant even if the external field and current are varying.
2. The magnetic flux Φ through the ring is quantized. This property is a demonstration of the essentially quantum mechanical effect at a macroscopic level.

We start with the conservation of the magnetic flux. The electric and magnetic fields are connected by the Maxwell equation

$$\mathrm{curl}\boldsymbol{E} = -\frac{\partial \boldsymbol{B}}{\partial t}. \qquad (9.54)$$

We integrate both sides of (9.54) over the surface S of the ring hole:

$$\int_S \mathrm{curl}\boldsymbol{E}\,d\boldsymbol{s} = -\frac{\partial}{\partial t}\int_S \boldsymbol{B}\cdot d\boldsymbol{s}. \qquad (9.55)$$

The left-hand side of (9.55) is reduced to the circulation of the electric field along the closed contour L on the surface of the ring and vanishes, since the electric field $\boldsymbol{E} = 0$ inside the superconductor, and its tangential component is also zero on the surface due to the boundary conditions

$$\int_S \mathrm{curl}\boldsymbol{E}\,d\boldsymbol{s} = \oint_L \boldsymbol{E}\cdot d\boldsymbol{l} = 0. \qquad (9.56)$$

The integral in the right-hand side of (9.55) is by definition the magnetic flux. Then we obtain that the total magnetic flux is kept constant:

$$\frac{\partial \Phi}{\partial t} = 0, \quad \Phi = \mathrm{const}. \qquad (9.57)$$

In the presence of the external magnetic field the total magnetic flux consists of two parts: the component Φ_I caused by the current I_s in the ring and the one Φ_e which is due to the external field. It has been shown in the previous chapter concerning magnetostatics that Φ_I can be expressed in terms of the self-inductance $\mathcal{L}$:

$$\Phi = \Phi_I + \Phi_e, \quad \Phi_I = \mathcal{L}I_s. \qquad (9.58)$$

Equations (9.57) and (9.58) describe the variation of the superconducting current I_s due to the change of the external magnetic field. Suppose that the ring became superconducting in the external magnetic field with the flux Φ_{e0}. Then the external magnetic field is removed. Using (9.57) and (9.58) we immediately find that the superconducting current I_s occurs in the ring:

$$I_s = \frac{\Phi_{e0}}{\mathcal{L}}. \qquad (9.59)$$

Equation (9.59) shows that the trapped magnetic flux is accompanied by an induced persistent current which represents a unique electromagnetic phenomenon. It has been proved experimentally that no decay of such a current occurs during several years. Persistent currents are used, for example, in superconducting magnet coils and in microscopic loops as memory elements for computer circuits.

The results obtained for one ring are valid also for multiply-connected superconducting bodies, including sets of rings. In the chapter *Magnetostatics* it has been shown that the total magnetic flux of a system of currents is determined by their self-inductances and mutual inductances. In the presence of the external magnetic field we have:

$$\Phi_{\text{tot},i} = \Phi_{e,i} + \sum_{k=1}^{N} \Phi_{ik} = \Phi_{0,i} = \text{const}, \quad \Phi_{ik,i \neq k} = M_{ik} I_{s,k}, \quad \Phi_{ii} = \mathcal{L}_i I_{s,i}. \tag{9.60}$$

Consider now the quantization of the magnetic flux [56,57]. This time we use (9.12) for the superconducting current density in the magnetic field. Carrying out the integration around the contour L we write

$$\oint_L \hbar \nabla \theta \cdot d\mathbf{l} = e^* \oint_L (\Lambda \mathbf{j}_s + \mathbf{A}) \cdot d\mathbf{l}. \tag{9.61}$$

The integral in the left-hand side of (9.61) is the wave function phase difference which can only be equal to $2\pi n$ where n is an integer, since the phase of the wave function must be unique. The first term in the integrand in the right-hand side of (9.61) can be reduced to zero by choosing the integration contour L in the bulk of the ring where the current vanishes. Then the integral in the right-hand side of (9.61) represents the circulation of the vector potential around the contour L, and it can be transformed according to the Stokes theorem as follows:

$$e^* \oint_L \mathbf{A} \cdot d\mathbf{l} = e^* \oint_S \text{curl} \mathbf{A} d\mathbf{s} = e^* \oint_S \mathbf{B} \cdot d\mathbf{s} = e^* \Phi. \tag{9.62}$$

Using these results we obtain the condition of the magnetic flux quantization:

$$2\pi n \hbar = e^* \Phi, \quad \Phi_n = \frac{nh}{e^*}, \quad n = 1, 2, 3, \ldots. \tag{9.63}$$

Taking into account that the effective charge of an electron pair $e^* = -2e$ we find that the magnetic flux is quantized in units $h/(2e)$. The basic quantum Φ_0 of magnetic flux has the value [56]

$$\Phi_0 = \frac{h}{2e} = 2.07 \times 10^{-15} \text{ Wb}. \tag{9.64}$$

It should be noted that the flux quantization can be comparatively easy observed, and the flux quantum is an important characteristic of superconducting state.

9.4 Examples of a DC Regime in a Superconductor

9.4.1 Junction between a Normal Conductor and a Superconductor

The London equations can be used in order to analyze phenomena in a superconductor in the dc regime. Some practically important examples are presented in [56]. Here we briefly discuss the dc current behavior in the vicinity of junction between a normal conductor and a superconductor following [56] and omiitting the detailed derivations which can be found there.

Evaluate the current distribution in the vicinity of the junction between a normal conductor and a superconductor in the dc regime [56]. The superconductor is assumed to be an equipotential, and the current density j distribution in the normal conductor is uniform. Consider the planar structure where the z-axis is chosen to be parallel and the z-axis is perpendicular to the junction respectively. Both the normal conductor and the superconductor are limited by the surfaces $y = \pm a$. The superconductor is situated in the half-space $z > 0$, and the normal conductor corresponds to $z < 0$. The geometry of the problem is shown in Fig. 9.2.

The influence of the junction on the density n_s of electron pairs is neglected. The distribution of the current density $\boldsymbol{j}_s = (0, j_{sy}, j_{sz})$ in the superconductor is determined by (9.39). It must be completed by the continuity equation which takes the form in the dc case:

$$\frac{\partial j_{sy}}{\partial y} + \frac{\partial j_{sz}}{\partial z} = 0. \tag{9.65}$$

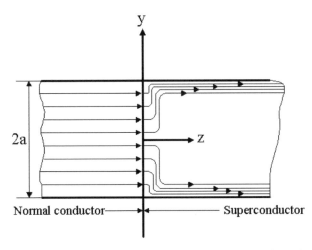

Fig. 9.2. The current distribution at the junction of the normal conductor and the superconductor

The boundary conditions are defined by the uniform distribution of the current density in the normal conductor, by the continuity of the current density at the junction $z = 0$ and by the finiteness of the current at the infinity:

$$j_{sz} = j \text{ for } -a \leq y \leq a, \ z = 0 \tag{9.66}$$

$$\frac{\partial j_{sz}}{\partial z} = 0, \ j_{sy} = 0 \text{ for } y = 0, \ z \geq 0 \tag{9.67}$$

$$j_{sy} = 0 \text{ for } y = \pm a, \ z \geq 0; \ -a \leq y \leq a, \ z \to \infty. \tag{9.68}$$

Equation (9.39) is solved by the separation of variables:

$$j_{sy} = f_1(y) f_2(z). \tag{9.69}$$

After the substitution of (9.69) into (9.39) the general solution takes the form:

$$j_{sy} = \sum_{n=1}^{\infty} A_n \sin \frac{\pi n}{a} y \exp\left\{ -z\sqrt{\lambda^{-2} + \left(\frac{\pi n}{a}\right)^2} \right\}. \tag{9.70}$$

The tangential component of the density current is obtained from (9.65).

$$j_{sz} = \mathcal{G}(y) + \sum_{n=1}^{\infty} A_n \frac{\pi n}{a} \left[\lambda^{-2} + \left(\frac{\pi n}{a}\right)^2 \right]^{-1/2} \cos \frac{\pi n}{a} y$$

$$\times \exp\left\{ -z\sqrt{\lambda^{-2} + \left(\frac{\pi n}{a}\right)^2} \right\}. \tag{9.71}$$

It can be found that [56]

$$\mathcal{G}(y) = \frac{ja \cosh(y/\lambda)}{\lambda \sinh(a/\lambda)} \tag{9.72}$$

and the coefficients A_n are given by [56]

$$A_n = -\frac{2j(-1)^n}{\pi n} \left(\frac{a}{\lambda}\right)^2 \frac{1}{\sqrt{(a/\lambda)^2 + (\pi n)^2}}. \tag{9.73}$$

The detailed solution shows that the periodically modulated components j_{sy} and j_{sz} decay rapidly with z and the transition from the uniform current distribution of the normal conductor to the surface distribution of the current flowing in the z direction in the superconductor usually occurs over the distance $0.1 - 0.12$ μm [56].

The model of the junction considered is too simplified for the two reasons [56]:

1. The proximity effect [68], i.e. the direct influence of the junction on the electron pair density, is neglected.
2. The fabrication of the junction between the normal conductor and superconductor close to the idealized model shown in Fig. 9.2 where the dimensions of the order of 0.1–0.12 μm are important is hardly possible.

Nevertheless, the example is instructive because it illustrates the application of the London equations and describes some general features of the junction [56].

9.4.2 Electrostatic Analogy

Establish an analogy between the magnetic field caused by currents on superconductors and the electrostatic fields caused by charges on the surfaces of normal conductors of identical form. In the absence of electric currents in a definite volume the Maxwell's equation yields

$$\mathrm{curl}\boldsymbol{B} = 0. \tag{9.74}$$

Introducing the vector potential $\boldsymbol{A}$ we get from equation (9.74)

$$\mathrm{curl}\mathrm{curl}\boldsymbol{A} = 0. \tag{9.75}$$

Let a differential current element $Id\boldsymbol{l}$ be a part of the surface current on a conductor lying near an infinite superconducting plane. If $Id\boldsymbol{l}$ is parallel to the z-axis, the magnetic vector potential possesses only a z component A_z, and in the region without currents equation (9.75) reduces to the Laplace equation

$$\nabla^2 A_z = 0. \tag{9.76}$$

Therefore we can define A_z as the analog of the electric potential Θ which by definition also satisfies the Laplace equation. The boundary condition on electric potential at the surface of the superconductor is that the tangential component of its gradient would vanish. In our case there are conditions

$$B_\perp = 0 \tag{9.77}$$

and

$$B_t = \mu_0 j_s \tag{9.78}$$

because the magnetic field vanishes inside a superconductor vanishes, and the normal induction component $B_\perp$ and the tangential field component should be continuous at the surface. Here j_s is the surface current density. Now we must relate $B_\perp$ and the tangential component of ∇A_z. Taking into account that

we have
$$A_x = A_y = 0$$

$$\boldsymbol{B} = \mathrm{curl}\boldsymbol{A} = \left(\frac{\partial A_z}{\partial y}, -\frac{\partial A_z}{\partial x}, 0\right), \quad B = \sqrt{\left(\frac{\partial A_z}{\partial y}\right)^2 + \left(\frac{\partial A_z}{\partial x}\right)^2} = |\nabla A_z|. \tag{9.79}$$

Obviously, the first of relations (9.79) yields:
$$\boldsymbol{B} \cdot \nabla A_z = 0. \tag{9.80}$$

Comparing (9.77), (9.79) and (9.80) one can see that the tangential component of ∇A_z vanishes at the superconductor surface:
$$(\nabla A_z)_t = 0. \tag{9.81}$$

We now can define the analogous magnetic and electrostatic quantities.. The electric field intensity $\boldsymbol{E}$ has a value
$$E = |\nabla \Theta|. \tag{9.82}$$

Consequently, the magnetic induction and the electric field have the same values if the values of A_z and Θ coincide. It is known that the electric field $\boldsymbol{E}$ is parallel to the gradient of the electrostatic potential while the magnetic induction is perpendicular to ∇A_z according to (9.80). We conclude that the analogous fields $\boldsymbol{E}$ and $\boldsymbol{B}$ are equal in absolute value and orthogonal at each point if $A_z = \Theta$. The value of $\boldsymbol{B}$ in the vicinity of the superconductor surface is determined by (9.78). The value of the electric field of the charged plane is given by
$$E = \frac{\rho_s}{\varepsilon_0 \varepsilon^r}, \tag{9.83}$$

where ρ_s is the surface charge density. Equating (9.78) and (9.83) we find that the analogy will be complete if ρ_s is given by
$$\rho_s = \mu_0 \varepsilon_0 \varepsilon^r j_s. \tag{9.84}$$

According to the image method in electrostatics, the image of a line charge in a conducting plane is a line of opposite charge situated at the same distance on the other side of the plane of the conductor surface. Then, it is seen from relation (9.84) that the image of a line current is the current of the opposite direction situated similarly to the image charge. It should be underlined, however, that the image method permits the determination of the fields only between current and the superconductor.

The principle of superposition makes it possible to generalize the results obtained for a differential current element to any arbitrary system of current elements. In order to determine the fields the distribution of currents on the conductor lying above the superconducting plate must be known. The method can be applied effectively only in the one-dimensional case where the vector potential satisfies (9.76). It should be also noted that all derivations are based on the assumption that $B = 0$ inside the superconductor which is a good approximation if the penetration depth λ is small compared with the thickness of the superconductors and the space separating them [56].

9.4.3 Inductance of a Superconducting Thin Strip

Consider now a practically important case of a thin superconducting strip of thickness h_1 and width W which is separated from a ground plane of thickness h_2 by a dielectric layer h_d. The x-axis is chosen to be along the wide side of the strip cross section. The geometry of the problem is shown in Fig. 9.3.

Such a configuration is typical in thin-film superconducting circuits where a film of width much greater than its thickness ($W \gg h_1$) is carrying current along its length which is much greater than its width. The strip is usually deposited on top of an insulating layer which separates it from the ground plane [56].

The analysis based on the electrostatic analogy method has shown [56] that if the current distribution is uniform with respect to x, the thickness h_1 can be neglected, the thickness of the ground plane is assumed to be infinite, and the strip is uniform in the current direction, then the magnetic field along the bottom and top surfaces of the strip peaks up at the edge over a distance about equal to the spacing h_d from the ground plane. The field at the ground plane decays considerably over a distance of a few times h_d away from the strip edge. The field B_0 in the center of the strip has the form:

$$B_0 = \mu_0 \kappa \frac{I}{W} \tag{9.85}$$

where $\kappa < 1$ is the parameter weakly depending on W/h_d and close to unity. It describes the field reduction due to the finite spacing h_d. In the limiting case $W/h_d \to \infty$ the parameter $\kappa \to 1$.

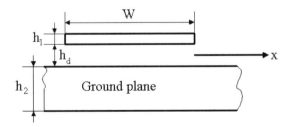

Fig. 9.3. Superconducting thin strip over a ground plane

The thin strip over a ground plane represents a strip transmission line since the currents in the strip and the ground plane flow in opposite directions. Calculate the inductance per unit length of such a line neglecting edge effects, and assuming the current distribution to be uniform and $\kappa = 1$. The magnetic field $B_{x1,2}(y)$ in the strip and ground plane has the form respectively [56]

$$B_{x1,2}(y) = \mu_0 \frac{I}{W} \frac{\sinh(y/\lambda)}{\sinh(h_{1,2}/\lambda)}, \qquad (9.86)$$

where the coordinate y is directed inwards the corresponding superconductor with origin at its outer edge. We must include in the total inductance the kinetic inductance (9.23) mentioned above and take into account the field energy $B^2/(2\mu_0)$. In the chapter *Magnetostatics* we have shown that the total magnetic energy of the system of currents can be expressed in terms of inductance (Eq. (2.53)). Consequently, we have [56]:

$$\frac{\mathcal{L}I^2}{2} = \frac{1}{2\mu_0} \int_V \left[B^2 + (\mu_0 \lambda)^2 j_s \right] dV, \qquad (9.87)$$

where j_s is the current density in the superconductors and λ is the penetration depth for the corresponding integration volume. The total volume of integration includes both superconductors and the intermediate space for a unit length in the direction of current flow. Substituting (9.85) and (9.86) into (9.87) we obtain the inductance per unit length [56]:

$$\mathcal{L} = \frac{\mu_0 h_d \kappa}{W} \left[1 + \frac{\lambda_1}{h_d} \coth \frac{h_1}{\lambda_1} + \frac{\lambda_2}{h_d} \coth \frac{h_2}{\lambda_2} \right], \qquad (9.88)$$

where $\lambda_{1,2}$ are the penetration depths in the two superconductors. It should be noted that the proper penetration depths should be used in (9.88) which account for the nonlocality effects. For the superconductors which are thick compared to their respective penetration depths expression (9.88) takes the form:

$$\mathcal{L} = \frac{\mu_0 h_d \kappa}{W} \left(1 + \frac{\lambda_1}{h_d} + \frac{\lambda_2}{h_d} \right). \qquad (9.89)$$

Expressions (9.88) and (9.89) are valid for the temperatures below $0.95 T_{cr}$ and for the frequencies below 1 GHz [56] where T_{cr} is the lower critical temperature of the ones for the strip and the ground plane.

The results obtained for the strip can be applied for a superconducting thin-film loop deposited on an insulator covering a superconducting ground plane as shown in Fig. 9.4 [56].

If the mean radius R_{loop} of the loop is much greater than its width W we can use expressions (9.88) and (9.89) as good approximations [56]. For the

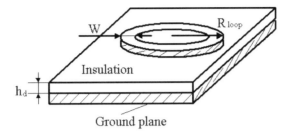

Fig. 9.4. Thin-film loop deposited on an insulated ground plane

loop and ground plane thick compared to their respective penetration depths we have for the total inductance of the loop similarly to (9.89)

$$\mathcal{L}_1 = \frac{2\pi R_{\text{loop}} \mu_0 h_d \kappa}{W} \left(1 + \frac{\lambda_1}{h_d} + \frac{\lambda_2}{h_d}\right). \quad (9.90)$$

The approximate expression for the inductance of an isolated loop is [56]

$$\mathcal{L}_0 \approx \mu_0 R_{\text{loop}} \left[\ln\left(\frac{16 R_{\text{loop}}}{W}\right) - 2\right]. \quad (9.91)$$

For the typical values of the parameters $R_{\text{loop}} = 0.1$ mm, $h_d = 300$ nm, $W = 0.01$ mm, $\lambda_1 = \lambda_2 = 50$ nm the comparison of (9.90) and (9.91) yields

$$\frac{\mathcal{L}_1}{\mathcal{L}_0} = 0.074.$$

The reduction of inductance is the primary reason for the use of ground planes [56].

9.5 Phenomena in a Superconductor under AC Regime

9.5.1 The Phenomenological Two-Fluid Model

The phenomenological two-fluid model was first introduced for the description of the superfluid properties of helium II [69]. It is assumed that helium II consists of a mixture of normal fluid and superfluid with different mass densities. The two fluids can move with different frequencies. In the absence of rotational motion in the superfluid there is no frictional interaction between the normal fuid and superfluid. The behavior of the normal fluid componentis described by the hydrodynamics of an ordinary viscous fluid while the superfluid component manifests the properties of an ideal classical fluid without viscosity [69, 70].

The two-fluid model is valid for a superconducting state as a superfluid of electrically charged particles (electrons) and can be applied to the description of a superconductor absorption of electromagnetic waves [70]. The two-fluid model for a superconductor presumes that the part of electrons in a superconductor is in superconducting state corresponding to the lowest energy while the other part is in the excited, or normal state [56, 70]. In such a case the current density j includes both a normal component j_n and a superconducting one j_s [56, 70].

$$j = j_s + j_n. \qquad (9.92)$$

The collisions between electron pairs do not occur as it was mentioned above. The behavior of the fluid of paired electrons is described by the London equation (9.16). The normal component of the electron fluid is governed by the momentum relaxation (9.18). Replacing the effective mass and charge of the electron pair by the electron mass and charge $-2e$ we can write these equations as follows [56]

$$m\frac{dv_s}{dt} = -eE \qquad (9.93)$$

and

$$m\frac{d\langle v_n \rangle}{dt} + m\frac{\langle v_n \rangle}{\tau} = -eE, \qquad (9.94)$$

where v_s and v_n are the particle velocities for the pair, or superconducting fluid and quasiparticle, or normal fluid respectively. The corresponding current densities are

$$j_s = -n_s e v_s, \quad j_n = -n_n e \langle v_n \rangle, \qquad (9.95)$$

where n_s and n_n are the number densities of electrons in paired and unpaired states respectively. The total temporal derivatives in (9.93) and (9.94) can be replaced by partial ones due to the slow spatial variations of the particle velocities. The time dependence of the electric field is assumed to be harmonic:

$$E = E_0 \exp i\omega t. \qquad (9.96)$$

Combining (9.93)–(9.96) and substituting the result in (9.2) we get

$$j = (\sigma_1 - i\sigma_2) E, \qquad (9.97)$$

where the real and imaginary parts of the superconductor conductivity σ_1 and σ_2 have the form respectively

$$\sigma_1 = \frac{n_n e^2 \tau}{m(1 + \omega^2 \tau^2)}, \quad \sigma_2 = \frac{n_s e^2}{m\omega} + \frac{n_n e^2 (\omega \tau)^2}{m\omega(1 + \omega^2 \tau^2)}. \qquad (9.98)$$

It is seen from (9.98) that the real part of the conductivity is determined by the normal fluid, while the imaginary part consists of two terms: the first

9.5 Phenomena in a Superconductor under AC Regime

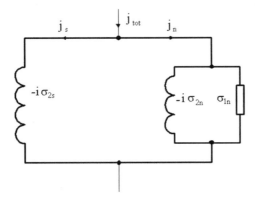

Fig. 9.5. Equivalent circuit for the admittance of a unit cube of superconductor in the two-fluid model

term is due to the superconducting fluid and the second one is caused by the normal electrons. For sufficiently low frequencies such that $\omega^2 \tau^2 \ll 1$ (9.98) reduces to the effective conductivity σ_{eff} which has the form

$$\sigma_{\text{eff}} = \sigma_1 - i\sigma_2 \approx \sigma_n \frac{n_n}{n} - i\frac{1}{\mu_0 \omega \lambda^2}, \qquad (9.99)$$

where $\sigma_n = ne^2\tau/m$ is the conductivity in the normal state. An equivalent circuit corresponding to the conductivity (9.98) is presented in Fig. 9.5 [56].

Consider now the behavior of the conductivity in the anomalous limit where the penetration length is much smaller than the intrinsic coherence length: $\lambda \ll \xi_0$ [56]. The real and imaginary parts of the conductivity have the form

$$\frac{\sigma_1}{\sigma_n} = \frac{2}{\hbar\omega} \int_\Delta^\infty [f(\mathcal{E}) - f(\mathcal{E} + \hbar\omega)]\, g(\mathcal{E})\, d\mathcal{E}$$

$$+ \frac{1}{\hbar\omega} \int_{\Delta - \hbar\omega}^{-\Delta} [1 - f(\mathcal{E} + \hbar\omega)]\, g(\mathcal{E})\, d\mathcal{E} \qquad (9.100)$$

and

$$\frac{\sigma_2}{\sigma_n} = \frac{1}{\hbar\omega} \int_{\Delta - \hbar\omega}^{\Delta} \frac{[1 - 2f(\mathcal{E} + \hbar\omega)]\left[\mathcal{E}^2 + \Delta^2 + \hbar\omega\mathcal{E}\right]}{[\Delta^2 - \mathcal{E}^2]^{1/2}\left[(\mathcal{E} + \hbar\omega)^2 - \Delta^2\right]^{1/2}} d\mathcal{E}, \qquad (9.101)$$

where Δ is the gap parameter, $f(u)$ is the Fermi function:

$$f(u) = \left[1 + \exp\frac{u}{k_B T}\right]^{-1/2} \tag{9.102}$$

and

$$g(\mathcal{E}) = \frac{\mathcal{E}^2 + \Delta^2 + \hbar\omega\mathcal{E}}{(\mathcal{E}^2 - \Delta^2)^{1/2}\left[(\mathcal{E} + \hbar\omega)^2 - \Delta^2\right]^{1/2}}. \tag{9.103}$$

The first integral in (9.100) represents the effect of the normal-state electrons, or the thermally excited quasiparticles. The second integral includes the contributions of the photon excited quasiparticles and is zero for $\hbar\omega < 2\Delta$. The integral (9.101) includes only the contribution of the paired electrons. It should be noted that its lower limit must be replaced by $-\Delta$ if $\hbar\omega > 2\Delta$.

In the following subsections we consider some practically important applications of the formalism developed using mainly the results presented in [56].

9.5.2 Surface Impedance

As we have seen in the chapter *Electromagnetic Waves*, the propagation properties of waveguides and the quality factors of cavity resonators strongly depend on the energy losses. The losses in conducting walls, in particular, can be expressed in terms of the surface impedance of the metals used in these constructions. We calculate the surface impedance of the superconductor using the results for the complex conductivity obtained above in the framework of the two-fluid model. Assume as usual that the superconductor and a free space are separated by a plane surface and determine the ratio E/H at this surface for spatially uniform fields. The displacement currents in a superconductor can be neglected just as in the case of ordinary conductors. A skin depth derived for the conductors can be also used with the replacement of the ordinary conductivity σ by the effective conductivity σ_{eff} (9.99). The skin depth δ is given by

$$\delta = \sqrt{\frac{2}{\mu_0 \omega \sigma}}. \tag{9.104}$$

Inserting σ_{eff} (9.99) instead of σ we get

$$\delta = \lambda \sqrt{\frac{2}{(\omega \tau n_n/n_s) - i}}. \tag{9.105}$$

Substituting the result (9.105) into the expression for the field in the case of the skin effect we find for $\omega = 0$

$$B = B_0 \exp\left(-\frac{1+i}{\delta}x\right) = B_0 \exp\left(-\frac{x}{\lambda}\right) \tag{9.106}$$

which coincides with the result (9.41) of the London equations. Obviously, the penetration depth decreases with the increase of a frequency. The superconductor penetration depth λ is much less than the skin depth δ for normal metals, even for microwave frequencies. The reduction of the penetration depth in a superconductor is due to the reactive current of the electron pairs [56].

The surface impedance of a normal metal $\mathcal{Z}_s$ is defined as the ratio of the tangential components of an electric and magnetic field intensities E_t and H_t near the surface. It has the form [4]:

$$\mathcal{Z}_s = \frac{E_t}{H_t} = \sqrt{\frac{\mu_0 \mu^r}{\varepsilon_0 \varepsilon^r}} = \sqrt{\frac{i\omega\mu_0}{\sigma}}, \qquad (9.107)$$

where we have taken into account that for a metal the real part of the dielectric constant can be neglected, $\mu^r = 1$ and the imaginary part $\varepsilon^{r''}$ of the dielectric constant and the conductivity are related by the formula

$$\varepsilon^{r''} = -\frac{\sigma}{\varepsilon_0 \omega}.$$

Once more we use the expression for σ_{eff} (9.99) given by the two-fluid model inserting it into (9.107). Then we obtain

$$\mathcal{Z}_s = i\omega\mu_0\lambda \frac{1}{\sqrt{1 + i\omega\tau n_n/n_s}} \approx i\omega\mu_0\lambda \left(1 - \frac{1}{2} i\omega\tau \frac{n_n}{n_s}\right). \qquad (9.108)$$

We separate the expression for the surface impedance into a real and imaginary parts corresponding to the surface losses R_s per unit area per unit surface current density amplitude and the surface inductive reactance $\omega \mathcal{L}_s$ of the superconductor [56]:

$$\mathcal{Z}_s = R_s + i\omega \mathcal{L}_s, \quad R_s = \frac{\omega^2 \mu_0^2 \lambda^3 n_n \sigma_n}{2n}, \quad \mathcal{L}_s = \mu_0 \lambda. \qquad (9.109)$$

It is easy to see that the phase angle of the surface impedance varies from $\pi/2$ at $T = 0$ when $n_n = 0$ to $\pi/4$ at $T = T_{\text{cr}}$ assuming additionally that $\omega\tau \ll 1$ [56].

9.5.3 Superconducting Transmission Lines and Microwave Cavities

Superconductors are important in the transmission of electric power because they reduce losses to negligible levels and permit the transfer of large amounts of electric power in compact geometries [56, 71]. At power frequencies of 50 or 60 Hz the losses calculated by virtue of the two fluid model are negligible, even close to the transition temperature. The dissipation is mainly due to

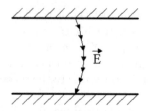

Fig. 9.6. Schematic representation of the electric field E in a parallel-plane superconducting transmission line

the hysteresis losses. In this frequency range it is worth to treat the superconductor as a magnetic material instead of the two fluid model [56, 70].

Hollow waveguides, cavity resonators, strip lines using normal metal conductors have been considered previously, in the chapter *Electromagnetic Waves*. They have comparatively low losses at radio frequencies and microwave frequencies up to 10 GHz. However, there exist two important applications where the normal loss level is unacceptable [56, 71]:

1. very long delay lines with the delay time $\tau_{\text{del}} > 1$ s;
2. wideband lines carrying short pulses.

Consider superconducting transmission line. We have shown that, unlike a "perfect conductor", the surface impedance of a superconductor at radio frequencies (RF) exists due to the inertia of electron-pair fluid. As a result, a RF electric field occurs within the skin depth at the superconductor surface. This field is necessary for the governing of the RF supercurrents flowing in a transmission line. The lowest mode of the two-conductor line shown in Fig. 9.6 is not a transverse electromagnetic (TEM) wave, as in the case for perfect conductors, but a transverse magnetic (TM) mode since a z component of the electric field exists in the superconductor even in the absence of losses [56]. These types of electromagnetic waves have been considered in details in chapter *Electromagnetic Waves*.

However, the z component of the electric field is small, and the expressions for TEM waves can be used [56]. The z-dependence of the TEM wave on a two-conductor line is described by the factor $\exp(-\gamma z)$ where the propagation constant γ and characteristic impedance $\mathcal{Z}_0$ have the form [56]:

$$\gamma = \alpha + i\beta = [(R + i\omega\mathfrak{L})(G + i\omega\mathcal{C})]^{1/2} \tag{9.110}$$

and

$$\mathcal{Z}_0 = \left[\frac{(R + i\omega\mathfrak{L})}{(G + i\omega\mathcal{C})}\right]^{1/2}, \tag{9.111}$$

where α is the attenuation constant, β is the phase constant, and $R, G, \mathfrak{L}$, and $\mathcal{C}$ are the distributed constants shown in Fig. 9.7. The series induction $\mathfrak{L}$

9.5 Phenomena in a Superconductor under AC Regime   363

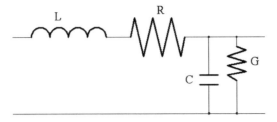

Fig. 9.7. General transmission line equivalent circuit

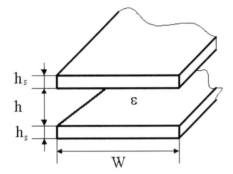

Fig. 9.8. Parallel-plane transmission line

includes the contributions from the magnetic flux in the space surrounding the conductors and from the surface impedance. The capacitance per unit length $\mathcal{C}$ is the ratio of electric charge on one conductor to the voltage between conductors. The line losses are represented by G and R; the losses in dielectric are represented by G and are usually less important than the losses in the conductors represented by R [56].

Consider a parallel-plane line presented in Fig. 9.8. It consists of the two superconducting plates of a thickness h_s and a width W separated by the spacing h_d that permits the simplification of the problem by neglecting the fringing fields. The thickness of the spacing is supposed to be small compared to the width: $h_d \ll W$. Usually, the lines of such a type are fabricated with thin films having the thickness substantially greater than the penetration length [56]. Consequently, we can use expression (9.109) for the line impedance. The total series impedance per unit length is given by [56]

$$R + i\omega\mathcal{L} = \frac{2\mathcal{Z}_s}{W} + i\omega\mathcal{L}_{\text{ext}}, \tag{9.112}$$

where $\mathcal{L}_{\text{ext}}$ is the inductance associated with the magnetic flux between the superconductors. This inductance is described by the first term of (9.90) with the parameter $\kappa = 1$. The other two terms are included in $\mathcal{Z}_s$. Combining (9.90), (9.109) and (9.112) we obtain for the impedance:

$$R + i\omega\mathcal{L} = \frac{\omega^2\mu_0^2\lambda^3 n_n\sigma_n}{nW} + i\frac{\omega\mu_0(h_d + 2\lambda)}{W}, \qquad (9.113)$$

and for the propagation constant [56]:

$$\gamma = \sqrt{-\omega^2\mu_0\varepsilon_0\varepsilon^r\frac{(h_d + 2\lambda)}{h_d} + i\varepsilon_0\varepsilon^r\frac{\omega^2\mu_0^2\lambda^3 n_n\sigma_n}{nh_d}}, \qquad (9.114)$$

where ε^r is the relative permittivity of the dielectric between the superconductors, and the conductance G is neglected. The attenuation and phase constants α and β can be evaluated directly from (9.114), but their explicit expressions are too complicated. They are simplified substantially in the case where the second term under the square root is small which is typical for low frequencies and temperatures well below T_{cr}. Then we get [56]:

$$\alpha = \frac{R_s}{h_d}\sqrt{\frac{\varepsilon_0\varepsilon^r}{\mu_0(1 + 2\lambda/h_d)}} \qquad (9.115)$$

$$\beta = \omega\left[\mu_0\varepsilon_0\varepsilon^r(1 + 2\lambda/h_d)\right]^{1/2}\left[1 + \frac{R_s^2}{2\omega^2\mu_0^2(h_d + 2\lambda)^2}\right] \qquad (9.116)$$

$$\mathcal{Z}_0 = \frac{h_d}{W}\left(\frac{\mu_0}{\varepsilon_0\varepsilon^r}\right)^{1/2}\left[1 + \frac{2\lambda}{h_d} - i\frac{2R_s}{\mu_0\omega h_d}\right]^{1/2}. \qquad (9.117)$$

The group velocity $v_{\text{gr}} = \partial\omega/\partial\beta$ and the phase velocity $v_{\text{ph}} = \omega/\beta$ can be calculated by using (9.116):

$$v_{\text{gr}} = \left[\mu_0\varepsilon_0\varepsilon^r(1 + 2\lambda/h_d)\right]^{-1/2}\left[1 - \frac{R_s^2}{2\omega^2\mu_0^2(h_d + 2\lambda)^2}\right]^{-1} \qquad (9.118)$$

and

$$v_{\text{ph}} = \left[\mu_0\varepsilon_0\varepsilon^r(1 + 2\lambda/h_d)\right]^{-1/2}\left[1 + \frac{R_s^2}{2\omega^2\mu_0^2(h_d + 2\lambda)^2}\right]^{-1}. \qquad (9.119)$$

The wave is propagated slower, if the superconductor film is thinner as it is seen from (9.118) and (9.119). This effect can be used in compact delay lines. Neglecting the surface losses connected with R_s and using expression (9.88) for the thin superconductor films one can see that the group and phase velocities coincide:

$$v_{\text{gr}} = v_{\text{ph}} = \left[\mu_0\varepsilon_0\varepsilon^r\left(1 + \frac{2\lambda}{h_d}\coth\frac{h}{\lambda}\right)\right]^{-1/2}. \qquad (9.120)$$

Slowing factors of at least 10 are feasible for very thin films separated by a thin film of a dielectric with high permittivity [56].

Consider briefly the specific features of superconducting cavity resonators. The possibility to reach a quality factor $Q \sim 10^{11}$ stimulated the interest to superconducting resonators. Similarly to the low-loss the surface-impedance function is a basis for the calculation of the highest possible values of Q. However, the calculation of surface resistance based on the two-fluid model becomes more less reliable as losses are progressively reduced due to the presence of the so-called residual losses [56]. The surface resistance R_s in microwave cavities has the form for the pure superconductors [56]:

$$R_s = \frac{C\omega^{3/2}}{T} \exp\left(-\frac{\Delta(T)}{k_B T}\right) + R_0(\omega), \quad (9.121)$$

where C is a constant, $\Delta(T)$ is the temperature dependent energy gap parameter, and $R_0(\omega)$ is the residual resistance. Enumerate the probable causes of the residual losses which have been determined experimentally [56]:

1. A trapped magnetic flux. Indeed, during the cooling of the superconductor through the transition temperature the presence of a weak external magnetic field can produce a trapped flux.
2. Impurities, and in particular the magnetic ones, result in the local variations of the energy gap, with a possibility of transformation of some isolated superconducting regions into normal ones.
3. Imperfection of the surface of the superconductor, which may not be perfectly flat or may contain mechanically damaged layers. The presence of roughness raises the level of a magnetic field locally above the average surface value, perhaps, beyond the critical value. Irreversible magnetic behavior occurs on a microscopic scale in these circumstances so that the maximum allowable stored energy for a given value of Q is reduced.

These factors limit the maximum possible value of Q for a given geometry. The small-signal Q values exceed those obtainable under high-power conditions. The Q values of resonant circuits are proportional to $\omega^{1/2}$ for normal metals in the classical skin-effect regime, since the surface resistance is proportional to $\omega^{1/2}$, while the stored energy density increases as ω. This means that the maximum possible value of Q corresponds to high frequencies. The superconducting losses, excluding residual losses, increase approximately as $\omega^{3/2}$, so that Q values are approximately proportional to $\omega^{-1/2}$. Consequently, in the superconducting resonators the highest values of Q occur at lower frequencies. The temperature dependence of the conduction losses shows that high-Q resonators must be operated as far below the critical temperature as possible. The lowest temperatures used in practice are just below the λ point for superfluid helium ($T = 2.17$K) where considerable refrigeration advantages can be obtained [56].

9.6 The Ginzburg–Landau Theory

9.6.1 The Ginzburg–Landau Equations

We mentioned above some deficiencies of the London theory of the superconducting state. The Ginzburg–Landau theory developed in 1950 makes it possible to overcome completely these difficulties. The remarkable fact about the Ginzburg–Landau theory is that it was created before the microscopic BCS theory but succeeded to describe adequately the phenomena typical for the superconducting state remaining in the framework of the phenomenological approach [56,57,72]. The Ginzburg–Landau theory permits the derivation of the complete system of nonlinear coupled differential equations, in general case, in partial derivatives, which are applicable in strong magnetic fields.

Ginzburg and Landau postulated the existence of a so-called order parameter which could be used as a measure of electron pairs condensation in a superconductor. It is a complex quantity which coincides with the macroscopic wave function (9.3). Unlike the London theory, the absolute value of this wave function $|\Psi(r)|$ is assumed to be varying spatially in response to applied magnetic fields which makes possible the analysis of systems of materials with the spatially variable physical properties [56]. Ginzburg and Landau constructed an expression for the free energy of a superconductor as a functional of the order parameter $\Psi(r)$. The minimization of such a functional yields the system of equations describing the distribution of the order parameter itself as well as the magnetic induction in the equilibrium state.

We start with the free energy $\mathcal{F}$ expression for a superconductor in the presence of a magnetic field which is determined by a vector potential $\boldsymbol{A}(r)$. This expression must obey some general conditions based on the physical considerations [57].

1. In the vicinity of the critical temperature the free energy density can be expanded in powers of the order parameter $\Psi(r)$ and its gradients. The term with the gradient of the order parameter occurs due to its spatial inhomogeneity. The order parameter itself is assumed to be sufficiently small, such that terms up to the fourth order should be kept in the power series.
2. According to the general requirements of the invariance with respect to the phase transformations of the type $\Psi'(r) = \Psi(r)\exp i\varphi$ all terms of the odd order in $\Psi(r)$ must be excluded.
3. The combination of two previous conditions requires that only two first terms proportional to $|\Psi(r)|^2$ and to $|\Psi(r)|^4$ should be kept.
4. In the presence of the magnetic field the term $B^2/(2\mu_0)$ describing the field energy density must be added to the free energy density.
5. The spatial variations of the order parameter are assumed to be sufficiently slow, and only first derivatives should be kept.

6. The gradient term would contain the gradient of the order parameter phase. In the presence of the magnetic field it must be of the type (9.12) in order to satisfy the gauge invariance conditions (9.8) for the vector potential and (9.13) for the order parameter.
7. For the sake of definiteness the symmetry of the metallic crystal is assumed to be cubic which permits using the gradient term squared instead of a more general quadratic form for the case of lower symmetry crystals.

Taking into account all these requirements we write [57]:

$$\mathcal{F} = \mathcal{F}_n + \int \left\{ \frac{B^2}{2\mu_0} + \frac{\hbar^2}{2m^*} \left| \left(\nabla - \frac{2ie}{\hbar} \mathbf{A} \right) \Psi(\mathbf{r}) \right|^2 + a \left| \Psi(\mathbf{r}) \right|^2 + \frac{b}{2} \left| \Psi(\mathbf{r}) \right|^4 \right\} dV, \quad (9.122)$$

where $\mathcal{F}_n$ is the free energy of a normal state, $2e = e^*$, $a = \varpi (T - T_{\mathrm{cr}})$, $\varpi > 0$, and the coefficient $b > 0$ depends only on the density of material. It should be emphasized that the complex order parameter $\Psi(\mathbf{r})$ is determined by two real quantities: $|\Psi(\mathbf{r})|$ and the phase θ. Therefore $\Psi(\mathbf{r})$ and $\Psi^*(\mathbf{r})$ must be considered as two independent functions. Then the minimization of the functional $\mathcal{F}$ (9.122) with respect to $\Psi^*(\mathbf{r})$ gives:

$$\delta \mathcal{F} = \int \left\{ -\frac{\hbar^2}{2m^*} \left(\nabla - \frac{ie^*}{\hbar} \mathbf{A} \right)^2 \Psi + a\Psi + b |\Psi|^2 \Psi \right\} \delta\Psi^* dV$$

$$+ \frac{\hbar^2}{2m^*} \oint \left(\nabla \Psi - \frac{ie^*}{\hbar} \mathbf{A} \Psi \right) \delta\Psi^* d\mathbf{S} = 0. \quad (9.123)$$

The second integral in (9.123) is the surface one, and it vanishes. The condition (9.123) is satisfied for any arbitrary $\delta\Psi^*$ if the integrand is zero which yields

$$\frac{1}{2m^*} \left(-i\hbar\nabla - e^* \mathbf{A} \right)^2 \Psi + a\Psi + b |\Psi|^2 \Psi = 0. \quad (9.124)$$

Similarly the minimization with respect to $\mathbf{A}$ gives

$$\frac{1}{\mu_0} \mathrm{curl} \mathbf{B} = -\frac{ie^* \hbar}{m^*} \left(\Psi^* \nabla \Psi - \Psi \nabla \Psi^* \right) - \frac{e^{*2}}{m^*} |\Psi|^2 \mathbf{A}. \quad (9.125)$$

The comparison of the Ginzburg–Landau equation (9.125) and the corresponding Maxwell equation shows that the right-hand side represents the superconducting current density $\mathbf{j}_s$:

$$\mathbf{j}_s = -\frac{ie^* \hbar}{m^*} \left(\Psi^* \nabla \Psi - \Psi \nabla \Psi^* \right) - \frac{e^{*2}}{m^*} |\Psi|^2 \mathbf{A}. \quad (9.126)$$

On the other hand, the comparison of (9.126) with (9.6), (9.7) and (9.12) of the London theory one can see that the expressions for the superconducting current density coincide if identify $|\Psi|^2$ according to (9.4). The current density (9.126) can be considered as a total one since the normal component is absent in the thermodynamic equilibrium [57]. Note that the continuity equation for the current density (9.126) follows immediately from (9.125). Relations (9.124)–(9.126) compose the complete set of the Ginzburg–Landau equations. The boundary condition is obtained from the fact that the surface integral in (9.123) must vanish for any $\delta\Psi^*$:

$$\boldsymbol{n} \cdot (-i\hbar \nabla \Psi - e^* \boldsymbol{A} \Psi) = 0, \tag{9.127}$$

where $\boldsymbol{n}$ is the normal to the surface of the superconductor. This simply means that the normal component of current density also vanishes at the surface [56, 57]

$$\boldsymbol{n} \cdot \boldsymbol{j}_s = 0. \tag{9.128}$$

The boundary conditions for the normal component of the magnetic induction remain the same as in the case of normal conductor. It is continuous at the surface of the superconductor. The tangential component of the magnetic induction is also continuous there due to the finite value of the current density. Hence, the magnetic induction $\boldsymbol{B}$ is continuous at the superconductor boundary [57].

The influence of weak magnetic fields on $|\Psi|^2$ can be neglected. Then, assuming the absolute value of the order parameter to be constant the Ginzburg-Landau equations reduce to the London equation (9.36) with the temperature dependent penetration depth $\lambda(T)$

$$\lambda(T) = \sqrt{\frac{m^* b}{\mu_0 e^{*2} \varpi (T_{\mathrm{cr}} - T)}} = \sqrt{\frac{m^* b}{\mu_0 e^{*2} |a|}}. \tag{9.129}$$

On the other hand, in the absence of a magnetic field the equilibrium value of Ψ is given by:

$$|\Psi_0|^2 = -\frac{a}{b} = \frac{\varpi (T_{\mathrm{cr}} - T)}{b}. \tag{9.130}$$

Consider the conditions of applicability of the Ginzburg–Landau equations. For low temperatures these equations are valid when

$$T_{\mathrm{cr}} - T \ll T_{\mathrm{cr}}, \tag{9.131}$$

since in this interval the order parameter is small, and the free energy density expansion is held. Condition (9.131) also provides the validity of the inequality $\xi_{\mathrm{GL}}(T) \gg \xi_0$ for the Ginzburg-Landau coherence length, or the temperature dependent coherence length $\xi_{\mathrm{GL}}(T)$

$$\xi_{GL}(T) = \frac{\hbar}{\sqrt{2m^*\varpi(T_{cr}-T)}} = \frac{\hbar}{\sqrt{2m^*|a|}}. \quad (9.132)$$

We should emphasize that this coherence length completely differs from the temperature independent Pippard coherence lengths mentioned above, since it characterizes the minimal distance over which the order parameter varies substantially [56]. The condition for the penetration depth $\lambda(T) \gg \xi_0$ requires the modification of inequality (9.131) for superconductors with small values of the Ginzburg–Landau parameter $\varkappa_{GL}$ which is given by

$$\varkappa_{GL} = \frac{\lambda(T)}{\xi_{GL}(T)} = \frac{\sqrt{2}\mu_0 e^* \lambda^2(T) H_{cr}(T)}{\hbar}, \quad (9.133)$$

where the critical magnetic field intensity $H_{cr}(T)$ is introduced as follows [56, 57]:

$$H_{cr}^2(T) = \frac{a^2}{\mu_0 b} = \frac{[\varpi(T_{cr}-T)]^2}{\mu_0 b} = \frac{|a|}{\mu_0}|\Psi_0|^2. \quad (9.134)$$

Then (9.131) takes the form

$$T_{cr} - T \ll \varkappa_{GL}^2 T_{cr}. \quad (9.135)$$

At $T \to T_{cr}$ it turned out that the Ginzburg–Landau equations are valid up to the critical point T_{cr} itself [57].

9.6.2 Examples of Ginzburg–Landau Theory Application

The solution of the coupled nonlinear Ginzburg–Landau equations (9.124)–(9.126) in general case is hardly possible. However, it is instructive to consider in details two relatively simple one-dimensional cases:

1. a superconducting half-space [56, 57, 72];
2. a thin superconducting film [56].

For the sake of simplicity we assume that the applied magnetic field H is sufficiently small, compared with the critical field H_{cr}, and only slightly affects the order parameter. The x-axis is chosen to be perpendicular to the boundary plane separating free space and a superconductor, filling a half-space $x > 0$. The magnetic field is applied parallel to the boundary. The vector potential gauge $\mathrm{div}\,\boldsymbol{A} = 0$ yields in our case

$$\frac{\partial A_x}{\partial x} = 0 \quad (9.136)$$

and consequently

$$A_x = 0, \; A_y = A, \; B = B_z = \frac{\partial A}{\partial x}. \quad (9.137)$$

For the simply connected superconducting bodies and in the absence of a transport current flowing the order parameter phase $\theta(r)$ can be reduced to zero, and Ψ becomes real [56]. Then (9.124) and (9.125) takes the form:

$$\frac{\hbar^2}{2m^*}\frac{\partial^2 \Psi}{\partial x^2} = \left\{a + b\Psi^2 + \frac{e^{*2}}{2m^*}A^2\right\}\Psi \qquad (9.138)$$

and

$$\frac{\partial^2 A}{\partial x^2} = \frac{e^{*2}}{m^*}\mu_0 \Psi^2 A. \qquad (9.139)$$

The solution is sought to be

$$\Psi = \Psi_0 + \Psi_1 + ..., \quad \Psi_0 = \text{const}, \quad \Psi_1 \ll \Psi_0. \qquad (9.140)$$

Substitution of (9.140) into (9.137) and (9.139) yields

$$\frac{\hbar^2}{2m^*}\frac{\partial^2 \Psi_1}{\partial x^2} = a\Psi_1 + \frac{e^{*2}}{2m^*}A^2 \Psi_0 \qquad (9.141)$$

$$\frac{\partial^2 A}{\partial x^2} = \frac{e^{*2}}{m^*}\mu_0 \Psi_0^2 A, \qquad (9.142)$$

where Ψ_0 is determined by (9.130). We start with the general solution of (9.142). It has the form

$$A = C_1 \exp\left(-\frac{x}{\lambda(T)}\right) + C_2 \exp\left(\frac{x}{\lambda(T)}\right), \qquad (9.143)$$

where $\lambda(T)$ coincides with (9.129), as it is seen from comparison of (9.129), (9.130) and (9.142). Obviously, $C_2 = 0$ since the vector potential must be finite at infinity. The constant C_1 is defined from the condition of the magnetic induction continuity at the superconductor boundary $x = 0$:

$$\frac{\partial A}{\partial x}\bigg|_{x=0} = -\frac{C_1}{\lambda(T)} = \mu_0 H, \quad C_1 = -\lambda(T)\mu_0 H \qquad (9.144)$$

and

$$A = -\lambda(T)\mu_0 H \exp\left(-\frac{x}{\lambda(T)}\right), \quad B(x) = \mu_0 H \exp\left(-\frac{x}{\lambda(T)}\right). \qquad (9.145)$$

Substituting A from (9.145) into (9.141) we obtain:

$$\frac{\hbar^2}{2m^*}\frac{\partial^2 \Psi_1}{\partial x^2} + 2a\Psi_1 = \frac{e^{*2}}{2m^*}\Psi_0 (\mu_0 H)^2 \lambda^2(T) \exp\left(-\frac{2x}{\lambda(T)}\right). \qquad (9.146)$$

The solution of (9.146) consists of the general solution of the homogeneous equation, i.e. of the left-hand side, and of the particular solution of the inhomogeneous equation. The latter solution is analogous to the function in the right-hand side of (9.146). We keep the term finite at $x \to \infty$ and get:

$$\Psi_1 = \Psi_{10} \exp\left(-\frac{x\sqrt{2}}{\xi_{GL}(T)}\right) + \Psi_{20} \exp\left(-\frac{2x}{\lambda(T)}\right), \quad (9.147)$$

where $\xi_{GL}(T)$ is the Ginzburg–Landau coherence length, as one can see from (9.146) and (9.132); Ψ_{20} is determined by the direct substitution of the second term of (9.147) into (9.146), and Ψ_{10} is found from boundary condition

$$\frac{\partial \Psi_1}{\partial x}\bigg|_{x=0} = 0. \quad (9.148)$$

After some algebra we obtain [56]:

$$\frac{\Psi}{\Psi_0} = 1 - \frac{\varkappa_{GL}}{2\sqrt{2}(2 - \varkappa_{GL}^2)} \frac{H^2}{H_{cr}^2(T)}$$

$$\times \left\{ \exp\left(-\frac{x\sqrt{2}}{\xi(T)}\right) - \frac{\varkappa_{GL}}{\sqrt{2}} \exp\left(-\frac{2x}{\lambda(T)}\right) \right\}. \quad (9.149)$$

The results (9.145) and (9.149) are illustrated in Fig. 9.9 and 9.10.

Consider now a thin film of a thickness d in a parallel magnetic field H. We assume that the film thickness is much less than the Ginzburg–Landau coherence length: $d \ll \xi(T)$, and therefore the order parameter Ψ depends on the magnetic field, but not on the position. The geometry of the problem is shown in Fig. 9.11.

In such a case (9.142) along with the boundary conditions of the type (9.144) at the surfaces of the film $x = \pm d/2$ has the solution:

$$A = \frac{\mu_0 H \lambda}{\varsigma} \frac{\sinh(\varsigma x/\lambda)}{\cosh(\varsigma x/2\lambda)}, \quad -\frac{d}{2} \leq x \leq \frac{d}{2}, \quad (9.150)$$

where $\varsigma = \Psi(H)/\Psi_0$, and

$$\lambda = \frac{m^*}{e^{*2}\mu_0 n_{seff}} \quad (9.151)$$

and the effective density of pairs $n_{seff} = |\Psi_0|^2$ depends on the film thickness. It can be shown that for film thickness less than the penetration length $d \leq \lambda$ and small field intensity H the order parameter has the form [56]

$$\frac{\Psi(H)}{\Psi_0} = 1 - \frac{1}{8}\left(\frac{H}{H_{cr}}\right)^2 \frac{\sinh(d/\lambda) - (d/\lambda)}{(d/\lambda)\cosh^2(d/\lambda)}, \quad (9.152)$$

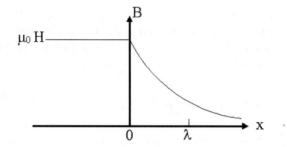

Fig. 9.9. The spatial dependence of the magnetic induction in a superconducting half-space

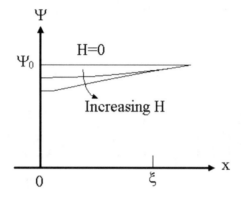

Fig. 9.10. The spatial dependence of the order parameter in a superconducting half-space

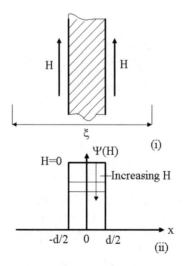

Fig. 9.11. (i) A thin superconducting film. (ii) The order parameter which is independent of position but dependent on magnetic field

where H_{cr} is the bulk critical field. The critical film $H_{\parallel}$ for the film has the form [56]:

$$H_{\parallel}^2 = \frac{H_{\mathrm{cr}}^2 \varsigma^2 \left(2 - \varsigma_{\mathrm{cr}}^2\right)}{1 - (2\lambda/\varsigma_{\mathrm{cr}} d) \tanh\left(\varsigma_{\mathrm{cr}} d/2\lambda\right)}, \qquad (9.153)$$

where ς_{cr} is the value of the quantity ς at the critical field, i.e. the field at which the free energy of the film in the superconducting state equals to that for the normal state. It turned out that for very thin films the critical field $H_{\parallel}$ is considerably larger than the bulk value H_{cr} for the same materials [56]. For example, for $d < \lambda\sqrt{5}$ it can be shown that [56, 57]

$$H_{\parallel} = \sqrt{24}\frac{\lambda}{d} H_{\mathrm{cr}}, \qquad (9.154)$$

and for films with $d/\lambda \to 0$

$$\frac{\Psi(H)}{\Psi_0} = \sqrt{1 - \left(\frac{H}{H_{\parallel}}\right)^2} = \sqrt{1 - \frac{H^2 d^2}{24 H_{\mathrm{cr}}^2 \lambda^2}}. \qquad (9.155)$$

9.6.3 Surface Energy at the Boundary between Normal Conductor and Superconductor

The Ginzburg–Landau equations make it possible to express the surface energy at the boundary separating the normal and superconducting phases in the same sample in terms of the bulk characteristics of the material. Experimentally, superconducting and conducting regions can coexist in a so-called intermediate state in metallic samples subjected to a magnetic field, where they are nonuniformly distributed through the sample. In the superconducting regions the order parameter $\Psi \neq 0$, and it tends to the equilibrium value, while in the normal ones it vanishes within a Ginzburg–Landau coherence length of the boundary [57]. The magnetic induction $\boldsymbol{B}$ being expelled from the superconducting parts can penetrate into normal regions. At the boundaries between normal and superconducting regions the magnetic induction $\boldsymbol{B}$ decays on a distance of an order of magnitude of the penetration depth λ.

For the sake of definiteness, consider the plane boundary between the normal and superconducting regions of the metallic sample. The boundary is chosen to be the yz plane, the superconductor is filling the half-space $x > 0$, and all quantities in both half-spaces depend on x only. In such a geometry we define the vector potential $\boldsymbol{A}$ and magnetic induction $\boldsymbol{B}$ according to (9.136) and (9.137). The quantitative picture of the magnetic induction and order parameter behavior at the boundary $x = 0$ is presented in Fig. 9.12.

The external magnetic field intensity $\boldsymbol{H}$ is assumed to be directed along the z-axis. As a result, the field intensity $\boldsymbol{H}_{\mathrm{int}}$ inside the superconducting region of the sample is everywhere equal to the field applied regardless of

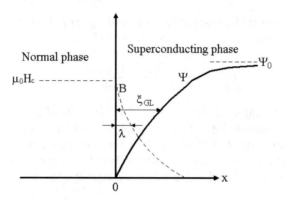

Fig. 9.12. Variation of the magnetic induction and order parameter at a plane interphase boundary in an infinite superconducting medium

the variating induction $\boldsymbol{B}$ [56, 57]. We consider the case when the external magnetic field $\boldsymbol{H}$ is equal to the critical field

$$H = H_z = H_{\mathrm{cr}}. \tag{9.156}$$

In the normal phase in the presence of the magnetic field 9.156, where

$$x \to -\infty, \ \Psi = 0, \ B = \mu_0 H_{\mathrm{cr}} \tag{9.157}$$

the free energy density $\mathcal{G}_n$ has the form [56, 57]:

$$\mathcal{G}(-\infty) = \mathcal{G}_n = \mathcal{F}_n - \frac{1}{2}\mu_0 H_{\mathrm{cr}}^2, \tag{9.158}$$

where $\mathcal{F}_n$ is the free energy density of the normal state without magnetic field. In the bulk of the superconductor,

$$x \to \infty, \ B = 0, \ |\Psi|^2 = |\Psi_0|^2 = -\frac{a}{b} \tag{9.159}$$

and the free energy density $\mathcal{G}_s$ is also given by 9.158:

$$\mathcal{G}_s(\infty) = \mathcal{F}_n - \frac{1}{2}\mu_0 H_{\mathrm{cr}}^2, \tag{9.160}$$

where relation (9.134) is used for the second term in the right-hand side. The surface energy $\mathcal{S}_{\mathrm{ns}}$ (per unit area of boundary) is defined as the difference of the energies of the superconducting and normal parts [56]:

$$\mathcal{S}_{\mathrm{ns}} = \int_{-\infty}^{\infty} dx \, \{\mathcal{G}_s - \mathcal{G}_n\}. \tag{9.161}$$

9.6 The Ginzburg–Landau Theory

Comparison of (9.158) and (9.160) shows that the essential contribution in the integrand of (9.161) is due to the region of the superconductor near the boundary. Therefore, we substitute into (9.161) the integrand of (9.122) reduced to the one-dimensional case and (9.158). For the sake of generality we must also include into $\mathcal{G}_s$ the term $-BH_{\mathrm{cr}}$ [56, 57]. Previously this term has been omitted since it is linear in the vector potential derivatives, and for this reason it does not contribute into the Ginzburg–Landau equations in the case of a uniform external magnetic field. Finally we get [56, 57, 73]:

$$\mathcal{S}_{\mathrm{ns}} = \int_{-\infty}^{\infty} dx \left[\frac{B^2}{2\mu_0} + \frac{\hbar^2}{2m^*} \left(\left| \frac{\partial \Psi}{\partial x} \right|^2 + \frac{e^{*2}}{\hbar^2} A^2 |\Psi(x)|^2 \right) + a |\Psi(x)|^2 + \frac{b}{2} |\Psi(x)|^4 \right]$$

$$- \int_{-\infty}^{\infty} dx \left[B H_{\mathrm{cr}} - \frac{1}{2} \mu_0 H_{\mathrm{cr}}^2 \right] \qquad (9.162)$$

Note that the term $i\mathbf{A} \cdot \nabla \Psi$ vanished because $A_x = 0$ according to (9.137). The integrand in (9.162) is real, and one can assume $\Psi(x)$ to be real.

In order to evaluate the integrals (9.162) we should solve the Ginzburg–Landau equations (9.138) and (9.139) with the boundary conditions (9.157), (9.159). We introduce the dimensionless quantities as follows [57]:

$$\tilde{x} = \frac{x}{\lambda}, \quad \tilde{\Psi} = \Psi \sqrt{\frac{b}{|a|}}, \quad \tilde{A} = \frac{A}{\lambda \mu_0 H_{\mathrm{cr}}}, \quad \tilde{B} = \frac{B}{\mu_0 H_{\mathrm{cr}}}. \qquad (9.163)$$

With these variables (9.138) and (9.139) take the form [57]:

$$\frac{\partial^2 \tilde{\Psi}}{\partial \tilde{x}^2} = \varkappa_{\mathrm{GL}}^2 \left[\left(\frac{\tilde{A}^2}{2} - 1 \right) \tilde{\Psi} + \tilde{\Psi}^3 \right] \qquad (9.164)$$

and

$$\frac{\partial^2 \tilde{A}}{\partial \tilde{x}^2} = \tilde{A} \tilde{\Psi}^2. \qquad (9.165)$$

In general case it is possible to get in a closed form only the first integral of these equations which has the form [57]:

$$\frac{2}{\varkappa_{\mathrm{GL}}^2} \left(\frac{\partial \tilde{\Psi}}{\partial \tilde{x}} \right)^2 + \left(2 - \tilde{A}^2 \right) \tilde{\Psi}^2 - \tilde{\Psi}^4 + \left(\frac{\partial \tilde{A}}{\partial \tilde{x}} \right)^2 = 1. \qquad (9.166)$$

Excluding by virtue of (9.166) the term with Ψ^4 in the integrand of (9.162) we obtain in the dimensionless variables (9.163) the following integral [57]:

$$\mathcal{S}_{ns} = \lambda \mu_0 H_{cr}^2 \int_{-\infty}^{\infty} d\tilde{x} \left[\frac{2}{\varkappa_{GL}^2} \left(\frac{\partial \tilde{\Psi}}{\partial \tilde{x}} \right)^2 + \frac{\partial \tilde{A}}{\partial \tilde{x}} \left(\frac{\partial \tilde{A}}{\partial \tilde{x}} - 1 \right) \right]. \qquad (9.167)$$

We start the analysis of (9.166) and (9.167) with the case of small Ginzburg–Landau parameter $\varkappa_{GL} \ll 1$. In such a case the magnetic field is varying substantially on a distance λ which is small compared to the characteristic length ξ of the order parameter variation. Consequently, in region of a strong magnetic field $\Psi \approx 0$, and conversely, in the region of the order parameter Ψ existence the vector potential A in (9.166) can be neglected. Then (9.166) reduces to the following equation

$$\frac{\partial \tilde{\Psi}}{\partial \tilde{x}} = \frac{\varkappa_{GL}}{\sqrt{2}} \left(1 - \tilde{\Psi}^2 \right) \qquad (9.168)$$

with the boundary condition $\tilde{\Psi} = 0$ at $x = 0$. It has the explicit solution

$$\tilde{\Psi} = \tanh \left(\frac{\varkappa_{GL} \tilde{x}}{\sqrt{2}} \right). \qquad (9.169)$$

The evaluation of the integral (9.167) with the solution (9.169) and $A = 0$ yields [57]:

$$\mathcal{S}_{ns} = \frac{1}{2} \mu_0 H_{cr}^2 \frac{1.9 \lambda}{\varkappa_{GL}}. \qquad (9.170)$$

It can be shown that the accuracy of expression (9.170) is of an order of magnitude $\sqrt{\varkappa_{GL}}$ [57]. The further analysis proves that the surface energy $\mathcal{S}_{ns}$ gradually decreases with the increase of $\varkappa_{GL}$, reduces to zero at $\varkappa_{GL} = 1/\sqrt{2}$ and then becomes negative [57]. The value of the Ginzburg–Landau parameter $\varkappa_{GL}$ plays key role in the calculation procedure since it relates two characteristic lengths of the theory, namely the penetration length λ and the Ginzburg–Landau coherence length ξ_{GL}. Generally, the penetration of the magnetic induction in superconductor lowers the free energy by approximately $\mu_0 H_{cr}^2 \lambda / 2$. The effect of the order parameter decay is opposite, since it raises the free energy by the amount close to $\mu_0 H_{cr}^2 \xi_{GL} / 2$. The interplay of these two contributions depending on λ and ξ_{GL}, respectively, determines the sign of the surface energy $\mathcal{S}_{ns}$.

The sign of the surface energy affects the properties of superconductors which are divided in the two types according to this feature. Superconductors of the type I possess $\varkappa_{GL} < 1/\sqrt{2}$, their surface energy is positive, it raises the free energy with respect to the uniform state, and, consequently, the existence of boundaries in such materials is energetically unfavorable. The division into superconducting and normal regions in the type I superconductors is possible only for the samples of some special shapes. However, the behavior of such materials is essentially different from the one of the superconductors with $\varkappa_{\mathrm{GL}} > 1/\sqrt{2}$. The latter materials, or superconductors of the type II have a negative surface energy which makes it possible to form an energetically favorable structure consisting of a large number of normal and superconducting regions separated by interphase boundaries [57,73]. The type I superconductors are, as a rule, pure superconducting metals. The type II superconductors are superconducting alloys [56].

9.7 Type II Superconductors

9.7.1 Mixed State

It was Abrikosov who first developed in 1956 the theory of the type II superconductors [74] based on the Ginzburg–Landau equations. The detailed investigation of the subject can be found, for example, in [52] and [75]. The state-of-art review of the microscopic theory is presented in [76]. Below we briefly discuss the basic properties of the type II superconductors using the phenomenological approach [56,57].

The negative surface energy of the superconductor corresponding to the Ginzburg–Landau parameter $\varkappa_{\mathrm{GL}} > 1/\sqrt{2}$ permits the formation of normal regions in a superconductor where the magnetic flux can penetrate thus lowering the total Gibbs free energy of the system [73]. As a result, a state occurs consisting of both superconducting and normal regions. Two possible distinct geometrical configurations of such a state have been studied [73]:

1. a laminar arrangement consisting of alternating normal and superconducting layers;
2. a filamentary arrangement consisting of cylindrical normal regions embedded in a superconducting matrix.

The analysis showed that the filamentary geometry has a lower Gibbs free energy than the laminar one, and for this reason it represents the equilibrium state whenever a magnetic flux partially penetrates into it. This filamentary system is known as a mixed state [57,73].

The specific feature of the type II superconductors is that they retain zero resistivity for steady currents in the presence of magnetic fields considerably greater than the corresponding critical magnetic fields H_{cr}. This practically important advantage is a result of the partial penetration of the magnetic flux into the bulk of the superconductor for the range of magnetic field intensities which determine the mixed state. The currents in the type II superconductors in the mixed state are larger than in the type I materials because the flow is distributed in the bulk rather than in the penetration layer near the surface.

The magnetic flux first penetrates the bulk of the superconductor at the lower critical field $H_{\mathrm{cr}1}$. Then, at the upper critical field $H_{\mathrm{cr}2}$ the order parameter vanishes and the transition to the normal state occurs. In the interval

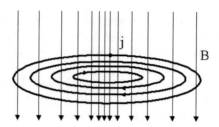

Fig. 9.13. Circulating current j and magnetic field B in a vortex

$H_{cr1} < H < H_{cr2}$ the superconductor exists in a mixed state. Its properties in such a state are gradually changing from the pure superconductivity at H_{cr1} up to the normal state at H_{cr2}.

Let us investigate the process of a magnetic flux penetration into the type II superconductor [56, 57, 73]. Consider a cylindrical sample of such a material subjected to the external magnetic field H parallel to the axis of the cylinder. We start the analysis with the fields which are slightly higher than H_{cr1}. In this situation in the bulk of the superconductor small regions of the normal phase emerge since the previous analysis showed that the significant variations of the order parameter can occur at least on a distance of a coherence length ξ_{GL}. The shape of the normal regions is determined by the energetical conditions. The normal phase regions possess the negative surface energy, and therefore the structure with a maximal possible surface area is energetically favorable. Consequently, they would be realized in a filamentary form where the order parameter is gradually reducing and finally reaches zero exactly at the center of the filament. The order parameter rises from zero at the center to an approximately constant value at a radius equal to two coherence lengths. That value depends on the external magnetic field and varies from roughly zero-field value at H_{cr1} to zero at the upper critical field H_{cr2}, where the transition to the normal state takes place. The magnetic field in the normal region is accompanied by a current which serves for a screening of the filament from the more strongly superconducting region outside. The direction of the current density is shown in Fig. 9.13.

The cylindrical filament with a distribution of Ψ, B and j is called a vortex. Consider the condition of the first vortex emergence. In a long thin superconducting sample parallel to the magnetic field the internal field intensity H is uniform and equal to the applied field. The magnetic induction B varies across the plane perpendicular to the vortex axis having a maximum along the axis. Let the energy of the vortex per unit length be $\mathcal{E}_v$ and the length of the vortex be l_v. Then the free energy of the sample with one vortex has the form [57]:

$$\mathcal{F} = \mathcal{F}_s + l_v \mathcal{E}_v - \int \boldsymbol{H} \cdot \boldsymbol{B} dV = \mathcal{F} = \mathcal{F}_s + l_v \mathcal{E}_v - H l_v \int B ds, \quad (9.171)$$

where $\mathcal{F}_s$ is the free energy of the superconducting state where the magnetic induction vanishes, the uniform magnetic field intensity H can be taken out of the integral, and the integral itself represent the magnetic flux Φ_v through the cross section of the vortex. It has been shown above that the magnetic flux through the ring with a current is quantized. At the lower critical field $H_{\text{cr}1}$ this condition can applied to the vortex. It can be shown that the energetically favorable vortex state corresponds to the minimal flux $\Phi_0 = h/2e$ [73]. Clearly, the vortex can occur when the addition to $\mathcal{F}_s$ becomes negative [57]. Then, the threshold condition is

$$l_v \mathcal{E}_v - H l_v \Phi_0 = 0 \qquad (9.172)$$

which yields the critical value of the external field $H_{\text{cr}1}$ corresponding to the first vortex emergence

$$H_{\text{cr}1} = \frac{\mathcal{E}_v}{\Phi_0} \qquad (9.173)$$

Above $H_{\text{cr}1}$ the state with individual magnetic flux quanta in vortices becomes energetically more favorable than the Meissner state with total magnetic flux exclusion [56]. In the ideal case, the vortices compose a regular two-dimensional array, or lattice. The lattice spacing is determined by the applied magnetic field and the mutual repulsion between neighboring vortices. The large-scale average of the magnetic induction B taken over many vortices approaches the external magnetic field value, as $H \to H_{\text{cr}2}$. When a transport current of density $\boldsymbol{j}_T$ passes through a superconductor in a mixed state, there exist a Lorentz force $[\boldsymbol{j}_T \times \boldsymbol{B}]$ acting on the vortices. The chemical and physical defects in the superconductor exert forces on the vortices to keep them trapped or pinned at the locations of the defects. If the Lorentz forces exceed the pinning forces, the vortices move and generate an electromotive force which results in dissipation.

9.7.2 London Model of the Mixed State

An understanding of the structure and behavior of vortices is necessary for an adequate macroscopic description of the mixed state. Consider a type II superconductor in the limiting case $\varkappa_{\text{GL}} \gg 1$ which means that $\lambda \gg \xi_{\text{GL}}$. The distance ξ_{GL} determines the radius of the vortex core where the order parameter Ψ is changing from zero at the vortex axis to the finite value corresponding to the surrounding superconducting region. On the other hand, the magnetic induction decays at large distances $r \sim \lambda \gg \xi_{\text{GL}}$, and the magnetic flux is mainly concentrated in the region out of the vortex core where $|\Psi|^2 = \text{const}$. The latter relation simplifies the problem and permits the application of the London equations to the analysis of the vortex because the relation between the current density $\boldsymbol{j}_s$ and the vector potential $\boldsymbol{A}$ becomes local [57]. Thus we return to equations (9.12), (9.25), (9.29) and (9.37) and write

$$\boldsymbol{A} + \lambda^2 \mathrm{curl} \boldsymbol{B} = \frac{\hbar}{e^*} \nabla \theta. \tag{9.174}$$

The factor in the right-hand side of (9.174) can be easily identified as $\Phi_0/(2\pi)$ where Φ_0 is the basic quantum of magnetic flux (9.64). Integrate (9.174) along the contour C enclosing the vortex at the sufficiently large distance $r \gg \xi_{\mathrm{GL}}$ from its axis. Transforming the integral over the left-hand side of (9.174) into the surface one according to the Stokes theorem we obtain:

$$\int_S \left(\boldsymbol{B} + \lambda^2 \mathrm{curl}\mathrm{curl} \boldsymbol{B} \right) d\boldsymbol{s} = \int_S \left(\boldsymbol{B} - \lambda^2 \nabla^2 \boldsymbol{B} \right) = \Phi_0, \tag{9.175}$$

where we have chosen in the right-hand side the minimal change of the phase θ equal to 2π. The equality (9.175) must be satisfied for any contour C which yields

$$\boldsymbol{B} - \lambda^2 \nabla^2 \boldsymbol{B} = z_0 \Phi_0 \delta(\boldsymbol{r}), \tag{9.176}$$

where $\delta(\boldsymbol{r})$ is the δ-function and $\boldsymbol{r}$ is the two-dimensional radius vector in the transverse cross section of the vortex. The presence of the δ-function in the right-hand side of (9.176) means that distances $\sim \xi_{\mathrm{GL}}$ are considered as zero [57]. The solution of (9.176) in cylindrical coordinates is given by the following expression:

$$B = B_z = \frac{\Phi_0}{2\pi \lambda^2} K_0 \left(\frac{r}{\lambda} \right), \tag{9.177}$$

where $K_0(x)$ is the modified Bessel function of zero order. The circulating in the vortex superconducting current has only a circumferential component $j_{s\varphi}$ and has the form

$$j_{s\varphi} = \frac{\Phi_0}{2\pi \lambda^3 \mu_0} K_1 \left(\frac{r}{\lambda} \right), \tag{9.178}$$

where $K_1(x)$ is the modified Bessel function of first order. Both functions $K_{0,1}(r/\lambda)$ decay exponentially for $r \gg \lambda$. In the opposite case where $r \ll \lambda$ the function $K_1(r/\lambda)$ varies approximate like $1/r$.

The London model of a vortex has the following limitations [56]:

1. It does not take into account the effect of decay of Ψ in the core region $r \lesssim \xi_{\mathrm{GL}}$ of the vortex.
2. It applies only where the Ginzburg-Landau parameter is large: $\varkappa_{\mathrm{GL}} \gg 1$.
3. It is valid in the interval of magnetic fields $(H - H_{\mathrm{cr1}}) \ll H_{\mathrm{cr1}}$ and for temperatures T not too close to T_{cr}.

A model of the mixed state can be constructed as the superposition of a large number of vortices of the form considered above. At the fields close to the lower critical field the vortices are widely separated and the interaction between them is weak. As the field increases, the vortices become densely packed and their interaction must be taken into account.

9.7.3 Vortex Energy

Evaluate the free energy $\mathcal{E}_v$ of the vortex [56, 57]. We take into account the contribution from the vortex region $\xi_{GL} \ll r \ll \lambda$ thus ignoring the core. The difference in the free energies of the superconducting and mixed states is due to the presence of the vortex. On the other hand, the free energy $\mathcal{F}$ of the superconductor (9.122) includes the identical terms for both states independent of the magnetic induction:

$$\mathcal{F}_{s0} = \mathcal{F}_n + \int \left\{ a\, |\Psi(\mathbf{r})|^2 + \frac{b}{2} |\Psi(\mathbf{r})|^4 \right\} dV. \tag{9.179}$$

The free energy $\mathcal{E}_v$ per unit length of the vortex is therefore given by the following formula [56]:

$$\mathcal{E}_v = \int_S \left\{ \frac{B^2}{2\mu_0} + \frac{\hbar^2}{2m^*} \left| \left(\nabla - \frac{2ie}{\hbar} \mathbf{A} \right) \Psi(\mathbf{r}) \right|^2 \right\} dS, \tag{9.180}$$

where S is the part of the cross section limited by the circumferences $r = \lambda$ and $r = \xi_{GL}$. The contribution of the first term in the integrand in (9.180) is negligibly small since the magnetic induction is mainly concentrated in the core of the vortex $r < \xi_{GL}$ [57]. In the region considered the absolute value of the order parameter $|\Psi(\mathbf{r})|$ is assumed to be constant. For this reason the second term in the integrand can be identified with superconducting current density $\mathbf{j}_s$ (9.12) which reduces (9.180) to

$$\mathcal{E}_v \simeq \int_S \frac{m^*}{2n_s e^{*2}} j_s^2 \, dS. \tag{9.181}$$

It can be shown that the evaluation of the integral (9.181) yields [56, 57]

$$\mathcal{E}_v = \frac{\Phi_0^2}{4\pi \lambda^2 \mu_0} \ln \varkappa_{GL}. \tag{9.182}$$

Substituting (9.182) into (9.173) we find the lower critical magnetic field

$$H_{\text{cr}1} = \frac{\Phi_0}{4\pi \lambda^2 \mu_0} \ln \varkappa_{GL}. \tag{9.183}$$

We can also express $H_{\text{cr}1}$ in terms of the critical magnetic field intensity H_{cr} for the type I superconductors. Equation (9.133) gives

$$H_{cr} = \frac{\varkappa_{GL}\Phi_0}{2\sqrt{2}\pi\lambda^2\mu_0}. \qquad (9.184)$$

Dividing (9.183) by (9.184) we get

$$H_{cr1} = H_{cr}\frac{1}{\varkappa_{GL}\sqrt{2}}\ln \varkappa_{GL}. \qquad (9.185)$$

It can be shown that the magnetic induction $B(0)$ at the center of an isolated vortex has the form [56]:

$$B(0) = \frac{2\mu_0 \mathcal{E}_v}{\Phi_0} = 2\mu_0 H_{cr1}. \qquad (9.186)$$

9.7.4 Vortex Lattice

In the mixed state of a type II superconductor, the magnetic field penetrates into the sample as Abrikosov vortices [77]. The emergence of the first vortex is followed by an influx of many others as the external magnetic field is slightly increasing above H_{cr1}. However, the average density of vortices n is small in comparison with that at the upper critical field H_{cr2} and will approach this limit when the magnetic flux penetrates the sample completely [56]. It is assumed that a regular array of vortices, or flux lines, represent the most favorable configuration of a type II superconductor [73]. Indeed, the calculation and minimization of the lattice free energy

$$\mathcal{F} = \mathcal{F}_v - \boldsymbol{B} \cdot \boldsymbol{H}, \qquad (9.187)$$

as a function of the lattice configuration with the account for the repulsive interaction between neighboring vortices showed that in the mixed state in the ideal pure crystalline material at the magnetic fields $H_{cr1} < H < H_{cr2}$ the vortices are always situated at the lattice sites of a regular array. The equilibrium lattice structure can be investigated by virtue of the London model. For n vortices equation (9.176) should be modified as follows:

$$\boldsymbol{B} - \lambda^2 \nabla^2 \boldsymbol{B} = z_0 \sum_j \Phi_0 \delta(\boldsymbol{r} - \boldsymbol{r}_j), \qquad (9.188)$$

where the summation is carried out over all vortices placed in the points $\boldsymbol{r}_j$. To construct the solution of (9.188) we use the superposition of the magnetic fields (9.177) for the isolated vortex neglecting the interaction effect. We have:

$$B = B_z = \frac{\Phi_0}{2\pi\lambda^2}\sum_j K_0\left(\frac{\boldsymbol{r}-\boldsymbol{r}_j}{\lambda}\right). \qquad (9.189)$$

The total free energy per unit volume $\mathcal{F}_v$ due to n vortices per unit area consists of two parts. The first contribution is the sum of self-energies $\mathcal{E}_v$

(9.182) of all vortices. The second part describes the repulsive interaction between vortices. Thus we get [56]:

$$\mathcal{F}_v = n\mathcal{E}_v + \frac{n\Phi_0^2}{4\pi\mu_0\lambda^2} \sum_j K_0\left(\frac{\boldsymbol{r}-\boldsymbol{r}_j}{\lambda}\right), \qquad (9.190)$$

where the summation is carried out over all vortices except the one situated at the origin. The equilibrium lattice structure can be found by minimizing the free energy $\mathcal{F}$ (9.187) with respect to the magnetic induction $\boldsymbol{B}$ averaged over all n vortices and using (9.190) for $\mathcal{F}_v$. This procedure [73] makes it possible to obtain a constitutive relation between $\boldsymbol{B}$ and the applied magnetic field intensity $\boldsymbol{H}$.

Assuming that for the magnetic fields slightly higher than $H_{\mathrm{cr}1}$ only nearest neighbors are important in the intervortex interaction and the distance between them is equal to d we can transform the interaction term from (9.190) [56, 73]:

$$\frac{n\Phi_0^2 N}{4\pi\mu_0\lambda^2} K_0\left(\frac{d}{\lambda}\right) \simeq \frac{BN\Phi_0}{4\pi\mu_0\lambda^2}\left(\frac{n\lambda}{2d}\right)^{1/2} \exp\left(-\frac{d}{\lambda}\right), \qquad (9.191)$$

where N is the number of the nearest neighbors and the asymptotic form of $K_0(d/\lambda)$ for $d \gg \lambda$ is used. The possible realizations of the array are either a triangular lattice, or a square one. The calculation for $z = 4$ (square array) and $z = 6$ (triangular array) shows that the triangular lattice has the lower Gibbs free energy just above $H_{\mathrm{cr}1}$ and represents the equilibrium state which corresponds to the experimental results [73]. It should be noted, however, that the difference in free energy for these two modifications is very small and can lead to anomalies caused by defects and other factors [56]. Finally, the constitutive relation for the triangular lattice takes the form [73]:

$$B \simeq \frac{2\Phi_0}{\sqrt{3}\lambda^2}\left\{\ln\left[\frac{3\Phi_0}{4\pi\mu_0\lambda^2(H-H_{\mathrm{cr}1})}\right]\right\}^{-2}. \qquad (9.192)$$

9.7.5 Upper Critical Field

At low magnetic fields the vortex cores occupy a negligibly small part of the superconductor volume. At higher magnetic fields, the entire superconductor contains a large number of weakly superconducting core regions through which the magnetic flux penetrates almost uniformly. At the upper critical magnetic field $H_{\mathrm{cr}2}$ the order parameter reduces to zero and the bulk of the sample passes to the normal state. It should be noted that a superconductivity may persist in a surface region up to a higher critical field $H_{\mathrm{cr}3}$. The continuous decrease of the order parameter Ψ to vanishingly small values with the increase of H up to $H_{\mathrm{cr}3}$ permits the application of the Ginzburg–Landau theory. Abrikosov solved the Ginzburg–Landau equations for the magnetic field values H close to H_{cr2} assuming that $|\Psi|$ is small compared to its value $|\Psi_0|$ with zero magnetic field [74]. It is also assumed that the perturbations of the magnetic field on the scale of the vortex screening current distribution are negligible in comparison with the field H applied in the z direction. In such a situation the Ginzburg–Landau equations can be linearized with respect to $\Psi(r)$. In particular, (9.124) takes the form:

$$\frac{1}{2m^*}(-i\hbar\nabla - e^*A)^2 \Psi + a\Psi = 0, \qquad (9.193)$$

where the vector potential is defined by (9.137):

$$A = A_y = \mu_0 H x. \qquad (9.194)$$

Substituting (9.194) into (9.193) and recalling the expressions for ξ_{GL} (9.132) and for the magnetic flux quantum Φ_0 (9.64) we obtain for the two-dimensional case:

$$\frac{\partial^2 \Psi}{\partial x^2} + \left[\frac{1}{\xi_{GL}^2} - \left(\frac{2\pi\mu_0 H x}{\Phi_0} + i\frac{\partial}{\partial y}\right)^2\right]\Psi = 0. \qquad (9.195)$$

The solution of (9.195) is sought in the form

$$\Psi = U(x)\exp(ik_y y). \qquad (9.196)$$

Inserting (9.196) into (9.195) we get the equation of a one-dimensional harmonic oscillator for the amplitude $U(x)$ [56]:

$$\frac{\partial^2 U}{\partial x^2} + \left[\frac{1}{\xi_{GL}^2} - \left(\frac{2\pi\mu_0 H x}{\Phi_0} - k_y\right)^2\right]U = 0. \qquad (9.197)$$

The eigenvalues of such an equation are determined by the following relation

$$H_m = \frac{\Phi_0}{2\pi\mu_0 \xi_{GL}^2 (2m+1)}, \quad m = 0, 1, 2, \ldots. \qquad (9.198)$$

The eigenvalue H_0 corresponding to $m = 0$ is the largest value for which the solution $\Psi \neq 0$ still exists. This eigenvalue can be identified as the upper critical field H_{cr2}

$$H_{cr2} = H_0 = \frac{\Phi_0}{2\pi\mu_0 \xi_{GL}^2} = \sqrt{2} H_{cr} \varkappa_{GL} \qquad (9.199)$$

The linear approximation equation (9.195) is valid for the fields in the vicinity of H_{cr2}, and its accuracy increases as H tends to H_{cr2}. The eigenfunction for $m = 0$ has the form

$$\Psi(x,y) = \Psi_0 \exp\left[-\frac{(x-x_0)^2}{2\xi_{GL}^2}\right]\exp\left(i\frac{2\pi\mu_0 H_{cr2} x_0}{\Phi_0}y\right), \qquad (9.200)$$

where

$$x_0 = \frac{k_y \Phi_0}{2\pi\mu_0 H_{cr2}}. \qquad (9.201)$$

The parameter x_0 is a continuous variable, so that solution (9.200) can be located at any interior point of the sample without a change of the free energy. Any linear combination of the vortex-type solutions may be assembled as a general solution with the values of x_0 arranged as the points of a lattice. The analysis showed that the account for nonlinear terms in the Ginzburg–Landau equations results in the triangular lattice of vortices corresponding to the minimum free energy configuration near H_{cr2} [56].

9.7.6 Vortex Motion

The structure of a vortex line is very rigid due to the quantization condition of a magnetix flux. For this reason, the vortex line reacts as a whole to any perturbation of the vortex system from equilibrium [77]. When a transport current passes through a superconductor with vortices, a Lorentz force gives rise to the vortices motion. Their motion causes a longitudinal potential gradient in the superconductor which results in the onset of a resistance. However, the vortices may be pinned by defects in the material with sufficient force to prevent continuous motion below a certain critical current density. Resistanceless operation of the type II superconductors can be achieved only in specially prepared so-called hard superconductors where the pinning force is sufficiently large to prevent flux motion [56].

A transport current applied by an external source passes through the bulk of material in contrast to the surface currents typical for the Meissner state. The transport current interacts with the magnetic induction in the vortices by the Lorentz force $\boldsymbol{f}_L$:

$$\boldsymbol{f}_L = [\boldsymbol{j} \times \boldsymbol{B}]. \qquad (9.202)$$

The transport current flows longitudinally, and because of the relation between the magnetic flux density in the vortex and the associated screening currents, the entire vortex is displaced laterally [56]. Calculate the magnitude of the force acting on an isolated vortex for large $\varkappa_{GL}$. Suppose that the transport current density is parallel to the x-axis and the magnetic induction is directed along the z-axis. Then the Lorentz force per unit volume for a stationary vortex possesses only the y component which has the form:

$$f_{Ly} = -n_s e^* v_x B_z, \qquad (9.203)$$

where v_x is the drift velocity associated with the uniform transport current that is superimposed on the circulating vortex currents. Using the vortex magnetic induction (9.177) and inserting the transport current density

$$j_T = j_{Tx} = n_s e^* v_x, \qquad (9.204)$$

we get for the Lorentz force (9.203)

$$f_{Ly} = -j_T \frac{\Phi_0}{2\pi\lambda^2} K_0\left(\frac{r}{\lambda}\right). \qquad (9.205)$$

The magnitude of the Lorentz force per unit length of the isolated vortex is obtained by integrating of (9.205) over the cross section area S extending to a distance of several penetration depths [56, 77]:

$$F_L = \int_S f_{Ly} ds = j_T \Phi_0. \qquad (9.206)$$

If the applied magnetic field is not perpendicular to the transport current but is oblique at the angle ϑ with respect to it then the Lorentz force according to (9.202) is given by

$$F_L = j_T \Phi_0 \sin\vartheta. \qquad (9.207)$$

In an ideal pure superconductor unimpeded vortex motion would occur for any value of the transport current and the mixed state could not be considered to be superconducting. However, in practice, inhomogeneities in interior and on the surface of the sample produce barriers to vortex motion and cause the vortices to be pinned locally at defect sites. Vortices that are not pinned are constraint to some extent by the intervortex interaction to stay with those that are. At the depinning current, the total Lorentz force on the array of vortices exceeds the total pinning force and the vortex lattice moves with a steady drift velocity. Thermally activated flux motion occurs as vortices are shaken free from their pinning sites. Both types of motion result in dissipation and heat generation [56].

Consider the time average motion of the vortices. In moving through the superconductor, the vortices experience a velocity dependent force $\boldsymbol{f}_v$ that is proportional to a velocity and can therefore be defined as a viscous force.

$$\boldsymbol{f}_v = -\eta \boldsymbol{v}, \qquad (9.208)$$

where η is a viscosity coefficient and $\boldsymbol{v}$ is the vortex velocity. The movement of magnetic flux induces an electromotive force in the core and in the surrounding region which drives current through the normal region. Bardeen and Stephen demonstrate that this current is just equal to the transport current when the vortices are moving. When the vortex is pinned or it is a part of a pinned lattice, there is no current flows in the core; the transport current flows around the core without energy loss [56].

Typical experiments results on voltages developed in type II superconductor films which are thinner than the penetration depth carrying a transport current showed that no measurable time-average voltage is developed until the current exceeds a certain value. For slightly higher currents, the voltage variation becomes linear. The maximum zero-voltage current depends on the density of defects in the samples. The onset of the voltage can be explained by virtue of the following phenomenological model [56]. The viscous force (9.208) is balanced by the difference between the Lorentz force (9.206) and a pinning force $\boldsymbol{f}_\mathrm{P}$ that averages the effect of the vortices' encounters with the potential wells at the defect pinning sites. Namely, we have [56, 77]

$$\eta v = F_\mathrm{L} - f_\mathrm{P}. \tag{9.209}$$

The forces in (9.209) can be written as scalars since they are either parallel, or anti-parallel to each other. An empirically useful picture of flux-flow resistance is based on the assumption that an electric field E is produced by the flux motion similarly to the Faraday's law of induction [56, 77]:

$$E = nv\Phi_0 = vB, \tag{9.210}$$

where n is the number of vortices per unit area and B is the average magnetic induction. Differentiating (9.209) with respect to F_L and combining the result with (9.206) we obtain

$$\frac{dv}{dj_T} = \frac{\Phi_0}{\eta}. \tag{9.211}$$

On the other hand, from (9.210) one can get

$$\frac{dE}{dv} = B \tag{9.212}$$

and finally combining (9.211) and (9.212) we find:

$$\rho_f = \frac{dE}{dj_T} = \frac{\Phi_0 B}{\eta}, \tag{9.213}$$

where ρ_f is the differential flow resistivity. The experimental results show that the differential flow resistivity ρ_f normalized to the normal state resistivity ρ_n at the same temperature linearly depends of on the magnetic field H. The linear field dependence becomes increasingly valid with the decrease of the temperature. At low temperatures $T \ll T_\mathrm{cr}$ the approximate relation has the form [56, 77]:

$$\frac{\rho_f}{\rho_n} = \frac{H}{H_{\mathrm{cr}2}(0)}. \tag{9.214}$$

Usually the normal resistivity of type II superconductors is relatively high (three orders of magnitude greater than that of copper at the same temperature). It is clear that no appreciable transport current is possible in the flux-flow state except brief transients. In practice, the superconductor must be surrounded by a normal metal of a good thermal conductivity to be able to withstand intermittent operation under flux-flow conditions.

9.8 The Josephson Effect

9.8.1 The Josephson Relations

In 1962, B.D. Josephson suggested that electron pairs can tunnel between closely spaced superconductors even without potential difference [62]. The predicted effect was later observed by Anderson and Rowell [78].

This phenomenon called the Josephson effect means that both superconductors separated by a thin layer of a dielectric become the unique system which is described by the common macroscopic wave function of the paired electrons liquid. Consequently, a superconducting current can flow through the junction even without any voltage applied. The superconducting current density in the junction depends on the wave function phase difference on the two sides of the junction similarly to the dependence of the superconducting current density in the bulk on the wave function phase gradient as was mentioned above. The phase difference across the junction in general depends both on time t and on spatial coordinates r. It is used as the variable in a Ginzburg–Landau theory of the electrodynamics of the junction [57, 79].

Consider two superconductors separated by a dielectric layer which represents a potential barrier. If separation between the superconductors 1 and 2 is large, the electron pairs in each of them are described by a macroscopic wave function $\Psi_{1,2}$ (9.3). The phases $\theta_{1,2}$ of the two wave functions are unrelated and can be defined with the accuracy to arbitrary additive constants. With the decrease of the separation layer thickness the functions can penetrate the barrier deeply enough to couple and the system energy is reduced by coupling. When the coupling energy exceeds the thermal fluctuation energy, the phases become locked and pairs can pass from one superconductor to the other without energy loss. In the limiting case of the impenetrable barrier the wave functions at the boundaries of both superconductors obey condition (9.127). Assume that all variables depend on the coordinate x which is perpendicular to the boundaries of superconductors. Then (9.127) takes the form for functions (9.3) [57]:

$$\frac{\partial \Psi_{1,2}}{\partial x} - \frac{ie^*}{\hbar} A_x \Psi_{1,2} = 0, \qquad (9.215)$$

where A_x is the vector potential. The penetration of the wave functions into the barrier results in the appearance of coupling terms in the right-hand part of (9.215) [57]:

$$\frac{\partial \Psi_{1,2}}{\partial x} - \frac{ie^*}{\hbar} A_x \Psi_{1,2} = \frac{\Psi_{2,1}}{v}, \qquad (9.216)$$

where v^{-1} is the coupling coefficient. Equations (9.216) must be invariant with respect to the transformation $\Psi \to \Psi^*$, $\boldsymbol{A} \to -\boldsymbol{A}$. Consequently, the constant v is real. The superconducting current density in the superconductor 1 has the form according to (9.126):

9.8 The Josephson Effect

$$j_s = -\frac{ie^*\hbar}{m^*\lambda_J}\left(\Psi_1^*\frac{\partial \Psi_1}{\partial x} - \Psi_1\frac{\partial \Psi_1^*}{\partial x}\right) - \frac{e^{*2}}{m^*}|\Psi|^2 A_x. \quad (9.217)$$

Substituting $\partial\Psi_1/\partial x$ from (9.216) into (9.217) we obtain:

$$j_s = -\frac{ie^*\hbar}{m^*v}\left(\Psi_1^*\Psi_2 - \Psi_2^*\Psi_1\right). \quad (9.218)$$

For the same superconductors on both sides of the contact the wave functions $\Psi_{1,2}$ differ only by the phases $\theta_{1,2}$:

$$\Psi_{1,2} = |\Psi_0|\exp i\theta_{1,2},\; |\Psi_0|^2 = n_s. \quad (9.219)$$

Then (9.218) yields

$$j_s = J_m \sin\theta_{21},\; J_m = \frac{2e^*\hbar}{m^*v}|\Psi_0|^2,\; \theta_{21} = \theta_2 - \theta_1. \quad (9.220)$$

Equations (9.220) show that the current flows through the contact between two superconductors without an external voltage. Near the transition point $|\Psi_0|^2 \to 0$ as $(T_{cr} - T)$ which means that the maximal current density J_m through the barrier tends to zero in the same manner.

Assume now that a voltage V is applied between the two sides. In this case the phases of the wave functions are not locked together but slip relative to each other at a rate precisely related to the voltage [56]. This time we should take into account the time dependence of the wave functions $\Psi_{1,2}$. The temporal evolution of these functions is described by the corresponding Schrœdinger equations which account for the coupling in the junction. We write [56]:

$$i\hbar\frac{\partial \Psi_{1,2}}{\partial t} = U_{1,2}\Psi_{1,2} + K\Psi_{2,1}, \quad (9.221)$$

where $U_{1,2}$ are the energies of the wave functions for the two superconductors and K is a coupling constant. The voltage applied is proportional to the energy difference:

$$U_2 - U_1 = e^*V. \quad (9.222)$$

For the sake of definiteness the zero level of energy can be chosen in the middle between the energies U_1 and U_2. Then (9.222) becomes

$$i\hbar\frac{\partial \Psi_{1,2}}{\partial t} = \mp\frac{e^*V}{2}\Psi_{1,2} + K\Psi_{2,1}. \quad (9.223)$$

Now returning to the definitions (9.3) and (9.4) and taking into this time both the absolute values $|\Psi_{1,2}|$ and the phases $\theta_{1,2}$ of wave functions $\Psi_{1,2}$ possess a temporal dependence we write real and imaginary parts [56,80]:

$$\frac{\partial n_{s1,2}}{\partial t} = \pm \frac{2K}{\hbar}\sqrt{n_{s1}n_{s2}}\sin\theta_{21} \qquad (9.224)$$

and

$$\frac{\partial \theta_{1,2}}{\partial t} = -\frac{K}{\hbar}\sqrt{\frac{n_{s2,1}}{n_{s1,2}}}\cos\theta_{21} \pm \frac{e^*V}{2\hbar}. \qquad (9.225)$$

Equations (9.224) show that the change rates of electron pair density in both superconductors are opposite. Actually, there cannot be a change of pair density since that would create a charge imbalance between the electrons and the background of ions. Such an imbalance is eliminated by the current in the external circuit. Assume that both superconductors are the same which means that the densities of the electron pairs on the both sides of the junction coincide $n_{s1} = n_{s2} = n_s$. Then, multiplying both sides of (9.224) by the effective charge e^* and the thickness of the junction electrodes d_J we get for the superconducting current density in the junction the expression coinciding with (9.220) if we identify the pair density n_s according to (9.4) and the coupling constant K as follows

$$K = \frac{\hbar^2}{m^* d_J v}. \qquad (9.226)$$

However, unlike the previous case, this time the phase difference θ_{21} is varying in time. In order to determine the temporal evolution of θ_{21} we subtract the first equation (9.225) from the second one which yields:

$$\frac{\partial \theta_{21}}{\partial t} = -\frac{e^*V}{\hbar}. \qquad (9.227)$$

Equation (9.227) must be invariant with respect to the gauge transformation of the scalar potential V

$$V' = V - \frac{\partial \chi}{\partial t}. \qquad (9.228)$$

Clearly, θ_{21} remains the same under the condition

$$\theta'_{21} = \theta_{21} + \frac{e^*}{\hbar}\chi. \qquad (9.229)$$

Equations (9.220) and (9.227) are the Josephson relations describing the behavior of electron pairs in the junction. Equation (9.227) can be easily integrated in the case of the constant voltage $V = $ const. In such a case we obtain

$$\theta_{21} = \theta_{21}^0 - \frac{e^*V}{\hbar}t. \qquad (9.230)$$

Equation (9.230) expresses the unique result that the constant voltage would cause the ac superconducting current

$$j_s = J_m \sin\left(\theta_{21}^0 + \omega_J t\right) \tag{9.231}$$

with the frequency ω_J which has the form

$$\omega_J = \left|\frac{e^* V}{\hbar}\right|. \tag{9.232}$$

The power P_J in the junction is given by

$$P_J = V j_s = V J_m \sin\left(\theta_{21}^0 + \omega_J t\right). \tag{9.233}$$

Obviously, its time average is zero, and the dissipation is absent which is appropriate to a superconducting current. It should be noted that in the case of an external voltage applied to the junction there exists some normal current which gives rise to a finite dissipation.

In the following subsections we investigate some practically important electrodynamic effects specific for a Josephson junction.

9.8.2 Spatial Variation of the Phase Difference in a Magnetic Field

Consider the effect of a magnetic field on the spatial variations of the current in the junction. For this purpose we calculate the phase difference θ_{21} between points 1 and 2 situated on the opposite sides of the junctions. The x-axis is chosen to be perpendicular to the junction boundaries, as in the previous section. The z-axis is parallel to the boundary and the yz plane coincides with the plane of the junction. The geometry of the problem is shown in Fig. 9.14.

The phase difference in the presence of the magnetic field characterized by the vector potential $\boldsymbol{A}$ has the form according to (9.12) [56]:

$$\theta_{21} = \theta_2 - \theta_1 + \frac{2e}{\hbar} \int_1^2 \boldsymbol{A}(x,t) \cdot d\boldsymbol{l}. \tag{9.234}$$

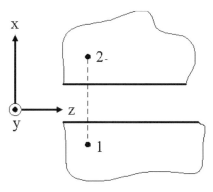

Fig. 9.14. Points on opposite sides of a tunneling junction

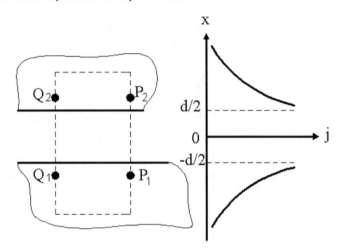

Fig. 9.15. The integration contour for the vector potential $\boldsymbol{A}$

Comparison of (9.234) with relations (9.8) and (9.13) shows that the phase difference θ_{21} remains gauge invariant. Then we can choose a gauge such that the order parameter Ψ is real [80], i.e. inside each of the two superconductors

$$\theta = 0. \tag{9.235}$$

Under such a condition the phase difference (9.234) is completely determined by the integral including the vector potential. Consider the rectangular contour $P_1 P_2 Q_1 Q_2$ in the xz plane shown in Fig. 9.15.

Then (9.234) takes the form for the phase difference $\theta_{21}(P)$ along the line $P_1 P_2$ [56]

$$\theta_{21}(P) = \frac{2e}{\hbar} \int_{P_1}^{P_2} \boldsymbol{A}(P,t) \cdot d\boldsymbol{l} \tag{9.236}$$

and, similarly, for the phase difference $\theta_{21}(Q)$ along the line $Q_1 Q_2$

$$\theta_{21}(Q) = \frac{2e}{\hbar} \int_{Q_1}^{Q_2} \boldsymbol{A}(Q,t) \cdot d\boldsymbol{l}. \tag{9.237}$$

The spatial variation of the phase difference along the z-axis is given by

$$\theta_{21} = \theta(P) - \theta(Q) = \frac{2e}{\hbar} \left\{ \int_{P_1}^{P_2} \boldsymbol{A} \cdot d\boldsymbol{l} - \int_{Q_1}^{Q_2} \boldsymbol{A} \cdot d\boldsymbol{l} \right\}. \tag{9.238}$$

On the other hand, the magnetic flux Φ through the area S of the rectangle $P_1 P_2 Q_1 Q_2$ is determined by the following expression:

9.8 The Josephson Effect

$$\Phi = \int_S \boldsymbol{B} \cdot d\boldsymbol{s} = \oint_{P_1 P_2 Q_1 Q_2} \boldsymbol{A} \cdot d\boldsymbol{l}$$

$$= -\int_{Q_1}^{Q_2} \boldsymbol{A} \cdot d\boldsymbol{l} + \int_{P_1}^{P_2} \boldsymbol{A} \cdot d\boldsymbol{l} + \int_{Q_1}^{P_1} \boldsymbol{A} \cdot d\boldsymbol{l} + \int_{P_2}^{Q_2} \boldsymbol{A} \cdot d\boldsymbol{l}. \tag{9.239}$$

The last two integrals in (9.239) are taken along the paths which can be chosen entirely in the superconducting regions. In the gauge (9.235) the vector potential $\boldsymbol{A}$ is proportional to the superconducting current density $\boldsymbol{j}_s$ according to (9.12) which vanishes in the bulk of the superconductor. As a result, both integrals can be neglected. Then the comparison of (9.238) and (9.239) shows that the phase difference θ_{21} and the magnetic flux Φ are related as follows

$$\theta_{21} = \frac{2e}{\hbar} \int_S \boldsymbol{B} \cdot d\boldsymbol{s}. \tag{9.240}$$

The distance between the points P_1 and P_2 can be evaluated as the sum of the barrier thickness d_J and of both penetration lengths $\lambda_{1,2}$ in the two superconductors. The width $P_1 Q_1$ of the rectangle $P_1 P_2 Q_1 Q_2$ in the limiting case tends to dz. Then the integral in the right-hand side of (9.240) reduces to [56, 80]:

$$\int_S \boldsymbol{B} \cdot d\boldsymbol{s} = \int_P^Q dz \int B_y dx = (d_J + \lambda_1 + \lambda_2) \int_P^Q dz B_y \tag{9.241}$$

and

$$\frac{\partial \theta_{21}}{\partial z} = \frac{2e(d_J + \lambda_1 + \lambda_2)}{\hbar} B_y. \tag{9.242}$$

If there exists also the magnetic induction component B_z then the similar procedure with respect to the contour in the xy plane results in the following equation [56, 80]:

$$\frac{\partial \theta_{21}}{\partial y} = -\frac{2e(d_J + \lambda_1 + \lambda_2)}{\hbar} B_z. \tag{9.243}$$

Combining (9.242) and (9.243) we obtain

$$\nabla \theta_{21} = \frac{2e(d_J + \lambda_1 + \lambda_2)}{\hbar} [\boldsymbol{n}_J \times \boldsymbol{B}], \tag{9.244}$$

where $\boldsymbol{n}_J$ is the unit vector directed from the superconductor 1 to the superconductor 2.

9.8.3 Current Dependence on a Magnetic Field

The results obtained can be applied to the study of the superconducting current in the junction on the external magnetic field. Taking into account the spatial dependence of the phase difference θ_{21} defined by (9.242)–(9.244) we rewrite the expression for the maximum zero-voltage current, or a critical current (9.220):

$$j_s(y,z) = J_m(y,z) \sin \theta_{21}(y,z). \tag{9.245}$$

For the sake of generality consider an arbitrary shape junction and evaluate the maximum zero-voltage current dependence on the uniform magnetic field induction applied in the junction plane yz. The geometry of the problem is shown in Fig. 9.16. In the further analysis we neglect a weak magnetic field created by the current. The y-axis direction is chosen along the magnetic field $B = B_y$. Then the spatial variations of the phase difference are governed by (9.242). The integrating of this equation with the account for the field uniformity yields

$$\theta_{21}(z) = \frac{2e(d_J + 2\lambda)}{\hbar} B_y z + \theta_{21}(0), \tag{9.246}$$

where $\lambda_1 = \lambda_2 = \lambda$ for the same superconductors on the both sides of the junction.

It is seen from (9.246) that all the current components along the magnetic field B_y are determined by the same phase. Hence we can express the current (9.245) in terms of the integral along the y-axis:

$$j_s(z) = \left[\int J_m(y,z)\,dy\right] \sin \theta_{21}(z), \quad J_m(z) = \int J_m(y,z)\,dy. \tag{9.247}$$

Combining (9.245)–(9.247) we calculate the total current in the junction $I(B_y)$

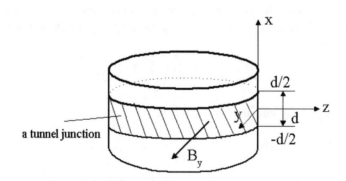

Fig. 9.16. A tunnel junction in a magnetic field

$$I(B_y) = \int_{-\infty}^{\infty} J_m(z) \sin\left[\frac{2e(d_J + 2\lambda)}{\hbar} B_y z + \theta_{21}(0)\right] dz. \qquad (9.248)$$

Consider in details the practically important case of the rectangular junction of the dimensions L and W in the z and y directions respectively and with a uniform critical current density $J_m = \text{const}$. In such a case (9.248) takes the form:

$$I(B_y) = J_m W \int_{-L/2}^{L/2} \sin\left[\frac{2e(d_J + 2\lambda)}{\hbar} B_y z + \theta_{21}(0)\right] dz$$

$$= \frac{\hbar J_m W}{e(d_J + 2\lambda) B_y} \sin\theta_{21}(0) \sin\frac{e(d_J + 2\lambda) B_y}{\hbar} L. \qquad (9.249)$$

The current (9.249) reaches the maximum value at $\theta_{21}(0) = \pm\pi/2$. It has the form

$$I_{\max}(B_y) = I(0) \left|\frac{\sin\left[e(d_J + 2\lambda) B_y L/\hbar\right]}{e(d_J + 2\lambda) B_y L/\hbar}\right|, \qquad (9.250)$$

where $I(0) = J_m W L$ is the total critical current without magnetic field. Since the current source has only one polarity, $\theta_{21}(0)$ flips from $\pi/2$ to $-\pi/2$ in order to keep $I_{\max}(B_y)$ positive [56]. Noting that

$$(d_J + 2\lambda) B_y L = \Phi, \qquad (9.251)$$

where Φ is the total magnetic flux through the junction and using expression (9.64) for the magnetic flux quantum Φ_0 we transform (9.250) as follows

$$I_{\max}(B_y) = I(0) \left|\frac{\sin(\pi\Phi/\Phi_0)}{(\pi\Phi/\Phi_0)}\right|. \qquad (9.252)$$

The maximum zero-voltage current $I_{\max}(B_y)$ reaches its largest value in the absence of the magnetic field when $\Phi = 0$ and $\sin(\pi\Phi/\Phi_0)/(\pi\Phi/\Phi_0) = 1$. In such a case this current is simply equal to the integral of the maximum zero-voltage current density over the junction. When the junction contains the magnetic flux the current diminishes. Finally, $I_{\max}(B_y)$ vanishes when the junction contain any integer number of magnetic flux quanta Φ_0. The phase differences are such that just as much current is flowing upward as downward, and the currents are circulating in the junction [56].

9.8.4 Wave Equation for a Josephson Junction

In this subsection we derive the equation which describes the spatial and temporal evolution of the phase difference θ_{21} and the Josephson tunneling current density j_s. For the sake of simplicity we assume that both superconductors are identical and their thickness in much larger than the penetration length λ. They are separated by a barrier layer of a thickness d. The geometry of the problem is presented in Fig. 9.17. The axes are oriented in the same way as in the previous section. The electric field $\boldsymbol{E}$ is normal to the superconductor and penetrates a negligible distance.

We start with the Maxwell equation in the integral form expressing the Faraday's law:

$$\int_S \mathrm{curl}\boldsymbol{E} d\boldsymbol{s} = \int_S \left(-\frac{\partial \boldsymbol{B}}{\partial t}\right) d\boldsymbol{s}. \tag{9.253}$$

We choose in the xz plane the integration surface $abcd$ with the infinitesimal width dz and length $(d+2\lambda)$. The left-hand side of (9.253) then takes the form

$$\int_S \mathrm{curl}\boldsymbol{E} d\boldsymbol{s} = \oint_{abcd} \boldsymbol{E} \cdot d\boldsymbol{l} = d\left[E_x(z+dz) - E_x(z)\right] = d\frac{\partial E_x}{\partial z} dz. \tag{9.254}$$

Integrating of the right-hand side of (9.253) on the same area yields

$$\int_S \left(-\frac{\partial \boldsymbol{B}}{\partial t}\right) d\boldsymbol{s} = -\frac{\partial B_y}{\partial t}(d+2\lambda) dz. \tag{9.255}$$

Equating (9.254) and (9.255) and applying to both sides the operation $\partial/\partial z$ we get

$$\frac{\partial^2 E_x}{\partial z^2} = -\left(\frac{d+2\lambda}{d}\right) \frac{\partial^2 B_y}{\partial t \partial z}. \tag{9.256}$$

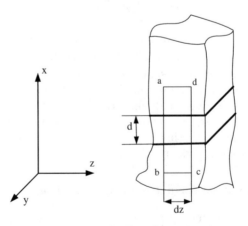

Fig. 9.17. The geometry of a Josephson junction waveguide

The similar procedure with respect to the xy plane gives

$$\frac{\partial^2 E_x}{\partial y^2} = \left(\frac{d+2\lambda}{d}\right)\frac{\partial^2 B_y}{\partial t \partial y}. \tag{9.257}$$

Taking the Maxwell equation

$$\operatorname{curl} \boldsymbol{H} = \boldsymbol{j} + \frac{\partial \boldsymbol{D}}{\partial t}$$

and operating with $\partial/\partial t$ on its x component we get

$$\frac{\partial^2 B_z}{\partial y \partial t} - \frac{\partial^2 B_y}{\partial z \partial t} = \mu_0 \mu^r \frac{\partial j_{sx}}{\partial t} + \mu_0 \mu^r \varepsilon_0 \varepsilon^r \frac{\partial^2 E_x}{\partial t^2}. \tag{9.258}$$

Substituting results (9.256) and (9.257) into the left-hand side of (9.258) we obtain the wave equation for the electric field E_x [56]

$$\frac{\partial^2 E_x}{\partial y^2} + \frac{\partial^2 E_x}{\partial z^2} - \frac{1}{v_{\text{ph}}^2}\frac{\partial^2 E_x}{\partial t^2} = \frac{1}{\varepsilon_0 \varepsilon^r v_{\text{ph}}^2}\frac{\partial j_{sx}}{\partial t}, \tag{9.259}$$

where the phase velocity v_{ph} is given by

$$v_{\text{ph}} = \sqrt{\frac{d}{\mu_0 \mu^r \varepsilon_0 \varepsilon^r (d+2\lambda)}} = c\sqrt{\frac{d}{\mu^r \varepsilon^r (d+2\lambda)}}. \tag{9.260}$$

If $\partial j_{sx}/\partial t = 0$ then (9.259) describes a TEM wave propagating between parallel plates with properties different from a normal metal structure. The structure is a kind of a parallel plane transmission line mentioned above. The inductance per unit length of such a line is proportional to the spacing through which the magnetic field passes, i.e. $\mathcal{L} \sim (d+2\lambda)$. The capacitance per unit length C is inversely proportional to the spacing for the electric field: $C \sim 1/d$. The phase velocity

$$v_{\text{ph}} = (\mathcal{L}C)^{-1/2} \sim \sqrt{\frac{d}{(d+2\lambda)}} \tag{9.261}$$

like in (9.259). The dielectric thickness and penetration depth typically are 2.5 and 50 nm, respectively, and the phase velocity is reduced by a factor of about 6 by the factor in (9.261) [56]. As $T \to T_{\text{cr}}$, the penetration depth increases considerably, and the slowing increases. Waves with $j_{sx} = 0$ are called Swihart modes [56].

Consider now the connection between the phase difference θ_{21} and the electric field E_x. Equation (9.227) can be rewritten as follows:

$$\frac{\partial \theta_{21}}{\partial t} = -\frac{e^* V}{\hbar} = -\frac{2eE_x d}{\hbar}, \quad E_x = -\frac{\hbar}{2ed}\frac{\partial \theta_{21}}{\partial t}, \tag{9.262}$$

where the upper plate is chosen as positive. Inserting E_x from (9.262) and $j_{sx} = -J_m \sin\theta_{21}$ into the wave equation (9.259) we get

$$\left(\frac{\partial^2}{\partial y^2} + \frac{\partial^2}{\partial z^2} - \frac{1}{v_{ph}^2}\frac{\partial^2}{\partial t^2}\right)\frac{\hbar}{2ed}\frac{\partial \theta_{21}}{\partial t} = \frac{J_m}{\varepsilon_0 \varepsilon^r v_{ph}^2}\frac{\partial (\sin\theta_{21})}{\partial t}. \qquad (9.263)$$

Assume that the initial conditions are

$$t = 0, \; \theta_{21} = 0. \qquad (9.264)$$

Then the first integral of (9.263) yields the basic barrier equation [80]

$$\left(\frac{\partial^2}{\partial y^2} + \frac{\partial^2}{\partial z^2} - \frac{1}{v_{ph}^2}\frac{\partial^2}{\partial t^2}\right)\theta_{21} = \frac{\sin\theta_{21}}{\lambda_J^2}, \qquad (9.265)$$

where the Josephson penetration depth λ_J is given by

$$\lambda_J^2 = \frac{\hbar}{2eJ_m\mu_0\mu^r(d+2\lambda)}. \qquad (9.266)$$

Using the typical values for $(d+2\lambda) \simeq 2\lambda = 90$ nm and $J_m = 10^4$ A/cm^2 the estimation yields $\lambda_J \simeq 5$ μm [56]. The left-hand side of (9.265) has the form of a wave equation.

Linearize (9.265) for small time-dependent variations of the phase difference in the vicinity of a spatially dependent time-average value $\theta_{21}^0(y,z)$.

$$\theta_{21}(y,z,t) = \theta_{21}^0(y,z) + \theta_{21}^1(y,z,t), \; \theta_{21}^1(y,z,t) \ll \theta_{21}^0(y,z). \qquad (9.267)$$

In the first approximation we set

$$\theta_{21}^0(y,z) = \theta_{21}^0 = \text{const}, \; \theta_{21}^1(y,z,t) \sim \exp i(\omega t - \mathbf{k r}), \qquad (9.268)$$

where $\mathbf{r}$ belongs to the plane of the junction. Substitution of (9.268) into (9.265) gives the following dispersion relation

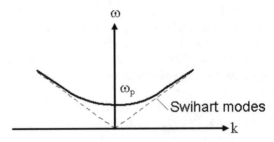

Fig. 9.18. The dispersion relation in a Josephson junction waveguide. The Swihart modes exist in the absence of Josephson tunneling current

$$\omega^2 = k^2 v_{\text{ph}}^2 + \omega_p^2, \quad \omega_p^2 = \frac{v_{\text{ph}}^2}{\lambda_J^2} \cos \theta_{21}^0 . \tag{9.269}$$

The dispersion relation (9.269) is presented in Fig. 9.18.

It is seen that propagating waves exist only for $\omega > \omega_p$. At a large phase difference $\theta_{21}^0 \sim \pi/2$ the frequency ω_p tends to zero. On the other hand, the dc current (9.220) reaches its maximum at $\theta_{21}^0 = \pi/2$ which means that the current is changing a little for small deviations θ_{21}^1. For $k = 0$ the high frequency magnetic field does not exist and only a periodic energy exchange between the electric field and the coupling energy occurs with the frequency ω_p which is called the plasma frequency of a Josephson junction. It is a natural frequency of oscillation of a perturbation with $k = 0$ in a junction [56].

A The List of Notations

$\boldsymbol{A}$	vector potential
$\boldsymbol{a}$	acceleration
a	geometric dimension
a_{jk}	components of a rotation transformation tensor
b	geometric dimension
$\boldsymbol{B}$	magnetic induction
B	elastic constant of a smectic liquid crystal
c	the light velocity in vacuum
$\mathcal{C}$	capacitance
$\boldsymbol{D}$	electric induction, or electric displacement
D	diffusion coefficient
e	electron charge
$\boldsymbol{E}$	electric field intensity
$\mathcal{E}$	energy of a particle
$\mathsf{E}(k)$	the complete elliptic integral of the second kind
$\boldsymbol{F}, \boldsymbol{f}$	force
F_{jk}	4-tensor of electromagnetic field composed of vectors $\boldsymbol{E}$ and $\boldsymbol{B}$
$\mathcal{F}$	free energy
$f(\boldsymbol{r}, \boldsymbol{v}, t)$	distribution function
G_{jk}	4-tensor of electromagnetic field composed of vectors $\boldsymbol{D}$ and $\boldsymbol{H}$
$\boldsymbol{G}, \boldsymbol{g}$	gyromagnetic vectors
G	conductance
g_f	filling factor
$\boldsymbol{H}$	magnetic field intensity
$\mathcal{H}$	hamiltonian
h	the Planck's constant
$\hbar$	$= h/(2\pi)$
$i = \sqrt{-1}$	the imaginary unit
I	electric current
$\boldsymbol{j}$	electric current density
J_i	4-vector of a current density
$J_m(x)$	the Bessel's functions
$\boldsymbol{k}$	wave vector
k_B	the Boltzmann constant

$K_{1,2,3}$	elastic constants of liquid crystals
$\mathcal{K}(k)$	the complete elliptic integral of the first kind
$\boldsymbol{L}$	angular momentum
$\mathcal{L}$	lagrangian
$\mathfrak{L}$	self-inductance
m, M	mass
$\boldsymbol{M}$	magnetization
$\mathcal{M}$	mutual inductance
n_e	concentration of electrons
n_i	concentration of ions
$\boldsymbol{n}$	vector director (in liquid crystals)
n	refraction coefficient
$\boldsymbol{P}$	polarization
p	concentration of holes
$\boldsymbol{p}$	momentum
$\boldsymbol{P}_g$	canonical, or generalized momentum
P	pressure
$\mathcal{Q}$	total energy flux in a waveguide
Q_p	quality factor of a cavity
$\boldsymbol{R}, \boldsymbol{r}$	radius vector
R	resistance
S	action
$\boldsymbol{S}$	Poynting vector
s	sound velocity
T	temperature
t	time
t'_{ij}	viscous stress tensor
u	layer displacement (in smectic liquid crystal)
$\mathcal{U}$	scalar potential of magnetic field
$\boldsymbol{v}$	velocity
V	electrostatic potential
$\mathcal{V}$	electromotive force
W	energy density
x, y, z	Cartesian coordinates
$Y_{mn}(\theta, \varphi)$	spherical harmonics
$\mathcal{Z}$	impedance
α_i	viscosity coefficients in liquid crystals
Γ	propagation constant
$\delta(x)$	Dirac delta-function
δ_{ik}	Kroneker's delta-function
ε_{ik}	dielectric constant (permittivity) tensor
ε_0	the dielectirc constant (permittivity) of free space
ε^r	relative dielectric constant (permittivity) of a medium
κ	wave number

λ	wavelength
μ	mobility of charge carriers
μ_0	the permeability of free space
μ^r	relative permeability of a medium
$\boldsymbol{\mu}_m$	magnetic moment
φ	scalar potential
Φ	magnetic flux
χ_e	electric susceptibility
χ_m	magnetic susceptibility
ρ	charge density
σ	electric conductivity
τ	relaxation time
ϱ_m	mass density
ϱ_0	resistivity
Ψ	order parameter
ω	frequency

B Formulae of Vector Analysis

B.1 Vector Operations in the Cartesian Coordinates

The Cartesian coordinates are (x, y, z). The operator ∇ in the Cartesian coordinates (x, y, z) has the form

$$\nabla = \boldsymbol{x}_0 \frac{\partial}{\partial x} + \boldsymbol{y}_0 \frac{\partial}{\partial y} + \boldsymbol{z}_0 \frac{\partial}{\partial z}, \tag{B.1}$$

where $\boldsymbol{x}_0, \boldsymbol{y}_0, \boldsymbol{z}_0$ are the unit vectors parallel to the x-,y-,z-axes, respectively.

$$\nabla f = \mathrm{grad} f = \boldsymbol{x}_0 \frac{\partial f}{\partial x} + \boldsymbol{y}_0 \frac{\partial f}{\partial y} + \boldsymbol{z}_0 \frac{\partial f}{\partial z}, \tag{B.2}$$

where f is a scalar function.

$$(\nabla \boldsymbol{F}) = \mathrm{div} \boldsymbol{F} = \frac{\partial F_x}{\partial x} + \frac{\partial F_y}{\partial y} + \frac{\partial F_z}{\partial z}, \tag{B.3}$$

where $\boldsymbol{F} = (F_x, F_y, F_z)$ is a vector function.

$$[\nabla \times \boldsymbol{F}] = \mathrm{curl} \boldsymbol{F} = \begin{vmatrix} \boldsymbol{x}_0 & \boldsymbol{y}_0 & \boldsymbol{z}_0 \\ \frac{\partial}{\partial x} & \frac{\partial}{\partial y} & \frac{\partial}{\partial z} \\ F_x & F_y & F_z \end{vmatrix}$$

$$= \boldsymbol{x}_0 \left(\frac{\partial F_z}{\partial y} - \frac{\partial F_y}{\partial z} \right) + \boldsymbol{y}_0 \left(\frac{\partial F_x}{\partial z} - \frac{\partial F_z}{\partial x} \right) + \boldsymbol{z}_0 \left(\frac{\partial F_y}{\partial x} - \frac{\partial F_x}{\partial y} \right) \tag{B.4}$$

$$\nabla^2 f = \Delta f = \frac{\partial^2 f}{\partial x^2} + \frac{\partial^2 f}{\partial y^2} + \frac{\partial^2 f}{\partial z^2} \tag{B.5}$$

$$\nabla^2 \boldsymbol{F} = \Delta \boldsymbol{F} = \boldsymbol{x}_0 \Delta F_x + \boldsymbol{y}_0 \Delta F_y + \boldsymbol{z}_0 \Delta F_z. \tag{B.6}$$

The element of volume dV is

$$dV = dx dy dz. \tag{B.7}$$

The element of length dl is

$$dl^2 = dx^2 + dy^2 + dz^2. \tag{B.8}$$

B.2 Vector Operations in Cylindrical Coordinates

The cylindrical coordinates are (r, φ, z).

$$\operatorname{grad} f = r_0 \frac{\partial f}{\partial r} + e_\varphi \frac{\partial f}{\partial \varphi} + z_0 \frac{\partial f}{\partial z}, \tag{B.9}$$

where r_0, e_φ, z_0 are the unit vectors parallel to the r, φ, z directions respectively.

$$\operatorname{div} \boldsymbol{F} = \frac{1}{r} \frac{\partial}{\partial r} (r F_r) + \frac{1}{r} \frac{\partial F_\varphi}{\partial \varphi} + \frac{\partial F_z}{\partial z}, \tag{B.10}$$

where F_r, F_φ, F_z are the components of a vector function $\boldsymbol{F}$ in cylindrical coordinates.

$$\operatorname{curl} \boldsymbol{F} = r_0 \left(\frac{1}{r} \frac{\partial F_z}{\partial \varphi} - \frac{\partial F_\varphi}{\partial z} \right) + e_\varphi \left(\frac{\partial F_r}{\partial z} - \frac{\partial F_z}{\partial r} \right) + z_0 \left(\frac{1}{r} \frac{\partial}{\partial r} (r F_\varphi) - \frac{1}{r} \frac{\partial F_r}{\partial \varphi} \right) \tag{B.11}$$

$$\Delta f = \frac{1}{r} \frac{\partial}{\partial r} \left(r \frac{\partial f}{\partial r} \right) + \frac{1}{r^2} \frac{\partial^2 f}{\partial \varphi^2} + \frac{\partial^2 f}{\partial z^2}. \tag{B.12}$$

$$\Delta \boldsymbol{F} = r_0 \left(\Delta F_r - \frac{F_r}{r^2} - \frac{2}{r^2} \frac{\partial F_\varphi}{\partial \varphi} \right) + e_\varphi \left(\Delta F_\varphi - \frac{F_\varphi}{r^2} + \frac{2}{r^2} \frac{\partial F_r}{\partial \varphi} \right) + z_0 \Delta F_z. \tag{B.13}$$

The element of volume dV is

$$dV = r \, dr \, dz \, d\varphi. \tag{B.14}$$

The element of length dl is

$$dl^2 = dr^2 + r^2 d\varphi^2 + dz^2. \tag{B.15}$$

B.3 Vector Operations in Spherical Coordinates

The spherical coordinates are (r, θ, φ).

$$\operatorname{grad} f = r_0 \frac{\partial f}{\partial r} + e_\theta \frac{1}{r} \frac{\partial f}{\partial \theta} + e_\varphi \frac{1}{r \sin \theta} \frac{\partial f}{\partial \varphi} \tag{B.16}$$

$$\operatorname{div} \boldsymbol{F} = \frac{1}{r^2} \frac{\partial}{\partial r} (r^2 F_r) + \frac{1}{r \sin \theta} \frac{\partial}{\partial \theta} (F_\theta \sin \theta) + \frac{1}{r \sin \theta} \frac{\partial F_\varphi}{\partial \varphi} \tag{B.17}$$

$$\text{curl}\mathbf{F} = \mathbf{r}_0 \frac{1}{r\sin\theta}\left[\frac{\partial}{\partial\theta}(F_\varphi \sin\theta) - \frac{\partial F_\theta}{\partial\varphi}\right] + \mathbf{e}_\theta\left[\frac{1}{r\sin\theta}\frac{\partial F_r}{\partial\varphi} - \frac{1}{r}\frac{\partial}{\partial r}(rF_\varphi)\right]$$

$$+ \mathbf{e}_\varphi \frac{1}{r}\left[\frac{\partial}{\partial r}(rF_\theta) - \frac{\partial F_r}{\partial\theta}\right] \qquad (\text{B.18})$$

$$\Delta f = \frac{1}{r^2}\frac{\partial}{\partial r}\left(r^2\frac{\partial f}{\partial r}\right) + \frac{1}{r^2\sin\theta}\frac{\partial}{\partial\theta}\left(\sin\theta\frac{\partial f}{\partial\theta}\right) + \frac{1}{r^2\sin^2\theta}\frac{\partial^2 f}{\partial\varphi^2} \qquad (\text{B.19})$$

$$\Delta\mathbf{F} = \mathbf{r}_0\left\{\Delta F_r - \frac{2}{r^2}\left[F_r + \frac{1}{\sin\theta}\frac{\partial}{\partial\theta}(F_\theta\sin\theta) + \frac{1}{\sin\theta}\frac{\partial F_\varphi}{\partial\varphi}\right]\right\}$$

$$+ \mathbf{e}_\theta\left\{\Delta F_\theta + \frac{2}{r^2}\left[\frac{\partial F_r}{\partial\theta} - \frac{F_\theta}{2\sin^2\theta} - \frac{\cos\theta}{\sin^2\theta}\frac{\partial F_\varphi}{\partial\varphi}\right]\right\}$$

$$+ \mathbf{e}_\varphi\left\{\Delta F_\varphi + \frac{2}{r^2\sin\theta}\left[\frac{\partial F_r}{\partial\varphi} + \cot\theta\frac{\partial F_\theta}{\partial\varphi} - \frac{F_\varphi}{2\sin\theta}\right]\right\}. \qquad (\text{B.20})$$

The element of volume dV is

$$dV = r^2\sin\theta dr d\theta d\varphi. \qquad (\text{B.21})$$

The element of length dl is

$$dl^2 = dr^2 + r^2 d\theta^2 + r^2\sin^2\theta d\varphi^2. \qquad (\text{B.22})$$

B.4 Vector Analysis Identities

Let f_1 and f_2 are the scalar functions, and $\mathbf{F}_1$ and $\mathbf{F}_2$ are the vector functions. Then the following identities are valid.

$$\text{grad}(f_1 + f_2) = \text{grad}f_1 + \text{grad}f_2 \qquad (\text{B.23})$$

$$\text{grad}(f_1 f_2) = f_1\text{grad}f_2 + f_2\text{grad}f_1 \qquad (\text{B.24})$$

$$\text{div}(\mathbf{F}_1 + \mathbf{F}_2) = \text{div}\mathbf{F}_1 + \text{div}\mathbf{F}_2 \qquad (\text{B.25})$$

$$\text{curl}(\mathbf{F}_1 + \mathbf{F}_2) = \text{curl}\mathbf{F}_1 + \text{curl}\mathbf{F}_2 \qquad (\text{B.26})$$

$$\operatorname{div}(f_1 \boldsymbol{F}_1) = (\boldsymbol{F}_1 \cdot \operatorname{grad} f_1) + f_1 \operatorname{div} \boldsymbol{F}_1 \tag{B.27}$$

$$\operatorname{grad}(\boldsymbol{F}_1 \cdot \boldsymbol{F}_2) = (\boldsymbol{F}_1 \nabla) \boldsymbol{F}_2 + (\boldsymbol{F}_2 \nabla) \boldsymbol{F}_1 +$$

$$+ [\boldsymbol{F}_1 \times \operatorname{curl} \boldsymbol{F}_2] + [\boldsymbol{F}_2 \times \operatorname{curl} \boldsymbol{F}_1] \tag{B.28}$$

$$\operatorname{curl}(f_1 \boldsymbol{F}_1) = [\operatorname{grad} f_1 \times \boldsymbol{F}_1] + f_1 \operatorname{curl} \boldsymbol{F}_1 \tag{B.29}$$

$$\operatorname{div}[\boldsymbol{F}_1 \times \boldsymbol{F}_2] = (\boldsymbol{F}_2 \cdot \operatorname{curl} \boldsymbol{F}_1) - (\boldsymbol{F}_1 \cdot \operatorname{curl} \boldsymbol{F}_2) \tag{B.30}$$

$$\operatorname{curl}[\boldsymbol{F}_1 \times \boldsymbol{F}_2] = \boldsymbol{F}_1 \operatorname{div} \boldsymbol{F}_2 - \boldsymbol{F}_2 \operatorname{div} \boldsymbol{F}_1 + (\boldsymbol{F}_2 \cdot \nabla) \boldsymbol{F}_1 - (\boldsymbol{F}_1 \cdot \nabla) \boldsymbol{F}_2 \tag{B.31}$$

$$\operatorname{curl} \operatorname{curl} \boldsymbol{F}_1 = \operatorname{grad} \operatorname{div} \boldsymbol{F}_1 - \Delta \boldsymbol{F}_1 \tag{B.32}$$

$$\operatorname{curl} \operatorname{grad} f_1 = 0 \tag{B.33}$$

$$\operatorname{div} \operatorname{curl} \boldsymbol{F}_1 = 0. \tag{B.34}$$

C The Physical Constants

Boltzmann constant	$k_B = 1.3806568 \times 10^{-23}$ joule/K
Dielectric constant (permittivity) of free space	$\varepsilon_0 = 8.854187817 \times 10^{-12}$ farad/m
Electron charge	$e = 1.60217733 \times 10^{-19}$ coulomb
Electron charge mass ratio	$e/m = 1.75881962 \times 10^{11}$ coulomb/kg
Electron mass	$m_e = 9.1093897 \times 10^{-31}$ kg
Permeability of free space	$\mu_0 = 1.2566370614 \times 10^{-6}$ henri/m $= 1.2566370614 \times 10^{-6}$ newton/ampere^2
Planck's constant	$h = 6.6260755 \times 10^{-34}$ joule s
Planck's hbar	$\hbar = 1.05457266 \times 10^{-34}$ joule s
Ration	$e/h = 2.41798836 \times 10^{14}$ ampere/joul
Speed of light in vacuum	$c = 2.99792458 \times 10^{8}$ m/s

D The MKS System of Units

The International System of Units (abbreviated to SI for Système International), which is also known as the MKS(A) (meter, kilogram, second, ampere) System of Units, or Giorgi system of units, was defined and given official status by the 11th General Conference on Weights and Measures, held in Paris in 1960 [81].

Generally, any absolute system of units should be based on the three fundamental physical quantities: mass, length and time, while all other physical quantities must be expressed in terms of the three former ones. This has been done in mechanics. For instance, in SI the units of mass, length and time are kilogram, meter and second, respectively. In 1901, Giorgi had shown that it is possible to derive the units and dimension of any electromagnetic quantity by an appropriate choice of the fundamental units. For this purpose, he chose a fourth fundamental unit, in addition to the three units mentioned above. This is a unit of an electric current, ampere. Then, the relations among the secondary quantities appeared to be the same as in the so-called practical system of units used in electrodynamics and therefore these quantities remain expressed in terms of the practical system units: ohm, coulomb and ampere. The permeability of a free space μ_0 has been chosen to be $4\pi \times 10^{-7}$ which permits the evaluation of the dielectric constant of a free space (permittivity) ε_0 from the definition of the vacuum light velocity c [3]. Thus, the SI, or MKS system of units is at the same time both absolute and practical. The dimensionality of the constants ε_0 and μ_0 which relate the vectors D and E, B and H in vacuum is arbitrary. The quantities ε_0 and μ_0 cannot be measured directly and independently. On the other hand, the quantity

$$c = \frac{1}{\sqrt{\varepsilon_0 \mu_0}} = 2.99790 \cdot 10^8 \text{m/s} \tag{D.1}$$

has the meaning of a light velocity in vacuum and its value is well established by virtue of experiments.

Consider briefly the derivation of the units and dimensions of electromagnetic quantities defined below [81].

Dielectric constant ε. The dielectric constant in farads per meter is the ratio of the force between two charged conductors measured in vacuum to that measured when the vacuum replaced by a homogeneous fluid insulating medium, multiplied by 8.85434×10^{-12}. In a homogeneous solid it is the product of 8.85434×10^{-12} by the ratio of the force on a given small charge measured at the center of a thin disk-shaped evacuated cavity placed normal to a uniform electric field to that on the same charge measured at the center of a thin needle-shaped evacuated cavity aligned with the same field.

Charge e. One coulomb is that charge which, when carried by each of two bodies whose distance apart r in meters is very large compared with their dimensions, produces in a vacuum a mutual repulsion of $8.98740 r^{-2} \times 10^9$ newton. A charge of one coulomb is transported by a current of one ampere in one second. There are two kinds of charge. Electrons carry a negative charge and protons a positive charge.

Current I. An ampere is that current which, flowing in the same direction in each of two identical coaxial circular loops of wire whose distance apart r in meters is very large compared with their radius a, produces in a vacuum a mutual attraction of $6\pi^2 a^4 r^{-4} \times 10^{-7}$ newton. A current of one ampere transports one coulomb of charge per second. Current direction is defined as that in which a positive charge moves.

Electric field intensity **E**. The electric field intensity in volts per meter is the vector force in newtons acting on a very small body carrying a very small positive charge placed at the field point, divided by the charge in coulombs. In a homogeneous solid the measurement is carried out at the center of a thin evacuated needle-shaped cavity aligned so that the force lies along the axis.

Electromotance or Electromotive Force $\mathcal{V}$. The electromotance in volts around a closed path is the work in joules required to carry a very small positive charge around that path, divided by the charge in coulombs.

Magnetic Induction or Magnetic Flux Density **B**. The magnetic induction in webers per square meter is a vector whose direction is that in which the axis of small circular current-carrying test loop that rests in stable equilibrium at the field point would advance if it were a right-hand screw rotated in the sense of the current circulation and whose magnitude equals the torque in newton meters on the loop when its axis is normal to the induction, divided by the product of loop current by loop area. In a homogeneous solid the measurement is carried out at the center of a thin evacuated disk-shaped cavity oriented so that the induction is normal to its faces.

Permeability μ. The permeability in henrys per meter is the ratio of the force between two linear circuits carrying fixed current measured in a homogeneous fluid insulating medium to that measured in a vacuum, multiplied by $4\pi \times 10^{-7}$. In a homogeneous solid it is the product of $4\pi \times 10^{-7}$ by the ratio of the magnetic induction at the center of a thin evacuated disk-shaped cavity oriented so that the induction is normal to its faces to that at the center of a thin evacuated needle-shaped cavity oriented so that the induction is directed along its axis.

D The MKS System of Units   411

Potential V. The potential in volts at a point in an electrostatic field is the work in joules done in bringing a very small positive charge to the point from a point arbitrarily chosen at zero potential, divided by the charge in coulombs.

The units and dimensions of electrostatic quantities are based on the Coulomb's law which states that the force F acting between two electric charges $q_{1,2}$ separated by a distance r is proportional to the product of their moduli and inversely proportional to the square of the distance between them [3]:

$$F = k_{\text{pr}} \frac{q_1 q_2}{r^2}, \tag{D.2}$$

where k_{pr} is the proportionality coefficient. It is chosen to be

$$k_{\text{pr}} = \frac{1}{4\pi\varepsilon_0}. \tag{D.3}$$

The dimension of the electric charge can be expressed in terms of ε_0 and the force dimension which is defined by the fundamental units of mass M, length L and time T. We denote the dimension of a physical quantity with square brackets. Then we can write [3]

$$[q] = \varepsilon_0^{1/2} M^{1/2} L^{3/2} T^{-1}. \tag{D.4}$$

By definition, an electric current I is an electric charge transported in a unit of time [3]. Consequently, the dimension of an electric current I has the form

$$[I] = \varepsilon_0^{1/2} M^{1/2} L^{3/2} T^{-2}. \tag{D.5}$$

It can be shown that introducing the unit of an electric current, ampere as a fundamental one, defining $\mu_0 = 4\pi \times 10^{-7}$, as mentioned above, and calculating ε_0 by virtue of the experimentally measured vacuum light velocity c (D.1) one can derive dimensions of all other electromagnetic quantities [3].

We present some examples of relations between the MKS units of essential electromagnetic quantities. Consider a conductor with an electric current I flowing in it and the resistance R. The energy W dissipating in the conductor during the time interval t is given by

$$W = I^2 R t. \tag{D.6}$$

Measuring the energy W in Joules, the current I in Amperes and the time t in seconds we can derive the resistance unit Ohm.

$$[R] = 1 \text{ ohm} = 1 \frac{\text{joule}}{\text{ampere}^2 \cdot \text{sec}} = 1 \frac{\text{watt}}{\text{ampere}^2}, \tag{D.7}$$

where we introduced a unit of power watt equal to the energy of 1 joule dissipated during 1 second. The joule is the unit of energy in the SI system, and by definition it is given by

$$[W] = 1 \text{ joule} = 1 \frac{\text{kilogram} \cdot \text{meter}^2}{s^2}. \tag{D.8}$$

The resistance unit can be easily expressed in terms of fundamental units of mass m, length l, time t and electric current I, i.e. kilogram, meter, second and ampere, respectively. It is known from mechanics that the dimension of power P is given by

$$[P] = 1 \text{ watt} = 1 \frac{\text{kilogram} \cdot \text{meter}^2}{s^3}. \tag{D.9}$$

Then, substituting (D.9) into (D.7) we get

$$[R] = 1 \text{ ohm} = 1 \frac{\text{kilogram} \cdot \text{meter}^2}{\text{ampere}^2 \cdot s^3}. \tag{D.10}$$

Similarly, we obtain the dimension of potential V. By definition

$$[V] = 1 \text{ volt} = 1 \frac{\text{joule}}{\text{coulomb}} = 1 \frac{\text{kilogram} \cdot \text{meter}^2}{\text{ampere} \cdot s^3}. \tag{D.11}$$

The unit of the electric field intensity $\boldsymbol{E}$ is

$$[\boldsymbol{E}] = 1 \frac{\text{volt}}{\text{meter}} = 1 \frac{\text{kilogram} \cdot \text{meter}}{\text{ampere} \cdot s^3}. \tag{D.12}$$

Note that the product of the electric field intensity dimension and the charge gives the dimension of force $\boldsymbol{F}$. Combining (D.12) and the we have

$$[\boldsymbol{F}] = 1 \text{ coulomb} \cdot 1 \frac{\text{volt}}{\text{meter}} = 1 \frac{\text{kilogram} \cdot \text{meter}}{s^2} = 1 \text{ newton}. \tag{D.13}$$

A newton is a unit of force in the SI system. The electric induction $\boldsymbol{D}$ is determined by the Poisson equation and its dimension is coulomb/m^2. Then the unit of the vacuum dielectric constant ε_0 is

$$[\varepsilon_0] = \frac{\text{coulomb} \cdot \text{m}}{\text{m}^2 \cdot \text{volt}} = \frac{\text{farad}}{\text{m}}, \tag{D.14}$$

where Farad is the unit of electric capacitance $\mathcal{C}$

$$[\mathcal{C}] = 1 \frac{\text{coulomb}}{\text{volt}} = 1 \text{ farad}. \tag{D.15}$$

Derive the dimension of the magnetic field characteristics. The dimension of magnetic flux $\varPhi$ can be defined from the Faraday's law. The electromotive force $\mathcal{V}$ and the magnetic flux $\varPhi$ are measured in volts and webers respectively. Then we have

$$[\mathcal{V}] = 1 \text{ volt} = 1 \frac{\text{weber}}{s}$$

and

$$[\Phi] = 1 \text{ weber} = 1 \text{ volt·s} = 1 \frac{\text{kilogram} \cdot \text{meter}^2}{\text{ampere·s}^2}. \qquad (\text{D.16})$$

The unit of a magnetic induction B, tesla is a magnetic flux Φ divided by an area S, i.e. it has the form

$$[B] = 1 \text{ tesla} = 1 \frac{\text{weber}}{\text{meter}^2} = 1 \frac{\text{kilogram}}{\text{ampere·s}^2}. \qquad (\text{D.17})$$

The dimension of the magnetic field intensity H is defined by virtue of the Ampère's law 2.9:

$$\oint_L \mathbf{H} d\mathbf{l} = I.$$

We obtain for the current in amperes and length in meters:

$$[H] = 1 \frac{\text{ampere}}{\text{meter}}. \qquad (\text{D.18})$$

In the case of the coil bound with n tours the total current is the product of the current in one tour I and n. Then the magnetic field intensity is measured in (ampere · tour) /meter. The unit of vacuum permeability μ_0 is by definition given by

$$[\mu_0] = \left[\frac{B}{H}\right] = 1 \frac{\text{tesla} \cdot \text{meter}}{\text{ampere}} = 1 \frac{\text{henry}}{\text{meter}}, \qquad (\text{D.19})$$

where relations (D.17) and (D.18) are used, and 1 henry is the unit of inductance $\mathcal{L}$ which has the form according to relation (2.48)

$$[\mathcal{L}] = 1 \text{ henry} = 1 \frac{\text{joule}}{\text{ampere}^2} = 1 \frac{\text{kilogram} \cdot \text{meter}^2}{\text{ampere}^2 \cdot \text{sec}^2}. \qquad (\text{D.20})$$

In order to obtain the number of the CGSE units of a physical quantity (the third column of the Table 4), one must multiply the number of the MKSA units of the same quantity (the first column of the Table 4) by the numerical factor (the second column of the Table 4).

Table D.1. Official SI (MKS) units names and symbols

Unit	Symbol	Unit	Symbol
meter	m	watt	W
kilogram	kg	coulomb	C
second	s	volt	V
ampere	A	ohm	Ω
kelvin	K	farad	F
candela	cd	weber	Wb
radian	rad	henry	H
steradian	sr	tesla	T
hertz	Hz	lumen	lm
newton	N	lux	lx
joule	J		

Table D.2. Official prefixes indicating decimal multiples and submultiples

Multiples and Submultiples	Prefix	Symbol
10^{12}	tera	T
10^{9}	giga	G
10^{6}	mega	M
10^{3}	kilo	k
10^{2}	hecto	h
10	deka	da
10^{-1}	deci	d
10^{-2}	centi	c
10^{-3}	milli	m
10^{-6}	micro	μ
10^{-9}	nano	n
10^{-12}	pico	p
10^{-15}	femto	f
10^{-18}	atto	a

Table D.3. Decimal multiples of SI units bearing coined names and acceptable in their special fields

Unit	Symbol	SI equivalent
angstrom	Å	10^{-10}m
bar	bar	10^{5}N/m
liter	l	10^{-3}m^{3}
poise	P	10^{-1}N$\cdot$s/m^{2}

Table D.4. Conversion of the MKSA units into the CGSE units

The MKSA unit	Factor	The CGSE unit
Ampere	3×10^9	Current I
Coulomb	3×10^9	Charge q
Farad	9×10^{11}	Capacitance $\mathcal{C}$
Henry	$(1/9) \times 10^{-11}$	Inductance $\mathfrak{L}$
Joule	10^7	Energy (erg) $\mathcal{F}$
Newton	10^5	Force (dyne) $\boldsymbol{F}$
Ohm	$(1/9) \times 10^{-11}$	Resistance R
Volt	$(1/3) \times 10^{-2}$	Potential V
Weber	$(1/3) \times 10^{-2}$	Magnetic flux Φ

References

1. A. Sommerfeld: *Electrodynamics* (Academic Press, New York 1964)
2. L.D. Landau and E.M. Lifshitz: *The Classical Theory of Fields* (Pergamon, New York 1975)
3. J.A. Stratton: *Electromagnetic Theory* (McGraw Hill, New York 1948)
4. L.D. Landau and E.M. Lifshitz: *Electrodynamics of Continuous Media* (2nd Edition) (Pergamon, Oxford 1984)
5. F.N.H. Robinson: *Macroscopic Electromagnetism* (Pergamon, Oxford 1973)
6. S. Ramo, J.R. Whinnery, T. van Duzer: *Fields and Waves in Communication Electronics* (Wiley, New York 1967)
7. W.L. Weeks: *Electromagnetic Theory for Engineering Applications* (Wiley, New York 1964)
8. K. Zhang, D. Li: *Electromagnetic Theory for Microwaves and Optoelectronics* (Springer, Berlin Heidelberg New York 1998) Chap. 4
9. A. Yariv: *Quantum Electronics* (2nd Edition) (Wiley, New York 1975)
10. I.D. Vagner, in: *Physical Phenomena in High Magnetic Fields*. Ed. by L.P. Gor'kov, Z. Fisk and J.R. Schrieffer (Addison-Wesley, 1991)
11. A.M. Fedorchenko, N.Ya. Kotzarenko: *Absolutnaya i Convectivnaya Neustoychivost v Plasme i Tverdich Telach (The Absolute and Convective Instability in Plasmas and Solid State)* (Moscow, Nauka 1981), in Russian
12. E.M. Lifshitz and L.P. Pitaevskii: *Physical Kinetics* (Pergamon, Oxford 1981)
13. M.C. Steele and B.Vural: *Wave Interactions in Solid State Plasmas* (McGraw-Hill, New York 1973)
14. I.D. Vagner and I.V. Ioffe: Soviet Physics Solid State **13**, 2926 (1972)
15. L.D. Landau and E.M. Lifshitz: *Mechanics* (Pergamon, New York 1960)
16. Y.L. Ivanov, S.M. Ryvkin: Sov. Phys. Tech. Phys. **3**, 722 (1958) [Zhournal Technicheskoy Physiki **28**, 774 (1958)]
17. V.V. Kadomtsev, A.V. Nedospasov: J. Nucl. Energy C **1**, 230 (1960)
18. M. Gliksman: Phys. Rev. **124**, 1655 (1961)
19. C.E. Hurwitz, A.L. McWhorter: Phys. Rev. **134**, A1033 (1964)
20. F.C. Hoh and B. Lehnert: Phys. Rev. Lett. **7**, (1961)
21. F.F. Chen: *Introduction to Plasma Physics and Controlled Fusion* (Plenum, New York London 1987)
22. P.H. Platzman and P.A. Wolff: Waves and Interactions in Solid State Plasmas. In: *Solid State Physics*. Vol. 13 (Academic Press, New York 1973)
23. V.T. Petrashov: Rep. Prog. Phys. **47**, 47 (1984)
24. I.D. Vagner, I.V. Ioffe, and A.A. Katanov: Sov. Phys. JETP **31**, 541 (1970)
25. B. Ancker-Johnson: Proc. Intern. Conf. Semiconductors Physics, Moscow, Vol. 2, 813 (1968)

26. L.É. Gurevich and B.L. Gel'mont: Sov. Phys. JETP **19**, 604 (1964)
27. S.J. Buchsbaum and P. Platzmann: Phys. Rev. **154**, 395 (1967)
28. P.D. de Gennes and J. Prost: *The Physics of Liquid Crystals* (2nd Edition) (Clarendon Press, Oxford 1993)
29. S. Chandrasekhar: *Liquid Crystals* (2nd Edition) (Cambridge University Press, New York 1992)
30. L. Lam, in: *Solitons in Liquid Crystals*. Ed. L. Lam, J. Prost (Springer, Berlin Heidelberg New York 1991)
31. B.I. Lembrikov: Sov. Phys. Tech. Phys **22(7)**, 908 (1977)
32. P.D. de Gennes: Solid State Comm. **6**, 163 (1968)
33. I.V. Ioffe and B.I. Lembrikov: Sov. Phys. Solid State **16**, 2304 (1975)
34. W.J.A. Goossens: Electrohydrodynamic Instabilities in Nematic Liquid Crystals. In: *Advances in Liquid Crystals*. Vol. 3. Ed. by G.H. Brown (Academic Press, New York 1978)
35. L.D. Landau and E.M. Lifshitz: *Theory of Elasticity* (3rd Edition) (Pergamon, Oxford 1986)
36. I.V. Ioffe and B.I. Lembrikov: Sov. Phys. Solid State **22**, 204 (1980)
37. M.J. Stephen, J.P. Straley: Rev. Mod. Phys. **46**, 617 (1974)
38. P.D. de Gennes. J. de Phys., Col. C**4**, 30, C4-65 (1969)
39. I.C. Khoo, and S.T. Wu: *Optics and Nonlinear Optics of Liquid Crystals* (World Scientific, Singapore 1993)
40. G.F. Kventsel and B.I. Lembrikov: Mol. Cryst. Liq. Cryst. **282**, 145 (1996)
41. M.C. Mauguin: Bull. Soc. Franc. Miner. Cryst. **34**, 71 (1911)
42. C.W. Oseen: Trans. Faraday Soc. **29**, 883 (1933)
43. H. de Vries: Acta Cryst. **4**, 219 (1951)
44. E.I. Katz: Sov. Phys. JETP **32**, 1004 (1971)
45. R. Nityananda: Mol. Cryst. Liquid Cryst. **21**, 315 (1973)
46. V.A. Belyakov: *Diffraction Optics of Complex-Structured Media* (Springer, Berlin Heidelberg New York 1992) Chap. **4**
47. J.D. Joannopoulos, R.D. Meade, J.N. Winn: *Photonic Crystals* (Princeton University Press, Princeton 1995)
48. H. Kamerlingh Onnes: Commun. Phys. Lab. Univ. Leiden **119b, 120b, 122b, 124c** (1911)
49. F. London: In: *Superfluids*, Vol. 1,2 (Wiley, New York 1950)
50. D. Shoenberg: *Superconductivity* (Cambridge, New York 1952)
51. J.R. Schrieffer: *Theory of Superconductivity* (Benjamin, New York 1964)
52. P.G. de Gennes: *Superconductivity of Metals and Alloys* (Benjamin, New York 1966)
53. M. Tinkham: *Superconductivity* (Gordon and Breach, New York 1965)
54. M. Tinkham: *Introduction to Superconductivity* (McGraw-Hill, New York 1975)
55. C.G. Kuper: *Theory of Superconductivity* (Oxford University Press, New York 1968)
56. T. Van Duzer, C.W. Turner: *Principles of Superconductive Devices and Circuits* (Elsevier, New York 1981)
57. E.M. Lifshitz and L.P. Pitaevskii: *Statistical Physics, Part 2* (Pergamon, Oxford 1981)
58. F. London and H. London: Proc. Roy. Soc. **A149**, 71 (1935)
59. F. London and H. London: Physica **2**, 341 (1935)
60. V.L. Ginzburg and L.D. Landau: Sov. Phys. JETP **20**, 1064 (1950)

61. W. Meissner, R. Ochsenfeld: Naturwiss. **21**, 787 (1933)
62. B.D. Josephson: Physics Letters **1**, 251 (1962)
63. B.S. Chandrasekhar: In: *Superconductivity*. Ed. by R.D. Parks. Vol. 1, pp. 1–49 (Marcel Dekker, New York 1969)
64. J. Bardeen, L.N. Cooper, and J.R. Schrieffer: Phys. Rev. **108**, 1175 (1957)
65. C.J. Gorter and H.B.G. Casimir: Physica **1**, 305 (1934)
66. A.B. Pippard: Proc. Poy. Soc. **A216**, 547 (1954)
67. L.P. Gor'kov: Sov. Phys. JETP **9**, 1364 (1959)
68. G. Deutscher and P.G. de Gennes. In: *Superconductivity*. Ed. by R.D. Parks. Vol. 2 (Marcel Dekker, New York 1969) p. 1005
69. I.M. Khalatnikov: *Introduction to the Theory of Superfluidity* (Benjamin, New York 1965)
70. W.F. Winen, in: *Superconductivity*. Ed. by R.D. Parks. Vol. 2 (Marcel Dekker, New York 1969) p. 1167
71. V.L. Newhouse, in: *Superconductivity*. Ed. by R.D. Parks. Vol. 2 (Marcel Dekker, New York 1969) p. 1283
72. N.R. Werthamer, in: *Superconductivity*. Ed. by R.D. Parks. Vol. 1 (Marcel Dekker, New York 1969) p. 321
73. A.L. Fetter and P.C. Hohenberg, in: *Superconductivity* Ed. by R.D. Parks. Vol. 2 (Marcel Dekker, New York 1969) p. 817
74. A.A. Abrikosov: Sov. Phys. JETP **5**, 1174 (1957)
75. D. Saint-James, G. Sarma, and E.J. Thomas: *Type II Superconductivity* (Pergamon, Oxford 1969)
76. T. Maniv, V. Zhuravlev, I. Vagner, P. Wyder: Rev. Mod. Phys. **73**, 867 (2001)
77. Y.B. Kim and M.J. Straley, in: *Superconductivity*. Ed. by R.D. Parks. Vol. 2 (Marcel Dekker, New York 1969) p. 1107
78. P.W. Anderson and J.M. Rowell: Phys. Rev. Lett. **10**, 230 (1963)
79. R. de Bruyn Ouboter, in: *Superconductor Applications*: SQUIDs and Machines. Ed. by B.B. Schwartz and S. Foner (Plenum, New York 1976) p. 21
80. B.D. Josephson, in: *Superconductivity*. Ed. by R.D. Parks. Vol. 1, (Marcel Dekker, New York 1969) pp. 423–448
81. American Institute of Physics Handbook (3rd Edition) (McGraw-Hill, New York 1972)

Index

absolute instabilities 193
adiabatic invariant 159
Alfven wave 265
Ampere's law 5
anisotropic medium 7

Biot–Savart law 34
birefringence 108
Boltzmann equation 273
Boltzmann–Vlasov equation 274

capacitance 75
cavity resonators VI
cholesteric liquid crystal 294
conductivity 69
– tensor 167
convective instabilities 193
Coulomb gauge 12
Coulomb law 5
critical field 305

dielectric
– constant 5
– waveguides 140
director 295
domain 296

eddy currents VI
electric
– charge 1
– current 3
– field 1
– induction 2
electromagnetic
– field V
– induction VI
– wave VI
electromotive force 9
electrostatic field 66

electrostatic ion waves 261
extraordinary waves 107

Faraday's law 5

gauge 11
– invariance 11
Ginzburg–Landau theory VIII
group velocity 101

Hall effect 165
helicons VII
hydrodynamic model 236

impedance VI
inductance 77
instabilities VII
ion acoustic waves 260

Josephson effect VIII

Kramers and Kronig's relations VI

Lagrangian VI
Landau damping 277
Larmor frequency 44
liquid crystal VII
Lorentz force 11
Lorentz gauge 11
Lorentz transformation 14

magnetic
– field V
– flux 8
– induction VI
– mirror 176
– moment 4
magnetosonic waves 266
magnetostatics VI
Maxwell equations VI
Meissner effect 337

nematic liquid crystals   293

order parameter   VII
ordinary waves   107

penetration depth   71
permeability   5
permittivity   VI
phase velocity   133
plane waves   VI
plasma   V
plasmon   242
polarization   5

quasineutral oscillations   213

resistance   75

scalar potential   10

second sound   327
self-inductance   76
semiconductor   69
skin effect   VI
smectic liquid crystals   293
spatial dispersion   VII
superconductivity   VIII
superconductors   V

uniaxial crystal   105

vacuum light velocity   1
vector potential   10

wave equation   13
wave vector   88
wave-guides   VI

Springer Series in Solid-State Sciences

Editors: M. Cardona P. Fulde K. von Klitzing H.-J. Queisser

1 **Principles of Magnetic Resonance**
   3rd Edition By C. P. Slichter
2 **Introduction to Solid-State Theory**
   By O. Madelung
3 **Dynamical Scattering of X-Rays in Crystals** By Z. G. Pinsker
4 **Inelastic Electron Tunneling Spectroscopy**
   Editor: T. Wolfram
5 **Fundamentals of Crystal Growth I**
   Macroscopic Equilibrium and Transport Concepts
   By F. E. Rosenberger
6 **Magnetic Flux Structures in Superconductors**
   2nd Edition By R. P. Huebener
7 **Green's Functions in Quantum Physics**
   2nd Edition By E. N. Economou
8 **Solitons and Condensed Matter Physics**
   Editors: A. R. Bishop and T. Schneider
9 **Photoferroelectrics** By V. M. Fridkin
10 **Phonon Dispersion Relations in Insulators** By H. Bilz and W. Kress
11 **Electron Transport in Compound Semiconductors** By B. R. Nag
12 **The Physics of Elementary Excitations**
   By S. Nakajima, Y. Toyozawa, and R. Abe
13 **The Physics of Selenium and Tellurium**
   Editors: E. Gerlach and P. Grosse
14 **Magnetic Bubble Technology** 2nd Edition
   By A. H. Eschenfelder
15 **Modern Crystallography I**
   Fundamentals of Crystals
   Symmetry, and Methods of Structural Crystallography
   2nd Edition
   By B. K. Vainshtein
16 **Organic Molecular Crystals**
   Their Electronic States By E. A. Silinsh
17 **The Theory of Magnetism I**
   Statics and Dynamics
   By D. C. Mattis
18 **Relaxation of Elementary Excitations**
   Editors: R. Kubo and E. Hanamura
19 **Solitons** Mathematical Methods for Physicists
   By. G. Eilenberger
20 **Theory of Nonlinear Lattices**
   2nd Edition By M. Toda
21 **Modern Crystallography II**
   Structure of Crystals 2nd Edition
   By B. K. Vainshtein, V. L. Indenbom, and V. M. Fridkin
22 **Point Defects in Semiconductors I**
   Theoretical Aspects
   By M. Lannoo and J. Bourgoin
23 **Physics in One Dimension**
   Editors: J. Bernasconi and T. Schneider
24 **Physics in High Magnetics Fields**
   Editors: S. Chikazumi and N. Miura
25 **Fundamental Physics of Amorphous Semiconductors** Editor: F. Yonezawa
26 **Elastic Media with Microstructure I**
   One-Dimensional Models By I. A. Kunin
27 **Superconductivity of Transition Metals**
   Their Alloys and Compounds
   By S. V. Vonsovsky, Yu. A. Izyumov, and E. Z. Kurmaev
28 **The Structure and Properties of Matter**
   Editor: T. Matsubara
29 **Electron Correlation and Magnetism in Narrow-Band Systems** Editor: T. Moriya
30 **Statistical Physics I** Equilibrium Statistical Mechanics 2nd Edition
   By M. Toda, R. Kubo, N. Saito
31 **Statistical Physics II** Nonequilibrium Statistical Mechanics 2nd Edition
   By R. Kubo, M. Toda, N. Hashitsume
32 **Quantum Theory of Magnetism**
   2nd Edition By R. M. White
33 **Mixed Crystals** By A. I. Kitaigorodsky
34 **Phonons: Theory and Experiments I**
   Lattice Dynamics and Models
   of Interatomic Forces By P. Brüesch
35 **Point Defects in Semiconductors II**
   Experimental Aspects
   By J. Bourgoin and M. Lannoo
36 **Modern Crystallography III**
   Crystal Growth
   By A. A. Chernov
37 **Modern Chrystallography IV**
   Physical Properties of Crystals
   Editor: L. A. Shuvalov
38 **Physics of Intercalation Compounds**
   Editors: L. Pietronero and E. Tosatti
39 **Anderson Localization**
   Editors: Y. Nagaoka and H. Fukuyama
40 **Semiconductor Physics** An Introduction
   6th Edition By K. Seeger
41 **The LMTO Method**
   Muffin-Tin Orbitals and Electronic Structure
   By H. L. Skriver
42 **Crystal Optics with Spatial Dispersion, and Excitons** 2nd Edition
   By V. M. Agranovich and V. L. Ginzburg
43 **Structure Analysis of Point Defects in Solids**
   An Introduction to Multiple Magnetic Resonance Spectroscopy
   By J.-M. Spaeth, J. R. Niklas, and R. H. Bartram
44 **Elastic Media with Microstructure II**
   Three-Dimensional Models By I. A. Kunin
45 **Electronic Properties of Doped Semiconductors**
   By B. I. Shklovskii and A. L. Efros
46 **Topological Disorder in Condensed Matter**
   Editors: F. Yonezawa and T. Ninomiya

Springer Series in Solid-State Sciences

Editors: M. Cardona P. Fulde K. von Klitzing H.-J. Queisser

47 **Statics and Dynamics of Nonlinear Systems**
Editors: G. Benedek, H. Bilz, and R. Zeyher
48 **Magnetic Phase Transitions**
Editors: M. Ausloos and R. J. Elliott
49 **Organic Molecular Aggregates**
Electronic Excitation and Interaction Processes
Editors: P. Reineker, H. Haken, and H. C. Wolf
50 **Multiple Diffraction of X-Rays in Crystals**
By Shih-Lin Chang
51 **Phonon Scattering in Condensed Matter**
Editors: W. Eisenmenger, K. Laßmann, and S. Döttinger
52 **Superconductivity in Magnetic and Exotic Materials**   Editors: T. Matsubara and A. Kotani
53 **Two-Dimensional Systems, Heterostructures, and Superlattices**
Editors: G. Bauer, F. Kuchar, and H. Heinrich
54 **Magnetic Excitations and Fluctuations**
Editors: S. W. Lovesey, U. Balucani, F. Borsa, and V. Tognetti
55 **The Theory of Magnetism II** Thermodynamics and Statistical Mechanics   By D. C. Mattis
56 **Spin Fluctuations in Itinerant Electron Magnetism**   By T. Moriya
57 **Polycrystalline Semiconductors**
Physical Properties and Applications
Editor: G. Harbeke
58 **The Recursion Method and Its Applications**
Editors: D. G. Pettifor and D. L. Weaire
59 **Dynamical Processes and Ordering on Solid Surfaces**   Editors: A. Yoshimori and M. Tsukada
60 **Excitonic Processes in Solids**
By M. Ueta, H. Kanzaki, K. Kobayashi, Y. Toyozawa, and E. Hanamura
61 **Localization, Interaction, and Transport Phenomena**   Editors: B. Kramer, G. Bergmann, and Y. Bruynseraede
62 **Theory of Heavy Fermions and Valence Fluctuations**   Editors: T. Kasuya and T. Saso
63 **Electronic Properties of Polymers and Related Compounds**
Editors: H. Kuzmany, M. Mehring, and S. Roth
64 **Symmetries in Physics** Group Theory Applied to Physical Problems   2nd Edition
By W. Ludwig and C. Falter
65 **Phonons: Theory and Experiments II**
Experiments and Interpretation of Experimental Results   By P. Brüesch
66 **Phonons: Theory and Experiments III**
Phenomena Related to Phonons
By P. Brüesch
67 **Two-Dimensional Systems: Physics and New Devices**
Editors: G. Bauer, F. Kuchar, and H. Heinrich

68 **Phonon Scattering in Condensed Matter V**
Editors: A. C. Anderson and J. P. Wolfe
69 **Nonlinearity in Condensed Matter**
Editors: A. R. Bishop, D. K. Campbell, P. Kumar, and S. E. Trullinger
70 **From Hamiltonians to Phase Diagrams**
The Electronic and Statistical-Mechanical Theory of sp-Bonded Metals and Alloys   By J. Hafner
71 **High Magnetic Fields in Semiconductor Physics**
Editor: G. Landwehr
72 **One-Dimensional Conductors**
By S. Kagoshima, H. Nagasawa, and T. Sambongi
73 **Quantum Solid-State Physics**
Editors: S. V. Vonsovsky and M. I. Katsnelson
74 **Quantum Monte Carlo Methods in Equilibrium and Nonequilibrium Systems**   Editor: M. Suzuki
75 **Electronic Structure and Optical Properties of Semiconductors**   2nd Edition
By M. L. Cohen and J. R. Chelikowsky
76 **Electronic Properties of Conjugated Polymers**
Editors: H. Kuzmany, M. Mehring, and S. Roth
77 **Fermi Surface Effects**
Editors: J. Kondo and A. Yoshimori
78 **Group Theory and Its Applications in Physics**
2nd Edition
By T. Inui, Y. Tanabe, and Y. Onodera
79 **Elementary Excitations in Quantum Fluids**
Editors: K. Ohbayashi and M. Watabe
80 **Monte Carlo Simulation in Statistical Physics**
An Introduction   4th Edition
By K. Binder and D. W. Heermann
81 **Core-Level Spectroscopy in Condensed Systems**
Editors: J. Kanamori and A. Kotani
82 **Photoelectron Spectroscopy**
Principle and Applications   2nd Edition
By S. Hüfner
83 **Physics and Technology of Submicron Structures**
Editors: H. Heinrich, G. Bauer, and F. Kuchar
84 **Beyond the Crystalline State** An Emerging Perspective   By G. Venkataraman, D. Sahoo, and V. Balakrishnan
85 **The Quantum Hall Effects**
Fractional and Integral   2nd Edition
By T. Chakraborty and P. Pietiläinen
86 **The Quantum Statistics of Dynamic Processes**
By E. Fick and G. Sauermann
87 **High Magnetic Fields in Semiconductor Physics II**
Transport and Optics   Editor: G. Landwehr
88 **Organic Superconductors**   2nd Edition
By T. Ishiguro, K. Yamaji, and G. Saito
89 **Strong Correlation and Superconductivity**
Editors: H. Fukuyama, S. Maekawa, and A. P. Malozemoff

Springer Series in Solid-State Sciences
Editors: M. Cardona P. Fulde K. von Klitzing H.-J. Queisser

90 **Earlier and Recent Aspects
of Superconductivity**
Editors: J. G. Bednorz and K. A. Müller

91 **Electronic Properties of Conjugated
Polymers III** Basic Models and Applications
Editors: H. Kuzmany, M. Mehring, and S. Roth

92 **Physics and Engineering Applications of
Magnetism** Editors: Y. Ishikawa and N. Miura

93 **Quasicrystals** Editors: T. Fujiwara and T. Ogawa

94 **Electronic Conduction in Oxides** 2nd Edition
By N. Tsuda, K. Nasu, F. Atsushi, and K. Siratori

95 **Electronic Materials**
A New Era in Materials Science
Editors: J. R. Chelikowsky and A. Franciosi

96 **Electron Liquids** 2nd Edition By A. Isihara

97 **Localization and Confinement of Electrons
in Semiconductors**
Editors: F. Kuchar, H. Heinrich, and G. Bauer

98 **Magnetism and the Electronic Structure of
Crystals** By V. A. Gubanov, A. I. Liechtenstein,
and A. V. Postnikov

99 **Electronic Properties of High-T_c
Superconductors and Related Compounds**
Editors: H. Kuzmany, M. Mehring, and J. Fink

100 **Electron Correlations in Molecules
and Solids** 3rd Edition By P. Fulde

101 **High Magnetic Fields in Semiconductor
Physics III** Quantum Hall Effect, Transport
and Optics By G. Landwehr

102 **Conjugated Conducting Polymers**
Editor: H. Kiess

103 **Molecular Dynamics Simulations**
Editor: F. Yonezawa

104 **Products of Random Matrices**
in Statistical Physics By A. Crisanti,
G. Paladin, and A. Vulpiani

105 **Self-Trapped Excitons**
2nd Edition By K. S. Song and R. T. Williams

106 **Physics of High-Temperature
Superconductors**
Editors: S. Maekawa and M. Sato

107 **Electronic Properties of Polymers**
Orientation and Dimensionality
of Conjugated Systems Editors: H. Kuzmany,
M. Mehring, and S. Roth

108 **Site Symmetry in Crystals**
Theory and Applications 2nd Edition
By R. A. Evarestov and V. P. Smirnov

109 **Transport Phenomena in Mesoscopic
Systems** Editors: H. Fukuyama and T. Ando

110 **Superlattices and Other Heterostructures**
Symmetry and Optical Phenomena 2nd Edition
By E. L. Ivchenko and G. E. Pikus

111 **Low-Dimensional Electronic Systems**
New Concepts
Editors: G. Bauer, F. Kuchar, and H. Heinrich

112 **Phonon Scattering in Condensed Matter VII**
Editors: M. Meissner and R. O. Pohl

113 **Electronic Properties
of High-T_c Superconductors**
Editors: H. Kuzmany, M. Mehring, and J. Fink

114 **Interatomic Potential and Structural Stability**
Editors: K. Terakura and H. Akai

115 **Ultrafast Spectroscopy of Semiconductors
and Semiconductor Nanostructures**
2nd Edition By J. Shah

116 **Electron Spectrum of Gapless Semiconductors**
By J. M. Tsidilkovski

117 **Electronic Properties of Fullerenes**
Editors: H. Kuzmany, J. Fink, M. Mehring,
and S. Roth

118 **Correlation Effects
in Low-Dimensional Electron Systems**
Editors: A. Okiji and N. Kawakami

119 **Spectroscopy of Mott Insulators
and Correlated Metals**
Editors: A. Fujimori and Y. Tokura

120 **Optical Properties of III–V Semiconductors**
The Influence of Multi-Valley Band Structures
By H. Kalt

121 **Elementary Processes in Excitations
and Reactions on Solid Surfaces**
Editors: A. Okiji, H. Kasai, and K. Makoshi

122 **Theory of Magnetism**
By K. Yosida

123 **Quantum Kinetics in Transport and Optics
of Semiconductors**
By H. Haug and A.-P. Jauho

124 **Relaxations of Excited States and Photo-
Induced Structural Phase Transitions**
Editor: K. Nasu

125 **Physics and Chemistry
of Transition-Metal Oxides**
Editors: H. Fukuyama and N. Nagaosa

Printing: Saladruck Berlin
Binding Lüderitz&Bauer, Berlin